Natural Science in Archaeology

Series Editors

Bernd Herrmann
Günther A. Wagner

For further volumes:
http://www.springer.com/series/3703

Claudio Margottini
Editor

After the Destruction of Giant Buddha Statues in Bamiyan (Afghanistan) in 2001

A UNESCO's Emergency Activity for the Recovering and Rehabilitation of Cliff and Niches

Springer

Editor
Claudio Margottini
Department of Geological Survey of Italy
ISPRA – Italian Institute for Protection and
Environmental Research
Rome
Italy

First published by Edition Viterbo – Italy February 2009
Union Printing S.p.A

ISBN 978-3-662-52215-8 ISBN 978-3-642-30051-6 (eBook)
DOI 10.1007/978-3-642-30051-6
Springer Heidelberg New York Dordrecht London

Printed on acid-free paper

Springer is part of Springer Science+Business Media (www.springer.com)

To Ines, Simone and Matteo

Foreword

Afghanistan's history can be traced back to when the land was called "Ariana". The native culture of Afghanistan was built on and influenced by Persian and Hellenistic cultures, and later from Indian culture. A mixture of indigenous and foreign cultures through the centuries created something unique, which deserves to be called "the crossroads of civilisation."

Bamiyan means "the place of shining light". There is great beauty in the variegated colours of its rugged lines in its folded hills. The central valley of Bamiyan is located at 2,500 meters, and is watered by two rivers flowing down from sources in the Kuh-e Baba: the Kakrak River to the east and the Foladi River to the west. The principle archaeological sites are located in the long east-west central valley of Bamiyan and in the Kakrak and Foladi river valleys.

The nucleus of Bamiyan's cultural legacy was formed by the two colossal Buddha images carved at the eastern and western ends of a high cliffs, which face the central valley. Some thousand caves are also cut into the cliff face and decorated with a rich variety of murals. The Buddhist art of Bamiyan, which enjoyed a renaissance in Central Afghanistan after the collapse of the earlier Gandharan culture, spread to and influenced various countries along the Silk Road.

Despite strong international demands for the protection of Bamiyan, the Taliban regime blew up the two giant Buddhas in March 2001. The world was stunned and shocked by this news.

Immediately after the collapse of the Taliban regime in December 2001, UNESCO responded quickly to the challenge of rehabilitating Afghanistan's endangered cultural heritage which, in some instances, had suffered irreversible damage and loss. Since 2002, an international project, funded by the UNESCO Japan Funds-in Trust for Preservation of the World Cultural Heritage and coordinated by UNESCO, has been operating in Bamiyan. Finally, in 2003, the site of Bamiyan is inscribed in the World Heritage List of UNESCO, as "Cultural Landscape and Archaeological Remains of the Bamiyan Valley, Afghanistan". Moreover, a Cultural landscape represents the "combined works of nature and of man", which express a long and intimate relationship between people and their natural environment. Bamiyan is certainly a flagship of this UNESCO World heritage List category.

To prevent the cliffs from collapsing where the Buddhas once stood and the crumbling of the niches because of the explosion in 2001, an international effort, at the border of a real adventure in science and restoration, was made by scientists from different part of world.

The present book is impressively reporting the successful work conducted to stabilize cliff and niches, otherwise at very high risk of collapse. The importance of the endeavour is clearly described as well as the cooperative efforts of all partners concerned, including the people living here. Under the essential economic support of the Government of Japan, this project was an example of a UNESCO initiative, in which high technologies and knowledge were applied in a remote part of the world, with a strong Afghan support and essential local cooperation. Also, the many disciplines involved and reported in this book, it clearly demonstrates the importance of a close collaboration among experts coming from different science sectors and magnifying the catalyst effect of these precious heritage of outstanding universal value.

Our endeavour, in close cooperation with the Afghans and international communities, is continuing today to preserve the precious site for future generations.

Francesco Bandarin
UNESCO
Assistant Director-General for Culture
Paris, France

Acknowledgments

The editor wishes to express his gratitude to all the people who made possible this relevant project and all the authors contributing to this book. I hope this experience will remain in their memory as a milestone in friendship, in cooperating for the renaissance of a marvelous Country, and in serving a very important issue: maintaining the traces of human beings on planet Earth, despite the many natural and human attacks they have to sustain.

A particular acknowledgment goes to those people who encouraged the development of this activity, also providing data and materials; a special thanks to Mr. Paul Bucherer-Dietschi, Director of the Swiss Afghan Museum in exile and the Bibliotheca Afghanica Foundation, Zurich (Switzerland) for providing many of the historical pictures used this book; and Mr. Kazuya Yamauchi, National Research Institute for Cultural Property (NRICP), head of the Regional Environmental Section, Tokyo (Japan).

Logistics in Afghanistan were quite difficult and the UNESCO local people provided excellent support in the last 12 years: to mention Brendan Cassar, Francois Langlois, Andy Miller, Masanori Nagaoka, Sara Noshadi, Emal Rasuli, Reza Shafiri and Jim Williams is more than a simple acknowledgement; they were really the engine of a complex system of relations and care for our safety.

The UNESCO World Heritage Centre in Paris demonstrated, once more, their high professionalism, organizing 12 years of very difficult tasks in a perfect manner: a special thanks is due to Roland Lin, Feng Jing, Junhi Han, Giovanni Boccardi, Christian Manhart, Sabrina Zannier, Francis Childe, Sarah Franke, and Junko Okahashi for their care in coordination, preparation, and support of the field missions.

Essential support was offered by the many colleagues and representatives of the Afghan Government who provided a great collaboration in all phases of this project. Thanks to the people of Bamiyan, to the Major, the Governor of Bamiyan Province, experts and officials from the Afghan Ministry of Culture and Information (MoCI), representatives of Afghan Government. Last but not least my friend Mr. Abdul Ahad Abbassy, Director, Department of Historical Monuments, MoCI.

The editor would like to thank the many ICOMOS colleagues who contributed to this important result with many discussions and brain storming.

The editor is extremely grateful to Prof. M. Pellegrini for continued suggestions and stimulation; unfortunately, he passed away in 2012. Many thanks to Prof. P. Canuti for reviewing the manuscript and for helpful advice.

The more practical aspects of this paper were elaborated with Dr. Eng. V. Manassero and Dr. Eng. B. Vingiani, with the collaboration of Mr. D. Fallini from Rodio (now TREVI S.p.A.). Their support was extremely important in defining the operational part of the proposed project.

A special thank is also due to my friend Dr. Eng. O. Pandolfi, President of Fondazione Marenostrum and manager of Carrera marble caves, who gave his support in new techniques for consolidating a highly unstable cliff.

Many colleagues gave their help and support in the development of laboratory investigations. It is a pleasure to thank Prof. Piero Manetti (University of Florence) for support in mineralogical and petrographical investigations, Prof. Tatiana Rotonda (University of Rome1) for petro-geophysical investigations, Dr. Cinzia Crovato and Dr. Giovanna Armiento (ENEA) for x-ray diffractometers, and Dr. Marcello Gioly Guidi and Mr. Fabrizio Pierdominici (ENEA) for use of the scanning electronic microscope.

A special thought to Prof. Ribacchi, who passed away in 2011, for his valuable comments.

A special gratitude goes to ISPRA – the Italian Insitute for Environmental Protection and Research Survey of Italy and its President Prof. Bernardo De Bernardinis and the Director of the Department Geological Survey of Italy Dr. Cladio Campobasso. A warm thank also to Dr. Renata Montesanti, and Dr. Cristina Pacciani who strongly encouraged the publication of this book.

Finally, last but not least my friend Eng. Daniele Spizzichino (ISPRA), who shared with us most of the brainstorming and interpretation of field data and laboratory tests.

This initiative has been implemented as a Coordinated Project of ICL (International Consortium on Landslides) within the framework of the International Programme on Landslides.

The present project has been possible thanks to the generosity of the people and Government of Japan.

This book is dedicated to the memory of Nicholas Ambraseys, who died 29 Dec 2012 leaving the world a five-decade-long legacy of studies of historical earthquakes, landslides and engineering. His expertise and his guidance was, for some of us, a fundamental step in our life and in our scientific career.

Contents

Contributors

Nicholas Ambraseys Department of Civil Engineering, Imperial College London, South Kensington Campus, London, UK

Roger Bilham Department of Geological Sciences, University of Colorado, Boulder, CO, USA

Andrea Bruno Studio Bruno, Torino, Italy

Vittorio Colombini Studio Colombini, Rome, Italy

Carlo Crippa Trevi Division Rodio, Casalmaiocco (Lodi), Italy

Giuseppe Delmonaco Department of Geological Survey of Italy, ISPRA – Italian Institute for Protection and Environmental Research, Rome, Italy

Roland Lin World Heritage Centre, UNESCO, Paris, France

Kosaku Maeda Japan Center for International Cooperation in Conservation, National Research Institute for Cultural Properties, Taito-ku, Tokyo, Japan

Christian Manhart Section of Museums and Cultural Objects, UNESCO, Paris, France

Claudio Margottini Department of Geological Survey of Italy, ISPRA – Italian Institute for Protection and Environmental Research, Rome, Italy

Guido Martini Department of UTPRA-PREV, ENEA – Italian Agency for New Technology Energy and Environment, Frascati (Rome), Italy

Salvatore Paolini ENEA – Italian Agency for New Technology Energy and Environment, Frascati (Rome), Italy

Bert Praxenthaler Epfenhausen, Penzing, Germany

Gedeone Tonoli Trevi Division Rodio, Casalmaiocco (Lodi), Italy

1 Introduction

The Buddhist art of the Hindu Kush mountain region, of which the Bamiyan Valley is a part, represents the final flowering of Buddhism in Afghanistan. The kingdom of Bamiyan was a Buddhist state positioned at a strategic location along the trade routes that for centuries linked China and Central Asia with India and the west. Bamiyan served as an important monastic and spiritual center, as well as a hub of intense commercial activity.

The site was constructed between approximately the fifth and ninth centuries A.D. during a distinctive phase in the history of Buddhist art, a period of intense cultural and religious exchanges between east and west, and a time of great cultural change within Buddhism itself. Bamiyan served as a ceremonial and spiritual center that attracted and comforted crowds of pilgrims and merchants traveling between Central and South Asia. The noted Chinese monk and traveler Hsuan Tsang, who visited the site in 634 A.D., reported that it was a thriving center of Buddhism and described in detail the ceremonies and rituals he witnessed there.

During this extended period of Bamiyan's Buddhist florescence two massive Buddha images were carved out of a high stretch of cliff facing the widest part of the valley. These colossal images are the largest Buddhist sculptures in the world. The greater of the two images stands 53 m/175 ft in height at the western end of the cliff-face (niche about 55 m); the second massive Buddha, at the eastern end of the cliff, is some 35 m/120 ft tall (height of the niche 41 m).

All along the cliff face between these monolithic images are carved hundreds of caves of varying size used as chapels for both private and communal worship. Ambulatories, off of which are further rock-cut chapels and image niches, surround the larger Buddha at the level of his feet and again at the level of his head high up the cliff face. Most of the rock-cut chapels and ambulatories at Bamiyan are covered with paintings over plastered walls that display an incredibly rich, varied, and important body of early Buddhist painting.

The Bamiyan style derives from the artistic traditions and iconography of both India and Central Asia. Generally speaking, the compositions and stylistic techniques displayed in the paintings at Bamiyan and neighboring sites consist of numerous separate figures represented frontally as iconic forms, rather than within narrative scenes. Buddhas and Bodhisattvas painted on the walls and ceilings at Bamiyan convey a vision of the Buddhist universe rather than events and moments in ordinary historical time.

Each of the many caves and niches at Bamiyan, including those that house the colossal Buddha figures, can be considered a complete and unified composition in which painting and sculpture work together to form a single symbolic configuration. At the center of each of these configurations is a sculpted Buddha, usually presented at ground level on the north wall.

C. Margottini (ed.), *After the Destruction of Giant Buddha Statues in Bamiyan (Afghanistan) in 2001*, Natural Science in Archaeology, DOI 10.1007/978-3-642-30051-6_1, © Springer-Verlag Berlin Heidelberg 2014

Surrounding this figure on the walls and ceiling of each cave there are innumerable painted images of further Buddhas, Bodhisattvas, and other heavenly figures.

The sculptured Buddha figures seem to have been painted originally, and are located in an axial and symmetrical position in relation to the paintings. The wall paintings thus appear to surround the central figure in numerous concentric circles or vertical rows that in both cases suggest a mandala, a conceptual representation of the Buddhist universe. The ceilings in particular are transformed into a "dome of heaven" through the manipulation of repeated Buddha forms around the central figure. Together, the wall and ceiling paintings work with the sculpture to create an entire heavenly environment, a symbolic representation of the universe.

In March 2001, the two great Buddhist statues were destroyed by the Taliban in the Bamiyan valley (Central Afghanistan), and great worldwide attention was paid to the possible reconstruction/rehabilitation of the sites, in order not to totally lose these important cultural heritage structures.

UNESCO was involved from the beginning, acting both before and after the destruction of the statues. Because the destruction was unavoidable, the main goal of UNESCO was to secure and preserve the remaining pieces that were not destroyed by the explosion, and to study the potential re-setting in place of the fragments that fell to the ground.

Nevertheless, preliminary to any further decision about the management of the fragments, since the first field inspections in 2002 the critical situation of the cliff in which the statues were carved was highlighted; and as a consequence of the explosion, its stability became precarious. The high risk of collapse of many parts of the cliff and niches was immediately and clearly established, which would produce the almost total erasure of the landscape and heritage hitherto known.

The first multidisciplinary UNESCO field mission to Bamiyan occurred in September/October 2002. Since that time, many other missions have been conducted to the site, investigating the overall stability of the niches and cliff, and providing a project for consolidating and correcting a most endangered situation. The purpose of this book is to tell a successful story in which, after devastation that astonished the world, the remaining falling part of the cliff was patiently fortified.

It is a success to share with the entire involved community: UNESCO officials and experts; Afghan officials; Japanese experts; the Japanese government, who provided the financial support for this work; the TREVI company; and last but not least, the local Bamiyan population. In fact, none of us will ever forget the amazingly warm welcome and support provided by the Bamiyan inhabitants. We saw many children grow during those 5 years; and now, when we are back in Bamiyan, we feel the unique emotion of being back home.

An important aspect of the successful result achieved in this project was the great care and professionalism offered by TREVI. TREVI is a company that specializes in special foundations and soil consolidation works. It belongs to the TREVI Group, a worldwide leader in foundation engineering and the design and production of relevant rigs and special equipment. The TREVI Group is active in the drilling field (oil, gas, and water), both for the production of systems and for the supply of services, and it is also involved in the emerging sector of renewable energies (wind and geothermal powers). It was established in Cesena, Italy in 1957 and achieved a remarkable reputation in the field of foundational engineering, hence becoming acknowledged as a worldwide leader for the execution of various types of interventions.

This is the case of special foundational works that are necessary to support complex projects such as the excavation of tunnels, the consolidation of soils (i.e., the restoration of the leaning Pisa Towers, among the many achievements), repair works for dams, the recovery of polluted sites, and the execution of automated underground car parks. Theirs is a transversal competence that finds application in every corner of the world.

The success of the TREVI Group is based on the integration and continuous exchange between technological and process innovation among the divisions making up the group: TREVI and Soilmec (producing and developing rigs and systems for the foundation engineering); Drillmec, which produces and develops drilling systems (oil, gas and water); and Petreven, which supplies oil drilling services.

2 The Foundation of the Colossal Buddha Statues

Kosaku Maeda

The Foundation of the Colossal Buddha Statues: The Eastern Buddha Statue

The Royal City and the Buddhist Sangharama

It is not clear when Buddhism started in Bamiyan. Around 305 B.C., Seleucus, who had succeeded the eastern area of the Empire of Alexander the Great ceded Afghanistan, the southeastern area of the Hindu Kush to the rising Maurya Dynasty of Chandraguputa and it was about 50 years later that his grandson, King Ashoka, sent a high Buddhist priest there to propagate Buddhism. That was just before the Greco-Bactrian Kingdom became independent (circa 250 B.C.) cutting the political bond with the Seleucid Dynasty. It is assumed that the people who accepted Buddhism were those who had been familiar with Persian culture since Achaemenian times and also had been deeply affected by Hellenistic culture. That is backed up by the bilingual rock edicts of King Priyadarsin (Ashoka) inscribed in Greek and Aramaic, which were discovered in Kandahar.

It became generally known through the excavations of the Buddhist sites that before long Buddhism gradually and quietly rooted in the northern and southern districts of the Hindu Kush and flourished at its most in the Kushan Dynasty of a nomad nation. It was also found that many Buddhist temples were built during the long period from the second century to the fourth century. In the north side district of the Hindu Kush, the nearest site to Bamiyan is the Buddhist site of Surkh Kotal discovered near the sanctuary for fire worship built by a king of the Kushan Dynasty. It was not assumed that the site was a Buddhist temple until finding "the turban of Boddhisattva" carved on one of the capitals. This might have been built in the third or fourth century. There is another rock cut temple in Haibak. This temple consists of five caves and one stupa cut out of rock. In the caves are the niches to place the seated Buddha statues, but now no Buddhas exist. No record of Chinese texts is left on the site. Probably Xian-zang and Wei Chao would not visit this temple.

In the south side of the Hindu Kush, the monasteries at Kapisa, that is, Begram, Shotorak, and Paitava had been in action already. That would be in the second century to the fourth century. Judging from the clay images excavated, it is thought that the Buddhist site of Tepe Maranjan in Kabul was built in the fourth and fifth centuries.

Those circumstances suggest that the creation of Buddha images and the building of Buddhist temples in Bamiyan started at the latest around the fifth century. Perhaps the period of establishing the royal city was not so far from that of building the first Sangharama in Bamiyan. According

K. Maeda
Japan Center for International Cooperation in Conservation, National Research Institute for Cultural Properties, Tokyo 13-43, Ueno Park, Taito-ku, Tokyo 110-8713, Japan
e-mail: isca-maeda@m2.dion.ne.jp

C. Margottini (ed.), *After the Destruction of Giant Buddha Statues in Bamiyan (Afghanistan) in 2001*, Natural Science in Archaeology, DOI 10.1007/978-3-642-30051-6_2, to Xian-zang, on the western side of Bamiyan valley exists the royal city "leaning on the mountains and cliffs" in the north and "extends as far as the valley" in the south. It would be built when the importance of the roads running through Bamiyan increased in economic and military importance.

The royal citadel of Shahr-i-Zohak located at the eastern entrance of the valley guards the east side of the kingdom and Shahr-i-Gorgora stands in the center of Bamiyan watching every corner of the valley as a whole. Surrounded by the high mountains, the holy Buddhist Sangharama and laic communities were divided by the river and the road.

The Great Image of Buddha as the Axis of Universe

In Bamiyan, before the colossal image of Buddha was cut on the rock cliff facing south, there had been a Sangharama built on the ground by "the former king of the country" as Xian zang described in his Buddhist Records of the Western World. Then the stupa was erected in front of the rock cliff and the cave Sangharamas were cut out in the cliff. In Haibak, Buddhist caves were also constructed in the same order. In most of the many neighborhood Buddhist temples, the stupas were still built not being separated from the monastery. Even in Bamiyan, Buddhist caves (F & G) with a small stupa placed in the center of the caves were discovered near the Eastern Colossal Buddha.

The colossal image of Buddha, to the contrary, is standing like the axis of the universe and many sanctuaries are arranged around it. It might be said that this decisive plan started in Bamiyan. The splendid idea of building the colossal image of Buddha above on the high mountain of 2,500 m as axis mundi connecting the earth and heaven must have some reasonable motives.

One may consider the change of Buddhist thought in the process of its propagation from east to west via Gandhara, Hadda, and Kapisa. However, why did the Buddhist priests of the school of Lokottaravadins belonging to the Little Vehicle "Hinayana" who had worshiped both the stupa and the images of Buddha hold the transcending motive for the construction of a colossal image of Buddha? It is difficult to find a decisive grounding in their canonical doctrine. Taking into consideration the huge cost of making the colossal image of Buddha, the strong political and economic motives of the patrons rather than religious ones cannot be negligible (Fig. 2.1).

When the stupa and the monasteries were built side by side, it is not a surprise that people were attracted to the visible Nirmanakaya, which represents the living (physical) Buddha much more than the abstract Dharmakaya. It is quite natural that longing for Nirmanakaya besides Sambhogakaya had been enhanced by the desire for transcendent existence. The desire for transcendent existence was accepted not only by the people but also by the reigning authority as a strong patron.

It might be to create an imago mundi, which involves the construction of a heaven-kissing colossal image of Buddha in Bamiyan that people coming here from all directions have never seen before. And at last, the patrons of Buddhism of Bamiyan came to think of carving the colossal Buddha statue in the center of the eastern steep and large cliff.

Many artisans came together from the southeastern areas such as devastated Gandhara and Hadda and also from northern areas such as Bactria to work on the big project, the new idea produced in this small valley. In Bamiyan, there is another valley called Foladi to the southwest of the royal city. It is said that Foladi means blacksmith. Probably the tools to cut the rock cliff and the caves were cast in this valley and supplied to the artisans. The forgers (smith) is "a unity of tinkers, musicians, physicians, and diviners" as Eliade said. These blacksmiths, who were the artists and artisans creating the idols, could thoroughly know the earth as well as their abilities and know which cliff to be chiseled at first. Unfortunately, no more than a few monasteries have been investigated fully in the valley of Foladi.

How Was the Colossal Buddha Statue Built?

I wonder how the artisans did the layout of the surface of the rock cliff standing sharply and perpendicularly. And I also wonder whether they designed

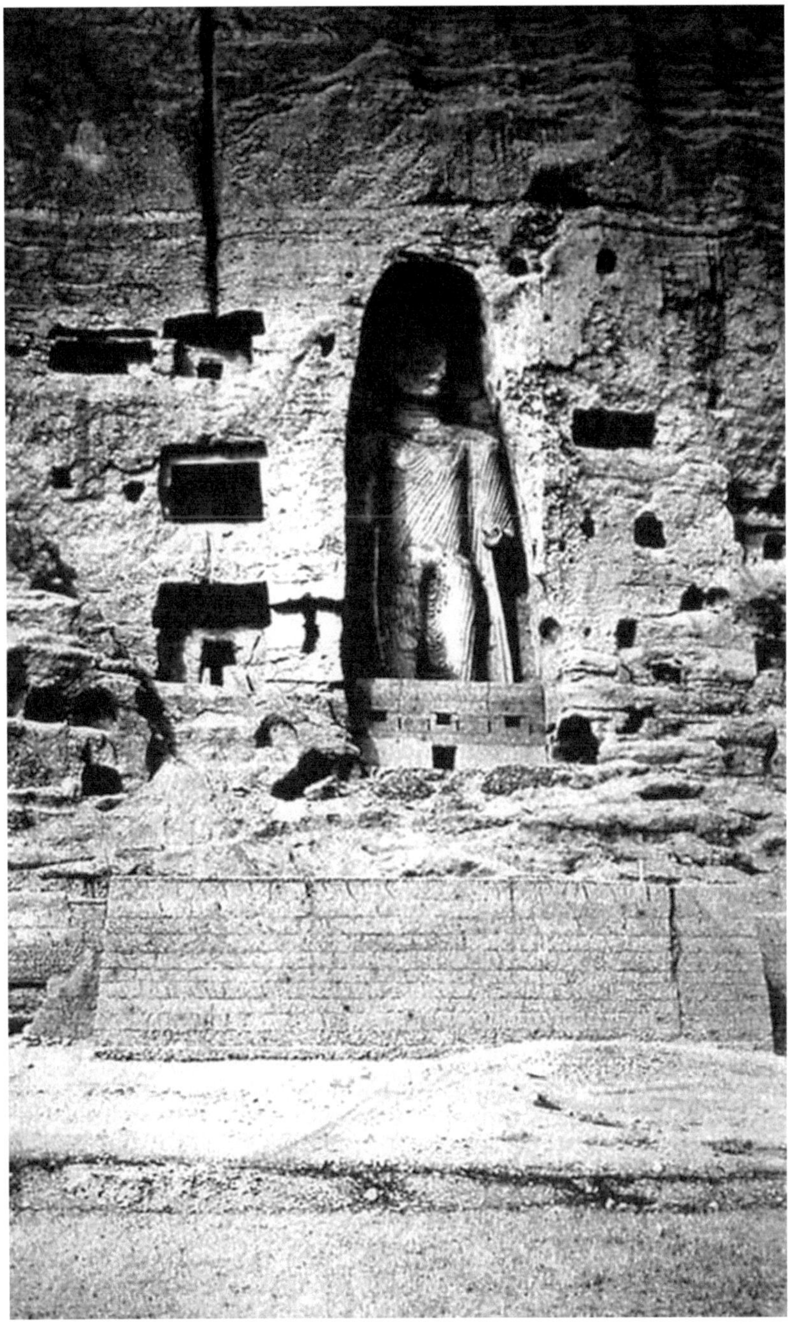

Fig. 2.1 Eastern Buddha statue

the surrounding rock caves and the passage-for-pradaksina at the same time, and then the group making the colossal Buddha statue (artisans of exterior carving) and the group making the monasteries (artisans of interior carving) advanced to carve together in parallel with the speed of cutting of the Buddha statue. Judging from the shallow cutting of the trefoiled niche of the colossal Buddha and the thin wall between the passage leading to the upper caves and the niche of the colossal Buddha, it is not necessarily considered that they started cutting the colossal Buddha at first.

As the model of the colossal Buddha statue, it is assumed that they selected the image of the Buddha statues of Gandharan style having clear and transcendent appearance created near the area of Kapisa. The hair is waving and the garment covers both shoulders completely. The three paths (lines) that symbolize the omniscience of Buddha are clearly shown on the neck and the comparatively long legs are designed to be seen under the garment. The Buddha was represented as Tathagata. Being suitable to the colossal Buddha statue, the insteps were made strong to stand on the earth, with the thick coating of mud. The finishing of the Buddha statue was to cut the niche into trefoil.

Probably it might be after the completion of the colossal rock cutting Buddha statue that eight caves, from cave I to cave VIII, were cut around the feet of the statue. The architectural design of the vault of five caves (I, III, IV, V, VI) are different from each other. However, out of three caves (II, VII, VIII) cave II might have been cut later for another purpose. Cave VII and cave VIII were planned also as the sanctuaries, but compared with the other five caves which had been cut before those two, there is less finesse in their workmanship.

Making use of the passage-for-pradakshina that encircles the colossal Buddha, they started to cut the three-story sanctuary (when it was investigated by French archaeological group, in 1923 and 1924, it was given the alphabetical sign of D1, D, C1, and C2 to each story from the bottom layer to the upper one). The passage would be also utilized in case of cutting the colossal Buddha statue from inside. It is thought that they planned to cut out this circumambulatory passage (pradakshina path) upward to lead to above the head of the colossal statue, starting from the side of cave VIII, which is positioned at the western side of the niche wall immediately below the colossal Buddha statue, so that this passage would meet another passage behind the protuberance on Buddha's head (ushunisa). The latter was also cut out to reach above the head of the statue passing through the eastern caves A and B.

Furthermore, large rectangular spaces are provided at the right and the left lateral walls above the head of the Buddha statue. In those opened spaces, seats were provided for the royal families who attended the celebration of the newly made Buddha image and for the Pancavarsika held once every 5 years (the Buddhist service to do offering, from kings to peasants and shepherds, regardless whether male or female, rich or poor, noble or common), and scattering flowers (one of the most important Buddhist rituals) also might be done making use of these spaces. A wooden balustrade was built at the front and the holy cloth might be hung on it as the murals painted on both sides of the niche had shown us. In this sense, this rectangular space would be in use as a topos (gr., common place) leading to the sacred perpetual time, or to the transcendental time for ceremonies.

The Mural Paintings in Bamiyan: The Great Composition on the Ceiling of the Eastern Colossal Buddha

The Sun God on the Golden Chariot Soaring in Heaven

The finishing touch of the making of the Eastern Colossal Buddha statue that was called Sakya Buddha by Xiang zang was to paint the vault and the lateral walls of the Buddha niche. When they looked up at the statue, the vault that decorated the overhead of Sakya was necessary to be the symbol of transcendence and at the same time, to reflect the philosophy of the world entertained by the royalty and nobility who promoted the big project to found the colossal Buddha statue. The design of the great composition that decorates the vault of the statue was probably entrusted to the artisans who could give full play to a remarkable presentation of imago mundi. The artisans finished the unparalleled mural paintings being helped by gods.

In the center of the great composition the great sun, which shines the every corner of the world, was painted. The blaze of the sun was represented with saw teeth on the edge of the large disk at the back of the central deity. The sun could absorb everything as the symbol of omnipotent invincible

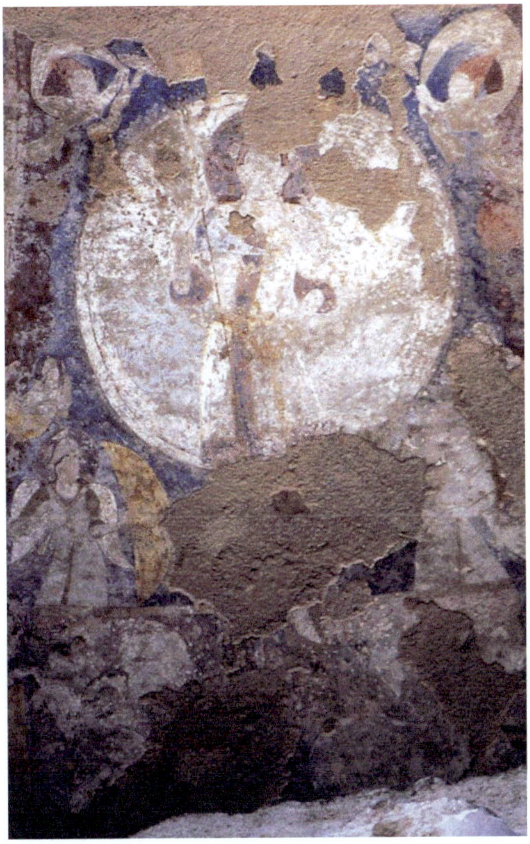

Fig. 2.2 The sun god

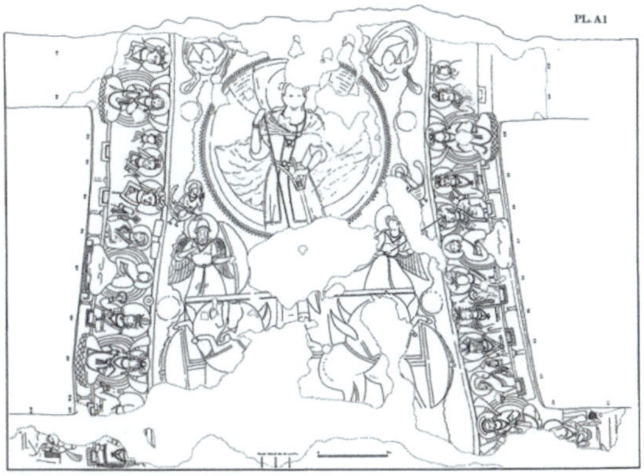

Fig. 2.3 Sketch of ceiling

divinity (Fig. 2.2). The traditional sun worship of the nomads and also Buddhist symbolism, which metaphorically compared the transcendence of Sakya to the sun, could both be represented here by the sun as the epiphany of supreme divinity. At the same time the artisans would be required to paint the image of the sun god doubled with the conception of the world that the royalty in this province required.

The sun god was painted with the figure who soars in heaven riding on the two-wheeled golden chariot pulled by four winged white horses. The sun god displays a nimbus and ribbons flying up from the both shoulders symmetrically, and wears a mantle, the skirt of which is waving in the wind, on a round-neck tunic, holding straight a rather slender spear in the right hand and also grasping the hilt of a sword hung from the waist belt in left hand. Two legs putting on boots would be painted beneath the solar disk, but the legs cannot be seen anymore, because that part of the painting came off. The hair is long enough to reach the shoulders and a round headdress can be seen, but the top part of it cannot be seen (Fig. 2.3).

The winged figure putting on boots, which can be seen scarcely on the wheel axle of the chariot will be a charioteer. The wheel is represented by a half circle with spokes. The divinity riding on a two-wheeled chariot has been painted in various styles in the Buddhist iconography from Bodh Gaya to Kizil and Dunhuang, but there is no example such as the sun god of Bamiyan, which was represented as the main theme by this independent great composition. There are some examples that show that four horses pulling a chariot are separated right and left symmetrically or that four horses make a pair. But the iconography that shows two pairs of horses separated into right and left and seeing each other, such as the white horses of Bamiyan, is a very rare example. Such difference in the details proves how creative the composition devised by the artists of Bamiyan was.

At both the right and left sides of the chariot on which the sun god is standing, two winged attendants are painted. The winged female attendant painted on the left side wears the Corinthian helmet with a feather and has a nimbus on her head and a shield in her left hand. It is impossible to be certain if she has something in her right hand or not. The attendant on the right side also wears a helmet, having a circular nimbus on his head and a bow in his left hand. He seems to be ready to fix an arrow to the bow. It is impossible to judge whether the helmet was the one of Ales without a feather.

In the center of the upper part of the great composition are several white birds flying over the sky and on both sides of them the wind gods flying with scarf floating in both hands are painted. Their hair is waving. Beneath the wind gods is a pair of half-man and half-bird figures flying with a torch in their left hand. And on the right and left edges of the composition, a bank of clouds are painted. In the clouds, four faint round shapes can be distinguished.

There is no precedent for such a great composition even in the iconography of India and Gandhara. The representation of the planet painted on the vault of a rock cave at Dunhuang, and those of the 38th cave at Kizil and the 46th cave at Kumutra seem to be slightly similar but they are far beyond the magnificent one of Bamiyan.

It is of no doubt anymore that the sun god enveloped by the radiant disk is a doubled figure with Sakya as the dharma-cakravarti raja, which was metaphorically said to ride the cosmic chariot. If it was so, the number of the spokes of the chariot soaring in heaven necessarily had been 6 or twice of it 12 symbolizing the circulating time.

This composition consisting of multiple concepts cannot be made clear only by Buddhism. In Hair Khane pass descending to the lower land called Koh-i-Daman, the suburbs of Kabul, a temple of the sun god was discovered by Jean Carl, one of the members of French Archaeology Mission, in 1938 and the figure of Surya made of marble (sixth to seventh century) was excavated there. This figure of the sun god Surya suggests the close relationship with the iconography of the figure of Surya in India, and it, accompanied by the attendants Danda and Pinghara on both sides, is represented by the posture of squatting down on the chariot pulled by two horses managed by a charioteer (the wheels are not shown). The long hair and personal ornaments and costume show similarity to the sun god of Bamiyan, but from the iconographic point of view the composition is quite different from the great composition of Bamiyan, and it seems that there is radical difference between the significance that the great composition of Bamiyan transmits and that of the sun god Surya.

Relations with the World of Avesta

The great composition of Bamiyan, which places the sun god in the center must be studied taking the relationship with the Iranian world of Avesta into consideration, as Benjamin Rowland, Professor of Fine Arts at Harvard University pointed out (Rowland 1938), to get the key for elucidation. The hymn "Mihr Yasht" (Gershevitch 1967) in the sacred book of Zoroastrianism "Avesta" (edited in the fourth to sixth centuries) tells that "Mithra manages the golden chariot, the four horses which pull the chariot are all white, they eat the foods of heaven to be immortal, and they put gold hoofs on their fore-podiums and platinum hoofs on their hind-podiums." It is natural that those horses soaring in heaven are winged. The composition of four horses that pull the chariot of the sun god Helios and the composition of two horses of the twin-god Dioscuri, which soar to the same direction face to face with each other, are common styles of Hellenism that have been conveyed by means of the coins of Bactria. And now in Bamiyan, the theological motif of Sassanian Iran was poured into this mold. As "Mihr Yasht" describes, the chariot is painted a yellow color indicating "the golden chariot."

"Mihr Yasht" continues the hymn and sings, "The charioteer guiding the chariot is tall and good Asi" and thus the name of the charioteer is disclosed. About this charioteer it was impossible to distinguish anything except the legs putting on boots in the great composition of Bamiyan. Asi is the goddess of luck, one of the subordinate yazata who attends Mithra. According to the comparative mythologist George Dumezil, it is said that Asi was equivalent to Bagha of India, and was always attributed magico-religious sovereignty (Dumezil 1947).

The attendant who is seen on the left side of the sun god Mithra holding a shield in the left hand and wearing a breastplate has been compared to the figure of the goddess Pallas Athena for a long time. Even though there are some differences in the attribute or in the position of the shield, it is undoubted that this attendant is a copy of the figure of Athena, which had been already known by the Bactrian coin. In the gold coin of King Huvishka of Kushan Dynasty, as Franz Gernet clearly pointed out in his excellent article "Bamiyan and the Mihr Yasht" (Grenet 1994), the name of the goddess Athena is engraved "Rishto." Rishto is "Arstat which enlarges the world" said by "Yasht." Arstat was the goddess of justice and one of the female yazatas who attend Mithra as well as Asi. This goddess who wears the Corinthian style helmet follows the iconography of the god Athena but the meaning of the iconography is drawn out of the theology of Avesta.

The attendant on the right side who wears a helmet and makes a pair with Athena = Arstat will be the attendant Vanainti Uparatat = "the excellent power to gain victory," who makes a pair with the god Sraosa = Mithra in "Yasna" the act of worship. This is the goddess who is represented by the figure of the goddess of victory Nike but engraved by the name of "Vanindo" on the coin of Huvishka of the Kushan Dynasty. The reason why all of the three goddesses, Asi, Athena = Arstat and Nike = Vanainti are winged would be to give them the same evaluation as the goddess of victory Nike and to show these figures as archangels who are worthy of the subordinate divinities of Mithra (Grenet 1994).

The standing figure of the central god Mithra was probably produced from the idea of polyvalent iconography that places the image of the Hellenistic figure Helios on the base and puts the Sassanian iconography of Mithra on it, and furthermore puts the iconography of the traditional god of Sogdiana over it. The bust of Mithra seen in the mark of the seal of Sassan Dynasty, which the British Museum possesses, tells clearly how deeply the image of the sun god of Bamiyan depends upon the presentation of the iconography in Iran. It can be said that either of the images is painted with the iconography that exactly fits the hymn of "Mihr Yasht" that tells of "a warrior who is excellent in martial arts, holding a long spear with a sharp head." Above the head of the god Mithra, four or five white birds are flying in the sky spreading their wings. They are Hamsa (geese), which tells of the circulation of seasons. Hamsa indicate circulating seasons and also it is thought that they symbolize the moon itself as the sacred missionary birds of the moon god. We discovered the image of the moon god, which was pulled by Hamsa, on the wall of a small Buddhist cave in Bamiyan (cave M) situated to the east of the colossal Buddha statue. The flying deities with something like a scarf floating in both hands on the right and left side of Hamsa will be the wind gods Vata seen in "Mihr Yasht." The hymn tells that "the wind smashes a devil, attending Mithra" and also "the wind blows off the spear thrown by the enemy of Mithra." It was the wind god who plays a role as a herald of the sun god Mithra who soars at full speed. Avesta tells that the wind god always leads as the metamorphoses of the god Verethraguna (Dumezil 1947). I am thinking that the wind gods in pair reveal, in secret, the dualistic feature of Avestan theology.

Half-human and half-bird figures situated on the right and left sides so as to surround the solar disk together with the wind gods are Kimnaras. In Buddhist texts, Kimnara is regarded as "the god of music" and also "the god of incense" but Kimnaras in Bamiyan wear a cap hanging long narrow ribbon and hold a burning torch in the left hand and something like an incense burner with a handle in the right hand. As "Mihr Yasht" says that "fire is flying in front of Mithra," Kimnara is probably presented here with an Iranian costume in style as "the torch holder" (dadophoros).

Furthermore, the banks of swelling cloud are seen on both edges of the composition, and if the two disks positioned two by two symmetrically in the cloud could be taken as the stars, it is concluded that the six elements, the sun, the moon, the stars, the clouds, the wind, and fire exist in this great composition. Then, it might be said that those are exactly the symbolic representations of six attending divinities Amesa Spentas being projected images of the unique god of Zoroastrian theology, Ahura Mazda.

The Double Image of Mithra and Dharma-Cakravarti Raja

The god Mithra who soars at dawn in the blue sky, managing a two-wheel chariot over the mountains

Hara (might be the Hindu Kush), will be a symbol of the religious cosmology of the people who founded the colossal Buddha statue in the middle of the eastern cliff of Bamiyan. At the same time it is thought that Mithra was accepted, being doubled with the image of Sakya as the dharma-cakravarti raja turning the law-wheel, by the people who were looking for the basic ground of recovery of Buddhism in Bamiyan, which had been newly opened as an important trading base and for Buddhism taking the place of declining Gandhara. For the royalty and nobility, the double image of the sun god and the dharma-cakravarti raja must have been a welcome ideal.

It was by the far-sightedness of Joseph Hackin, who investigated the Buddhist sites of Bamiyan on a full scale for the first time, that he characterized the Buddhist arts of Bamiyan as Irano-Buddhist arts. In 1969 we discovered two long narrow ribbons flying upward from both shoulders of the sun god in the composition to reinforce his opinion.

In the extraordinarily important document (Da Tang Xi Yu Ji), Xuan Zang (Hiang tsang) refers to "celestial deva, showing signs or omen to indicate good fortune or evil according to the amount of the merchants' donation." This "celestial deva" should be the god of contract Mithra, I assume, who "is always awake and watches over" from the head of the colossal Buddha statue. Mithra was the god of truth and faith, and also the god who guarantees contracts and punishes persons who disobey the oath. For people who are engaged in trade, the sign of luck brought by the god who guarantees credit must be very important tos their life.

Mithra was also the god that as "the owner of a large pasture, brings good harvest, flock of livestock, and posterity and lives". It seems that Mithra could merge with Buddhism quite well because of its multilateral aspects as the old Aryan god.

The Dramatic Scene Painted on Both Side of the Wall

On the eastern and western wall beneath the great composition, the sitting images of Buddha and Boddhisattvas, and the scene of the royal family procession led by a priest in profile are painted, facing the laterals of the Buddha statue. Being separated from the royal procession, Boddhisattvas situated on both sides of the sitting Buddha, which wears a garment in a way leaving the right shoulder and breast bare, are painted in images sitting in the air outside of the balustrade to divide the space into two, one for saints and another for laics. Buddha does not wear a ribbon but the Boddhsattvas wear it. Each Boddhsattva wears a shawl that has two parts cut out mountain-shape, and has a sacred cloth in hand. Hackin called the unique Boddhisattva who wears a headdress and a ratnavali (necklace) "the bejeweled Buddha," and according to Paul Mus, he concluded that this was one of the peculiar representations of Buddhist paintings of Bamiyan. Rowland took the Buddha and the Boddhisattvas as the representation of seven past Buddhas and he regarded them as the representation of "Sambhoga-Kaya" (image of Buddha that has the perfect benevolence) developed by the thoughts of the Mahasanghika (Rowland 1938).

The royals bearing the nimbus might be regarded as the saints. Most of the royal families painted on the eastern wall have nimbus and ribbons on their shoulders but any of the royal families painted on the western wall do not have them. This might indicate the differences in rank order among them. The worship procession of royal families is led by a priest from right to left, facing to the wall.

The person just behind the priest will be the king of Bamiyan who wears a characteristic crown, a round-neck tunic and a Sassanin cross band, holding a sacred ribbon in his right hand. Someone wears the Central Asian nomad clothes with the one-side-turned-down neck opened to the right, which are seen quite often in Toharistan. They are attending the ceremony for worship respectively having a sacred object, flower, and a ring (khwarnah) in their hands. The royals, nobles, and benefactors who pushed forward the building of the colossal Buddha statue and now attend the ceremony to commemorate the completion of the project, perhaps took their seats at the balustrades provided just under the mural painting on both side walsl.

They who dressed pompously must have ascended to the balustrades on the occasion of "pancavarsika" held every 5 years. The donation or charity, that is "dana" in Sanscrit, to be done in pancavarsika was one of the very important moral deeds for the worshipers of the Buddhism. And dana, gorgeous dana by the king would be, as Emile Benveniste explicated in his wonderful book (Benveniste 1969), the religious and social demonstrative activity. In that sense, I think that the mural paintings of the vault of the Eastern Colossal Buddha were painted to represent as an unforgettable moment, the national monumental festival that was held periodically in the form of a politico-religious assembly rather than to represent the sublime Buddhist world of Sakya Tathagata.

These masterpieces of the mural paintings of the Eastern Colossal Buddha completely disappeared in 2001.

References

Benveniste E (1969) Don et echange. Le Vocabulare des Institutions, Indo-Europeennes. Paris
Dumezil G (1947) Tarpeia. Gallimard, Paris
Gershevitch I (1967) The Avestan hymn to Mithra (translation). Cambridge
Grenet F (1994) Bamiyan and the Mihr Yasht. Bulletin of the Asia Institute, 7
Rowland B Jr (1938) Buddha and the sun god, Zalmoxis. Paul Geuthner, Paris

The History of the Buddhas of Bamiyan

3

Guido Martini and Salvatore Paolini

If we wanted to compile a rating of the most mysterious monuments, the Buddhas of Bamiyan would certainly occupy a position of great importance. No information is available about the patron, the artist who coordinated the cutting of the cliff, its duration, nor any other information useful for dating.

It almost seems that during the history of humanity a hypothetical curtain was raised on the Valley of Bamiyan, allowing the world to admire its gigantic statues. These statues, commissioned to glorify Buddhism, mark the geographical limits of Buddhist westward expansion; and from 1222, which marks the destruction by Genghis Khan of the housing installation, they represent the only witnesses of the greatness and economic-political importance that the valley once assumed in the region.

Lack of news about the statues is reflected in the literary sources. One of the first literary sources available is the beautiful travel account written by the Chinese monk Hiuan Tsang in the seventh century, followed by a long period of silence as a consequence of the Mongolian destruction. The colonial period, with its race for predominance in the East, occasioned new interest in this region, particularly by the United Kingdom. Consequently, from the beginning of the nineteenth century a great number of studies, drawings, surveys, associations, and photographs began to be published, which spread the knowledge and beauty of the Bamiyan Buddhas to the vast European public.

G. Martini • S. Paolini (✉)
Department of UTPRA-PREV,
ENEA – Italian Agency for New Technology Energy and Environment, CR Frascati Via Enrico Fermi, 45, 00044 Frascati (Rome), Italy
e-mail: guido.martini@enea.it; salvatore.paolini@enea.it

Ancient Authors

In 629 A.D., a Chinese Buddhist monk named Hiuan-Tsang started a journey to India to improve his study of sanskrit books. Following the Silk Route with risk to his life, he traveled westward, passing through Afghanistan. He returned to China in 645, was received with many honors, and began to transcribe the memoirs of his travels.

The literary work of Hiuan-Tsang (Fig. 3.1), despite its fantastic aspects, is quite precise, especially pertaining to the places visited. His descriptions are so detailed that they will be useful for future French archaeological expeditions.

Describing the colossal statues, Hiuan-Tsang noted that near the two statues in a standing position, there was a third statue of a "sleeping Buddha." In another version of these memoirs, the sleeping Buddha is described as being adorned with gold and jewels. This statue may be identified according to Yule (Talbot et al. 1886), with those described by Kaye (Kaye 1879)

> In the North East of the Royal City, in the flank of the mountain, there is one statue of a standing Buddha made in stone; it is tall between 140 to 150

C. Margottini (ed.), *After the Destruction of Giant Buddha Statues in Bamiyan (Afghanistan) in 2001*,
Natural Science in Archaeology, DOI 10.1007/978-3-642-30051-6_3, © Springer-Verlag Berlin Heidelberg 2014

Chronology

Fourth century B.C.	The site is not mentioned by the historians and geographers of Alexander Magno, who invades the region in 329 B.C.
250 B.C.	Conversion to Buddhism of the great king Ashoka of the Maurya dynasty
First century A.D.	Presumed dating of the Buddhist settlement in the Bamiyan Valley
First to fourth century A.D.	Reign of the Kushana, with its ancient capital Kapisia (Begram). Beginning of the spread of Buddhism. The Afghan king Kanishka converts to Buddhism.
241	The Sassanids from Iran conquer Bamiyan.
371–420	Conquest by the Huns of Eftaliti tribe
560	Turkish invasion at Balkh
630	First account of the reign and city of Bamiyan by Hiuan-tsang
727	The Korean pilgrim Houei-tch'ao visits the Buddha of Bamiyan
781	Yakib ben Laith, an Iranian prince, plunders the sanctuary at Bamiyan
876	Kabul is conquered by the reign of Khorassam. Start of Hindu rule.
977–997	Turkish (Gaznavidi) army defeats the Hindu troops
1222	Bamiyan is conquered and completely destroyed by Genghis Khan
1318	Supposed crossing of the Bamiyan Valley by the Italian monk Odorico da Pordenone
1678–1707	Aurangzeb is the first Muslim governor who shoots guns at the Bamiyan statues
1798	First detailed description of Bamiyan site by Francis Wilford
August 25–27, 1824	William Moorcroft and George Trebeek, officials of the East India Company, cross the Bamiyan Valley
December 26, 1828	Edward Stirling reaches the Bamiyan Valley, coming from Turkestan
1832	Alexander Burnes visits the Bamiyan Valley; he draws the firsts sketches of the statues
1830–1837	Charles Masson conducts archaeological excavations in the Bamiyan Valley
1840	Lt. Gen. Kaje passes through the Bamiyan Valley with English troops
1839–1842	Vincent Eyre visits the site of Bamiyan
1880–1901	Abdur Rahman is proclaimed King. The Afghan reign is founded.
November 13, 1885	Description of the Valley by Talbot and Maitland
1919–1928	Reign of Amanullah
1922	Exclusive rights are granted to France for archaeological research in Afghanistan after an agreement between the French and Afghan governments
1931–1973	Afghanistan becomes a constitutional monarchy
1931	The Bamiyan Valley is crossed by the Citroen Asiatic Expedition
1934	The Italian reporter Arnaldo Cipolla visits the Bamiyan Valley
1960s	Agreement is reached between Afghanistan and the Indian government for archaeological excavations and consolidation works in the Bamiyan Valley
Summer 1969	Bruce Chatwin and Peter Levi visit the Valley
1973–1978	Afghanistan becomes a republic
1978	President Daud is killed
December 27–28, 1979	Afghanistan is invaded by Soviet troops; the Afghan War begins
February 15, 1989	Soviet troops are completely retired from Afghanistan
September 1998	Taliban troops occupy the Bamiyan Valley
February 26, 2001	Taliban leader Mullah Mohammed Omar orders the destruction of all statues, including the Buddhas of Bamiyan. The destruction of the statues is submitted to the Minister for the Development of Virtue and the Repression of Vice and to the Minister of Culture
March 2, 2001	Taliban troops begin the destruction of the first statue (the big one), in spite of international protests
March 8, 2001	The big statue is destroyed
March 9, 2001	The second smaller statue is destroyed

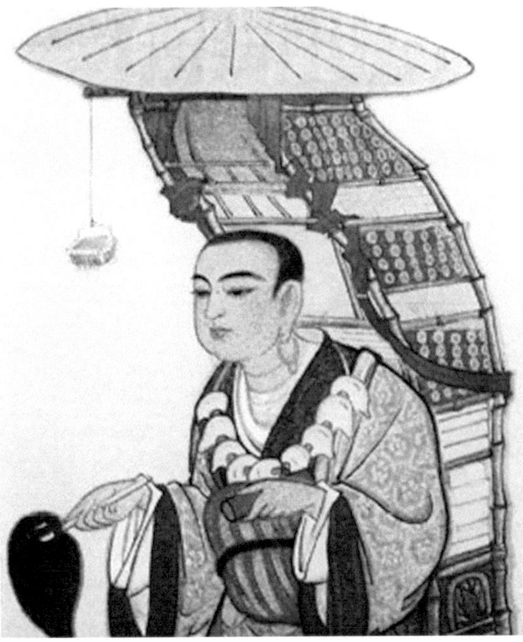

Fig. 3.1 Hiuan-Tsang (http://www.btrts.org.sg/venerable-xuan-zang)

> feet, its golden colour is dazzling and the precious ornaments are shinning. Eastward [of this statue] there is another statue of standing Buddha named "Che-kia", that is tall more than 100 feet. The body has been constructed in assembled blocks, subsequently covered and adorned. In two or three "li" [an ancient measure unit] in the direction of the Royal City [...], there is a "sleeping Buddha" statue, length more than 1000 feet. Here the King of the Country organizes the great assembly of "Wuo-tcho" [a great five-year assembly] (Hackin et al. 1928).

After quite a century, another Buddhist monk, Houei-tch'ao, coming from Korea, reached the Bamiyan Valley in 727 A.D. We do not know the reasons for his travel, but like Hiuan-tsang, he wrote memoirs entitled "Wang wou t'ien tchou kuouo" (account of a travel in the five Indias). Three chapters were written in a synthetic style and, like the previously mentioned account, he referenced "three jewels," referring to the Buddha's images, which confirms the presence of a third statue.

> Then, seven days north of Sie-yu, he reached Fan-yin (Bamiyan). The king was a Hou (Iranian). This kingdom was independent. Its foot-soldiers and horse-soldiers were strong and numerous, and other kings did not dare to attack it. In dressing, the people wore cotton shirts, padded shirts, and padded clothes. In this country there were horse-breeding and ram-breeding, the cultivation of cotton, and abundant vineyards. It was snowy and cold; many houses were constructed in the mountains. The king, the ministers, and the people were devoted to the three jewels; monks and monasteries were numerous (Hackin et al. 1928).

At the end of tenth century the Bamiyan Valley fell under Muslim rule. The Buddhist religion began to lose its influence and the colossal statues of Buddha begin to decline in importance and suffer the offenses of the new religion. In spite of this, the Bamiyan Valley maintained its importance in communication and trade, increasing its prestige over the surrounding area, as testified by the accounts of Arab geographers.

Ibn Abdallah El Roumi El Hamawi Yaqout (1179–1229 A.D.), a geographer, theologian, and erudite trader and traveler, who lived in that period, describes Bamiyan in these terms:

> Bamiyan. This is the name of a city and an important district between Balkh and Ghaznah, settled between mountains, with one fortress. This city is small, but it is the main town in an extensive territory. Ten days of march separate it from Balkh and eight from Ghaznah. There you can admire a building supported by big pillars and covered of frescoes representing all the birds created by God. [In Bamiyan] there are two statues carved in the stone cliff and high from the base of the mountain until its top. One is called Surkhbud, the red statue, and the other Khinkbud the white statue. Nothing can be compared with these statues in the entire world (Yaqout 1970).

Abou'l Feda, a geographer born in Damascus 672–1273 A.D., confirmed the importance of Bamiyan:

> The cities under the jurisdiction of Bamiyan were Baghschour, Sakawand, Kabul, Djara, Farwan Ghaznah and Pandjhir (El Feda 1848).[1]

Hamd Allah Mustawfi El Qazwin (1340 A.D.) remembered the history of the fall of Bamiyan and provided information on the natural resources of the Valley.

> Bamiyan. At the time of the Mogol invasion Prince Mutukin, son of Jaghatay Khan, met his death, and for this reason, to avenge his grandson, Chenghiz Khan ordered the town to be laid in ruins, renaming the place Mav Baliq ("Bad Town" in Mongol), and commanding that no one should ever build or

settle there. And so it remains a ruin even to the present time.

Iron. Of this metal there are many mines, and in the kingdom of Bamiyan is a district called The Iron Foundry (Ahan Kar).

Sulphur. There is also sulphur in the mine at Bamiyan, where there is a spring; and here the water bursts forth with such violence that the sound of it may be heard at some distance away, but when the water has run for a certain distance it petrifies, and then forms brimstone (Qazwin 1919).

The decline of the Bamiyan Valley coincides with the advance of the Mongolian troops. In 1222 Genghis Khan, coming from the East with an impressive army, swept away the Muslim defenses one after another, imposing its "Pax Mongolica." After this invasion, the entire Valley of Bamiyan knew one of its most mournful days:

[Coming] from Enderab, Genghiz Khan besieged Bamiyan, where the inhabitants sheltered themselves in its walls and fought with great courage. A son of Djaghatai Khan [Mutugan] was killed from an arrow during the battle.

Genghiz Khan, full of anger for the loss of this young and particularly appreciated prince, ordered a general attack. The city was stormed and all its inhabitants were massacred without distinction in age or sex. The fortress was razed to the ground. The Khan gave to this place the name of Maw-Baligh, that is Yaman Qal'a. Evil Place (Aboul Ghazi 1871).

After all these events, the Valley of Bamiyan did not reach its previous splendor. The new trade sea routes brought about a further decline to the area (Mehendale 1996). The colossal Bamiyan statues remained unique dumb witnesses of a definitively passed magnificence. The ruins of the ancient city of Bamiyan were named Shar-I-Gholgola, "the city of the laments," those unlucky inhabitants massacred by Genghis Khan.

A further Arabic author, Abu' l Fazl, provided more information about the state of the Bamiyan fortress and its statues.

In the Tuman of Zohak and Bamiyan, the fortress of Zohak is a monument of great antiquity, and in good preservation, but the fort of Bamiyan is in ruins. In the mountain-side caves have been excavated and ornamented with plaster and paintings. Of these there are 12.000 which are called Sumaj and in former times were used by the people as winter retreats. Three colossal figures are here: one is a statue of a man, 80 yards height; another that of a woman 50 yards high, and the third is that of a child measuring 15 yards (Fazl-Abu-Fazl 1891).

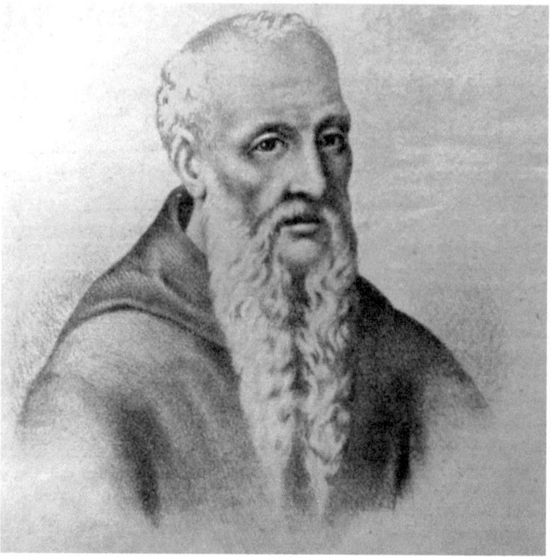

Fig. 3.2 Friar Odorico da Pordenone (http://en.wikipedia.org/wiki/Odoric_of_Pordenone)

A vivid description of the sense of desolation, abandonment, and death that wrapped this region, is reported in the memories of an Italian traveler, the friar Odorico da Pordenone (Fig. 3.2).

The friar Odhoricus, started his travel toward China with a precise scope: *di questo alcuno frutto che sia utile all'anime nostre guadagnare*; that is, to convert the greatest possible number of souls in the East. His travel began in 1318 and ended in 1330. In the successive year, January 14, 1331, he died in Udine, and in 1755 he was canonized.

Unfortunately, in his memoirs Odorico did not report the name of the valley he crossed.[2] The reference to a valley: *da un lato del monte vidi sulla stessa pietra della parete una faccia d'uomo assai terribile* (in a side of the mountain I saw on the same stone of the wall a very terrible face of a man), could be connected with a rough description of the Buddhas of Bamiyan (Yule and Cordier 1913, Pullè 1931).[3] Here follows the Odorico account:

De valle quadam in qua terribillia vidi

Aliud terribile magnum vidi nam cum irem per vallem, que est posita supra flumen deliciarum in ea multa corpora vidi morta in qua etiam audiebam diversa genera musicorum maxime autem nachara que ibi mirabilliter pulsabantur quare tantum clamor erat ibi quod michi timor maximus inferebat & est vallis longa septem vel VIII milliaribus in qua si aliquis infidelium intrat nunquam egreditur sed statim moritur sine mora & tamen non dubitavi in illam intrare ut viderem finaliter quid hoc effet & cum intrassem vidi ibi tot corpora mortuorum, quod nisi quis vidisset incredibille putaretur a latere etiam vallis in ispo saxo vidi faciem hominis maximam & terribillem que in tantum terribillis erat quod pro nimio timore spiritum penitus perdere me videbam qua propter cum signo crucis verbum caro factum est oretenus proferebam ad ipsam tamen facies numquam ausus fui propinquare sed ab ipsa VII vel VIII passibus distans ivi tandem ad aliud capud vallis & ascendi montem quemdam arenosum in quo undique circumspiciens nichil videbam nisi quod audiebam nachara illa pulsare, que mirabilliter pulsabantur.

Cum autem in capite montis fui illic aurum & argentum reperi in maxima quantitate quasi squame pisscium congregatum de quo accepi in gremio& quia de ipso non curabam & timens etiam ne tali illusione forte michi denegare exitus illud totaliter in terra proieci & sic deo dante inde illexus exivi.

Tunc omnes saraceni qui hoc sciverunt reverebantur me multum dicentes me esse baptizatum & sanctum. Illos autem qui in illa valle mortui erant dicebant esse homines demonis infernalis (Benoffi 1761).[4]

The Death of the Old Man of the Mountain

There was another terrible thing which I saw there. Passing by a certain valley, which is beside a pleasant river, I saw many dead bodies, and in the valley also I heard divers sweet sounds and harmonies of music, especially the noise of citherns. I was greatly amazed. This desert valley is in length seven or eight miles at the least, into which any one who enters dies presently, and can by no means pass alive through it. Moreover, I was tempted to go in, and to see what it was.

At length making my prayers, and recommending myself to God in the name of Jesus, I entered, and saw such swarms of dead bodies there was no man would believe unless he were an eye-witness thereof.

At the one side of the valley in a certain rock, I saw the face of a man, which beheld me with such a terrible aspect that I thought verily I should have died in the same place.

But always this sentence, "The Word became Flesh, and Dwelt amongst us", I began to pronounce, making the sign of the cross, and nearer than seven or eight paces I dared not approach to the face in the rocks. But I departed and fled to another place in the valley, ascending up to a little sandy mountain, where looking round about, I saw nothing but heard the citherns, which continued sounding and playing by themselves without the help of musicians.

And being upon the top of the mountain, I found silver there like the scales of fishes in great abundance; and I gathered some into my bosom to show for a wonder, but my conscience rebuking me, I cast it up. And so, by God's grace, I departed without danger. And when the men of the country knew that I was returned out of the valley alive, they reverenced me, saying that I was baptized and holy, and the bodies were men subject to the devils infernal, who used to play upon citherns, to the end they might allure people to enter, and so murder them. Thus much concerning those things which I beheld most certainly with mine eyes, I, Friar Odoric, have here written. Many strange things also I have of purpose omitted, because men will not believe them unless they seen them (Komroff 1928).[5]

Modern Authors

In modern times, the explorations and successive colonial periods gave a decisive impulse regarding knowledge of these sculptures. After the uncertain memories of friar Odorico, the first European account regarding the colossus of Bamiyan was provided by Thomas Hyde (1636–1703), a teacher at Oxford University and a librarian in the Bodleian library (Hyde 1760). The work of Thomas Hyde, written in Latin and based on the collection of several Islamic sources, did not receive great interest.

More than a century after the death of Thomas Hyde, the United Kingdom became interested in Afghanistan. The Afghan region was crossed by many English visitors with a variety of interests. Explorers with scientific interests, civil employees of the East India Company, soldiers during the movements of troops, as well as spies and travelers, all were attracted by the adventure.[6] The curtain once lowered over the valley and its statues began to raise once more.

In 1819, two explorers, William Moorcroft and George Trebeck, crossed the Bamiyan Valley to find new trade routes and zones for the East India Company. Moorcroft and Trebeck, whose travel memoir was published posthumously by the Royal Asiatic Society in 1841, reported the difficulties and dangers faced during these expeditions.[7] Moorcroft died on August 27, 1825, of a disease contracted during travel, and the same fate awaited George Trebeck a few days later.[8]

> The figures stood in niches or recesses cut out of the rock, the upper part of which was arched so as to form an alcove or vaulted canopy over the head of the figure. The sides advanced so as to form wings, in which were staircases ascending to a gallery behind the neck of the statue, whereas other galleries ran off from their sides, right and left, into the rock. The flights of steps of the larger image were so much decayed as to be inaccessible, but one of those on the side on the smaller was mostly intact, and led to the head of the figure. It was said that both figures were constructed by order of Aurangzeb (Moorcroft and Trebeck 1842).

About 1830 another English traveler, Charles Masson, who was a real adventurer, crossed the Valley of Bamiyan. Masson, whose real name was James Lewis, was a British Army deserter, and from 1827 he began to tour through Afghanistan, as traveller, archaeologist, spy and writer.[9]

Masson's published report on the two Buddha statues is important for two reasons: first because this was the first time a sketch of the valley with the Buddha niches and caves was published; and second because he described the complex access structure around the colossus by which it was possible to reach to the head of the statues.

A nearly hidden sentence was written in coal on the head of the great Buddha: *If any fool this high samootch explore, Know Charles Masson has been here before.*[10]

> The idols were cut or hewn into the rock, and were covered with a surface of cement. They were erect figures, with their hands extended, supporting the folds of drapery in which they were clothed. Their features were destroyed by removing half of their heads, or as far as the lips, and leaving the hinder halves, with enormously large ears, appended.
>
> The work of mutilation required considerable labor, was executed with precision, and must have been directed by authority, possibly that of the Arabian conqueror. A subsequent and less systematic mutilation was also carried out on the idols, by breaking off their hands and fracturing their legs, the merits of which Jenghiz, Timu'r, Aurengze'b, and even Timur Shah, who are all accused, may dispute.
>
> The idols stand in vast niches formed in the rock, whose sides have been embellished with paintings on a level with their necks. These consist of busts and seated figures, both male and female. The niche of the superior idol has a line of twelve female figures on each side, and there is an important inscription at the summit of the idol's head that is obviously intended to reveal the mystery. The niche in which the second idol stands has no inscription, but has twelve male and female busts on either side, among which is one so valuable that its loss is regretted. In the absence of a literal testimony, over the head of this idol, a full-length female figure has been painted. The niches of the other idols are also embellished with paintings.
>
> On either side of the niches are series of stairs, cut in the rock, which lead to the heads of the idols. Each series of steps leads to a small square apartment, and these apartments have been superbly decorated with gilding and lapis lazuli. To illuminate the passages, apertures have been cut through the rock toward the idols. We ascended to the summit of the second idol by the passage on one side, and walking round the hinder part of its head, descended by the steps on the other side. Near the summit or above the lines of paintings, the niches widen, and on either side a Takhit, or sofa, obviously for the convenience of sitting upon, has been formed. The superior idol has or had the same facilities of ascent to the summit, but at the time of our visit the lower caves near it

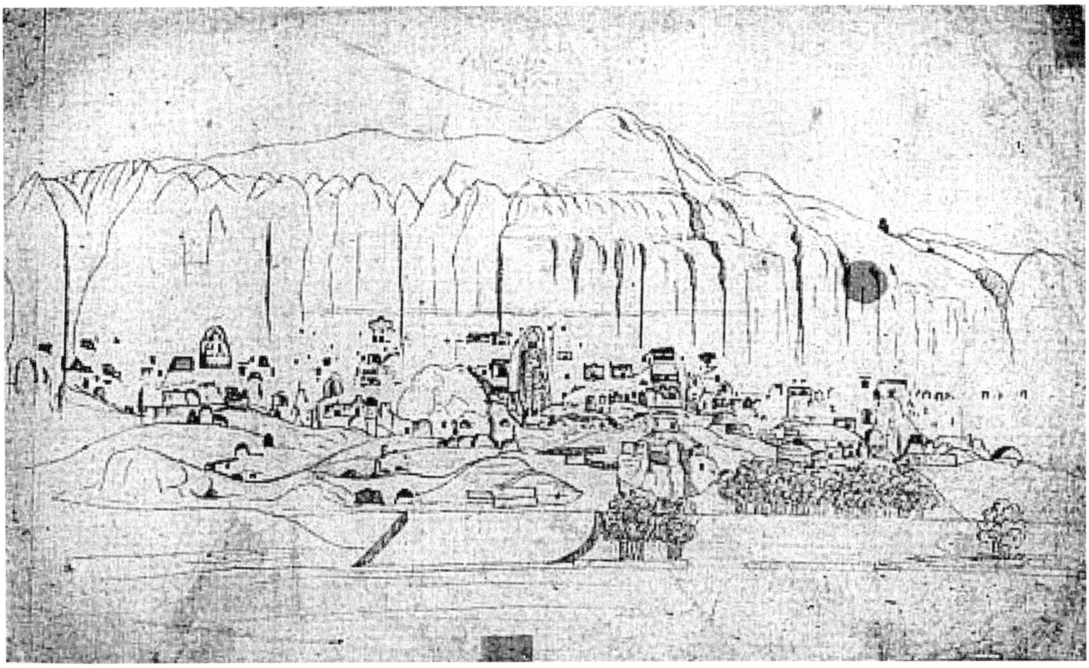

Fig. 3.3 Charles Masson, Notes on the Antiquities of Bamiyan, in JASB, V, 1836

were occupied by an unaccommodating Tajik, who had stowed his stock of provender in the passage. We could not prevail upon him by menace or entreaty to open the path, and he evasively affirmed that he had never heard of one. We did not insist, hoping to make a future visit, which until now has not happened. It is important to gain entrance to the upper stations so we may inspect the paintings.

Between the legs of the superior idol are entrances that conduct one into spacious apartments surmounted by domes, which many imagine to have been temples. In common with the other caves, they were covered with cement, with which the moulding surrounding their circumferences also was formed, as well as the ornaments at the summits of the domes. The interiors are glossy black from the smoke of fires that were kept up in them. Many of the caves at Bamiyan are remarkable for their dimensions, and have other peculiarities in their form and embellishments. Most are found above the superior idol, but in another cliff rising backward, so that in walking from them to the front or south, we reach the edge of the perpendicular wall of rock in which that sculpture is carved. They are gained by a slight ascent to the left or west of the idol (Fig. 3.3). In these caves we saw written in charcoal the names of W. Moorcroft, W. Trebeck, and G. Guthrie! (Masson 1836)

In the same period another British traveler, Edward Hamilton Stirling, without specific military training or great experience as an explorer, was the first European to cross the Hindu Kush mountains, passing through Turkistan. Stirling, a member of the Scottish upper class, was a civil employee of the English tax collector's office in the Indian district of Agra. His travel was personally motivated. In his notes, Stirling describes the hazards he encountered during his travels: the sourness of the geography, the climate, diseases, marauders, and the hostility of the local people.[11] The following is how Stirling described the Bamiyan Valley:

> 26th December 1828. After making a slight halt, or rather unloading our animals for the night, we again marched and reached the valley of Bamian at the close of the evening. We saw there two enormous images cut in the sides of the vertical face of the indurate clay hills on the left of the valley. These no doubt existed prior to the introduction of the Mahomedan religion. They are of gigantic size and were probably, as tradition gives out, worshipped as the gods of Bamian. They are simple images with no extra limbs. Their garments appear to have been loose and flowing, like those of the

Hindoos and Bengalis of the present day. The Mahomedans have disfigured them by cutting off their features, eyes and nose, and the face is sawed off as far as the mouth. They stand erect. They have nothing to recommend them except their singular appearance in the wall, and the tradition the natives tell of them in praise of their present religion and the efforts of Shah Mardan and Murtaz Ally Sheer; but it may be remarked as extraordinary that from the valley of Sighan the hills are perforated for the purpose of offering dwelling places. Indeed, this practice seems to have prevailed in all the country between Kooloom and Bamian, and perhaps is of a most extensive practice in the hills on the north of the Cabool dominions, and in no place that I have seen are the hills more perforated then in those on the sides of the valley now the subject of observation. An indurate clay formation seems particularly adapted for this purpose, and hundreds of holes or dwelling place are seen far and near to these images. This custom of burrowing in the hills is undoubtedly one of the most ancient modes of securing the human person from the severities of the weather, but it must be acknowledged that it proves the race unremoved from a savage state, and that civilization has never made progress among them.

I expected to enter a large town at least on reaching Bamian, and was much surprised at finding that we encamped under the walls of a square fort whose side scarcely extended two hundred feet, and was told that this was Bamian, or rather that the whole valley was so called, and this was merely the residence of the chief of it.

On the opposite side of the stream, placed at some distance on the eminence of an isolated hill of considerable magnitude and of a conical shape, are seen the ruins of an ancient extensive city which bore the name of Bamian, or the city of Goolgoolah. It is now entirely uninhabited. The valley of Bamiyan contains about twelve killahs. Besides the inhabitants of these, a number of people inhabit the burrows before mentioned excavated in the sides of the hills, which I found on closer inspection to be composed generally of clay or sandstone conglomerate (Lee 1991).

In September 1840, after the occupation of the Afghan territory by the British Army, Colonel E. Kaye commander of a contingent of British soldiers, crossed the Valley of Bamiyan. After some years Kaye, during a meeting of the Royal Geographic Society, remembered the valley this way (Kaye 1879):

At night, on the 5th October, our camp was formed on the left bank of the Kalu stream, and on the right of that flowing from Bamian, at the junction of the two waters. The gorge of the defile leading from the Kalu Pass was on our left, as we looked to the west; and at its entrance on the summit of a loftily insulated rock, with perpendicular faces, frowned down the ruined fortress of the Emperor Zohauk, whence the place takes its name. At this extremity the valley is about a quarter of a mile in breadth, and well-cultivated, the immediate hills bordering on it of no great height. The Bamian River, rapid and of some volume, though generally fordable, hides itself in a rocky passage, dividing the spurs branching down from the southern and northern ranges. It takes its rise from the foot of the westernmost of Koh-I-Baba, as do also, I conjecture, the rivers which water the parallel valleys of Saighan and Kamard.

The great image cut in the face of the cliff bounding the valley on the north in 9 miles from Zohauk. The valley winds much, varying in width, generally not more than a quarter of a mile broad, until Bamian is reached, where it opens out considerably. It is well cultivated, but there are few trees. Several narrow glens, the channels of streams flowing from the Koh, on the south, fall into this valley; one of them is at Topchi, or Ahinghar, a small fort about half-way from Zohauk. But the largest, which is indeed a valley itself, having numerous terraces of fields on either side of the rivulet, joins that of Bamian nearly opposite the images: this is the Fouladi Vale. The two streams converge in the main valley, and unite near a small fort situated in the fork itself.

Higher up, near the idols (of which I have little to say except that they are very large and very ugly), there are, or more correctly there stood at the time whereof I write, three forts, forming together a triangle: two of these, having four towers (one at each corner of the square), were assigned to the infantry; and the third, a double fort, with six towers and a dividing wall in the middle, to the artillery, for the accommodation, one part, of the men, the other of the gun horses: the cavalry were sent back to Kabul. The political officers and their escort were lodged at the confluence of the rivers.

Near the foot of the great image, on some rising ground, there were the ruins of a fort, which must have been of considerable magnitude, much stronger and more capacious than any of those still existing in the valley. Between the images and at their sides, peeping over their shoulders, and some even above their heads, were many caves in the cliff side, having intricate connecting approaches, and galleries cut within the rock. These formed dwellings for many Bamianchis, and also for some camp followers of the British.

On the opposite side of the valley, about a mile to the west, a stony gully leads into the hills: a short way up this, there is a nearly insulated rock, on the

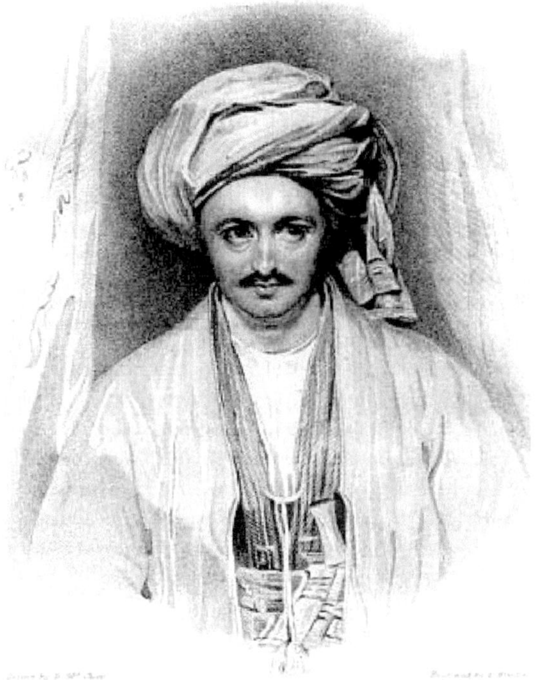

Fig. 3.4 Burnes Alessandro, Viaggi di Alessandro Burnes, tip. Giachetti, Prato, 1842

flat summit of which there is in relief a recumbent figure bearing a rude resemblance to a huge lizard, and near the neck of the reptile there is a red splash, as of blood. This is called the Azdahar, or the dragon, said to have been slain by Ali or some Mahomedan saint of by-gone days, and an indication in the rock close by is held to be the gigantic footprint of the slayer.

A great success in terms of public interest was conferred on the travels of another Englishman (Fig. 3.4), Alexander Burnes (1805–1841).

Burnes, at that time an officer in the British Army, was in Afghanistan as a diplomat. In 1832 he began to travel through the Hindu Kush coming from Kabul. His emotion at the sight of the two colossi was so strong that he decided to publish the memoirs of his travels in a book. Unfortunately, Burnes did not enjoy success from his books, *Travels into Bokhara* and *On the Colossal Idols of Bamiyan* (Burnes 1833)." In the meantime his diplomatic career was broken. At the beginning of the Afghan rebellion,[12] Burnes and his brother were killed in Kabul at the Khyber Pass slaughter in 1841.

The work of Alexander Burnes has been appreciated not only in England, but also in foreign countries. Thanks to Burnes' books, Western society received new information about the Bamiyan Valley and, for the first time, Europeans appreciated the rituals, traditions, and cultures of Afghanistan, which were—and are—greatly various and different from their own (Burnes 1834).[13]

Nothing could be more grand than the scenery which we met in this valley. Frightful precipices hung over us; and many a fragment beneath informed us of their instability. For about a mile it was impossible to proceed on horseback, and we advanced on foot, with a gulf beneath us. The dell presented a beautiful section of the mountains to the eye of the geologist; and, though a by-path, appeared to have been fortified in former years, as innumerable ruins testified. Some of these were pointed out as the remnants of the post-houses of the Mogul emperors; but by far the greater number were assigned to the age of Zohak, an ancient king of Persia. One castle in particular, at the northern termination of the valley, and commanding the gorge, had been constructed with great labour on the summit of a precipice, and was ingeniously supplied with water. It would be useless to record all the fables of the people regarding these buildings.

Bameean is celebrated for its colossal idols and innumerable excavations, which are to be seen in all parts of the valley, for about eight miles, and still form the residence of the greater part of the population. They are called "Soomuch" by the people. A detached hill in the middle of the valley is quite honeycombed by them, and brings to our recollection the Troglodites of Alexander's historians. It is called the city of Ghoolghoola, and consists of a continued succession of caves in every direction, which are said to have been the work of a king named Julal.

The hills at Bameean are formed of indurated clay and pebbles, which renders their excavation a matter of little difficulty; but the great extent to which it has been carried, excites attention. Caves are dug on both sides of the valley, but the greater number lie on the northern face, where we found the idols: altogether they form an immense city. Labourers are frequently hired to dig in them; and their trouble is rewarded by rings, relics, coins & c. They generally bear Cufic inscriptions, and are of a later date than the age of Mahommed.

These excavated caves, or houses, have no pretensions to architectural ornament, being no more than squared holes in the hill. Some of them are finished in the shape of a dome, and have a carved frieze below the point, from which the cupola springs. The inhabitants tell many tell many remarkable tales of the caves of Bameean; one in particular – that a mother had lost her child among them, and recovered it after a lapse of twelve years! The tale need not be believed; but it will convey an idea of the extent of the works.

There are excavations on all side of the idols; and below the larger one, half a regiment might find quarters. Bameean is subject to Cabool: it would appear to be a place of high antiquity; and is, perhaps, the city which Alexander founded at the base of Paropamisus, before entering Bactria.

The country, indeed, from Cabool to Balkh, is yet styled "Bakhtur Zumeen", or Bakhtur country. The name of Bameean is said to be derived from its elevation, – "bam" signifying balcony, and the affix "eean" country. It may be so called from the caves rising one over another in the rock.

There are no relics of Asiatic antiquity which have roused the curiosity of the learned more than the gigantic idols of Bameean. It is fortunately in my power to present a drawing of these images. They consist of two figures, a male and a female; the one named silsal, the other Shahmama. The figures are cut in alto relievo on the face of the hill, and represent two colossal images. The male is the larger of the two, and about 120 feet high. It occupies a front of 70 feet; and the niche in which it is excavated, extends about that depth into the hill.

This idol is mutilated; both legs having been fractured by cannon; and the countenance above the mouth is destroyed. The lips are very large; the ears long and pendent; and there appears to have been a tiara on the head. The figure is covered by a mantle, which hangs over it in all parts, and has been formed of a kind of plaster; the image having been studded with wooden pins in various places, to assist in fixing it.

The figure itself is without symmetry, nor is there much elegance in the drapery. The hands, which held out the mantle, have been both broken. The female figure is more perfect than the male, and has been dressed in the same manner. It is cut in the same hill, at a distance of 200 yards, and is about half the size. It was not to be discovered whether the smaller idol was a brother or son of the Colossus, but from the information of the natives.

The sketch which is attached will convey better notions of these idols than a more elaborate description. The square and arched apertures which appear in the plate represent the entrance of the different caves or excavations, and through these there is a road which leads to the summit of both the images. In the lower caves, the caravans to the upper ones are used as granaries by the community.

I have now to note the most remarkable curiosity in the idols of Bameean. The niches of both have been at one time plastered, and ornamented with paintings of human figures, which have now disappeared from all parts but that immediately over the heads of the idols. Here the colours are as vivid, and the paintings as distinct, as in the Egyptian tombs. There is little variety in the design of these figures; which represent the bust of a woman, with a knob of hair on the head, and a plaid thrown half over the chest; the whole surrounded by a halo, and the head again by another halo. In one part, I could trace a group of three female figures following each other.

The execution of the work was indifferent, and not superior to the pictures which the Chinese make in imitations of an European artist.

The traditions of the people regarding the idols of Bameean are vague and unsatisfactory. It is stated, that they were excavated about the Christian era, by ta tribe of Kaffirs (infidels), to represent a king, named Silsal, and his wife, who ruled in a distant country, and was worshipped for his greatness. The Hindoos assert that they were excavated by the Pandoos, and that they are mentioned in the great epic poem of the Mahaburat. Certain it is, that the Hindoos, on passing these idols, at this day, hold up their hands in adoration: they do not make offerings; and the custom may have fallen into disuse since the rise of Islam. I am aware that a conjecture attributes these images to the Boodhists; and the long ears of the great figure render the surmise probable. I did not trace any resemblance to the colossal figures in the caves of Salsette, near Bombay; but the shape of the head is not unlike that of the great trifaced idol of Elephanta.

At Manikyala, in the Punjab, near the celebrate "tope", I found a glass or cornelian antique, which exactly resembles this head. In the paintings over the idols I observed a close resemblance to the images of the Jain temples in Western India, on Mount Aboo, Girnar, and Politana in Kattywar. I judge the figures to be female; but they are very rude; though the colours in which they are sketched are bright and beautiful. There is nothing in the images of Bameean to evince any great advancement in the arts, or what the most common people might not have easily executed. They cannot, certainly, referred to the Greek invasion; nor are they mentioned by any of the historians of Alexander's expedition. I find, in the history of Timourlane, that both the idols and excavations of Bameean are described by Sherif o deen, his historian. The idols are there stated to be so high that none of the archers could strike the head. They are called Lat and Munat; two celebrated idols which are mentioned in the Koran: the writer also alludes to the road which led up to their summit from the interior of the hill.

There are not inscriptions at Bameean to guide us in their history; and the whole of the

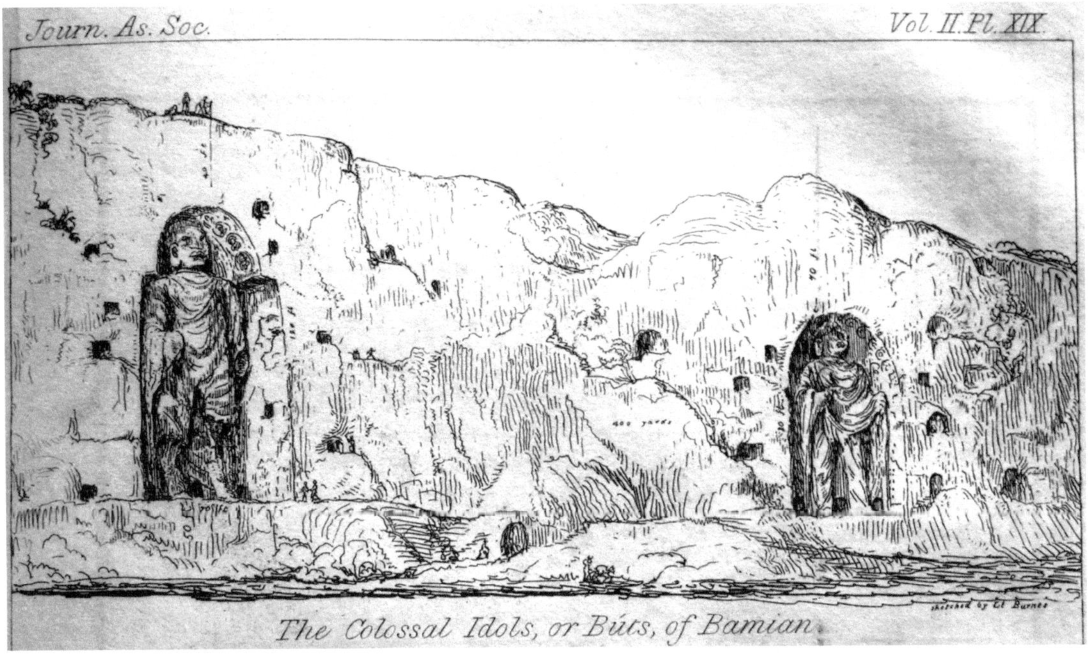

Fig. 3.5 Burnes Alexander, *On the Colossal Idols of Bamian*, in the *Journal of Asiatic Society*, 23, 1833

> later traditions are so mixed up with Ali, the son-in-law of Mohammed, who, we well know, never came into this part of Asia, that they are most unsatisfactory. It is by no means improbable that we owe the idols of Bameean to the caprice of some person of rank, who resided in this cave-digging neighbourhood, and sought for immortality in the colossal images which we have now described (Burnes 1834).

The work of Burnes is also important because a landscape with the two statues was reproduced in two engravings for the first time in his publication.

In particular, in these prints are represented the flank of the mountain with niches, the statues, and the entrances to the caves. However, such drawings can give us only a rough idea of the real aspect of the site because more restricted access to the caves is reported in both. Furthermore, the proportions between persons and statues are not respected and maintained; the traces of erosion are almost absent, especially if compared with, for example, their representation in the sketches made by Talbot and Maitland; the upper side of cliff appears less irregular than in reality. There is an interesting contradiction regarding the larger statue. In the sketch executed by Burnes and published in *On the Colossal Idols of Bamian* (Fig. 3.5), he reproduced the Buddha with a face, whereas in reality it had been erased, as correctly shown in the print executed by Haghe Louis and published in *Travels into Bokhara* (Figs. 3.6 and 3.7).[14]

In his drawings, Burnes reported technical data such as the dimensions of the two colossi, and chose a lateral point of view to show the digging of the cliff and the rest of the frescoes. It appears to have been a good choice, as in the print by Haghe the arch that surpasses the head of the larger statue appears flattened to stress the presence of frescoes. Nonetheless, both prints are pleasant and well realized. An interesting aspect of Haghe's work is the presence of camels and personages in traditional dresses. These details give a natural Oriental aspect to the whole scene.

After the Khyber Pass slaughter, many British soldiers and citizens were captured by Afghan rebels. During one of the numerous forced marches imposed on prisoners, Lieutenant Vincent Eyre obtained permission to visit the

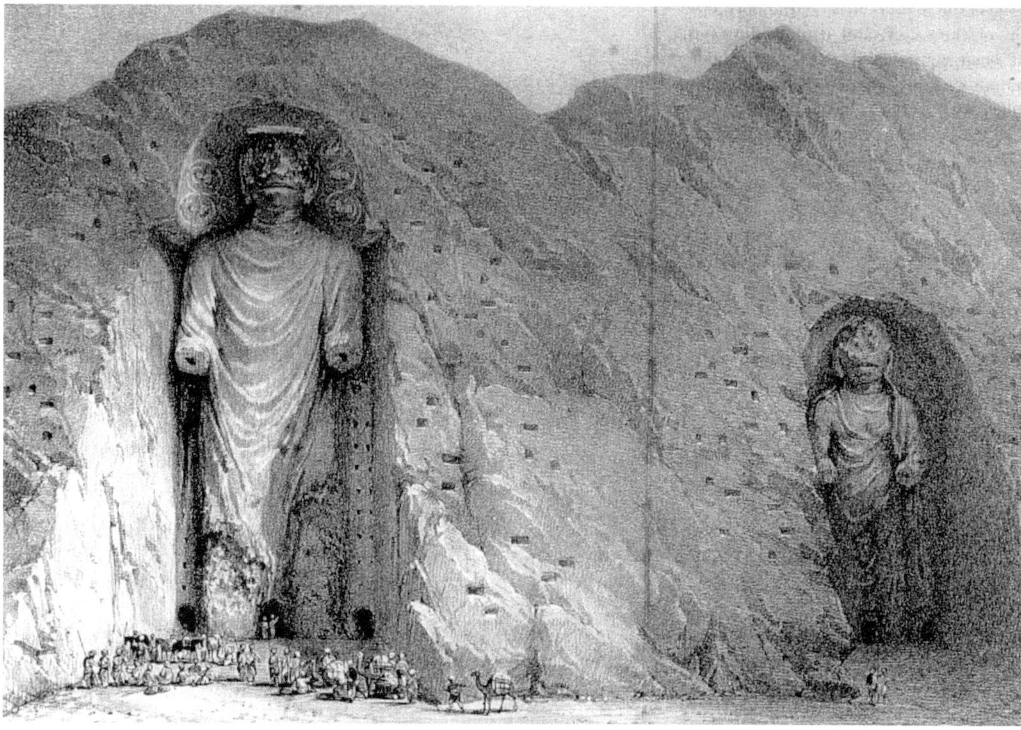

Fig. 3.6 Burnes Alexander, *Travel into Bokhara,* 1834, London

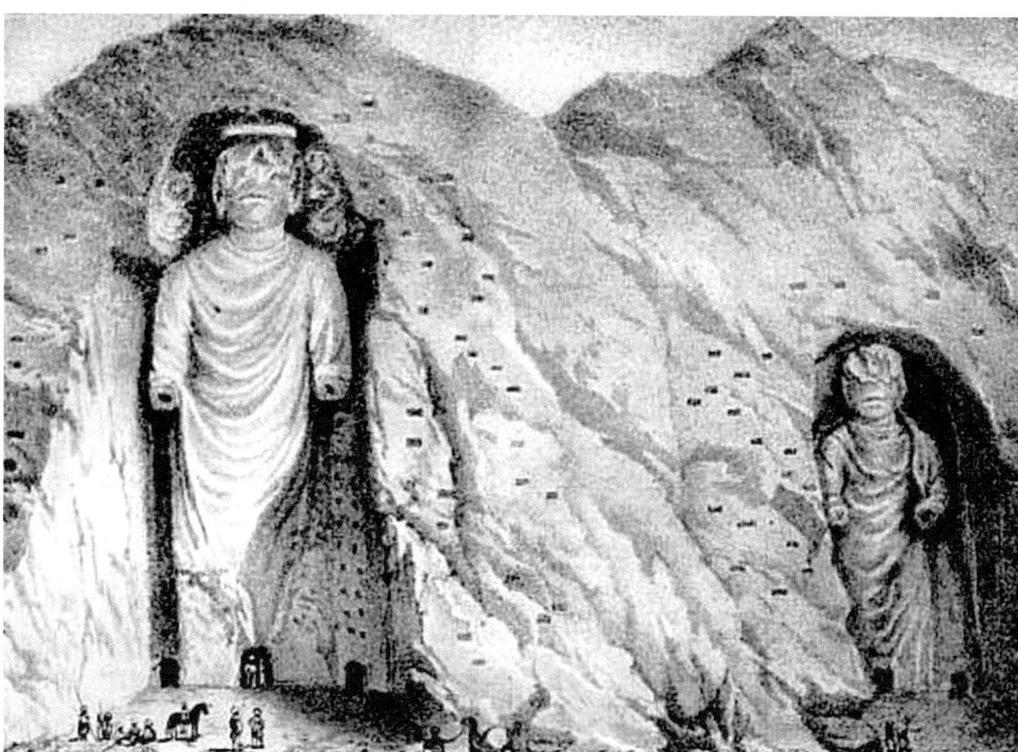

Fig. 3.7 "Die Kolosse zu Bamiyan n. Al-Burnes Zeichnung". In: Carl Ritter, Die Stupa's (Topes) oder die architectonischen Denkmale an der Indo-Baktrischen Königsstraße und die Colosse von Bamiyan, Berlin: Nicolaische Buchhandlung, 1838

Bamiyan site with its statues and cells dug in the cliff. On this occasion, besides the guards, Lady Sale and her daughter also were present (Fig. 3.8).[15]

Vincent Eyre, in spite of the situation, sketched the site (Figs. 3.9, 3.10, and 3.11), some of which are now in the Cambridge University, Royal Commonwealth Society Library. A copy of the original drawings was executed by William Simpson, who edited the memories of Maitland and Talbot.

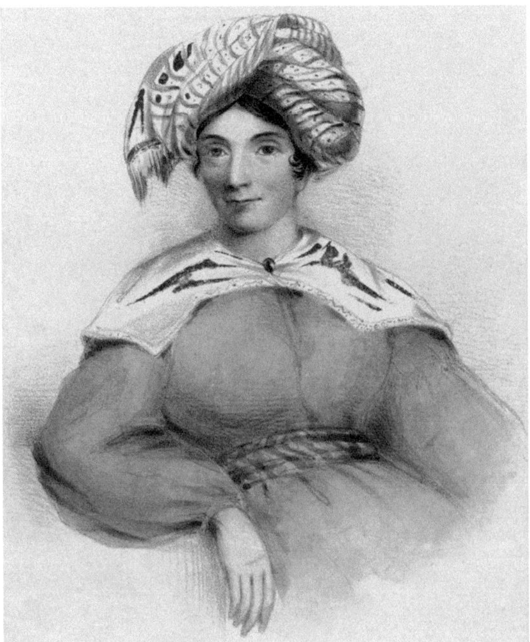

Fig. 3.8 Lady Florentia Sale (http://ozebook.com/wordpress/archives/26555)

Here follows the account by Lieutenant Vincent Eyre:

> The vast assemblage of caves, for which the place is celebrated, became visible at a great distance, and the ancient citadel of Gulguleh, with its lofty ruined towers, crowning an isolated pyramidal hill, rising behind the scene, formed a striking and imposing object. We passed several Afghan forts, and halted at one somewhat better than the rest, where we had just got our tents pitched and our breakfast in a state of forwardness, when we were peremptorily ordered to remount and move a few miles higher up the valley. We were very lot to obey; but neither sulky looks, nor open remonstrance, had any effect on Saleh Mahomed, and onward we travelled in that sort of temper peculiar to hungry men disappointed to their morning's meal. On our way we passed close by the famous colossal images, sculptured in a cliff, which was about 300 feet high; the body covered with a clever representation of thin drapery; the position easy and natural. The upper part of the face is entirely destroyed. About 400 yards further on, in the same cliff, stands the male figure, about 160 feet high, and clothed in a similar fashion with a light drapery. The mouth alone remains a perfect feature of the face, and is well formed. Each figure is sculptured in a deep recess, having an ornamented ceiling, on which are painted what seem intended to represent ancient kings and queens and a variety of emblematic personages, whose mystery it would be difficult, at this remote date, to clear up to the satisfaction of the Society of Antiquaries.
>
> There are openings at the head and feet of both images, which are connected, by galleries and stairs cut inside the rock. The whole long line of excavations forms a wonderful scene, and carries the fancy back thousands of years, to the date at which a widely different race peopled the country from any now existing.

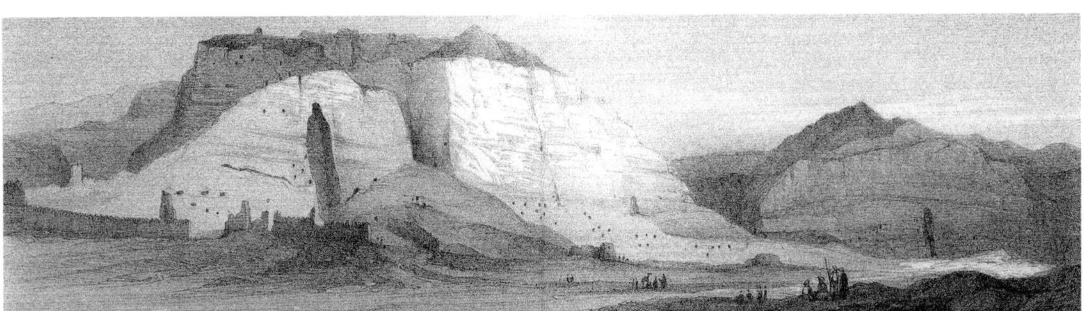

Fig. 3.9 Vincent Eyre, *Portraits of the Cabul Prisoners*. Cambridge University Library: Royal Commonwealth Society Library, *Portraits of the Cabul Prisoners*, RCMS 101, 1842 (Courtesy of D. Bukerer)

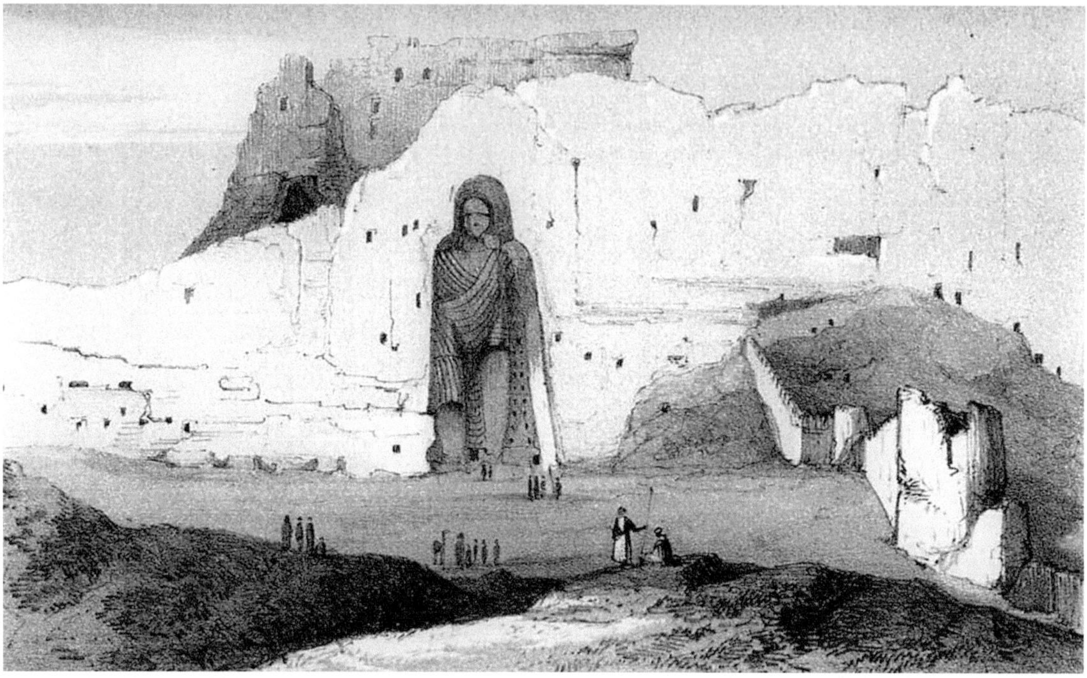

Fig. 3.10 Vincent Eyre, *Portraits of the Cabul Prisoners*. Cambridge University Library: Royal Commonwealth Society Library, *Portraits of the Cabul Prisoners*, RCMS 101, 1842 (Courtesy of D. Bukerer)

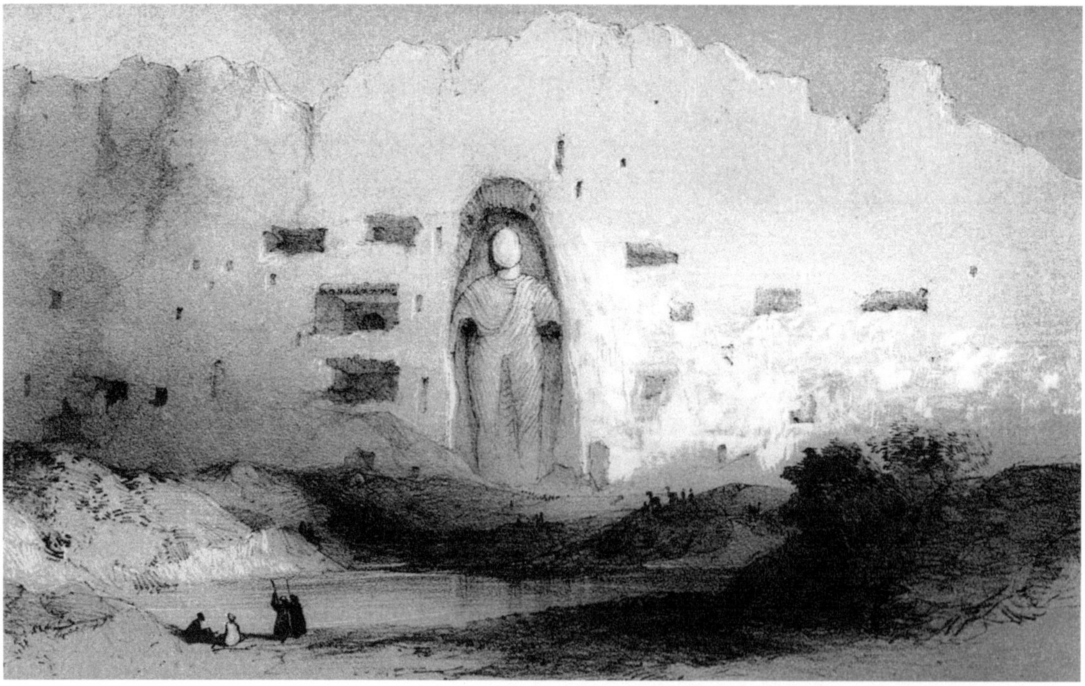

Fig. 3.11 Vincent Eyre, *Portraits of the Cabul Prisoners*. Cambridge University Library: Royal Commonwealth Society Library, *Portraits of the Cabul Prisoners*, RCMS 101, 1842 (Courtesy of D. Bukerer)

Sept. 5th. – I obtained leave to visit the caves and images, one of the guards accompanying me. A walk of one mile brought me to the base of the male figure, the size of which on a near approach seemed really stupendous. Along the sides of the recess, in which it is carved, are ranges of large circular chambers with vaulted ceilings. The image has been greatly mutilated by cannon shot, for which act of religious zeal credit is given to Nadir Shah. One circumstance struck me as remarkable, which was, that in all those parts where the limbs are deficient, there are regular rows of small holes in which pieces of wood have been stuck, for the evident purpose of making plaster adhere. From this it would appear either that an attempt had been made to restore the mutilated parts by these means, or that the figure was originally only partially sculptured in the rock, and the deficiencies made up with plaster in the manner I have mentioned. From the apparent facility with which from the softness of the rock, the image might have been chiselled perfect at the first, I incline to the belief than an attempt has been since made to repair the work of destruction, during some temporary success of the hearten inhabitants against their Mahomedan invaders. The cliff is composed of that species of conglomerate known by the name of pudding-stone, consisting of very hard clay, thickly studded with various kinds of rounded pebbles.

September 6th. – I accompanied Lady Sale on a second trip to the caves. Her Ladyship, who is well skilled in numismatics, at the first glance pronounced the figures painted on the ceiling over the images to be identical with those on many Sassanian coins. This, if really the case, it may throw considerable light on the history of these curious relics of antiquity (Hackin et al. 1928).

And here is a short note by Lady Florentia Sale.

From the 5th to the 9th we made excursions to see the caves, etc. At first some difficulty was made: but the General sent about thirty men to guard us and our pencils; for several went intent on sketching. I only copied the frescoes that were on the walls and ceiling near the large image; but Mr Eyre made some very pretty and correct sketches of Ghoolghoola (the ancient city) (Hackin et al. 1928, Lady Florentia 2002).

Towards the end of the nineteenth century two British soldiers, Captain M. G. Talbot and Captain P. J. Maitland, crossed the valley of Bamiyan in order to realize a complete topographic survey for military purposes.

On that occasion, Talbot wrote a detailed report regarding the caves and the two colossi, and Maitland was the author of a series of drawings showing the cliff and statues. These drawings were reproduced by Mr. W. Simpson, who was present during a previous government mission, and published by W. Simpson himself in 1886 (Figs. 3.12 and 3.13) (The Illustrated London News 1886).

During their survey, for the first time Talbot and Maitland measured the exact height of the two Buddhas: one was 173 ft. and the other was 120 ft. On this subject, talking about the first statue, W. Simpson remarked: "The Nelson Column in Trafalgar-square is 176 ft., just three feet higher than the Bamian figures, and thus giving almost an exact counterpart of its height."

Next is a letter by Captain M. G. Talbot to W. Simpson and some notes by Captain P. J. Maitland on their activities in the Bamiyan valley:

Camp Haibak, Nov. 13th 1885 – Letter –

My Dear Simpson,

Maitland and I have made the long-talked-of trip through the Hazara country to Bamian, and we are now on our way to Turkistan... I spent four days at Bamian, but could give only a very small portion of that time in visiting the antiquities.

To begin with the figures: there are five.

1. First the big idol, male. The passage up to the top of this is broken away, so I measured it with my theodolite and found it to be 173 feet high. It is sunk in a niche, so as to be protected from the weather (...).
2. A female figure 120 feet high, measured by Maitland with a tape. The passage up to the top is still accessible. There are paintings on the roofs of the niches of both these figures. In the case of the latter some have been copied. Both figures are hewn out of the conglomerate rock, but the finishing, drapery, etc., was all added by putting on stucco. The niche of the female figure is irregular, and looks as if it had been left unfinished.
3. A smaller figure, 50 or 60 feet high, estimated. This figure has almost entirely disappeared. (...)
4. A seated figure about 25 to 30 feet high, in a niche. This figure looks as if it had been cut out and prepared for stucco, but the stucco had never been applied.
5. A standing figure about one mile from the others. Unfortunately, owing to a misunderstanding, we never visited this. (...)

No. 4 also has paintings, some of which have been copied. The caves are innumerable, they extend for miles. The best ones are close to the

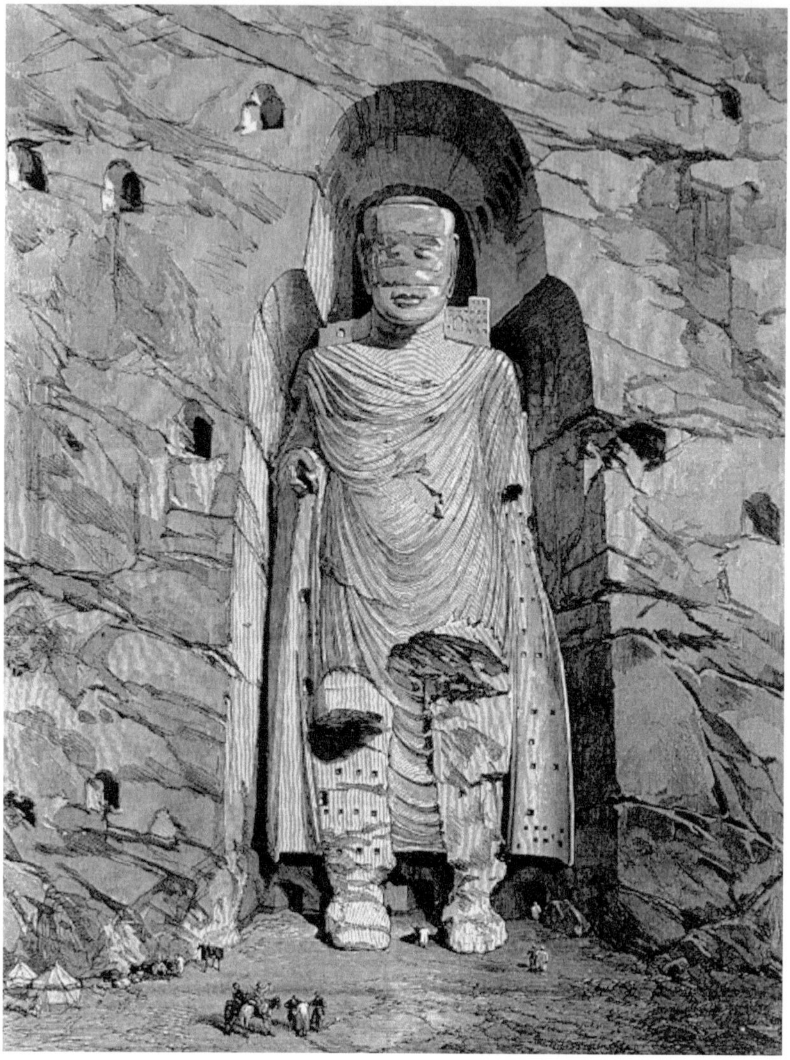

Fig. 3.12 The largest statue. *The Illustrated London News*, Nov. 6, 1886, pag. 491 (Courtesy of D. Bukerer)

female figure. The doorways are mostly sunk well 10 or 15 feet into the rock, with a porch excavated outside. (…)

Most of the caves in good order are now inhabited, so I could not visit them; of those I did visit most had domed roofs, the floor being square. The conversion of the square into a circle, preparatory to the springing of the dome, is effected or rather indicated in the manner used in the present day with Kacha bricks, that is, by a succession of arches at the corners. Looking at the corner from the centre of the cave it appears thus.

The caves near the idols are all connected by rambling passages and staircases cut in the rock. Many of the caves have paintings, but we could not see them as the caves were full of grass, etc., stored for winter.

In Shahri Ghulghulah we saw nothing remarkable. Zohak is a wonderful fort as regards situation and multiplication of lines of defence. It is fairly well preserved in parts, so much so that the natives say it was not captured but abandoned. Some domed chambers in it are the exact facsimiles of the domed caves above mentioned…

Additional Notes on Bamyan by Capt. P. J. Maitland

The Bamian valley is about half a mile broad and well cultivated, but there is no town or even central agglomeration of houses, only small villages scattered up and down the valley. To the north is a fairly continuous wall of cliffs averaging about 300 feet in height; to the south is a central plateau separated by the glens called Dahaneh-i-Tajik and Dahanehi-Saidabad from the cliffs limiting the western and eastern part of the valley. On the edge

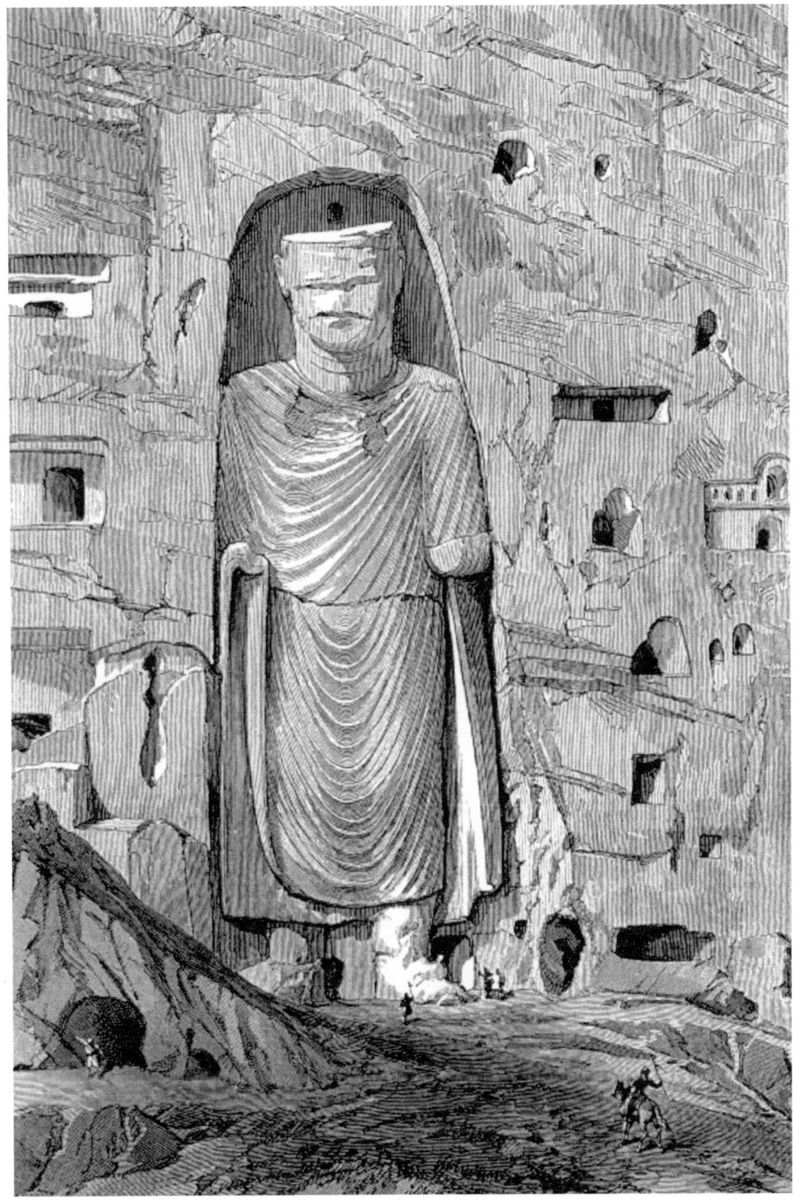

Fig. 3.13 The second great statue. *The Illustrated London News,* Nov. 13, 1886, pag. 535 (Courtesy of D. Bukerer)

of the central plateau is a small, conical, clayey hill, covered with the ruins of Ghulgulah. This is probably the ancient Bamian. The cliffs are everywhere pierced with numerous caves, but the greatest number is found on the north side of the valley, and here are also the famous idols, The But-i-Bamian. The cliffs round these are literally honeycombed with caves, which are found even in the debris slope at the bottom. They are almost inhabited by Tajiks, or used as store rooms, and the entrance is frequently protected by a low mud wall.

Facing the cliff the larger of the two big idols is to the left, the other to the right. They are about a quarter of a mile apart and supposed to be male and female, and their heights are respectively 180 and 120 feet. Their names are, as reported by former travellers, Sal Sal for the male and Shah Mameh for the female figure.

The idols are standing figures, sculptured in very bold relief in deep niches. Between the two large idols are, or rather were, two smaller ones, also in niches. These are equidistant from the large idols and from each other, that is to say, there is about 150 yards between each of the niches, large and small. One of the smaller niches is about 60 to 70 feet high, and is now empty, though a close inspection shows fragments of the idol that once filled it The second small niche is still occupied by a sitting figure, which is about 40 feet high and known as the Bacheh, or child. The general shape of the niches is the same in all cases, but that of the large female figure is evidently unfinished, and the shoulders are not marked, nor the edges smoothed off.

The depth of the niches of the two large idols is about twice the thickness of the figures standing in them: the latter are therefore fairly well protected from the weather, and this accounts for their preservation, nearly all the damage done to them being due to the hand of man.

The whole interior of the niches, and particularly the arches over the heads of the idols, have been painted with what appears to be allegorical designs. Although much damaged, in fact, obliterated, where they could be easily got at, enough remains to show the general style of the work, which is exceedingly well executed, and forcibly reminds one of what is generally understood to be Byzantine art.

The idols themselves are rather clumsy figures, roughly hewn in the tough conglomerate rock, and afterwards thickly overlaid with stucco, in which all the details are executed. The whole arrangement clearly shows that this was done at a later period, but is part of the original design of the figures. The stucco appears to have been painted, or at least paint was used in some places. The features of the figures have been purposely destroyed, and the legs of the larger one have been partly knocked away, it is said by cannon-shot fired at it by Nadir Shah. Both idols are draped in garments reaching to below the knee. The limbs and contour of the body show through, and the general effect of Muslim is excellently imitated in the stucco. The arms of both are bent at the elbow, the forearms and the hands projecting but the latter are now broken off. The feet have also been battered out of shape. Narrow stairways hewn in the interior of the rock lead up from cave to cave to the heads of the idols, and even to the summit of the hill.

The caves, though so numerous are not large. By far the greater portion of them are chambers 12 or 14 feet square, with domed roofs. I think as a rule several chambers open into each other, and have a wide portico in front by which light is admitted to the doorways. These have generally round arches. There are certainly no pointed arches anywhere, but some of the openings may be square headed. The domes are set on the four sided chambers in a remarkable manner, the square being reduced to an octagon by cornices springing by tiers from the angles in unmistakeable imitation of brick-work (…). The largest cave of all is said to be between the feet of the great idol, but like several others it is used for government stores, and was filled with lucerne, etc.… As above mentioned, the majority of the caves are inhabited. A few years ago it is said they all were. The interior are plain, without sculptural ornamentation, and now smoke-blackened.

The whole, however, were plastered with stucco and painted. In one of the upper caves, near the head of the female idol, some designs are still visible. A short distance east of the female idol, near the foot of the cliff, is a mound, which seems to be the remains of Buddhist tope.

A design on the arch over the female idol can, to a certain extent, be made out with a field-glass. Within a circle is a figure in a long robe with a spear, apparently slaying something. The two upper corners without the circle are filled with figure of angels or cherubs, waving scarves at each other. They might well date from the last century. On either side is a border with male and female busts or half-figures in circles, and all adorned with halos.

Outside the central design on the left-hand side is a very curious figure of a human-headed bird. On the east side of the Saidabad glen is an idol, which does not appear to have been noticed by former travellers. It is somewhere up the cliff, which is pierced with numerous caves. The niche is 40 to 60 feet high, and the figure in it has its head covered with a sort of cap or tiara. The two big idols may possibly have been adorned in like manner; the top of their heads is now unnaturally flat, suggesting the idea that something has been cut off.

It should be mentioned that the caves are very dark, only a small doorway admits light, and without candles, or rather good lanterns, nothing can be seen. The stairways are always very narrow and steep. There is almost invariably a shallow recess opposite the doorway. No traces of doors were seen (Figs. 3.14, 3.15, and 3.16) (Talbot et al. 1886).

In the same period another article on Bamiyan was published in a London magazine, *The Graphic*. This article stressed not only the artistic value of the statues, but also reported information about the dangerous situation in the region caused by continuous rebellions of local tribes.

> To give you some ide.a of their size. I may say that I saw a man on horseback ride up to the male figure. Man and horse together were not so high as the toe…

The drawings are all very interesting and detailed and the general view of the valley with niche of the smaller statue is particularly important for some elements like: the use of light and

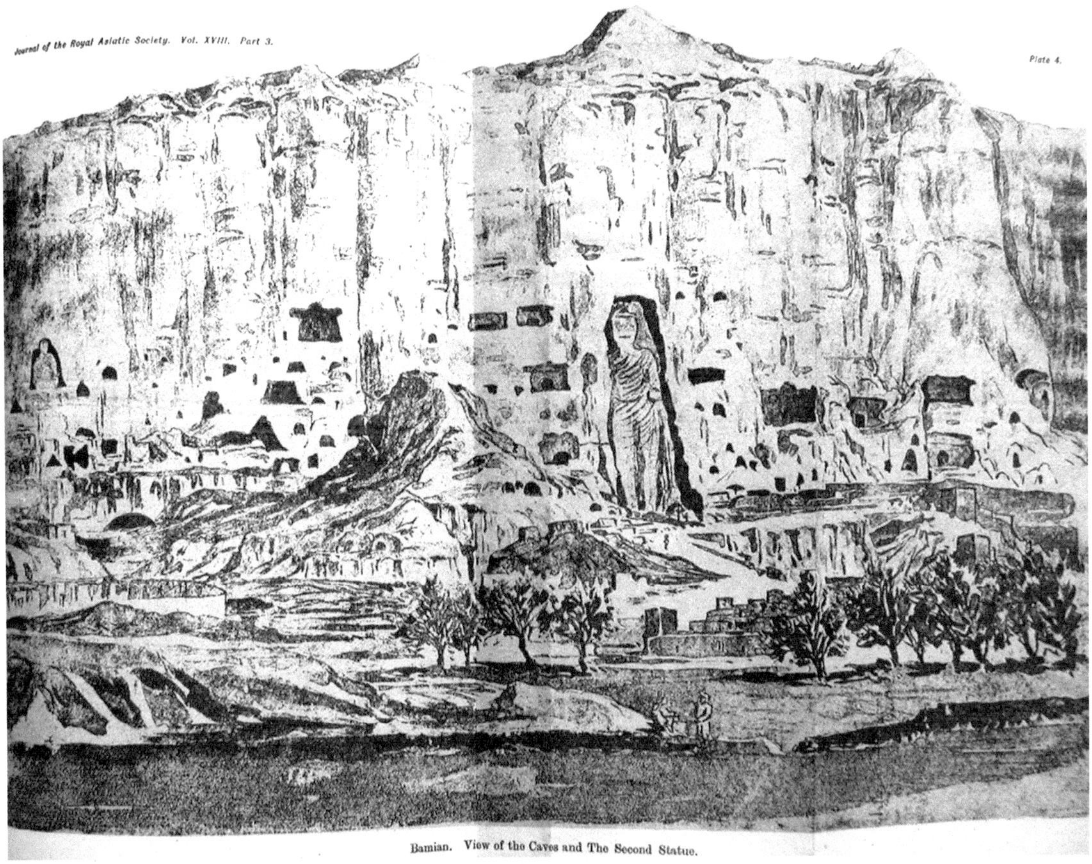

Fig. 3.14 Talbot M.G. and Maitland P.J., *The Rock-Cut Caves and Statues of Bamian,* in *The Journal of the Royal Asiatic Society, XVIII,* 1886

shade technique, the view in perspective, the attention in details, the vegetation and the small village at the feet of the statue.

Moreover in the drawings are well represented the cliff, the debris accumulation area at its feet and the entrance to the caves, with an accuracy nearly photographic.

> The ancient rock-cut temples in the cliff (caverns hollowed out by the side of the colossal figures), had been turned into storehouses, and were full of grain and military stores, while groups of tents and long lines of horses were visible in every direction, and large new barracks have been built (Fig. 3.17) (The Graphic 1894).

For this reason it isn't a surprise to see the military context in which one of the colossi is represented. The drawing is a typographic reproduction probably realized by photography. It shows a particular moment during one of the many military training parades that were executed in the valley. At the beginning of the twentieth century the colossus of Bamiyan and surrounding caves were well known Asian monuments. However, there is a real lack of archaeological study about the statues and the valley as a whole. The first scientific research conducted was the topographic work by Talbot and Maitland, as mentioned. Successively, a new step in this direction was taken by H. H. Hayden, an British geologist, who published a short essay.

> There are at present altogether five statues in Bámián three of which to are in the main valley at Taibut. Plate XII gives to general view of the cliff of Tertiary conglomerated on the left side of the valley and shows the two chief niches, one at either end, and two smaller niches between them. Only one of the latter contains to statues and that has suffered, like the two principal figures, from systematic mutilations as well as from the effects of Cannon-ball fired at it by Mahommedan invaders (Hayden 1910).

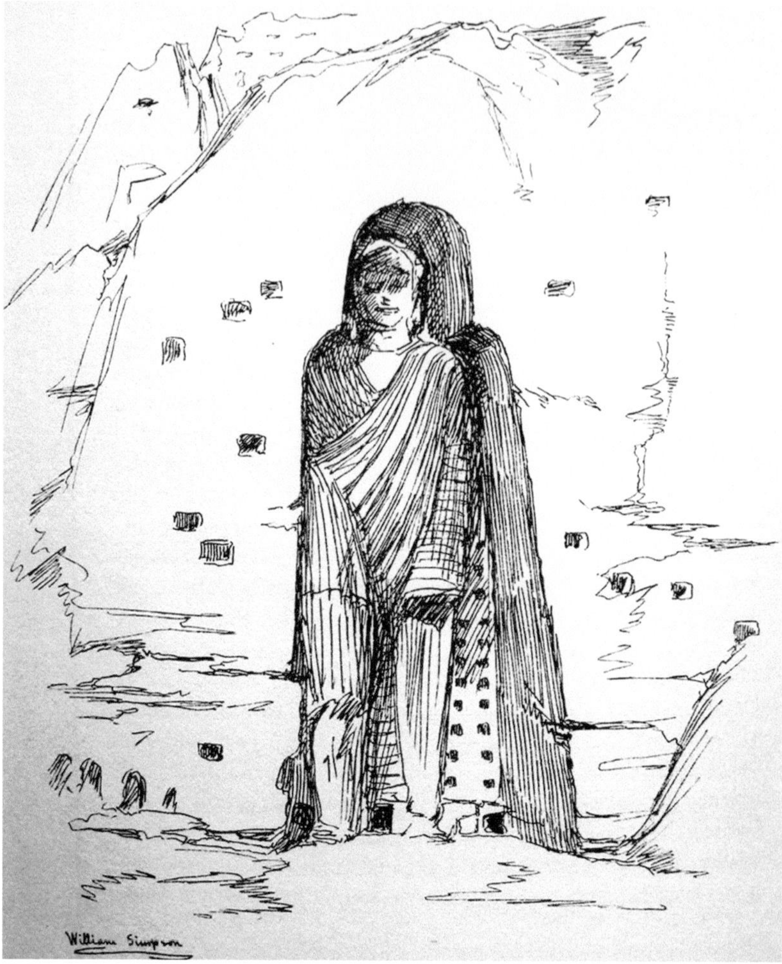

Fig. 3.15 Talbot M.G. and Maitland P.J., *The Rock-Cut Caves and Statues of Bamian*, in *The Journal of the Royal Asiatic Society, XVIII*, 1886

In his article, Hayden published a series of photos of the Bamiyan Valley and its colossi for the first time (Figs. 3.18 and 3.19).

The use of early photographic equipment and the rapidity of duplication and publication of his photographs rapidly replaced the use of drawings in its descriptive function of objects and landscapes. In this way a perfect likeness of the real image was afforded by photography, and cameras became indispensable for travelers, especially in countries such as Africa and Asia, which up to then were not well known. With their artistic value and mystical fascination, the Bamiyan colossi and the entire valley became a preferred destination for travelers crossing Afghanistan in that period. Through the publication of travel diaries and reports of archaeological excavations, it became clear that the production of monumental masterpieces such as the statues of Abu Simbel or obelisks and pyramids were not the exclusive expression of only one civilization. The twentieth century afforded new opportunities for archaeological research in Afghanistan. In 1922, the excavations and studies conducted by the French archaeologist Alfred Foucher (Fig. 3.20) were realized through an agreement between the French and Afghan governments. The DAFA (Delegation Archeologique Francaise en Afghanistan) was created by this agreement, with a monopoly on all archaeological researches in the Afghan

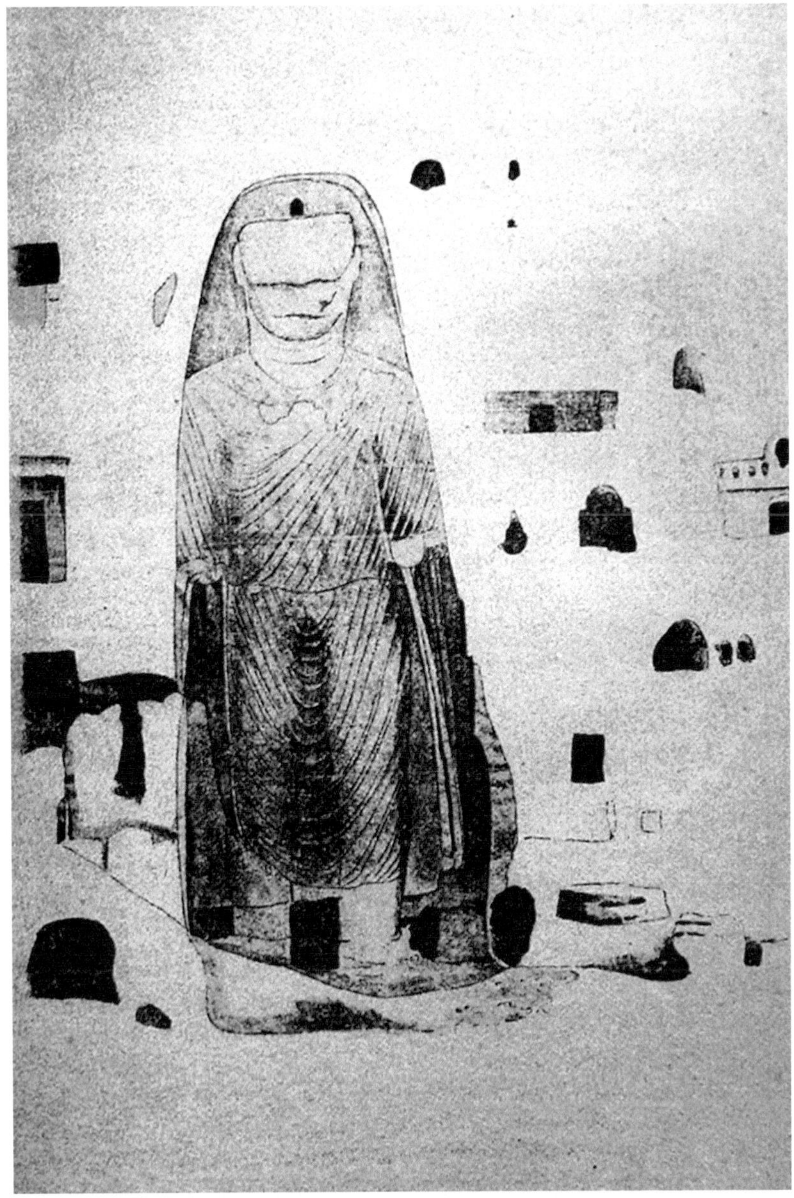

Fig. 3.16 Talbot M.G. and Maitland P.J., *The Rock-Cut Caves and Statues of Bamian*, in *The Journal of the Royal Asiatic Society, XVIII*, 1886

territory (Figs. 3.21, 3.22, 3.23, 3.24, 3.25, 3.26, 3.27, 3.28, 3.29, 3.30, 3.31, 3.32, 3.33, 3.34, 3.35, and 3.36) (Foucher 1905).[16]

In the same period, another French geographical expedition crossed the valley, the Citroen Trans Asiatic expedition, which started in April 1931 and finished in February 1932.

The members of this expedition, on board 14 Citroen halftracks, covered a total of 12,000 km from Beirut to Beijing. The expedition entered and crossed Afghanistan in the south, passing through the cities of Herat, Farak, Girishk, Kandahar, Mukur, Ghazni, Kabul, and Jalalabad, and left the country through the Khyber Pass toward Peshawar. Their arrival in Kabul, and the following rest stop, were used to visit the Valley of Bamiyan, as testified by the usual souvenir photos (Fig. 3.37) (Hackin 1934).

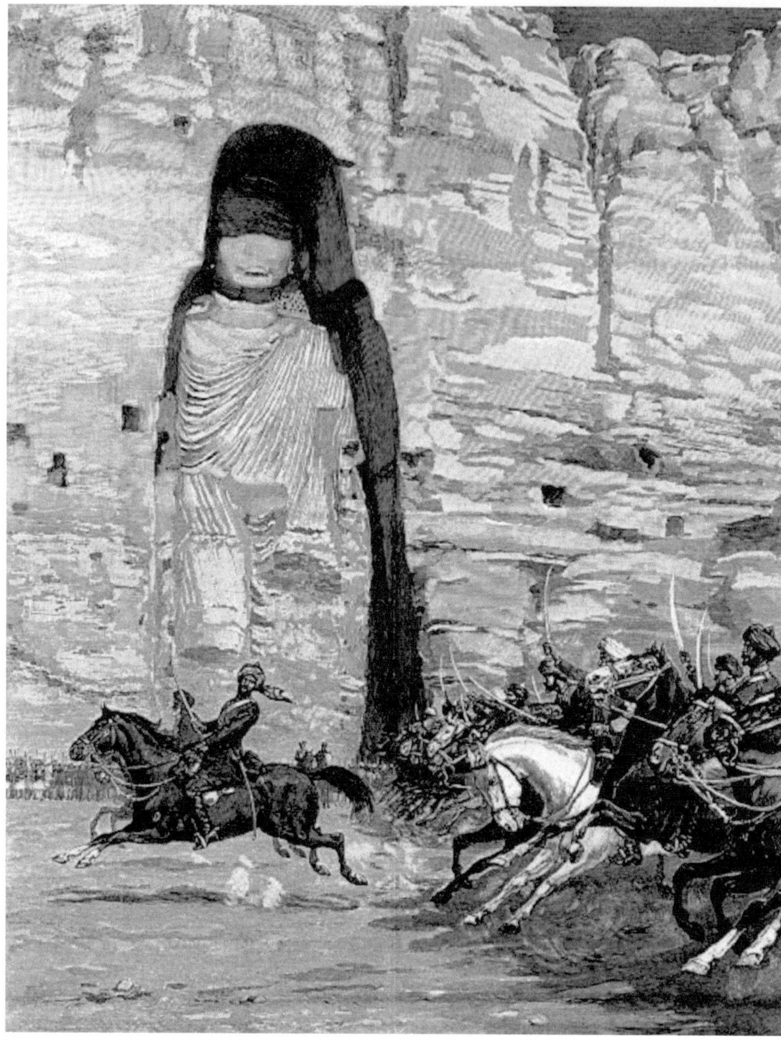

Fig. 3.17 *Graphic,* January 6, 1894, p. 9 (Courtesy of D. Bukerer)

A few years before World War II, an Italian traveler visited the Bamiyan Valley.

Arnaldo Cipolla, a writer and reporter for the Italian newspaper *La Stampa*, together with his friend Conte Barbellini Amidei, made a grand tour of the Afghan tracks between the spring and summer of 1934 (Fig. 3.38).

Cipolla wrote a day-to-day diary containing his observations on the places, villages, and people encountered during his journey. Like previous visitors, Arnaldo Cipolla also was fascinated by the "*sfingi grattacielo*" (skyscraper-sphinx), which is how he defined the Buddhas. In his book, the author summarized Hiuan Tsang memories and gave interesting indications of the state of preservation of the statues and caves.

> About the Sleeping Buddha there is not any trace today; about those standing, as they were used as target for mountain batteries of all the Afghans Emirs before Amanullah, their faces are in pitiful state, limbs and folds of the frock are mixed with the original matter from which they originate; not to mention the right arm expressing "the reassuring gesture," which probably has been the first part crumbled by the iconoclastic shots of the Muslims guns; nevertheless, especially when the sun draws shadows of their figures, they can still provoke a sensation comparable with the view of the colossus of Abù Simbel, there on the Nile river, between Aswan and Wadi Alpha.

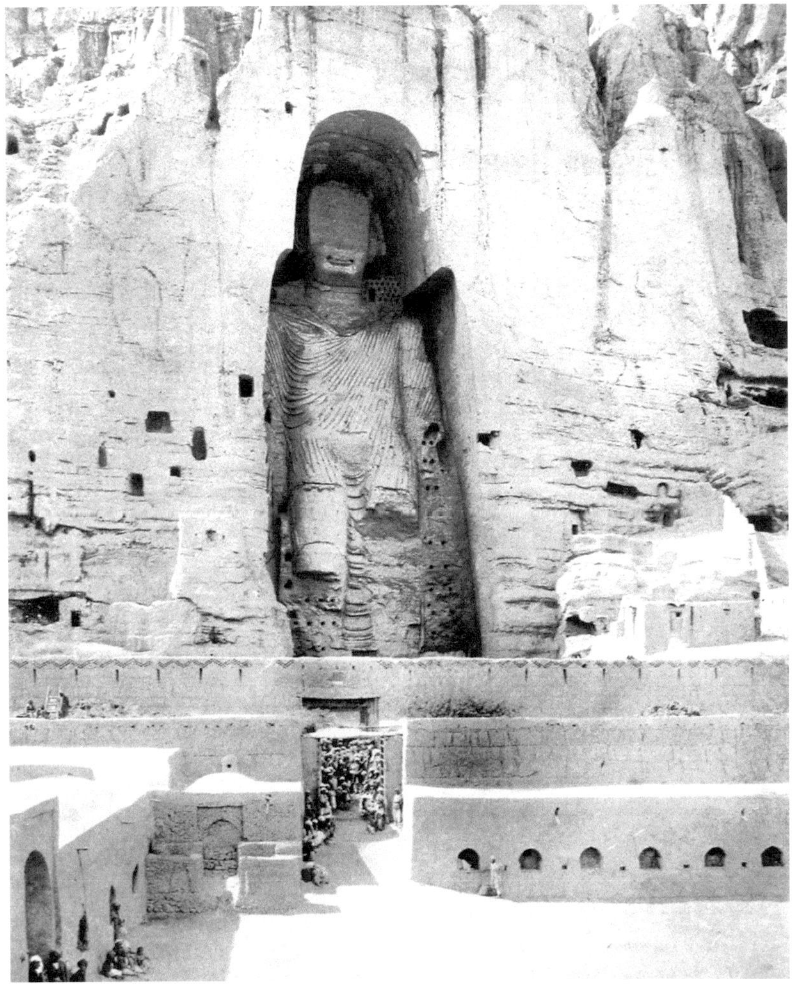

Fig. 3.18 Hayden H.H., *Notes on Some Monuments in Afghanistan,* in *Memoirs of the Asiatic Society of Bengal, II,* 1910

But we must take in mind that these Asian Sphinxes have been seen by a number of Europeans certainly less than fifty, while the funeral monument of Ramsete is too much easily accessible. … [the most easily accessible caves] are used by the Afghans peasants as residences and granaries; the most ancient ones are close to the 35 meters high Buddha—within its deep niche, the enormous statue has a monstrous large head, a too big torso, too much wide shoulders, legs fixed on the body like columns and its garment drape is obtained by a mixture of lime, earth and crushed straw—and, thanks to a mass of debris produced by a landslide fallen in front of the entrance, coloured frescos and bass-relief which adorned [the caves] preserve their self in a good condition.

There you can see many images of Buddha, minor divinities, demons, monks and groups of donors with aureole. Some caves are located close to the head of the statue, and, by means of inner scales, they communicate with large halls, some of them with an octagonal plant, that probably have been devoted as reunion centre by the Buddhist monks assigned to the colossus, and with particular ceilings formed by embedded squares that looks like those that are nowadays constructed in the Badakshan and Caschemire. Moreover, there are sanctuaries with other frescos of moving Buddha, moon divinities and others floral decorations.

The archaeologists seem to be quite satisfied by the styles showed in these types of frescos: an expression of the Sassanian Iran art with Greek influences. Regarding the 53 meters high Buddha, archaeologists agree in recognizing its greater artistic value compared with the lower one. Niche and statue made a pleasant whole. Around this greater colossus there are too caves, halls, recesses, sanctuaries and cells more or less decorated by figures, symbols, birds inspired by the Sassanian Iran and oriental-Hellenic art (Cipolla 1935).

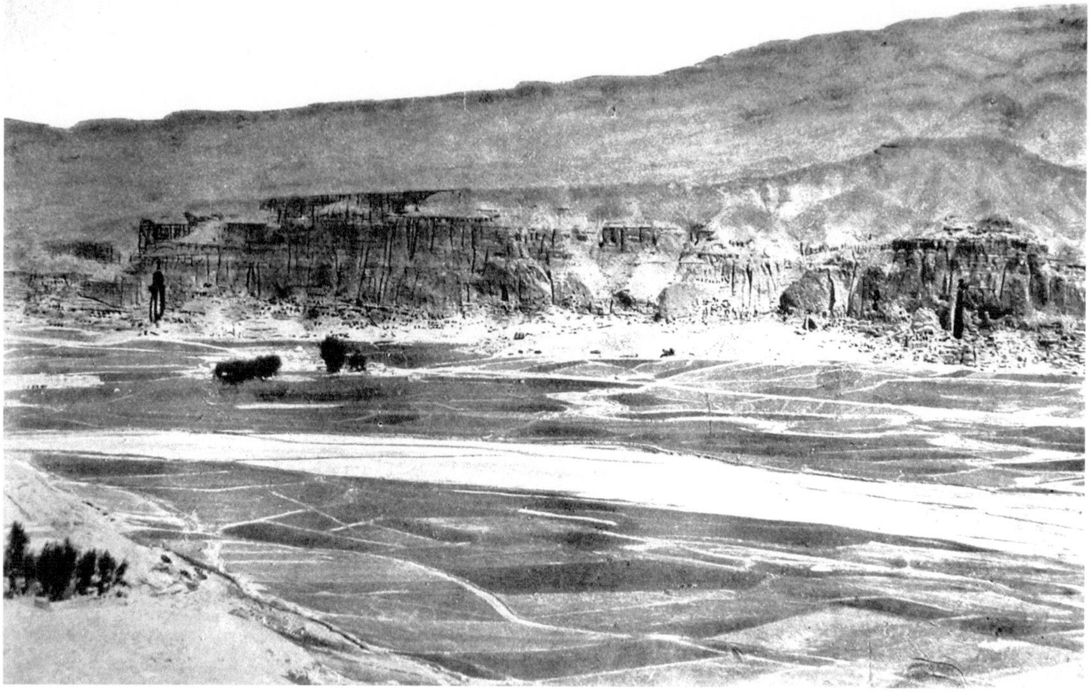

Fig. 3.19 Hayden H.H., *Notes on Some Monuments in Afghanistan,* in *Memoirs of the Asiatic Society of Bengal, II,* 1910

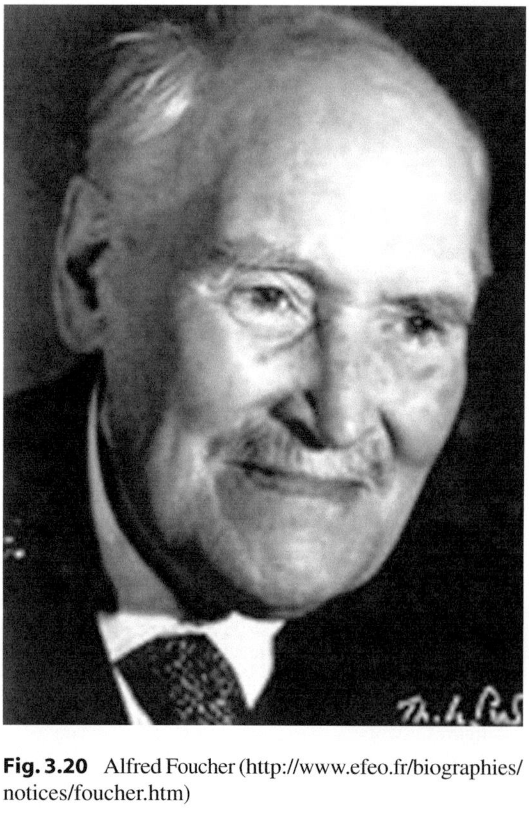

Fig. 3.20 Alfred Foucher (http://www.efeo.fr/biographies/notices/foucher.htm)

After World War II, following an agreement between the Afghan and Italian governments, archaeological excavations were conducted by the ISMEO (Italian Institute for the Middle and the Far East); in the same period the Afghan government made archaeological agreements with India and the United States.

In the 1960s the Indian government executed some works on the cliff of the Bamiyan Valley, just over the niches of the Buddhas, for water drainage and erosion risk reduction.

In the same period a philatelic series dedicated to the colossi of Bamiyan was printed, showing that the Afghan government acknowledges the importance of its cultural heritage without any religious distinctions (Fig. 3.39).

In chronological order, some of the last "travelers" who visited the Bamiyan Valley were two British men, Peter Levi and Bruce Chatwin (Fig. 3.40). They had different personalities. Bruce Chatwin was a traveller and writer and had visited Afghanistan at least twice before; Peter Levi was a Jesuit student of archaeology, a learned and refined poet. In the summer of 1969,

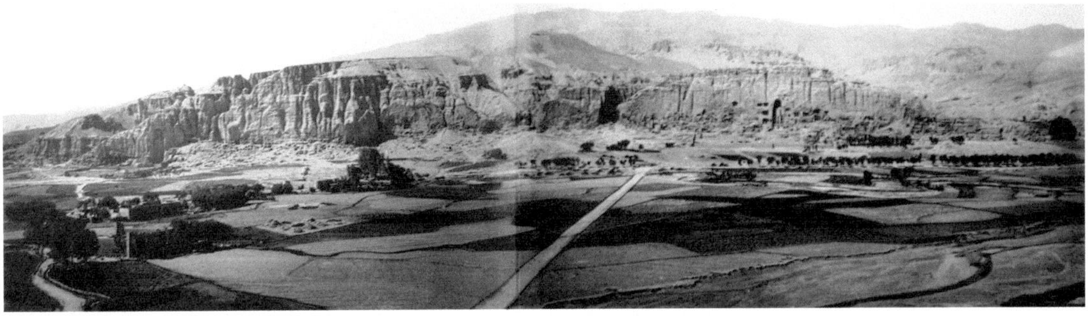

Fig. 3.21 Hackin M., Godard A., and Godard Y. *Memoires de la delegation Archeologique Francais en Afghanistan,* Tome II, *Les Antiquites Bouddhiques de Bamiyan,* Paris et Bruxelles, 1928

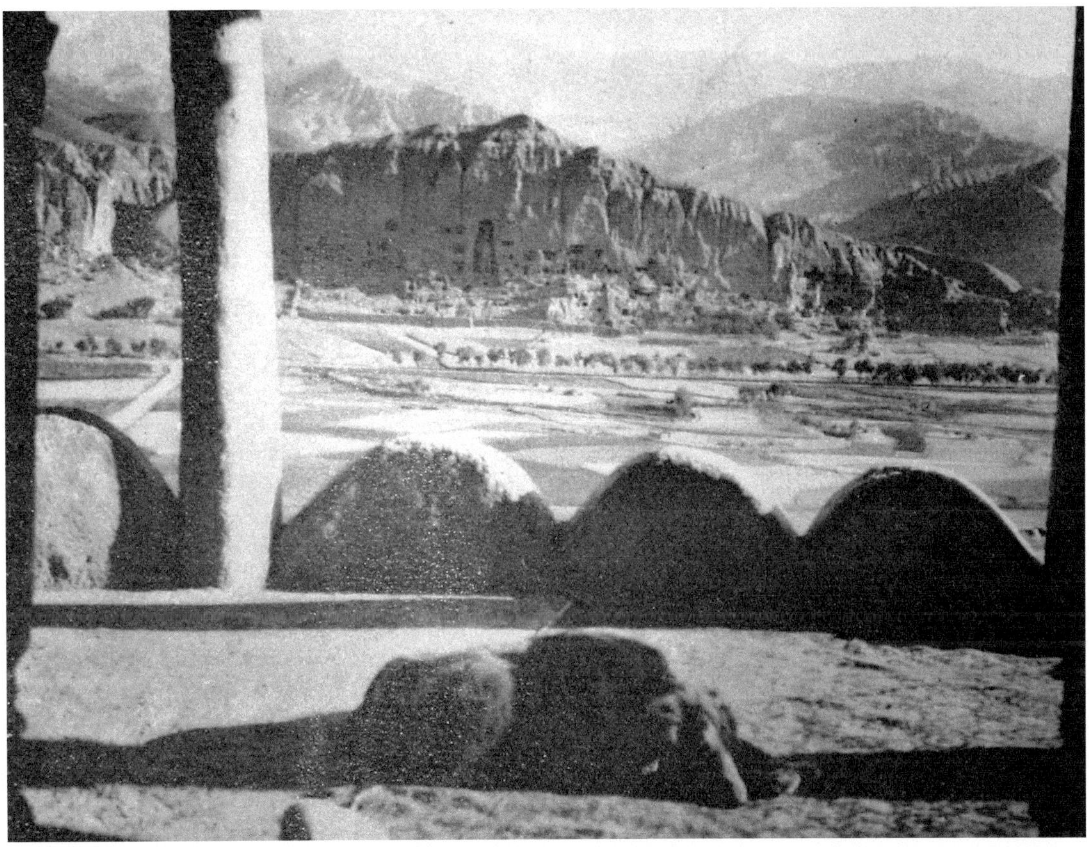

Fig. 3.22 Hackin M., Godard A., and Godard Y. *Memoires de la delegation Archeologique Francais en Afghanistan,* Tome II, *Les Antiquites Bouddhiques de Bamiyan,* Paris et Bruxelles, 1928

the two travelers, following the indications of Italian archaeologist Maurizio Tosi, who was a friend of Bruce Chatwin, began a three-month tour. Levi and Chatwin, rather than being tourists, can be better thought of as descendants of the romantic travelers of the nineteenth century, who were diligent and acute observers of places and persons, enjoying the adventure and

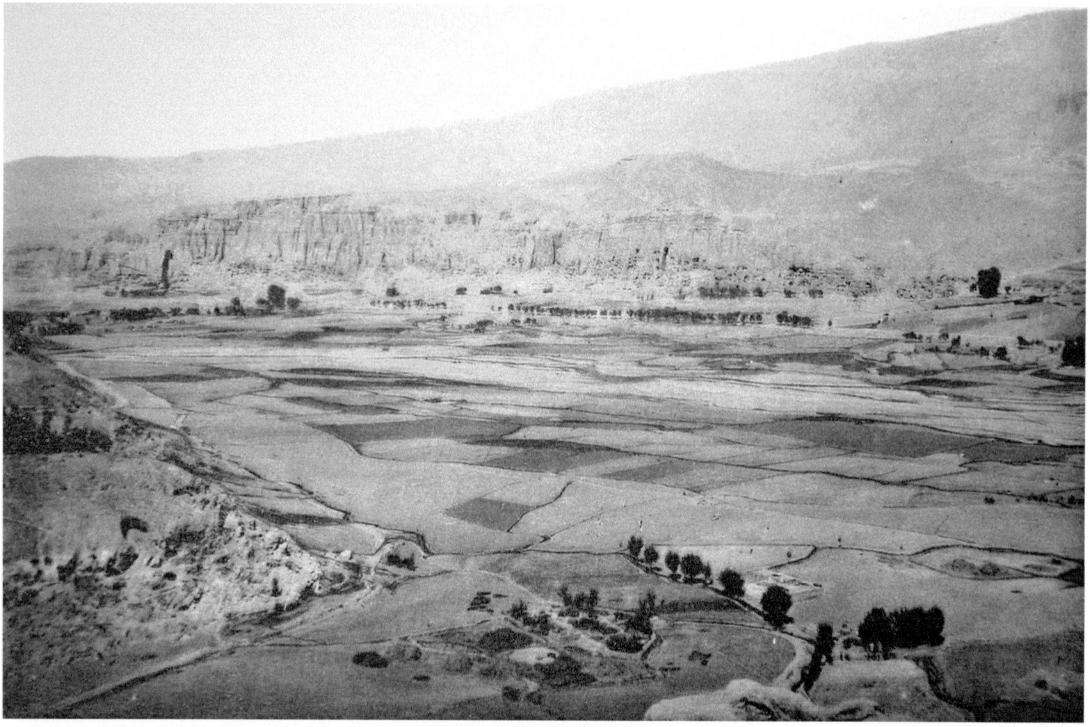

Fig. 3.23 Hackin M., Godard A., and Godard Y. *Memoires de la delegation Archeologique Francais en Afghanistan,* Tome II, *Les Antiquites Bouddhiques de Bamiyan,* Paris et Bruxelles, 1928

admiring the monuments, which was the basis of their inspired literary and photographic expressions (Bruce 1993).

They crossed the entirety of Afghanistan, sharing risks and dangers, by the use of various means of transport such as taxies, horses, small planes, and buses.

Upon returning to England, Peter Levi wrote a book on his travels (Levi 2002).[17] Here follows his memories of Bamiyan:

> There are trees, rivers and caves in Bamiyan. The small hotel on a slope southwards; on its back there is a short airfield; a little far away you can see the snowed massif of Koh-i-Baba. The sun set exactly at the end of the valley; it seemed a light yellow liquid dropping from a pale blue sky. The valley was obscure and pacific. In two enormous niches on the northern slope, as many Buddhas more than thirty meters high are towering. Red and gilded they were once, but now they have the same clay colour of the mountain.
>
> They stare at the snow and the sky behind the hotel, like meek and magical robots. The wind blew noisily and the river roared even more loud. The opposite cliffs were swallowed up by the darkness, riddled with hundreds of black cells and chapels of the monastic whole.
>
> In the morning the Buddhas remain in the darkness; a sequence of bare and dark hills, only just dusted in green and accurately embroidered by shadows, are offered to the view. The Buddhas are not depicted in a natural manner; they remember instead the image of the Virgin with the eyes of gull described by Robert Lowell, non est species, neque décor. The faces have been damaged and their restored parts are cold; the position assumed by feet communicates solidity.
>
> No statue whose face has been removed the can express rectitude, authoritativeness, intelligence, mercy, nevertheless the disquieting presence of the two giants succeeds in transmitting something. I don't know, maybe "the lack of expression, expresses God".
>
> Caves looks really like cells in a honeycomb or nests of birds. The valley is meticulously cultivated; small woods of white poplar, intense greens and a great number of small fortresses containing small farms. Almost every day, early in the morning, there was a boy who was playing his flute in a field.
>
> The instrument has two octaves: I thought that it was a double flute but during the travel we often

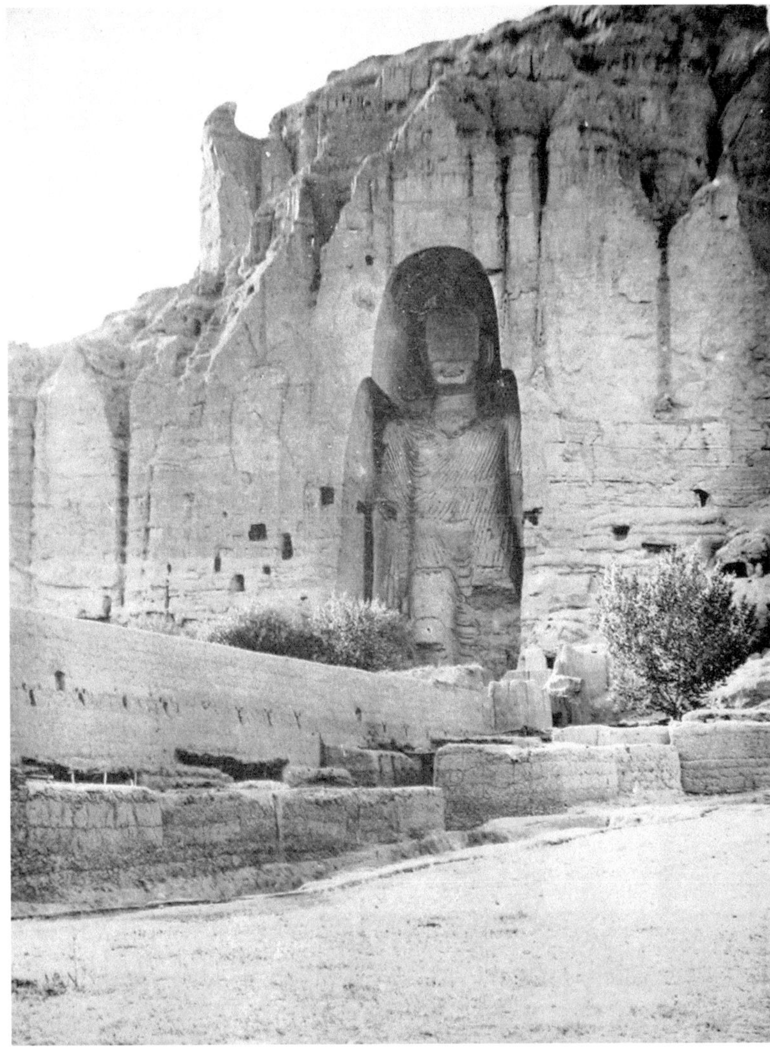

Fig. 3.24 Hackin M., Godard A., and Godard Y. *Memoires de la delegation Archeologique Francais en Afghanistan,* Tome II, *Les Antiquites Bouddhiques de Bamiyan,* Paris et Bruxelles, 1928

saw boys who was playing and their instruments was always a simple wooden tube, poplar wood I suppose. Like all tourists we ascended to the head of the Buddha climbing us gallery by gallery and cave by cave; up there we heard the swallows chirping and flying around to the faces without face.

The Buddha was here to contemplate or to be contemplated? Probably to contemplate, as both monks from their caves, as the Enlightened watched toward the same mountains and same snows. To cut in the rock architectonic shapes so large and elaborated should require enormous hard work and, probably, many slaves (Fig. 3.41).

The visits of Bruce Chatwin and Peter Levi ended the golden period of Bamiyan Valley "grand tours"—archaeological researches characterized by drawings, photos, and technical and literary reports—in other words, the peaceful era.

In 1979 the Red Army invaded Afghanistan, starting a 10-year war, and Soviet soldiers became the new tourists to Bamiyan, as testified by souvenir photographs (Figs. 3.42 and 3.43). With the Afghan War, the country became an enormous battlefield, with several alternating governments and violent events. In such a situation, both the civil population and its cultural heritage, such as the Bamiyan Buddhas, pay a high price.

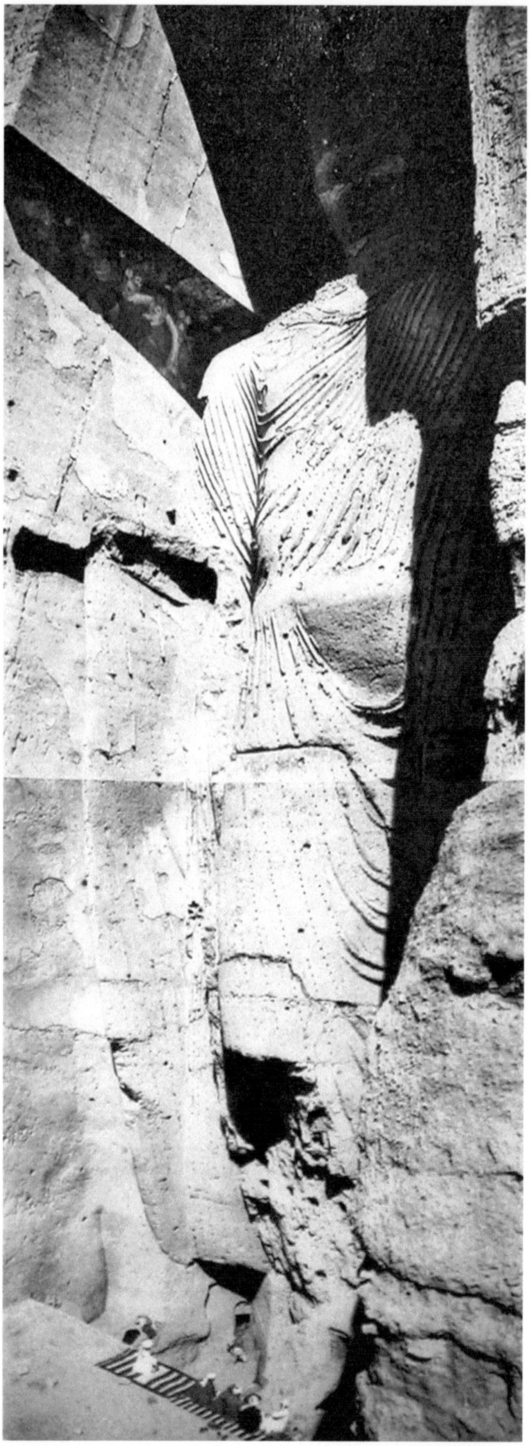

Fig. 3.25 Hackin M., Godard A., and Godard Y. *Memoires de la delegation Archeologique Francais en Afghanistan,* Tome II, *Les Antiquites Bouddhiques de Bamiyan,* Paris et Bruxelles, 1928

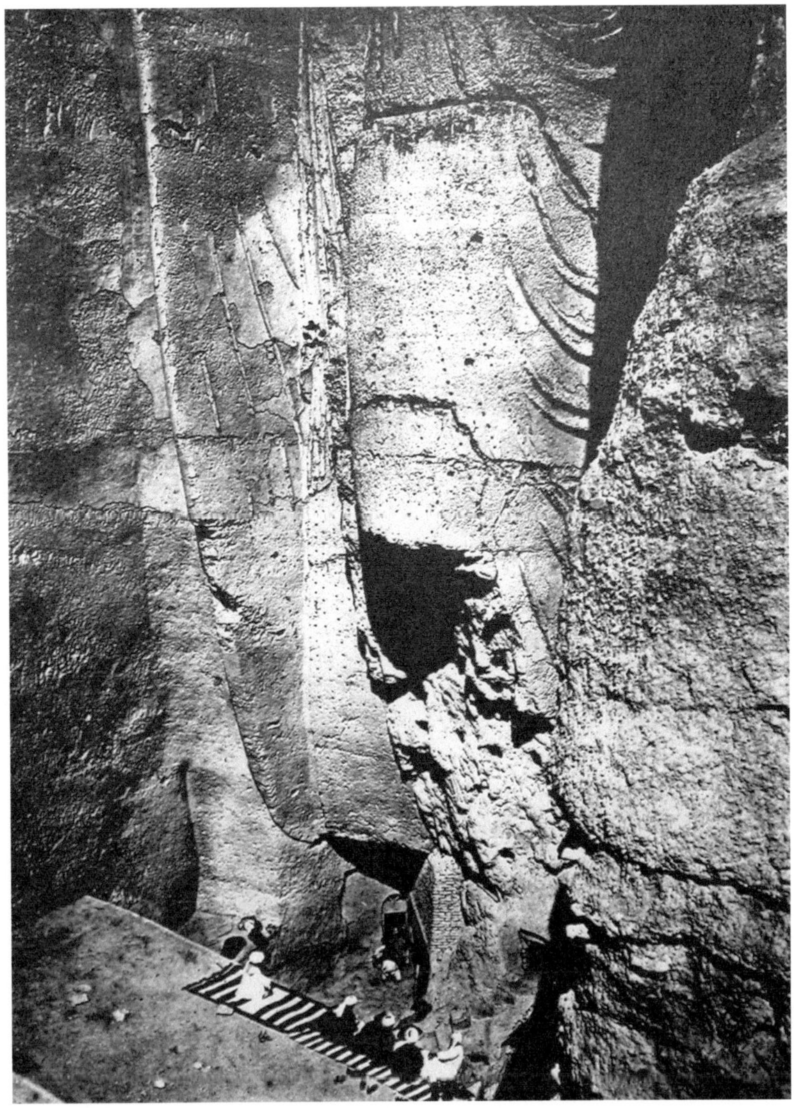

Fig. 3.26 Hackin M., Godard A., and Godard Y. *Memoires de la delegation Archeologique Francais en Afghanistan,* Tome II, *Les Antiquites Bouddhiques de Bamiyan,* Paris et Bruxelles, 1928

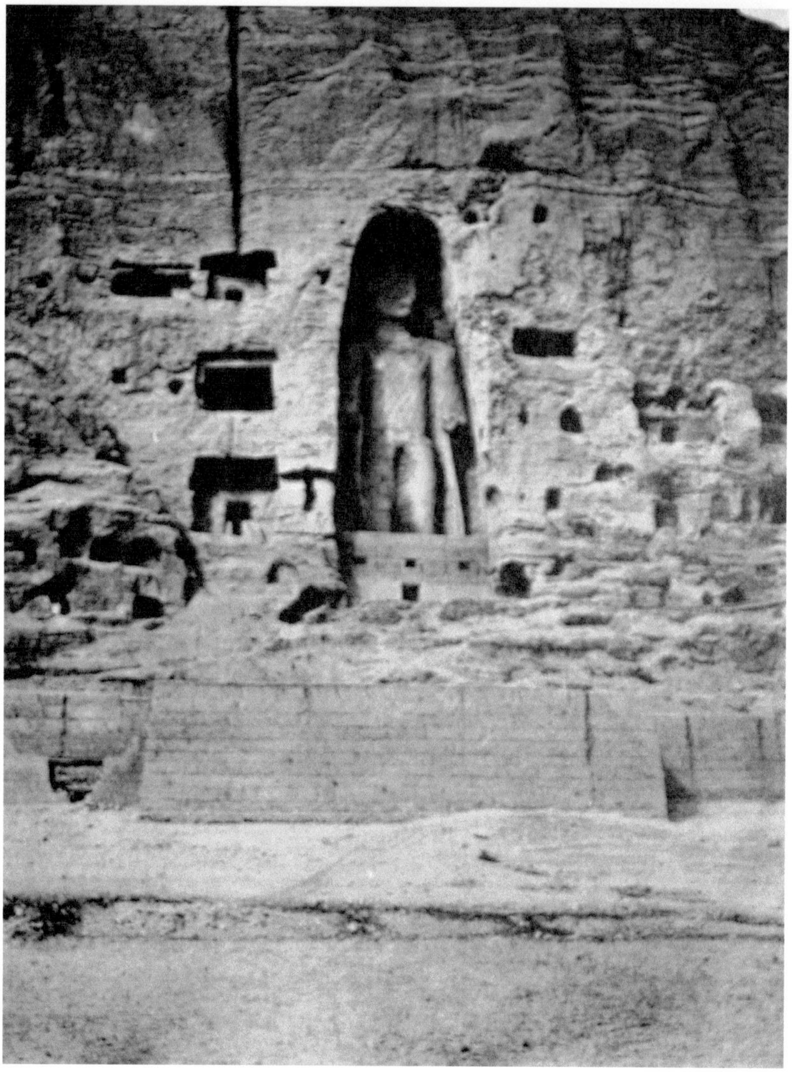

Fig. 3.27 Hackin M., Godard A., and Godard Y. *Memoires de la delegation Archeologique Francais en Afghanistan,* Tome II, *Les Antiquites Bouddhiques de Bamiyan,* Paris et Bruxelles, 1928

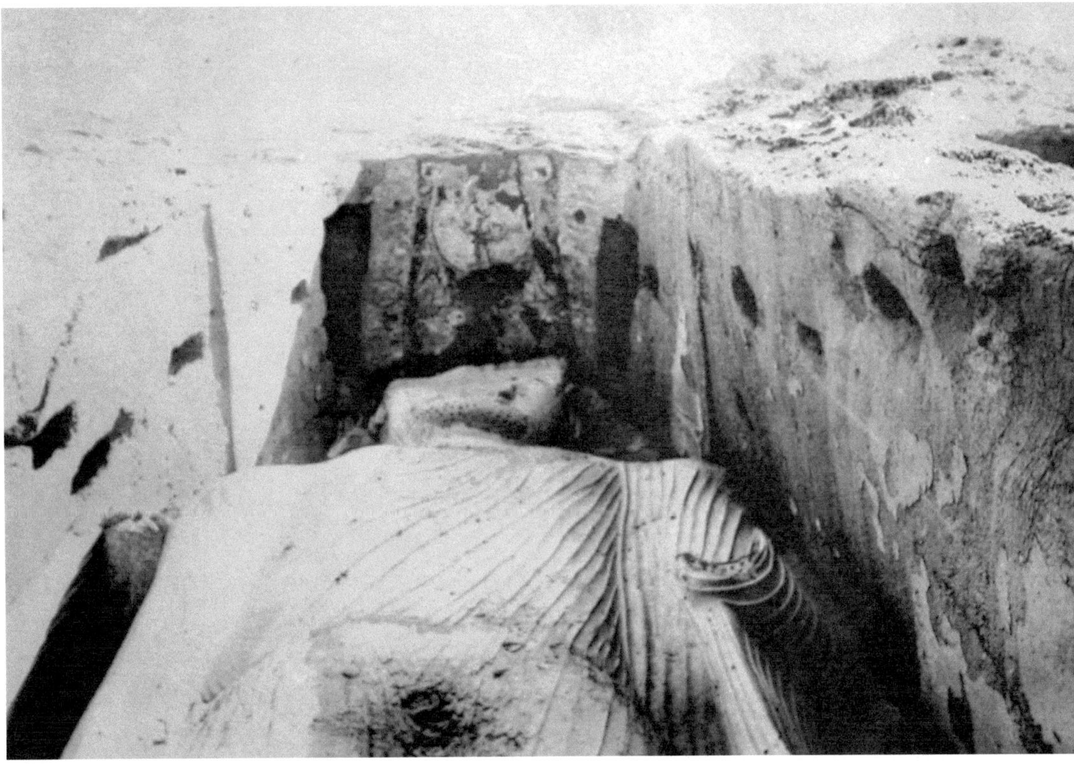

Fig. 3.28 Hackin M., Godard A., and Godard Y. *Memoires de la delegation Archeologique Francais en Afghanistan,* Tome II, *Les Antiquites Bouddhiques de Bamiyan,* Paris et Bruxelles, 1928

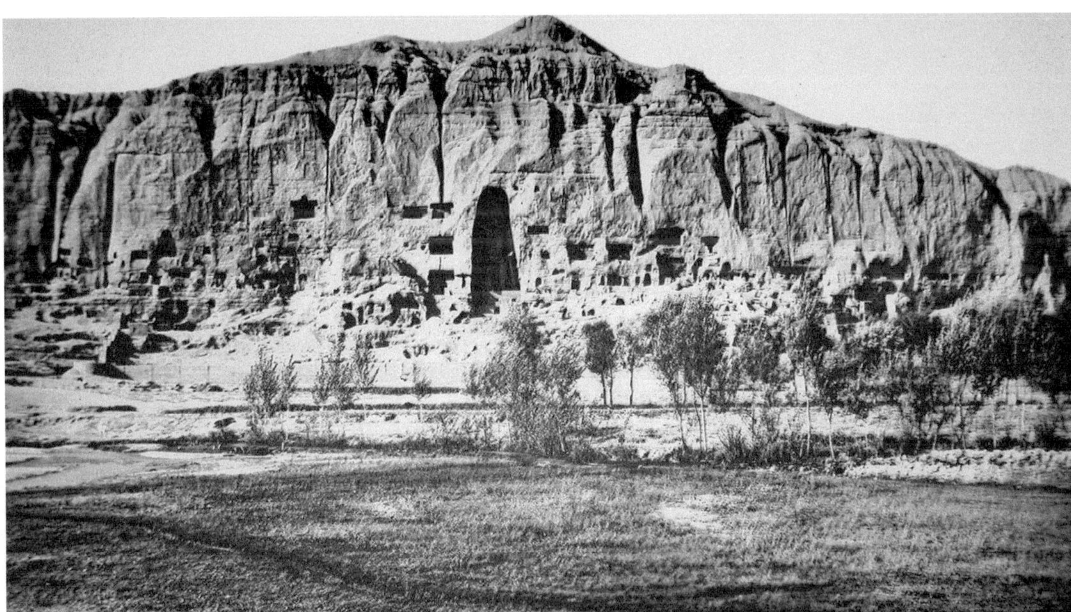

Fig. 3.29 Hackin M., Godard A., and Godard Y. *Memoires de la delegation Archeologique Francais en Afghanistan,* Tome II, *Les Antiquites Bouddhiques de Bamiyan,* Paris et Bruxelles, 1928

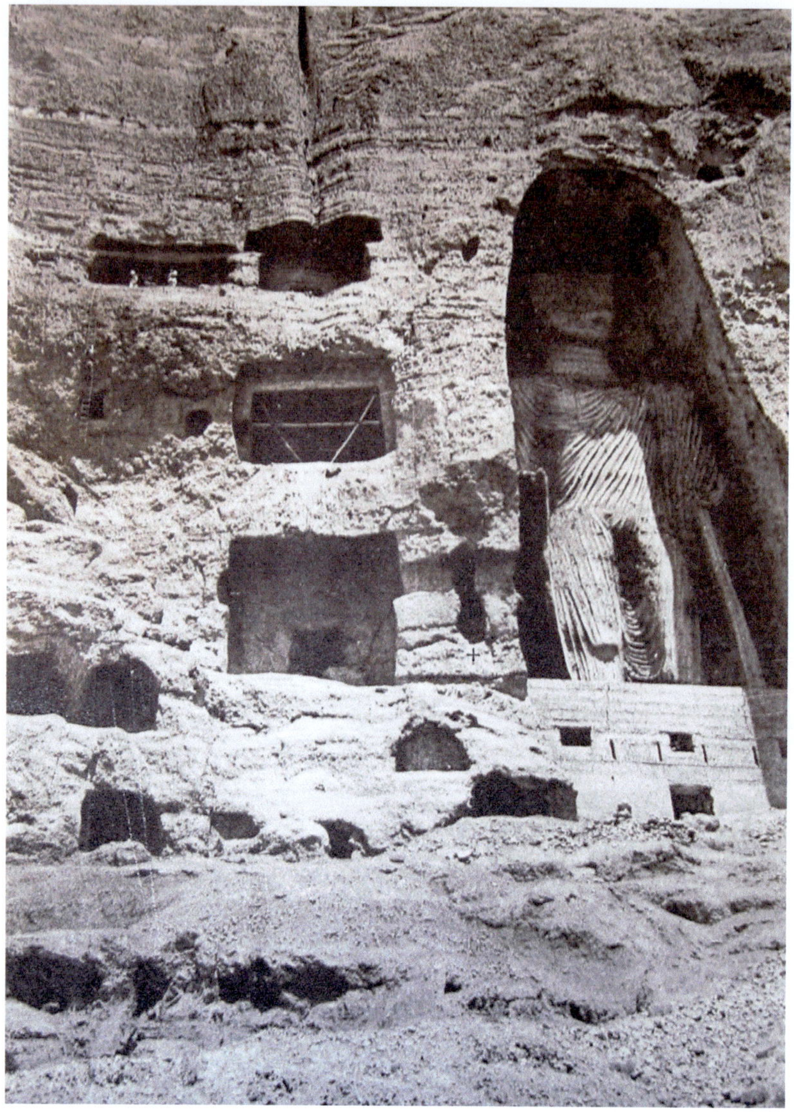

Fig. 3.30 Hackin M., Godard A., and Godard Y. *Memoires de la delegation Archeologique Francais en Afghanistan,* Tome II, *Les Antiquites Bouddhiques de Bamiyan,* Paris et Bruxelles, 1928

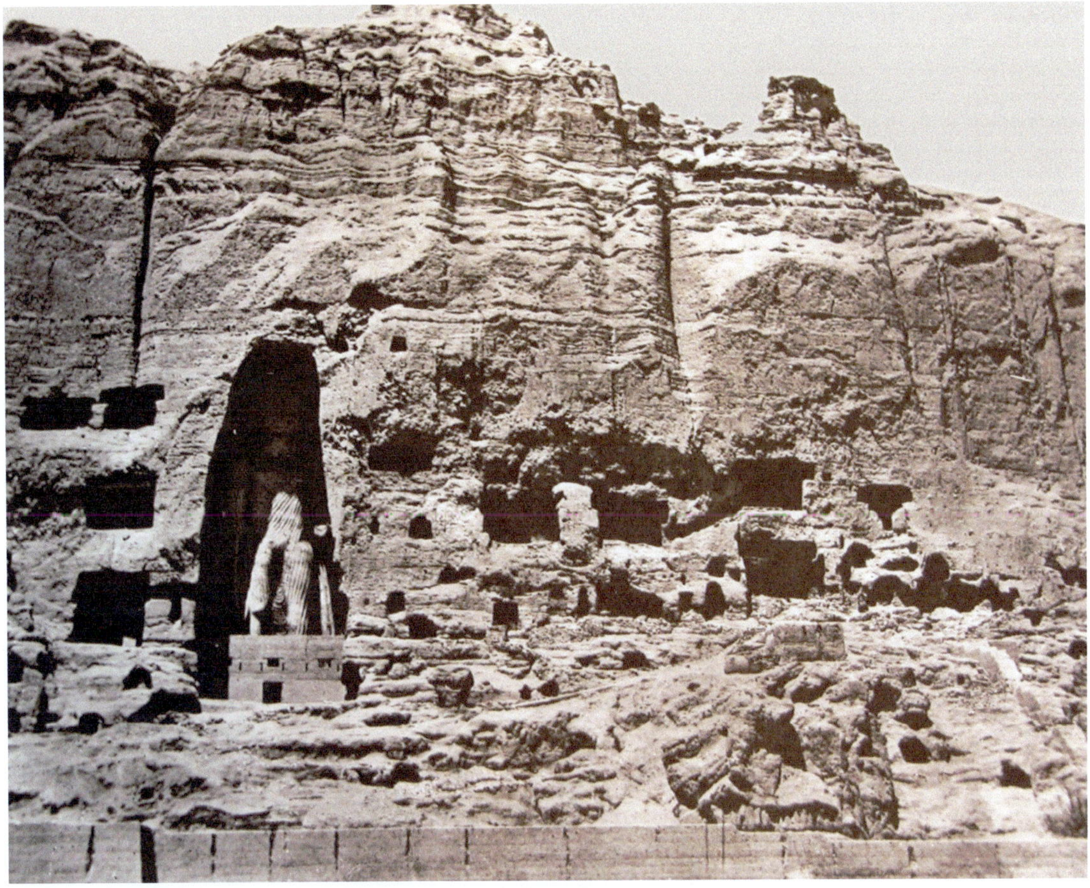

Fig. 3.31 Hackin M., Godard A., and Godard Y. *Memoires de la delegation Archeologique Francais en Afghanistan,* Tome II, *Les Antiquites Bouddhiques de Bamiyan,* Paris et Bruxelles, 1928

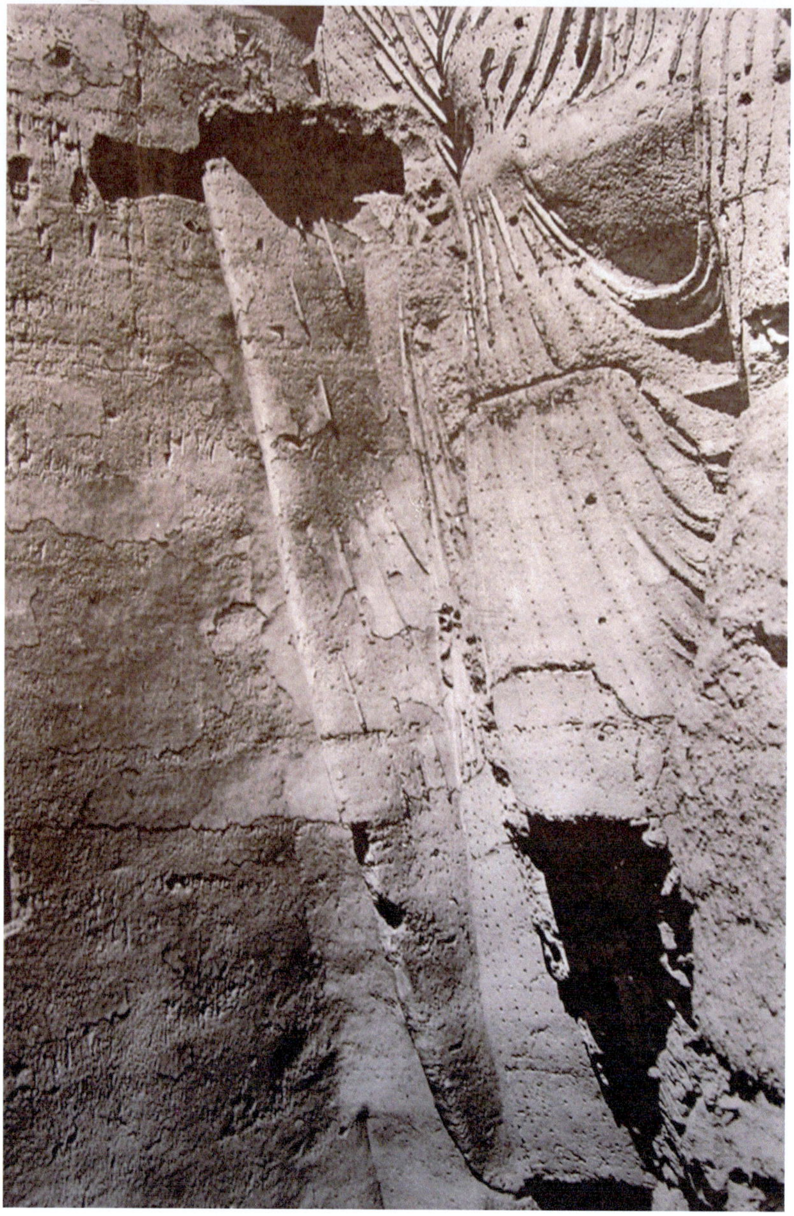

Fig. 3.32 Hackin M., Godard A., and Godard Y. *Memoires de la delegation Archeologique Francais en Afghanistan,* Tome II, *Les Antiquites Bouddhiques de Bamiyan,* Paris et Bruxelles, 1928

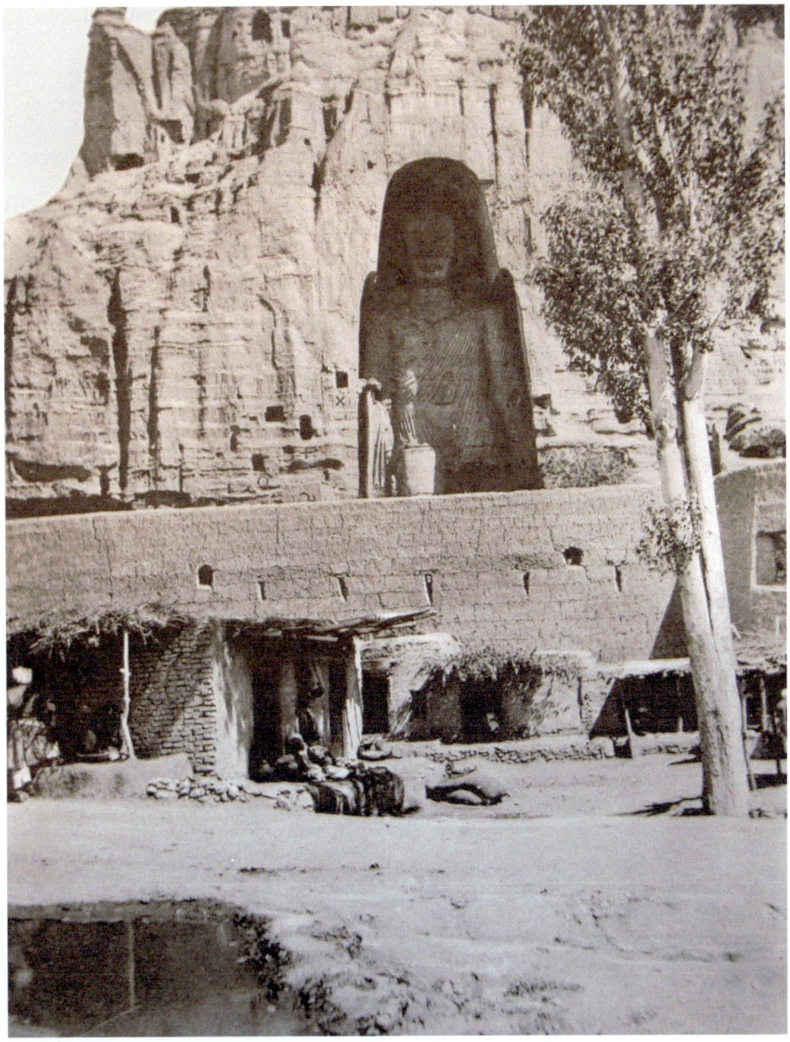

Fig. 3.33 Hackin M., Godard A., and Godard Y. *Memoires de la delegation Archeologique Francais en Afghanistan,* Tome II, *Les Antiquites Bouddhiques de Bamiyan,* Paris et Bruxelles, 1928

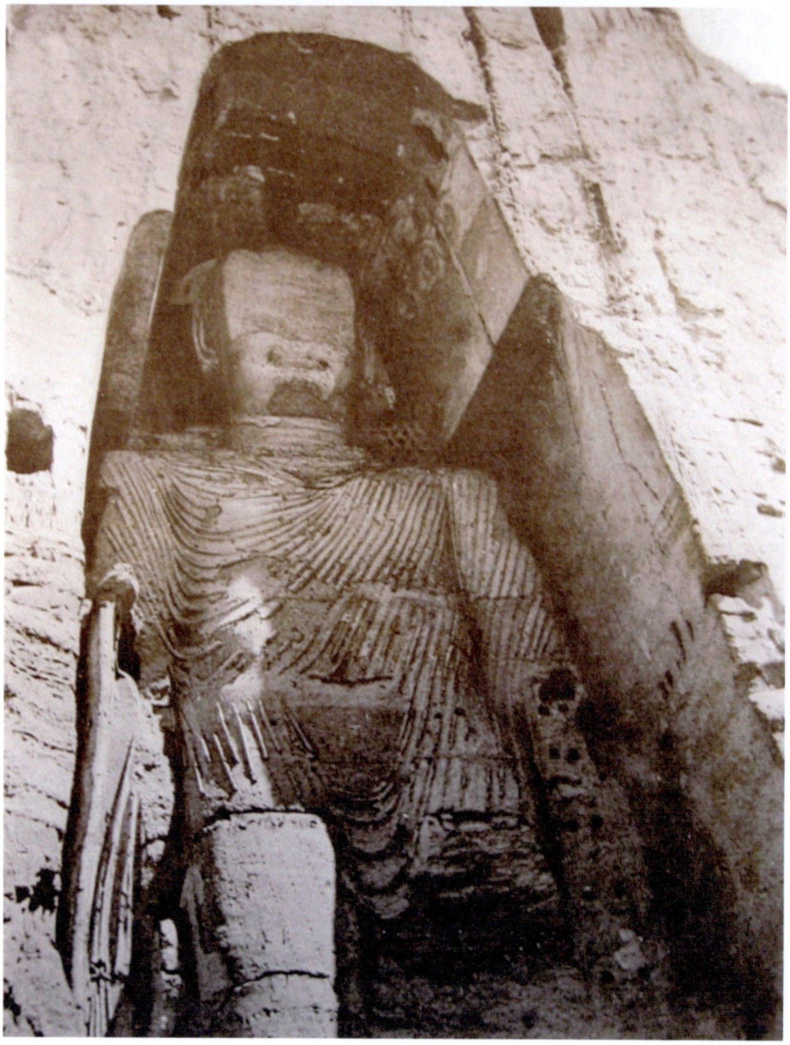

Fig. 3.34 Hackin M., Godard A., and Godard Y. *Memoires de la delegation Archeologique Francais en Afghanistan,* Tome II, *Les Antiquites Bouddhiques de Bamiyan,* Paris et Bruxelles, 1928

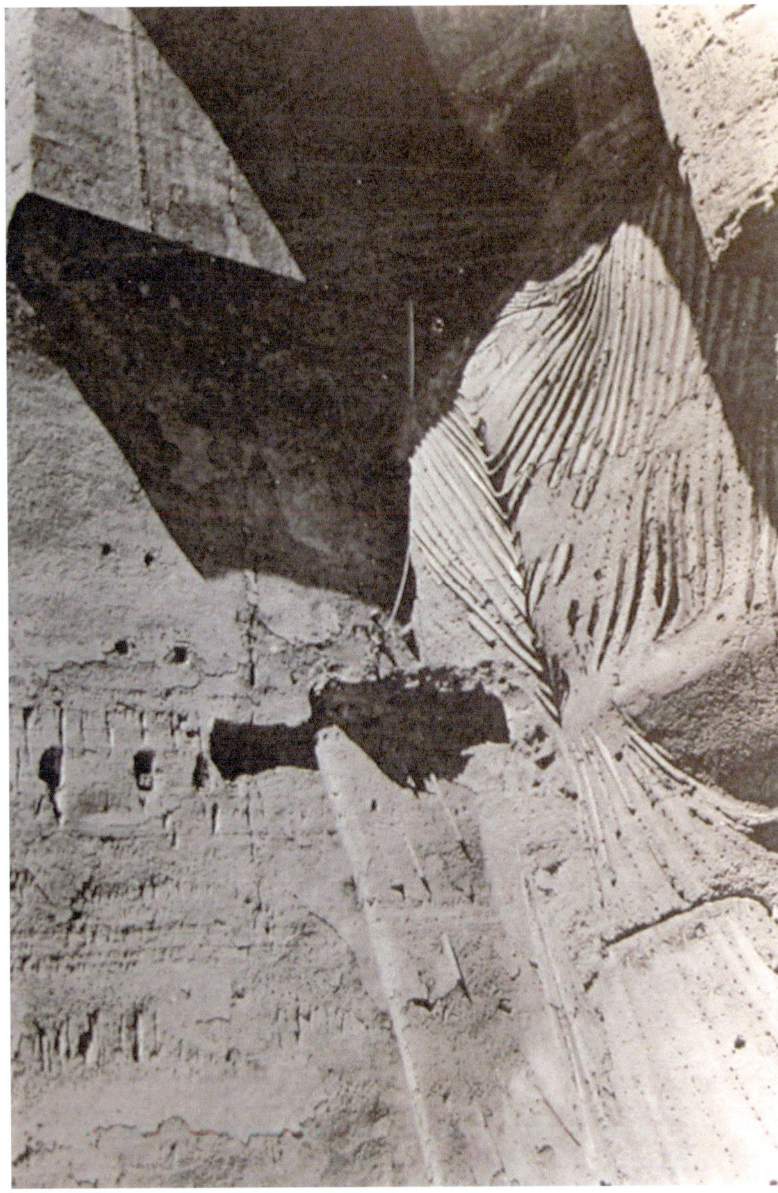

Fig. 3.35 Hackin M., Godard A., and Godard Y. *Memoires de la delegation Archeologique Francais en Afghanistan,* Tome II, *Les Antiquites Bouddhiques de Bamiyan,* Paris et Bruxelles, 1928

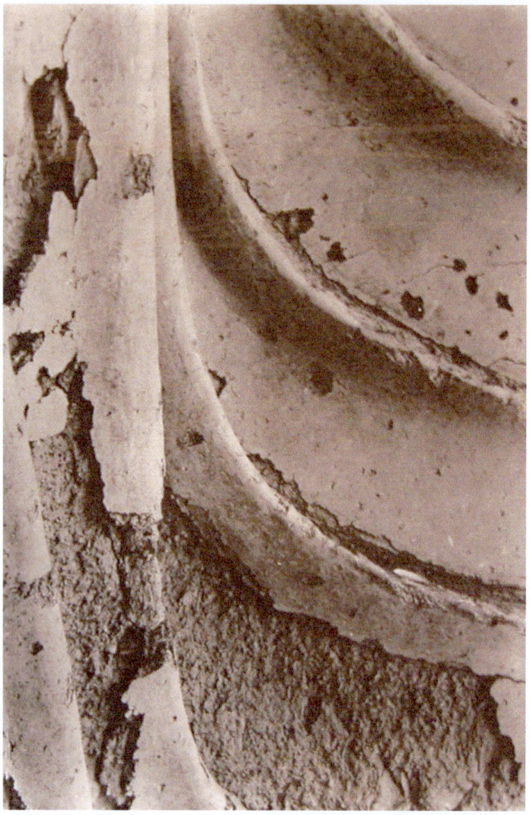

Fig. 3.36 Hackin M., Godard A., and Godard Y. *Memoires de la delegation Archeologique Francais en Afghanistan,* Tome II, *Les Antiquites Bouddhiques de Bamiyan,* Paris et Bruxelles, 1928

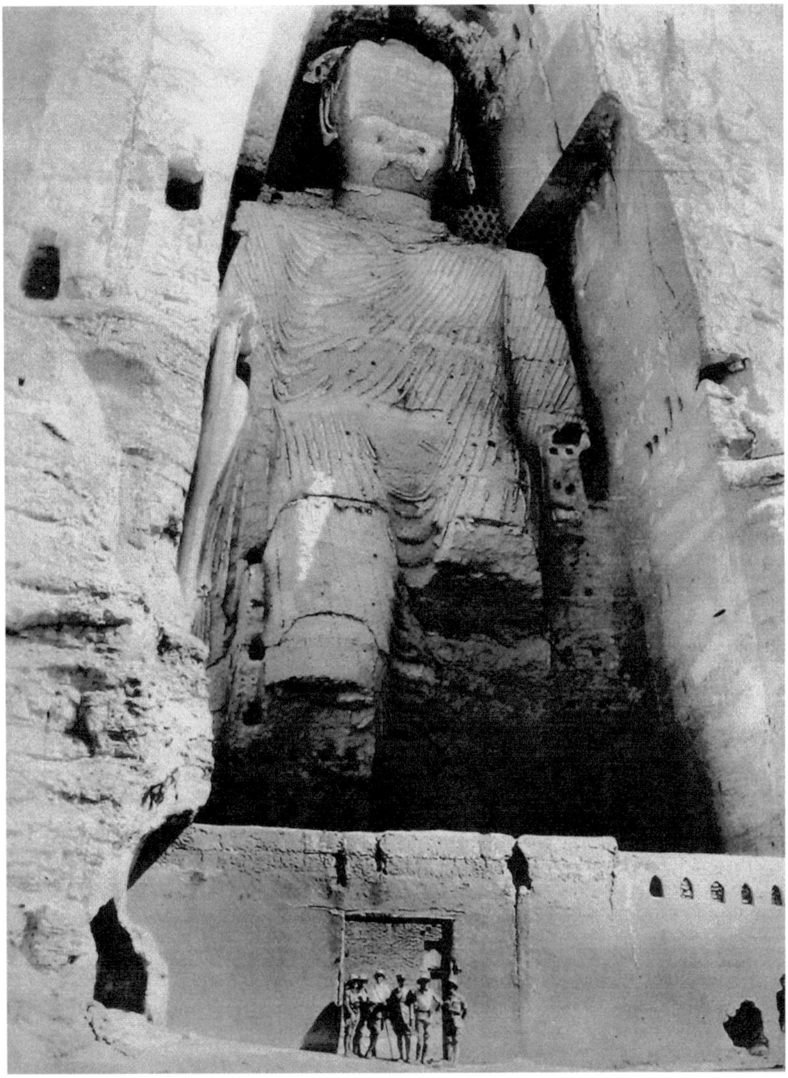

Fig. 3.37 The Giant Buddha of Bamiyan: rock-sculpture in Afghanistan. A photograph of the bigger statue, taken by members of the Citroen Trans-Asiatic Expedition. *The Illustrated London News, 179,* October 10, 1931, p. 557 (Courtesy of D. Bukerer)

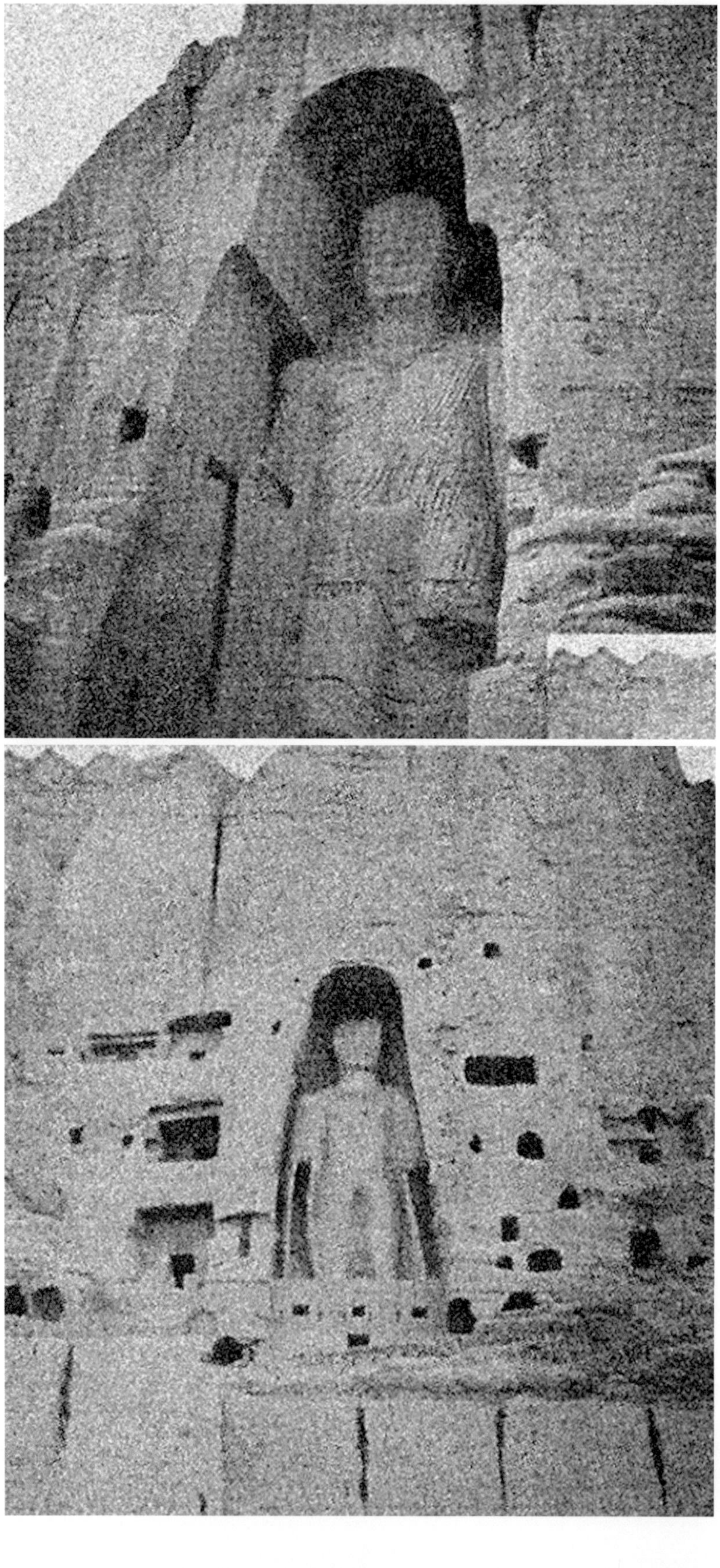

Fig. 3.38 Cipolla Arnaldo, Asia Centrale Sovietica contro India, Milano 1935

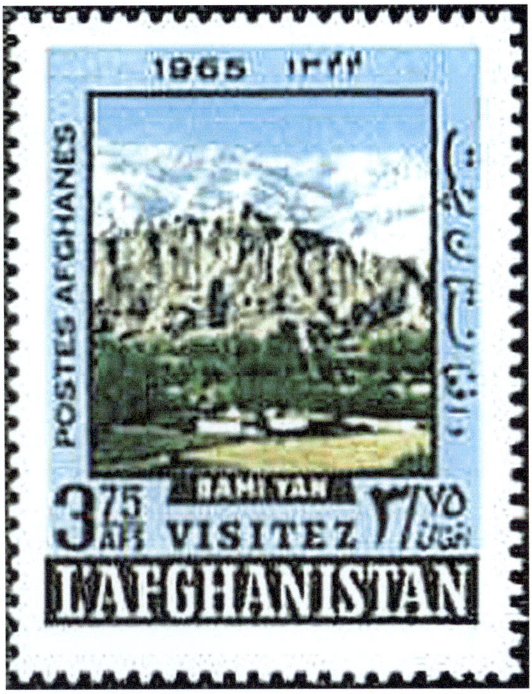

Fig. 3.39 Some Afghan philatelic series (Images from several web sites)

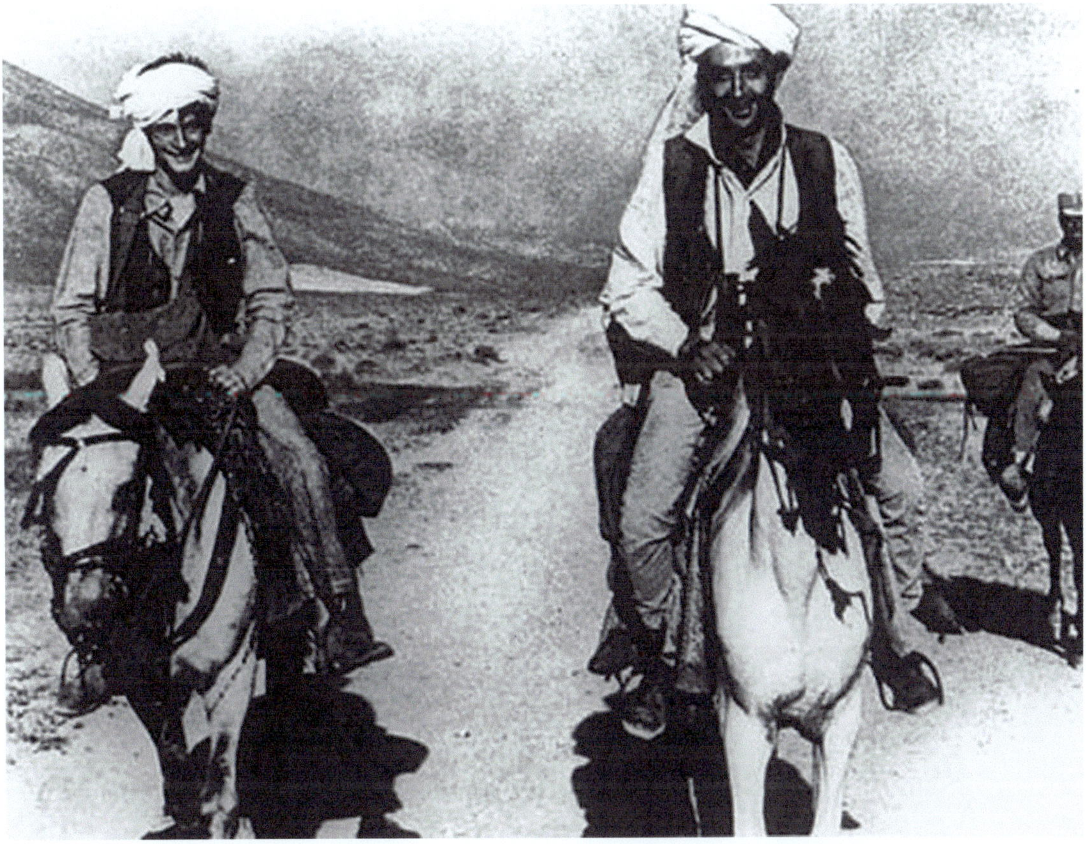

Fig. 3.40 Bruce Chatwin and Peter Levi. Levi Peter, Il giardino luminoso del Re Angelo, Torino, 2002

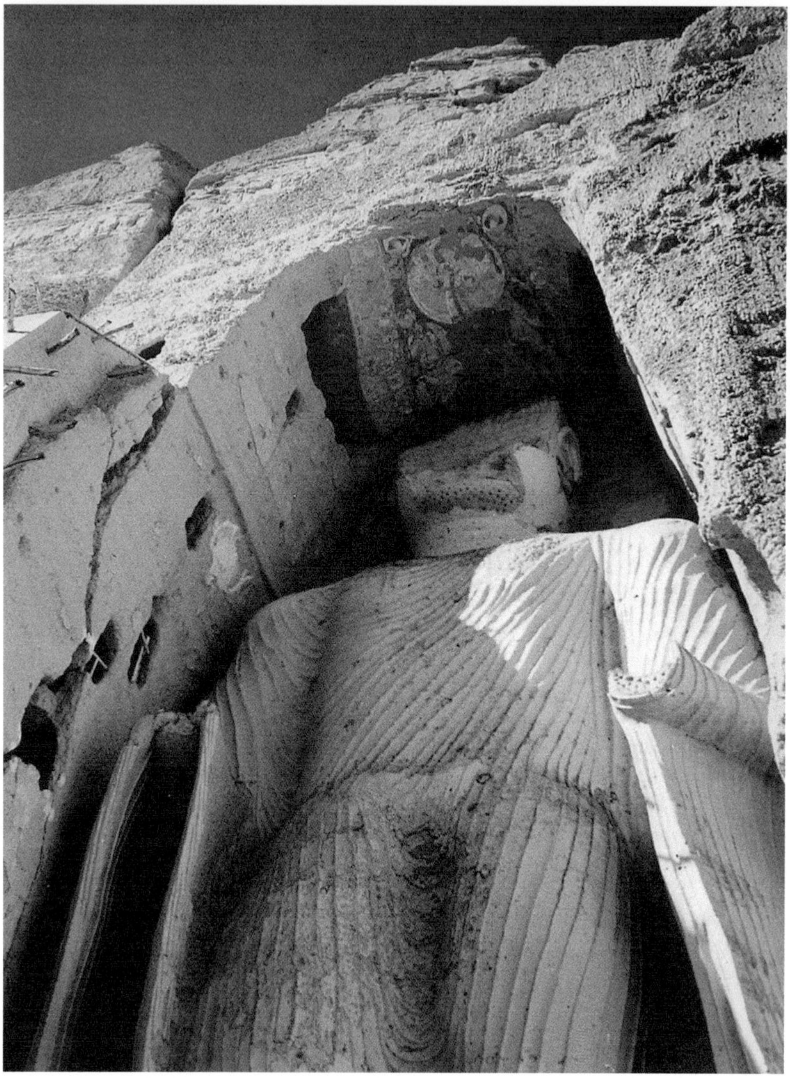

Fig. 3.41 Levi Peter, Il giardino luminoso del Re Angelo, Torino 2002

Fig. 3.42 Soviet troops during the Afghanistan War 1979–1989 (www.boser.chat.ru)

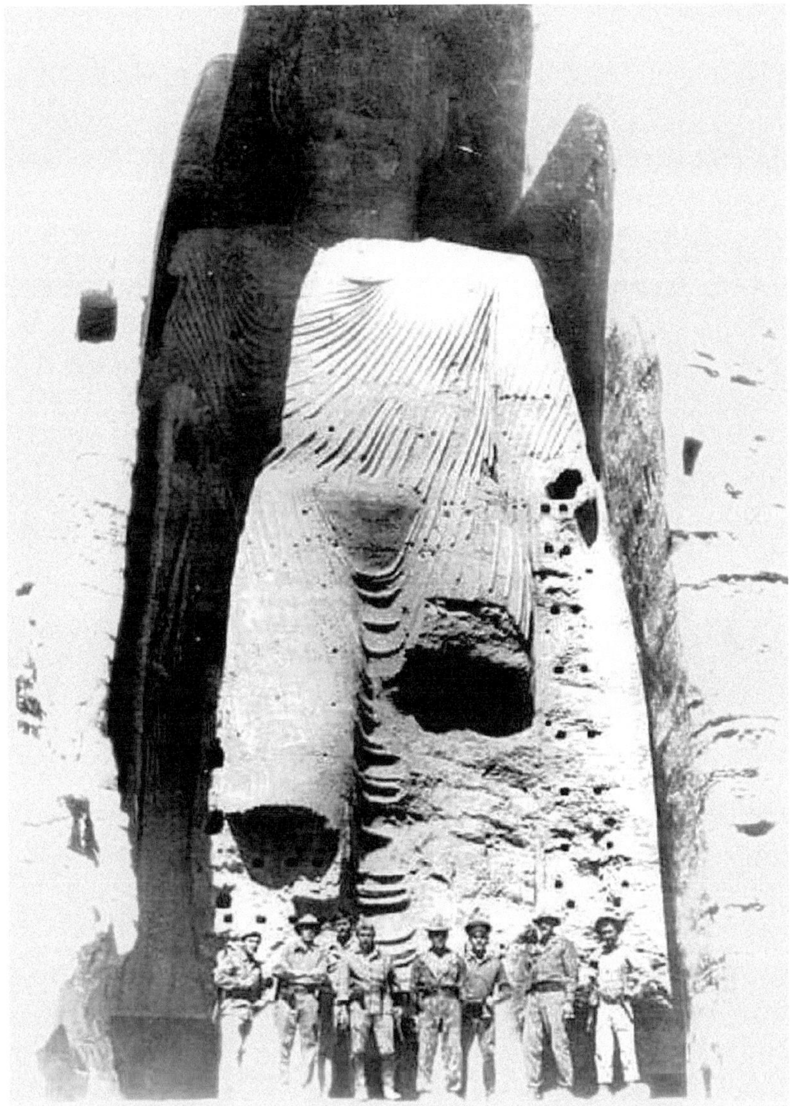

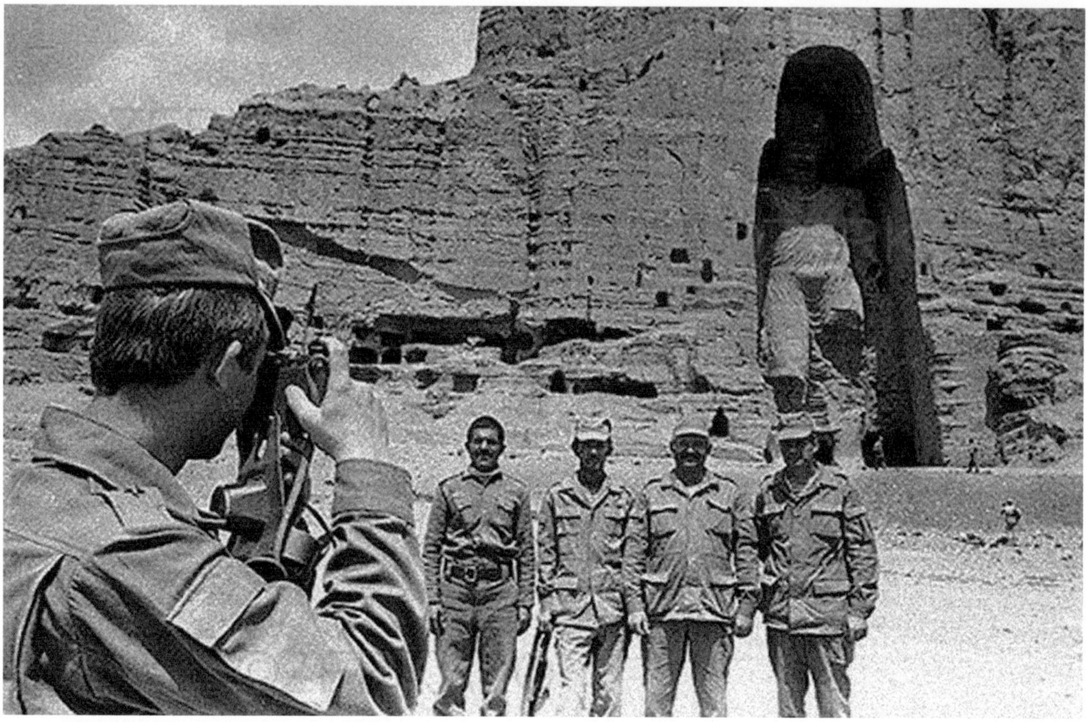

Fig. 3.43 Soviet troops during the Afghanistan War 1979–1989 (www.zerkalopress.ru)

Notes

1. This Arab geographer repeats some information coming from the work of another Arab geographer, Ibn Hauqal. In 1321, when Abou'l Feda finished its treaty on geography, the Valley of Bamiyan had already been destroyed by Mongolian troops.
2. Odorico dictated his travel memories to another friar, Guglielmo da Solagna, when he was tired and sick. These memories were therefore written not in chronological or geographic order, but in summary form.
3. Regarding the noise of nachara (castanet) reported by Odorico, an Arab writer, Hamd Allah Mustawfi El Qazwin, talking about a water source, said: "… here the water bursts forth with such a violence that its sound may be heard at some distance away."
4. Other interesting edition of the Odorico account is Monaco and Testa (1986). In the 1513 edition, the episode of the "face of stone" is omitted.
5. In his translation from Latin, the author reports the amazing sound produced by citherns, but in the Latin text, Odorico talks about castanets that produce a horrible noise, inducing great fear.
6. For a spy's point of view, see Ferrier (1860), where the valley is examined only from a military point of view.
7. Afghanistan, yesterday as today, is an extremely dangerous land to cross. In their notes, Burnes (1834), Moorcroft and Trebeck (1842), and Cipolla (1935) remember the epidemics, marauders, and bandits.
8. Other authors report that William Moorcroft reached Lhasa in 1826, and lived there for 12 years. He was assassinated in 1838 during his back travel to India.
9. Charles Masson collected nearly 3,000 Greek or Kushan period coins, which are now conserved in the British Museum (Masson 1841).
10. This sentence was discovered by the following French Archaeological Mission (DAFA).
11. Ten years later, two other Englishmen, Colonel Stoddard and Lieutenant De Mooyerost, followed the same long way during a diplomatic mission. Both were imprisoned by the Boccara's Emir Nars Ullah, because Stoddard introduced himself in the presence of the Emir while remaining on his horse. After some time Captain Conolly reached the Emir to negotiate the release of the prisoners, but he was imprisoned as well. After suffering torture, the three prisoners were beheaded by order of the same Emir Nars Ullah.
12. The British expedition was entirely destroyed. The account of the battle reports that only a doctor, Major Dr. William Brydon, was able to reach Jalalabad.
13. This book was translated in Italian with the title, Viaggi di Alessandro Burnes (1842). In France a large extract of Burnes's work was included in Xavier Raymond (1848). In Germany this report was mentioned in Carl Ritter (1838).
14. Haghe Louis (1833–1865), artist and engraver, worked in London and specialized in landscapes.
15. Lady Florentia Sale was the wife of British general Sir Robert Sale. When the rebellion began, Lady Sale and her daughter remained separated from her husband, who was in India. During this period Lady Sale wrote a diary, which was published in London in 1843 and became a bestseller.
16. Alfred Foucher (1905) was the first director of the French delegation, followed by his assistant Joseph Hackin. Foucher was the author of a treaty that became basic for archaeological studies in Afghanistan. The archaeological research conducted by DAFA were collected in Hackin et al. (1928).
17. The original notes on this travel, written by Bruce Chatwin, are today conserved in the Bodleian Library, Oxford University.

References

Aboul Ghazi B (1871) Histoire des Mogols et des Tatares, Desmaisons (trad.). St. Petersbourg

Benoffi FA (1761) Elogio storico delle gesta del Beato Odorico. Venezia

Bruce C (1993) L'occhio assoluto. Adelphi, Milano

Burnes A (1833) On the colossal Idols of Bamian. J Asiat Soc 2(23):561–564

Burnes A (1834) Travel into Bokhara. J. Murray, London

Burnes A (1842) Viaggi di Alessandro Burnes, tip. Giochetti, Prato

Cipolla A (1935) Asia Centrale Sovietica contro India. Casa Editrice EST, Milano

Fazl-Abu-Fazl (1891) The ain i Akbari, trad. Jarrett, vol II. Asiatic Soc. of Bengal, Calcutta

El Feda F-A (1848) Geographie d'Aboulfeda, M. Reinaud (trad). Impr. Nationale, Paris

Ferrier JP (1860) Voyage en Perse dans l'Afghanistan, le Bélouchistan et le Turkestan. E. Dentu, Paris

Foucher A (1905) Art the Greek-buddhique du Gandhara. Paris

Hackin J (1934) In Persia and Afghanistan with the Citroen Trans Asiatic expedition. Geogr J LXXXIII(5). p. 353–361

Hackin J, Godard A, Godard Y (1928) Memoires de la delegation Archeologique Francais en Afghanistan, Tome II, Les Antiquites Bouddhiques de Bamiyan. Van Oest, Paris et Bruxelles

Hayden HH (1910) Notes on some monuments in Afghanistan, in: Memoirs of the Asiatic Society of Bengal, Vol II

Hyde T (1760) Veterum Persarum et Parthorum et Medorum religionis historia, Editio secunda, Oxonii, Typographeo Claredoniano

Kaye E (1879) The Mountain Passes leading to the Valley of Bamian (Read at the Evening Meeting, February 24th, 1879). In: Proceedings of the Royal Geographic Society, new monthly series, London. vol I, n. 4 April

Komroff M (1928) Contemporaries of Marco Polo. Jonathan Cape, New York

Lady Florentia S (2002) A journal of the disasters in Afghanistan. Tennessee

Lee JL (ed) (1991) The journals of Edward Stirling in Persia and Afghanistan (1828–1829), Dip. Di Studi Orientale, Serie Minor XXXVI. Istituto Universitario Orientale, Napoli

Levi P (2002) Il giardino luminoso del Re Angelo. Einaudi, Torino

Masson C (1836) Notes on the antiquities of Bamiyan. J Asiat Soc Bengal V. p. 707–720

Masson C (1841) A descriptive account of the Antiquities and Coins of Afghanistan with a Memori on the buildings called Topes. London

Mehendale S (1996) Begram: along ancient central Asian and Indian trade routes. Cahiers d'Asie Centrale IFEAC 1–2

Monaco L. e Testa G.C. (a cura di) (1986) Odorichus de rebus incognitis – Odorico da Pordenone nella prima edizione a stampa del 1513. Pordenone

Moorcroft W, Trebeck G (1842) Travels in the Himalayan Provinces of Hindustan and the Panjab; in Ladakh and; in Peshawar, Kabul, Kunduz, and Bokhara from 1819 to 1825, vol 2. London

Pullè G (a cura di) (1931) Viaggio del Beato Odorico da Pordenone. Alpes, Milano

Qazwin-Hamd Allah Mustawfi of Qazwin (1340 AD) (1919) The geographical part of the Nuzhat al Qulub, G. Le Strange (trad). London

Ritter C (1838) Die Stupa's (Topes) oder die architectonischen Denkmale an der Indo-Baktrischen Konigsstrasse und die Colosse von Bamiyan. Berlin

Talbot MG, Maitland PJ, William S (1886) The rock-cut caves and statues of Bamiyan. J Royal Asiat Soc 18:323–350

The Graphic (1894) Jan 6, London

The Illustrated London News (1886). Nov. 6, p. 490; and Nov. 11, p. 525

Xavier R (1848) Afghanistan. Paris

Yaqout-Ibn Abdallah El Roumi El Hamawi Yaqout (1179–1229 AD) (1970) Du Mo'Djem El-Bouldan – Dictionnaire geographique, historique et litteraire de la Perse et des contrees adjacentes, Casimir Barbier De Meynard (translation), Amsterdam

Yule H, Cordier H (1913) Cathay and the way thither, etc. Hakluyt Society, London

4 UNESCO's Activities for the Safeguarding of Bamiyan

Christian Manhart and Roland Lin

On February 26, 2001 (3/12/1421), the Islamic Emirate of Afghanistan issued the following edict from the City of Kandahar: "On the basis of consultations between the religious leaders of the Islamic Emirate of Afghanistan, the religious judgments of the Ulema and the rulings of the Supreme Court of the Islamic Emirate of Afghanistan, all statues and non-Islamic shrines located in the different parts of the Islamic Emirate of Afghanistan must be destroyed. These statues have been and remain shrines of infidels and these infidels continue to worship and respect these shrines. Allah almighty is the only real shrine and all false shrines should be smashed. Therefore, the Supreme Leader of the Islamic Emirate of Afghanistan has ordered all the representatives of the Ministry of the Promotion of Virtue and Suppression of Vice and of the Ministry of Information and Culture to destroy all the statues. As ordered, by the Ulema and the Supreme Court of the Islamic Emirate of Afghanistan, all statues must be annihilated so that no one can worship or respect them in the future."

C. Manhart (✉)
Section of Museums and Cultural Objects, UNESCO,
7, Place de Fontenoy, 75352 Paris 07 SP, France
e-mail: c.manhart@unesco.org

R. Lin
World Heritage Centre, UNESCO,
7, Place de Fontenoy, 75352 Paris 07 SP, France
e-mail: r.lin@unesco.org

As soon as this order was made public, UNESCO issued appeals to the Taliban leaders, exhorting them to preserve the Afghan cultural heritage. These appeals were widely carried by the international press. The Director-General addressed a personal letter to the Taliban leader, Mullah Omar, on February 28. The Director-General also obtained the full support of the Islamic countries for UNESCO's activities to save the Afghan cultural heritage. On March 1, Mr. Pierre Lafrance, Special Representative of the Director-General, left for Islamabad, Kandahar, Kabul, the United Arab Emirates, Qatar, and Saudi Arabia. A number of Muslim religious leaders (*Ulema*) from Egypt, Iraq, and Pakistan intervened at the request of UNESCO, issuing fatwas against the Taliban's order. UNESCO launched an international petition for the safeguarding of the Afghan cultural heritage on its web site, and a special Funds-in-Trust account has been created for this purpose. The crisis gained a great deal of international media attention and UNESCO also received many letters of support for its actions in this matter from heads of state, ministers, other international organizations, and individuals. The Director-General personally contacted the Presidents of Egypt and of Pakistan, as well as of the Organization of the Islamic Conference, all of whom tried to use their influence to persuade the Taliban to cancel the order. Following these interventions, a delegation of 11 international Muslim leaders went to Kandahar in order to try to convince Mullah Omar and his advisors that

C. Margottini (ed.), *After the Destruction of Giant Buddha Statues in Bamiyan (Afghanistan) in 2001*,
Natural Science in Archaeology, DOI 10.1007/978-3-642-30051-6_4,

the Koran does not prescribe the destruction of statues. The Emir of Qatar generously provided his private airplane to enable this group of *Ulema* to travel directly to Kandahar. The meeting with Mullah Omar and his advisors lasted 4 h and as we all know unfortunately did not achieve any result.

All these political and religious interventions proved to be in vain, and in March 2001, the Taliban destroyed not only the Buddhas in Bamiyan but also a large number of statues throughout Afghanistan.

In December 2001, an international conference of *Ulema* was organized in Doha, Qatar, jointly with the OIC, ISESCO, and ALECSO, to examine the position of the Muslim world toward the preservation of Islamic and non-Islamic heritage. This conference resulted in a clear declaration of principles in favor of the protection of cultural heritage, including statues, that can be appealed to in the future and which can be found on the UNESCO website http://unesdoc.unesco.org/images/0014/001408/140834m.pdf.

UNESCO, the UN specialized agency having responsibility for culture, has a duty to take action in order to try to avert future cultural disasters of this type. It responds firmly to the challenge of rehabilitating Afghanistan's endangered cultural heritage, which has suffered irreversible damage and loss during the past two decades of war and civil unrest. The safeguarding of all aspects of cultural heritage in this country, both tangible and intangible, including museums, monuments, archaeological sites, music, art, and traditional crafts, is of particular significance in terms of strengthening cultural identity and a sense of national integrity. Cultural heritage can become a point of mutual interest for former adversaries, enabling them to rebuild ties, to engage in dialogue, and to work together in shaping a common future. UNESCO's strategy is to assist in the reestablishment of links between the populations concerned and their cultural history, helping them to develop a sense of common ownership of monuments that represent the cultural heritage of different segments of society. This strategy is therefore directly linked to the nation-building process within the framework of the United Nation's mandate and concerted international efforts for rehabilitating Afghanistan.

With reference to the UN Secretary-General's dictum, "Our challenge is to help the Afghans help themselves," policies and activities for the safeguarding of Afghanistan's cultural heritage focus on training and capacity-building activities related to the preservation of this cultural heritage.

Entrusted by the Afghan Government to coordinate all international efforts aiming to safeguard and enhance Afghanistan's cultural heritage, the Organization coordinates and carries out various activities. As the UN Programme Secretariat for Culture, Youth and Sports, UNESCO is supporting the Afghan Ministry of Information and Culture and related government agencies by coordinating all activities in the field of culture (Figs. 4.1 and 4.2).

In May 2002, UNESCO organized, in cooperation with the Afghan Ministry of Information and Culture, the first International Seminar on the Rehabilitation of Afghanistan's Cultural Heritage, held in Kabul, which gathered 107 specialists in Afghan cultural heritage, as well as representatives of donor countries and institutions. Under the chairmanship of H. E. Dr. Makhdoum Raheen, Minister of Information and Culture of the Afghan Government, the participants gave presentations on the state of conservation of cultural sites across the country and discussed programs and coordination for the first conservation measures to be taken. This seminar resulted in more than US$7 million being pledged for priority projects, allocated through bilateral agreements and UNESCO Funds-in-Trust projects. An 11-page document containing concrete recommendations for future action was adopted, in which the need to ensure effective cooperation was emphasized. Bearing in mind the enormous need to conserve sites in immediate risk of collapse, it was clearly stated, and approved by the Afghan Government, that the Bamiyan statues should not be reconstructed.

To this end and following the Afghan authorities' request to UNESCO to play a coordinating role in all international activities aimed at the safeguarding of Afghanistan's cultural heritage,

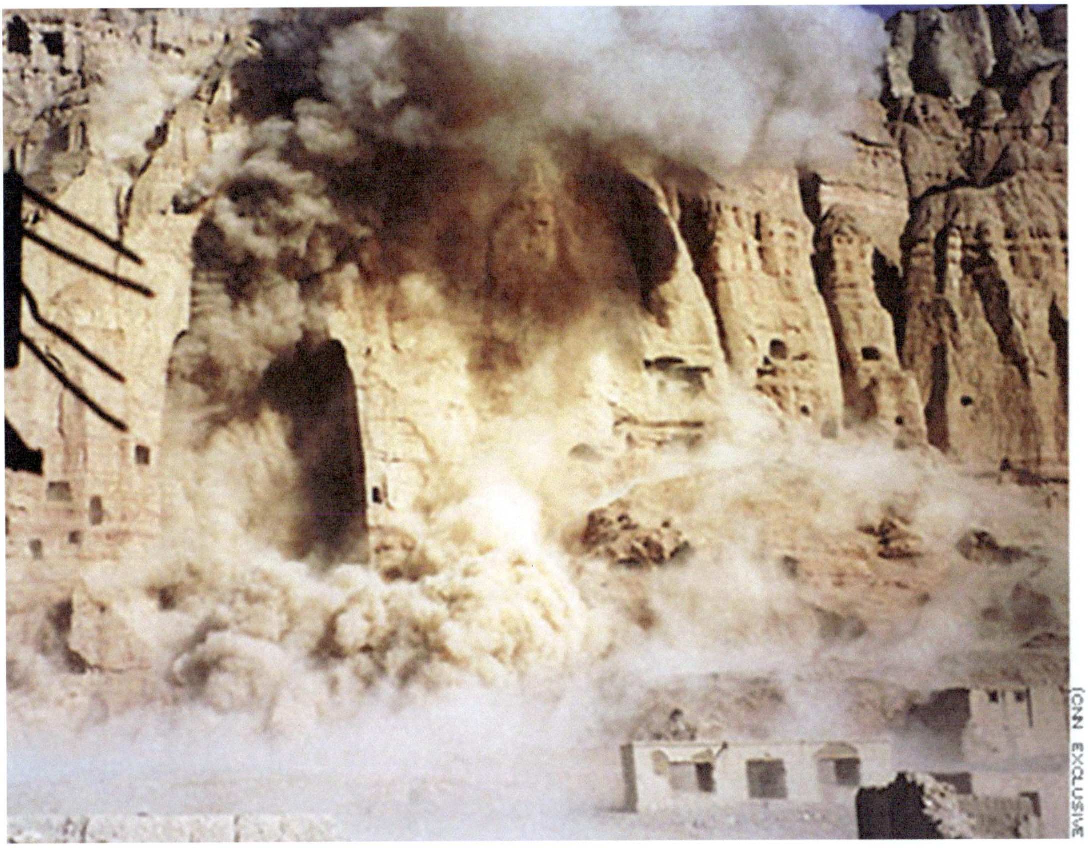

Fig. 4.1 The destruction of the Western Giant Buddha statue (From www.cnn.com)

UNESCO established an International Coordination Committee. The statutes of this Committee were approved by the 165th session of the Organization's Executive Board in October 2002. The Committee consists of Afghan experts and leading international specialists belonging to the most important donor countries and organizations providing funds or scientific assistance for the safeguarding of Afghanistan's cultural heritage. It meets on a regular basis to review ongoing and future efforts to rehabilitate Afghanistan's cultural heritage.

In June 2003, the First Plenary Session of this Committee was organized at UNESCO Headquarters. The meeting was chaired by H. E. Dr. Makhdoum Raheen, Minister of Information and Culture, in the presence of His Highness Prince Mirwais, seven representatives of the Afghan Ministry of Information and Culture, and more than 60 international experts participating as members of the Committee or as observers. The meeting resulted in concrete recommendations, which allowed the efficient coordination of actions to safeguard Afghanistan's cultural heritage to the highest international conservation standards. These recommendations concern key areas such as the development of a long-term strategy; capacity building; the implementation of the World Heritage Convention and the Convention on the Means of Prohibiting and Preventing the Illicit Import, Export and Transfer of Ownership of Cultural Property; national inventories and documentation, as well as the rehabilitation of the National Museum in Kabul; and the safeguarding of the sites of Jam, Herat, and Bamiyan. Several donors pledged additional funding for cultural projects in Afghanistan during and following the meeting.

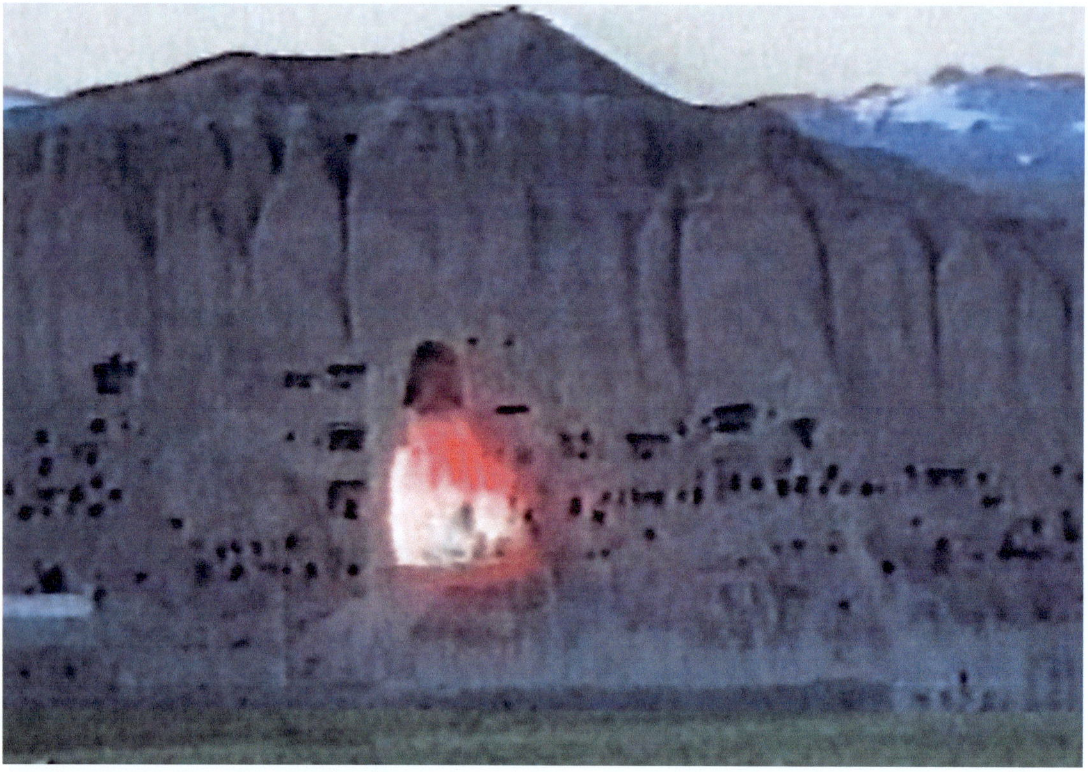

Fig. 4.2 The destruction of the Eastern Giant Buddha statue

Bamiyan Early Conservation Measures Following the Destruction of the Giant Buddhas

Immediately after the collapse of the Taliban regime in December 2001, UNESCO sent a mission to Bamiyan to assess the condition of the site and to cover the remaining large stone blocks with fiberglass sheets protecting them from harsh climatic conditions during winter. In July 2002, a second UNESCO mission jointly organized with the International Council on Monuments and Sites (ICOMOS) and directed by its president, Professor Michael Petzet, was undertaken in order to prepare conservation measures at the Bamiyan site.

In response to this situation, a contract was concluded through the Afghan Ministry of Information and Culture with the local commander, who immediately provided ten armed guards to be responsible for the permanent surveillance of the site, and no further thefts were noted since. It was also noted with concern that large cracks have appeared in and around the niches where the Buddha statues had previously been situated, which could lead to the collapse of parts of the niches and inner staircases.

In October 2002 a mission of Japanese experts noted that over 80 % of the mural paintings dating from the sixth to the ninth century A.D. in the Buddhist caves have disappeared since their last evaluation in the 1970s. The losses occurred through neglect or looting. In one cave, experts even found tools of the thieves and the remains of freshly removed paintings. Since 2003 a large number of missions of specialists from the Japanese National Research Institute for Cultural Properties have been fielded to Bamiyan for safeguarding of the mural paintings and for preparation of a master plan for the long-term preservation and management of the site. A Japanese company was contracted for the preparation of a topographic map of the valley and of a 3-D model of the niches and the cliffs.

After careful analysis, the UNESCO experts advised on appropriate measures to consolidate the cliffs and the niches and the Japanese Foreign Ministry generously approved a UNESCO Funds-in-Trust for the Safeguarding of the Bamiyan site. Following some delays caused by the security situation because of the war in Iraq, the first activities under this project started in June 2003. In addition, ICOMOS financed the restoration of a Sunni mosque and another building, both of which are located in close proximity to the niche of the Large Buddha. The aforementioned building is now used to accommodate the guards and to store the project equipment.

During the International Coordinating Committee plenary session in June 2003, a number of recommendations were made for the safeguarding of the Bamiyan site. It was notably recommended to consider the consolidation of the extremely fragile cliffs and niches, the preservation of the mural paintings in the Buddhist caves, as well as the preparation of an integrated master plan a priority.

In July 2003, the Bamiyan site was inscribed simultaneously on the UNESCO World Heritage List and List of World Heritage in Danger as the Cultural Landscape and Archaeological Remains of the Bamiyan Valley.

A large scaffolding, given free of charge by the German Messerschmidt Foundation, was transported by the German Army to Afghanistan in August 2003. With the help of this scaffolding and of additional imported specialized equipment, the internationally renowned Italian firm RODIO has successfully implemented the first phase of the emergency consolidation of the cliffs and niches.

The International Effort for the Preservation of the Cultural Landscape and Archaeological Remains of the Bamiyan Valley

The Japan Funds-in-Trust project for the Safeguarding of the Cultural Landscape and Archaeological Remains of the Bamiyan Valley is now in its fourth phase and has received a generous US$6,345,807 in funding. The activities have been coordinated and implemented under the supervision of UNESCO in close collaboration with Afghan authorities (the Ministry of Culture and Information, the Ministry of Urban Development, and the Provincial authorities of Bamiyan), the Japanese National Research Institute for Cultural Properties, the Italian experts from ISPRA (Italian Institute for Protection and Environmental Research) and RODIO, Aachen University, Munich Technical University, and ICOMOS Germany. Progress has been facilitated by regular meetings of members of these institutions forming the Expert Working Group on the Preservation of the Cultural Landscape and Archaeological Remains of the Bamiyan Valley, which met 11 times between 2002 and 2012. It has been proposed to hold the 12th Bamiyan Expert Working Group meeting in early November 2013 either in Orvieto, Italy, or in Bamiyan, Afghanistan.

The first, second, and third phases of the Japan Funds-in-Trust project were successfully completed in December 2004, in March 2008, and in September 2012, respectively. The most important outcomes of phases I, II, and III are as follows:

- Stabilization and consolidation of the cliffs and two Buddha niches that were threatened with collapse following the destruction of the two Buddha statues in 2001
- Conservation, monitoring, and documentation of the fragments of the two destroyed Buddha statues (implemented with additional funding of over one million euros from the government of Germany by ICOMOS Germany)
- Safeguarding the remains of the mural paintings in the Buddhist caves along the cliff as well as in the adjacent valleys of Kakrak and Foladi
- Archaeological research to define the exact boundaries of the various archaeological zones and register and document historical monuments, traditional settlements, irrigation systems, etc.
- Elaboration of a Cultural Master Plan consisting of protective zones with corresponding regulations to control urban development that could have a destructive effect on cultural heritage

- Management Plan preparation for the sustainable management of the World Heritage Site
- Fighting against illicit excavations through site surveillance
- Training Afghan experts (conservators, archaeologists, architects, heritage professionals) in heritage management and conservation

The World Heritage Committee, at its 31st session (New Zealand, July 2007), adopted several benchmarks (criteria for a "Desired State of Conservation") which, if met, will indicate that the site no longer faces immediate threats to its conservation and can be removed from the List of World Heritage in Danger. These include ensured site security, structural stability of the two Giant Buddha niches, adequate state of conservation of archaeological remains and mural paintings, and implementation of the Management Plan and Cultural Master Plan (the protective zoning plan).

The main objectives of the current Bamiyan Phase IV (US$ 1,564,070 with the implementation period 2012–2014) are:

- To achieve the Desired State of Conservation for the removal of the property from the List of World Heritage in Danger and to facilitate a feasibility study for the long-term presentation of the Buddha niches
- To promote institutional and community capacity building in conservation, monitoring, planning, management, and community awareness for a cultural heritage-based sustainable development program, including the linkage of this phase IV project with the concept of Culture for Development for Bamiyan and the creation of a Bamiyan Museum for Peace
- To ensure scientific documentation of site findings and operations

The current work at Bamiyan focuses on progress toward the abovementioned Desired State of Conservation and its benchmarks and the eventual removal of the property from the List of World Heritage in Danger. While substantial progress has been made, a number of threats to the site remain. Development pressure is an imminent threat to the cultural landscape of the Bamiyan Valley and its archaeological remains. The Foladi road construction – which develops a critical road system for Bamiyan – poses a danger to the property, especially in the areas where the road comes near to or inside its boundaries. The Hajigak mining project near the Shari Zorak monument zone of the property, equally has the potential to impact it. The State Party has reported increasing urban pressure on the property from the growing resettlement of people within the valley, complicated by the private ownership of some land within the property's boundaries. Development anticipating an influx in tourists has also contributed to urban growth and exerted greater pressure on the cultural landscape.

Proposals for the Future Presentation of the Buddha Niches

While there has been much debate about the possible reconstruction of the two Buddha statues, UNESCO is not pursuing this objective, nor is the Organization involved in any initiative to achieve this aim. UNESCO's actions in Bamiyan focus on undertaking urgent conservation measures and protecting as much of the Bamiyan heritage as possible. Through UNESCO's projects, international experts and the Afghan authorities will elaborate possible technical options for the future use of the Buddha fragments, based on international conservation standards and in consultation with the World Heritage Committee. Such projects aim to support the Afghan Government in taking its future decision regarding the archaeological remains of the Bamiyan Valley.

At the 9th Expert Working Group Meeting in March 2011 at UNESCO in Paris, four international teams made proposals for remedial measures and the future presentation of the two Buddha niches. The 11th Expert Working Group meeting (Aachen, Germany, December 2012) revisited these proposals, concluding again that based on available scientific data and financial resources, a total reconstruction of either Buddha sculpture cannot be considered at present. Rather, in light of the policy of the Afghan Ministry of Information and Culture, the Western Buddha niche should be consolidated and left empty at present as a testimony to the tragic act of its destruction.

Nevertheless, the government of Afghanistan expressed its strong wish for the partial reassembly of the Eastern Buddha to be an option in the coming years, and a feasibility study should explore whether or not this is possible. Such a study should be in line with the World Heritage Committee's Decision 36 COM 7A.26 at its 36th session (Saint Petersburg, 2012), which stated that feasibility studies on partial reassembly of the Bamiyan Eastern Buddha should consider an overall approach to conservation and presentation of the property, an appropriate conservation philosophy based on the Outstanding Universal Value of the property, as well as the technical and financial possibilities for the implementation of the project proposals. The study should be further examined by the Expert Working Group and the International Coordination Committee for the Safeguarding of Afghanistan's Cultural Heritage – as well as by the World Heritage Committee – before final submission to the Afghan authorities for their consideration. Should the partial reassembly of the Eastern Buddha be considered, more scientific and technical data must be acquired, and the process should be undertaken in a flexible step-by-step approach. Finally, safeguarding and preserving the entire cultural landscape of the Bamiyan Valley is paramount, with due regard to all of archaeological and architectural components necessary for the future development of Bamiyan.

UNESCO has assisted Afghanistan in safeguarding this important World Heritage property since 2002 when we reestablished an office in Kabul. We have concentrated on identification, conservation, and management of existing sites. This has also involved the production of a Cultural Master Plan and zoning plan for the valley to help balance heritage safeguarding and infrastructure development to improve the lives of people (supply of electricity, water, roads, etc.). For UNESCO, the main concern for the Buddha sculptures is to consolidate and stabilize their remains and the niches themselves which have been in danger of collapse since the tragic events of March 2001. Since 2002, in cooperation with ICOMOS Germany, we have removed and safely housed the remaining fragments of the sculptures in temporary structures at the site. We have consolidated the Eastern Buddha niche (Small Buddha) and remaining sculpture on the rear wall and we are now concentrating on the immense task of the Western Buddha niche and surrounding cliff. Funding is available to begin this process, but there is a gap of at least a million dollars to see its completion over the next few years. This emergency conservation work on the Buddha site, urban planning, and safeguarding the remaining fragments and the other parts of the property in danger from further deterioration continue to be our focus.

UNESCO would like to take this opportunity to thank all our generous donors for their indispensable contributions to the safeguarding of Afghanistan's cultural heritage. It should also be emphasized that these funds come from specific cultural budgets. As such, they are in no instances taken from humanitarian funds but rather constitute an addition to them.

General Environmental Condition of the Bamiyan Valley

5

Giuseppe Delmonaco and Claudio Margottini

Meteoclimatic Setting

Afghanistan is a mountainous country in a dry part of the world, which experiences extremes of climate and weather. Winters are cold and snowy, and summers hot and dry. The wet season generally runs from winter through early spring, but the country on the whole is dry, falling within the Desert or Desert Steppe climate classification. Very little snow falls in the lowland deserts of the southwest, but the snow season averages roughly October to April in the mountains and varies considerably with elevation.

Climate normals were computed by Afghanistan 10 years ago and provided to the World Meteorological Organization for the global standard normals project. The Afghan normals cover the period roughly from 1956 to 1983. Based on stations for which climate normals were provided (NCDC 2003):

- The average annual precipitation ranges from 51.6 mm at Zaranj in southwest Afghanistan to 992,124 mm in the northeast mountains at North Salang.
- Temperatures can vary widely, from as cold as 51 °F below zero at Chakhcharan (in the north central mountains at an elevation of 2,183 m) to as hot as 124 °F in the southwestern deserts at Zaranj.
- The mountain valleys can experience, on average, 10–30 days per year with snowfall, but the higher passes receive much more snow.
- At an elevation of 3,366 m, North Salang receives snow, on average, 98 days out of the year with depths reaching as high as 4,496 mm. Snow has been observed on the ground there as early as August and as late as June.

The precipitation pattern is reported in the following figure.

Strong winds can blow any time of the year, spawning blinding dust storms in the summer and raging blizzards in the winter. During the period 1961–1983, wind gusts were clocked as high as 98 mph in February at Farah, in western Afghanistan. Sunny skies characterize the dry summers, whereas low clouds bring rain and snow to much of the country during the winter wet season (Fig. 5.1).

Based on data from 4 weather stations (in Kabul, Mazari-Sharif, Herat, and Qandahar):

- Average cloud cover runs from 0 to 1/8 of sky coverage in summer to 1/2 to 5/8 in the winter months.
- The percentage frequency of occurrence of low ceiling and/or visibility (cloud height 3,000 ft or less, or visibility 3 miles or less) ranges from 1 to 5 % of the time in the summer months to 11–24 % of the time in January and February.

G. Delmonaco (✉) • C. Margottini
Department of Geological Survey of Italy,
ISPRA – Italian Institute for Protection and Environmental Research,
Via Brancati, 60, 00143 Rome, Italy
e-mail: giuseppe.delmonaco@apat.it;
claudio.margottini@isprambiente.it

C. Margottini (ed.), *After the Destruction of Giant Buddha Statues in Bamiyan (Afghanistan) in 2001*,
Natural Science in Archaeology, DOI 10.1007/978-3-642-30051-6_5, © Springer-Verlag Berlin Heidelberg 2014

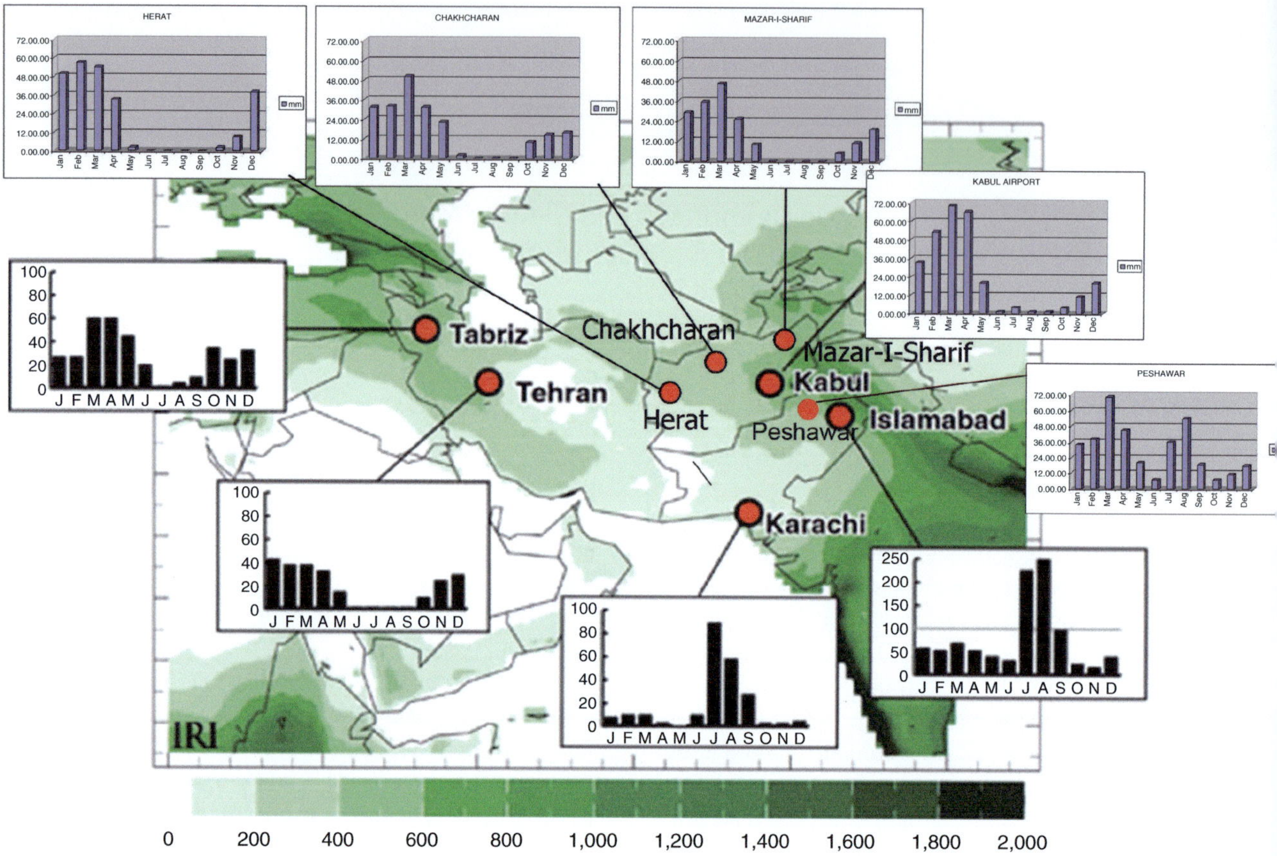

Fig. 5.1 Mean rainfall precipitation (mm) in Central Asia (IRI 2003, modified with data from NCDC 2003)

- Average number of days per month with fog ranges from zero in the summer months to 4 days per month in the winter.
- Average number of days per month with blowing dust or blowing sand ranges from 1 to 2 in the winter months to 6 days per month in July.

Specific data for Afghan meteorological stations are available at Operational Climatic Data Summary (2002). Reported data include temperature, precipitation, snowfall, mean relative humidity/vapor pressure/dew point, surface wind, mean cloud cover/thunderstorm/fog/blowing sand & dust, as well as other statistical interpretation. The site of Bamiyan is included in the list because there was a meteorological station belonging to the Afghan Air Authority Meteorology Department, Climate Section, operating for 8–9 years in the period 1958–1977 (added October 2001) (Fig. 5.2).

Mean annual precipitation in Bamiyan can be estimated at 162.56 mm and mean annual temperature at 7.40 °C.

Topographical Setting

At the beginning of this work, in 2002, a detailed topographical map, at a scale suitable for engineering purposes, was not available for the study area. A 1/50,000 map was produced by the Russians in the past whereas, a more reliable map, though referring to a frontal view of the cliff and caves, is reported in Higuchi (1984–1985).

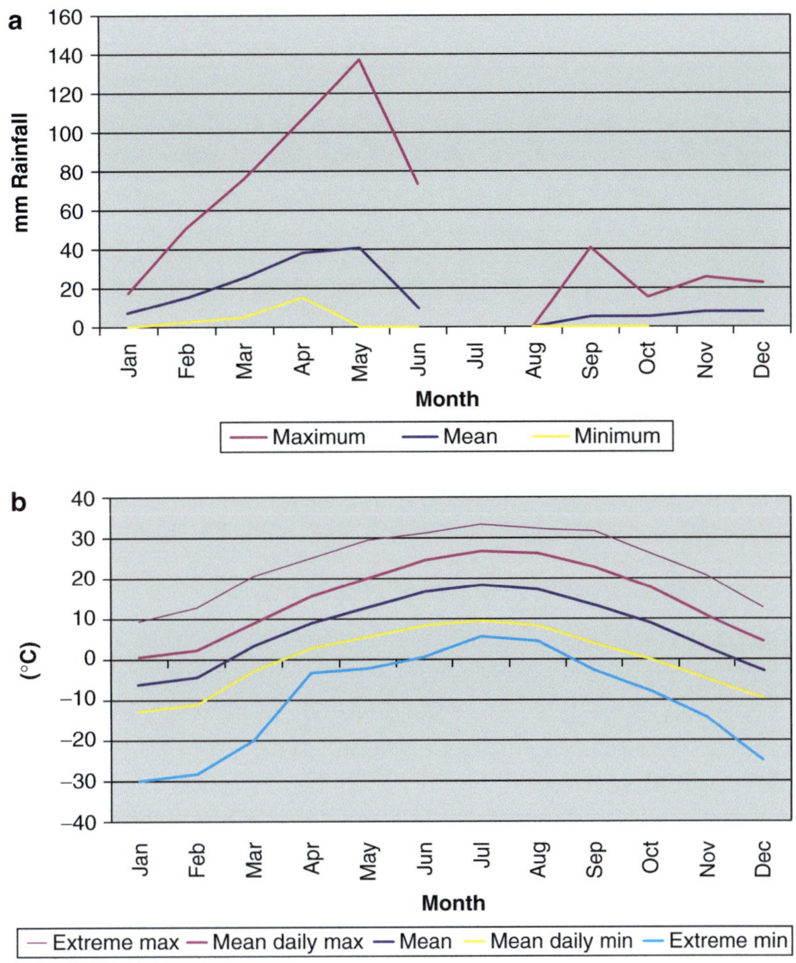

Fig. 5.2 Meteorological data for the Bamiyan Station in the period 1958–1977 (added October 2001) (Source: Operational Climatic Data Summary 2002)

In 2002, a preliminary topographic investigation was performed by Santana (2002) under UNESCO coordination.

Recently, the Japanese company PASCO (2003a) has provided a detailed survey of the area, based on already existing stereo couple aerial photographs. The work was excellent and very useful for many purposes but, unfortunately, the photos do not cover the northern side of the valley where the niches are located (Fig. 5.3).

For this reason, and considering the need for a detailed topography of the cliff, suitable to be used for stability analysis with Finite Element Models, all the above information was integrated with direct topographic measurements and a portable GPS survey. The results, in terms of topographic profile of the niches and related cliff, are shown in Fig. 5.4.

Uncertainties still remain in the upper part of the topographic profile where an accurate field survey was not undertaken because of the presence of land-mines in the area.

Also PASCO (2003) provided the laser scanner survey of both niches (Figs. 5.5 and 5.6). This information, available to all the involved teams, highly supported the stabilization work. All the planned and executed consolidation work was investigated and designed with the help of such maps. Before these maps, Gruen et al. (2002) and Gruen et al. (2004) provided useful data on the topographic feature of niches and statues based on photogrammetric techniques.

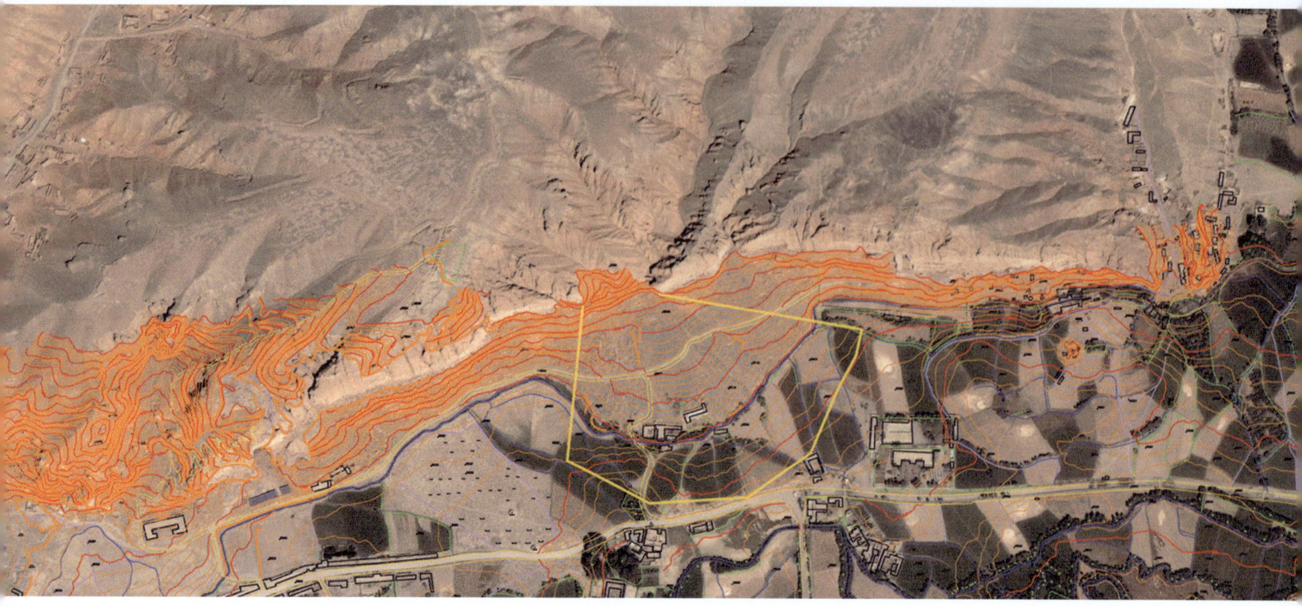

Fig. 5.3 The topographic survey developed by PASCO (2003a) in the whole Bamiyan valley, displayed only for the northern part of the valley where the niches are settled, superimposed on the QuickBird satellite image of the area

Geological Setting

In both Western and Eastern Buddha areas, continental lithotypes, probably coming from dismounting of surrounding morphological peaks and deposition in a flood plain and small lagoon are seen as outcropping. The subsequent river erosion produced the present morphology that, for the part where the rupestrian settlement is located, could also be affected by human excavation in order to obtain a vertical cliff suitable to host the monastic civilization. As possible confirmation on the hypothesis of some human influence on present morphology, it can be pointed out that such a steep morphology is not recognized in other parts of the valley.

In general terms (Lang 1968, 1972; Reineke 2006) the Bamiyan Valley is an intramountainous basin, subsequently filled with debris material originating from the surrounding mountain ranges (Lang 1971). The neogenic, more or less horizontally bedded sediments can be distinguished into four strata, which are shown in Figs. 5.7 and 5.8.

Starting with the Eocene Dokani Formation (>80 m of sandy carbonates and anhydrite) and the Zohak-Formation (>1,000 m of red conglomerates), the so called Buddha Formation is deposited in the Oligocene and is built up by over 70 m of yellow-brown pelites, sandstones, conglomerates, and some volcanic material. On the top are the Miocene Ghulghola Formation (>200 m of sandstone, clay, and lacustrine carbonates) and the Pliocene Khwaja-Ghar Formation (approx. 200 m of travertine, sandstone, and conglomerate) lie. Almost contemporary to the Buddha Formation is the Qal'acah one, which reflects a detritic facies on the slope of a volcano (Lang 1972).

At the northern and southern fault lines of the tectonic graben, red clayey soils formed by metamorphic contact can be found. Along these fault lines volcanic activity can be recognized as well, which may have modified (fritted) the surrounding sediments and changed their color to red.

In each of these layers the variability of substrata in size, structure, and color is enormous; this is an indicator for the vicinity of origin areas, where Mesozoic and Paleozoic lithology including volcanic, intrusive, and other sedimentary components can be found. This is typical for a continental sedimentation milieu which is near the origin (erosion) area.

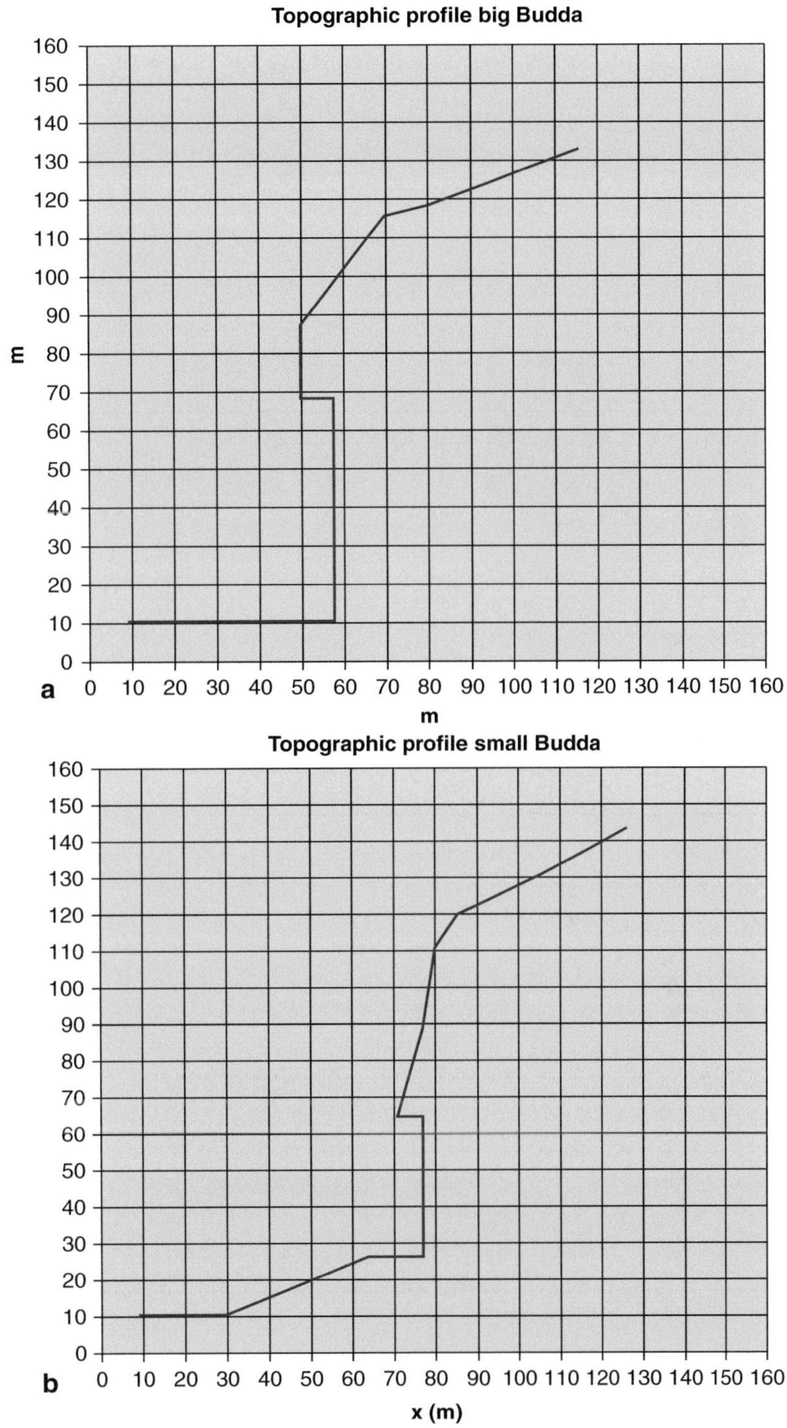

Fig. 5.4 Topographic profile of Western and Eastern Buddha sites

During the time from the late Pliocene to the end of the Pleistocene epoch (Reineke 2006), these neogenic sediments were incised by fluvio-glacial erosion. Alternating warm and cold periods led to changing conditions between accumulation and erosion, so that different

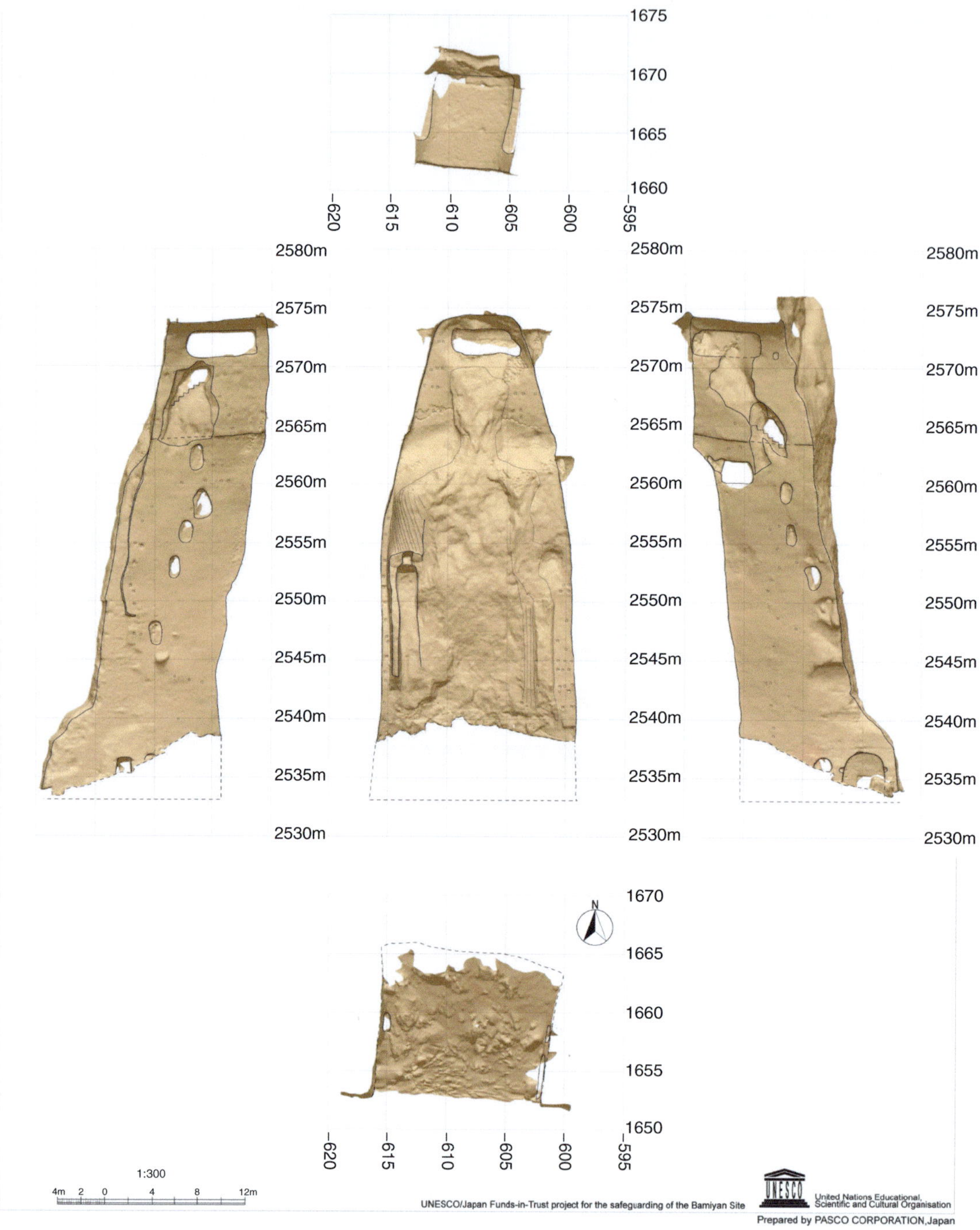

Fig. 5.5 Laser scanning map of Eastern Giant Buddha niche (PASCO 2003b)

quaternary terraces developed, which can be found at distinct levels according to Lang (1972) (respectively 5, 30, 50, 105, 115, 220, and 240 m above the river).

The cliff and niches have been excavated into the so called Buddha Formation and are composed by an alternation of conglomerate and siltstone (yellow at the bottom and red in the

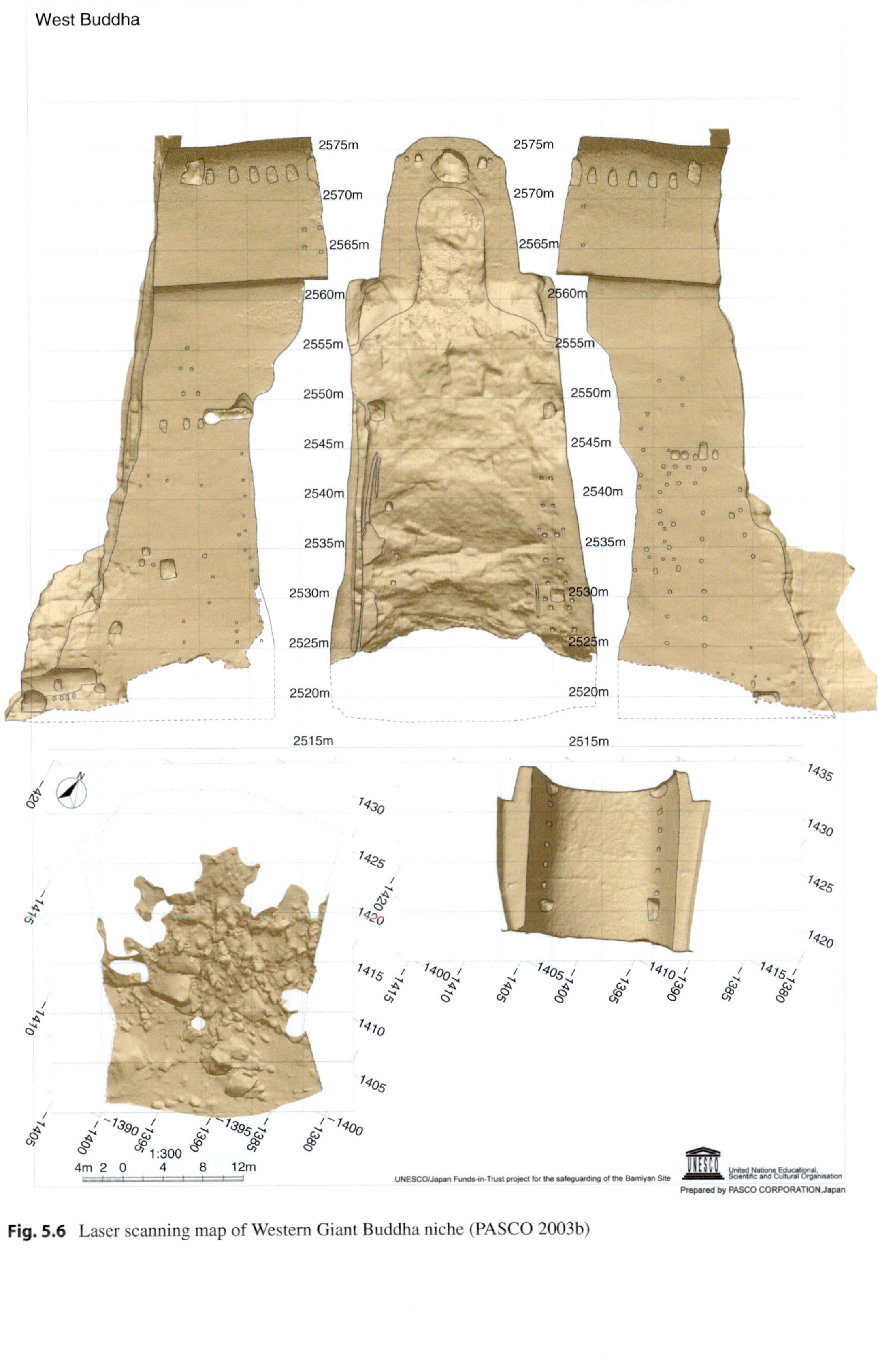

Fig. 5.6 Laser scanning map of Western Giant Buddha niche (PASCO 2003b)

1–3: Quaternary alluvial and fluvio-glacial accumulation		
4: Khwaja-Ghar Formation	10: Dokani Formation	16: Old volcanic rock
5: Ghulghola Formation	11: Cretaceaous Formations	17: Chaidan-granite
6: Buddha Formation	12: Permian Formations	18: Contact-metamorphic rock
7: Volcanics	13: Devonien Formation	
8: Qal'acah Formation	14: Recrystallized limestone (not dated)	
9: Zohak Formation	15: Schist, meta-schist	

middle of the cliff) with some pelites, sandstones, and volcanic material; the conglomerate is the predominant material in the cliff and exhibits a moderate cohesion (Fig. 5.9). The varying grain size distribution (from conglomerate to clay) clearly demonstrates a nonselective depositional environment, characterized by a high energy flood plain. Large discontinuities have been recognized in the conglomerate, with a tendency to enlarge in the upper part.

The siltstone exhibits an apparent moderate cohesion and two major pattern of discontinuities that are isolating small blocks, with volumes that generally are <1 m^3. The fine grain size distribution and the higher percentage of clay mineral and phyllosilicate in the siltstone with respect to conglomerate, demonstrate for siltstone a lower energy depositional environment (small lake or lagoon?), probably corresponding to dry periods.

The depositional environment and diagenetic conditions probably affected the present day characteristics of lithotypes, because the conglomerate exhibits a permanent cohesion whereas the siltstone

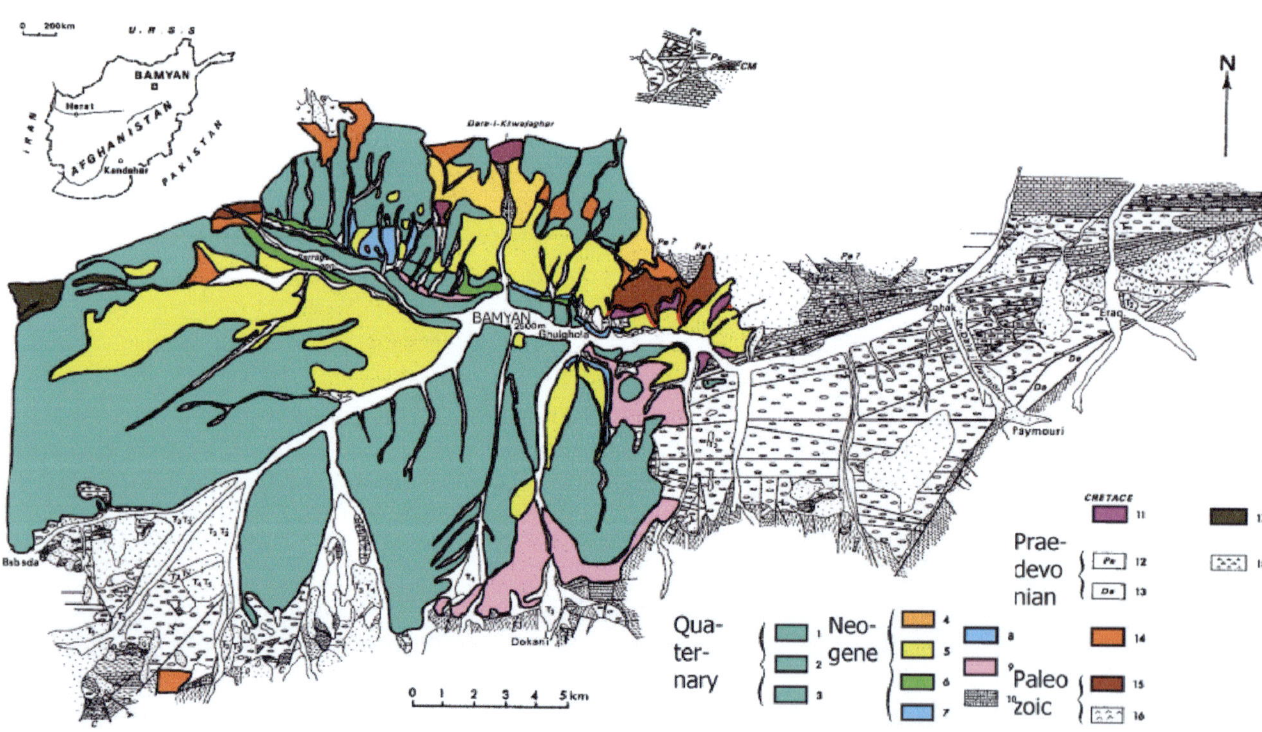

Fig. 5.7 (*Top*) Geological map of Bamiyan after Lang 1971, (Redrawn from Reineke (2006))

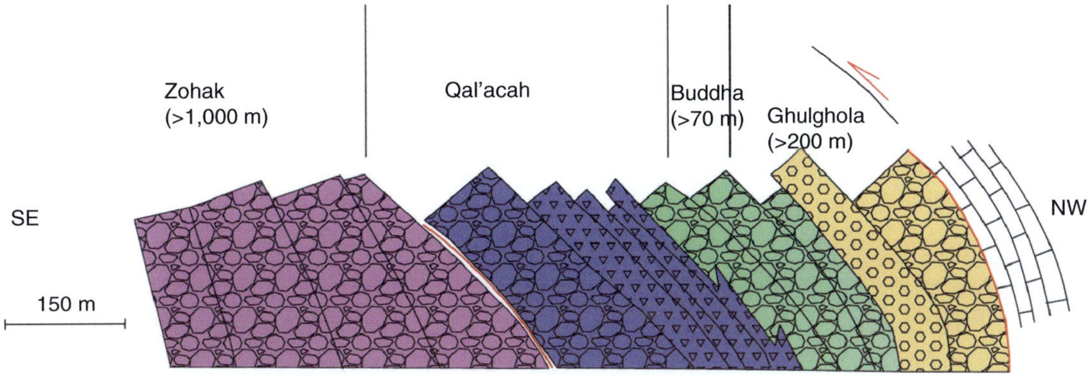

Fig. 5.8 (*Left*) Schematic SE-NW-Profile through neogenic layers after Krumsiek1980 and Urbat 2003 (From Reineke 2006, redrawn)

easily loses the apparent cohesion when saturated, as in Fig. 5.10. The mechanism for this behavior will be tentatively explained in the following chapters by means of x-ray diffractometry, microscopic thin section, and petro-geophysical analysis, scanning electronic microscope investigations, as well as the determination of physical and mechanical properties in soil and rock via mechanical laboratory tests.

Geomorphological Setting

Introduction

The reconstruction of the geomorphological activity in the Bamiyan Valley was developed by means of a detailed field survey integrated with a kinematic analysis performed along 18 sections of the cliff. These data were compared with a multitemporal sequence of ground-based pictures, covering about 100 years of observations (Delmonaco and Margottini 2007).

In general terms, the following active processes have been recognized in the area (Fig. 5.11):

- Water infiltration from the upper part of the cliff
- Gully erosion caused by rainfall and snowmelt in the upper part of the cliff
- Accumulation of debris materials at the toe
- Occurrence of mud flow, in the upper part, probably caused by sandstone saturation
- Toppling affecting some isolated blocks
- Sliding in the large portion of the slope, mainly when discontinuities from the top are reaching the lower highly jointed sandstone formation
- Progressive opening of discontinuities that mainly affects the external part of the cliff

The explosion of March 2001, apart from the collapse of the statues, produced a deterioration of the slope stability conditions, mainly in the shallower part of the niches (Fig. 5.12). In the Eastern Giant Buddha niche, besides the collapse of statue, three minor rock falls occurred in the top of the structure. In the meantime the blasting produced a degradation of the back side of the niche, especially in the right part where a stairs is located inside the cliff and the section between the stairs and the niche is quite thin (about 30–50 cm).

This part is presently the most critical for future stability. The left side, as a consequence of the protection offered by the existing buttress, is remarkably not exhibiting substantial displacement. A rock fall occurred only in the upper part and some instability phenomena are now evident.

In the Western Giant Buddha site, major effects were the collapse of the statue and the consequent instability of the back side of the niche. A small rock fall occurred on the top of the niche, in the left side. Probably the large thickness of the section between the stairs going up into the cliff and the niche (about 1 m) did not allow a large propagation of the effects

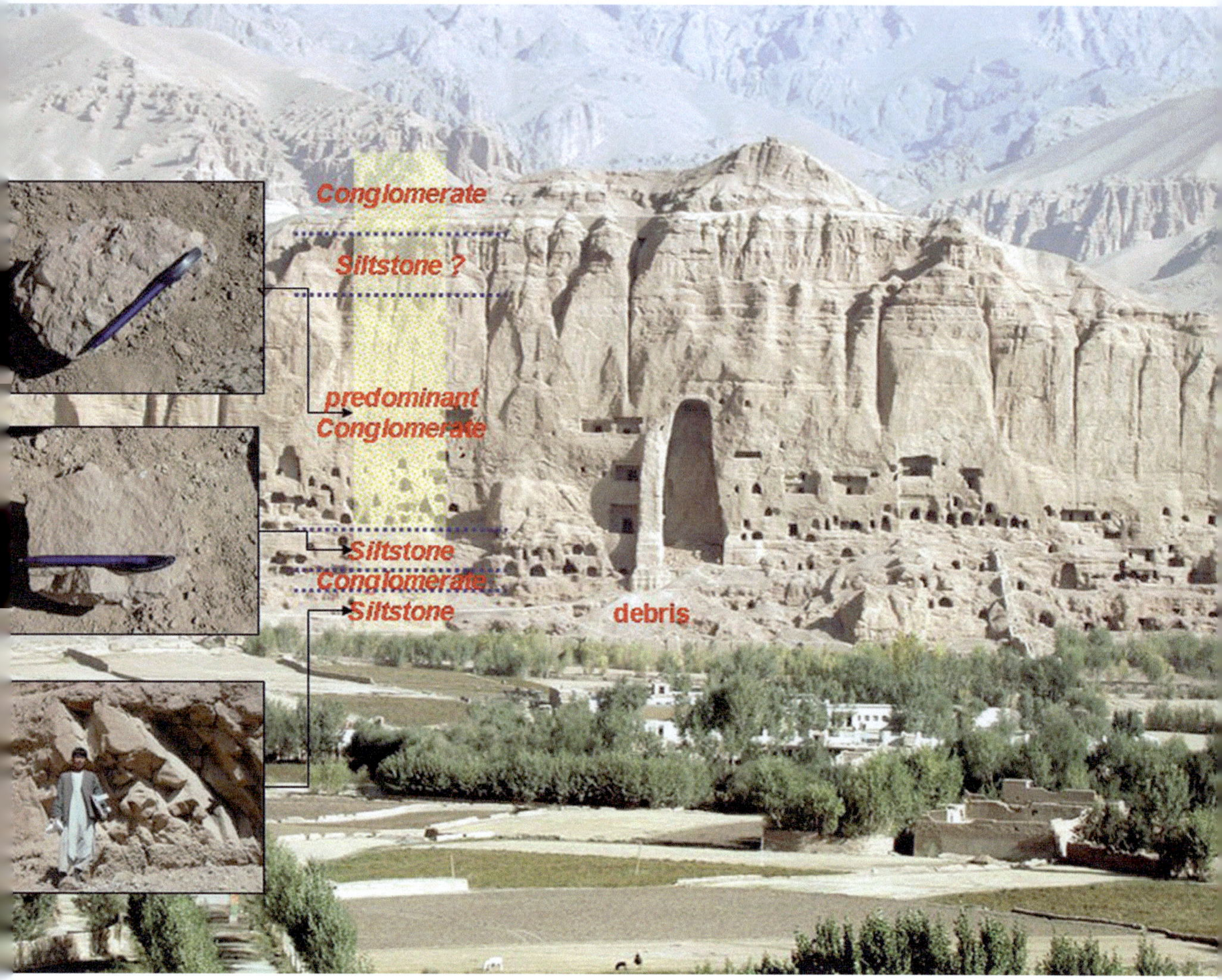

Fig. 5.9 Major features of the local geological series in the Bamiyan cliff

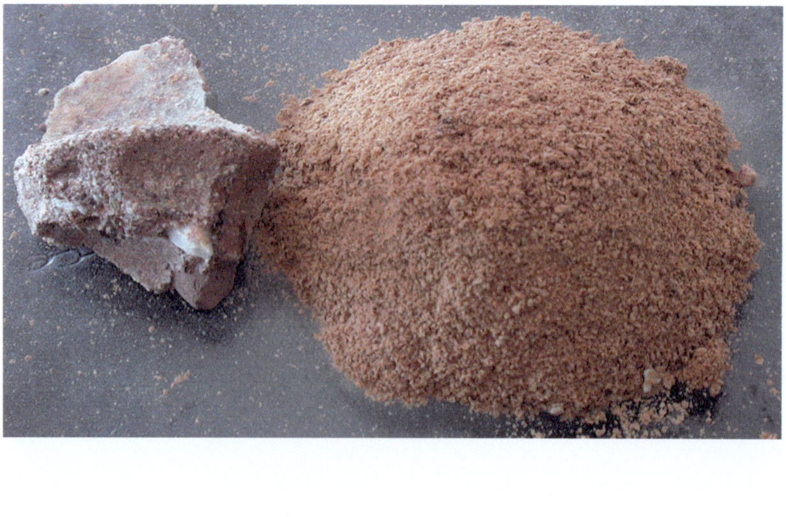

Fig. 5.10 Stability of conglomerate (*left*) and degradation of the siltstone (*right*) when weathered

of blasting and, consequently, severe damage to the structure.

The Geomorphological Survey

The Bamiyan Valley is located at 2,500 m elevation on the northern edge of the 600-km-long E–W valley along the Herat fault, one of the largest and most prominent faults in Afghanistan (Fig. 5.13). The fault represents a suture zone between the Farad and Tadjik blocks occurred during the Cretaceous periods.

From a geomorphological point of view, the Buddhas' cliff presents an average slope inclination around 80°; that is probably due, as previously mentioned, to human excavation of niches and caves for religious reasons. Such a vertical shape, as a matter of fact, is almost unique in the whole area of the region and cannot be explained as a result of geological and structural causes.

The river valley in Bamiyan is quite large because of the confluence of three different rivers, with a local elevation of about 2,540 m. The flood plain is mainly formed by fluvial sediments (alluvial and alluvial fans coming from N torrents) and by slope sediments (landslide deposits, slope deposits, eluvial and colluvial coverings). The valley formation is based on a variety of factors such as lithological characteristics, tectonic activity, paleoclimatic events, fluvial activity along the river valley, and slope evolution in the cliff.

The cliff presents a general East–West orientation, although it can be divided into two distinct sectors: the western side, where the Western Giant Buddha statue is located, shows a N65 E orientation with a length of approx. 820 m, whereas the eastern portion, where the Eastern Giant Buddha is placed, exhibits a N95 E orientation and a length of ca. 525 m. The two segments are separated by a large alluvial cone area, generated by two distinct torrents flowing into the Bamiyan river, still very active, that have diverted the river flow towards SSE. The change of orientation from E–W to NNE–SSW occurs in correspondence with the torrent located at E.

This configuration can be mainly caused by the Herat fault system and to local faults oriented NE–SW where major tributaries of the Bamiyan river are likely developed and evolved with typical straight lines. Because of this, tectonic evolution the valley profile may be likely asymmetric with left banks, where the cliff outcrops, presenting higher slope inclination with respect to the opposite slopes. At present there is no evidence from direct and indirect investigations to confirm this assumption, except visual observation of morphological indications.

Slope debris and clasts are elements mainly of various nature (i.e., siltstone, sandstone, granite), mostly well rounded and variable in size, immersed in a silty-sandy matrix.

Landslide deposits are widely present along the slope toe, generated by fall and toppling of large conglomerate and siltstone blocks that display a modest runout also evidenced by their typical sharp-edged shape. Some blocks, especially in the most active sectors of the cliff and in correspondence with active debris cone areas, have been translated to the river valley. Block volumes are quite variable from <1 m^3 to over 10 m^3. Planar sliding deposits are diffusely outcropping at the base of the cliff and somewhat immersed and/or partially covered by the debris.

The top of the cliff, as well as the outer walls, are largely affected by diffuse and intense erosion of conglomerate and siltstone, especially in the western side of the cliff. This produces gully erosion that is the typical landform that outcrops and is widely extended in the slope face and in the upper parts of the Buddhas' cliff. The concentration of gullies is very high in the very small basins located on top of the cliff area, especially along the steep slopes of those tributary creeks that form active debris cones when flowing into the Bamiyan Valley.

The easily erodible soils with a weak structure like those forming the Bamiyan cliff, the absence of vegetation, as well as climatic conditions of the area are prominent factors in accelerating this type of phenomenon in the catchments located on the back of the cliff, with a typical retrogressive activity.

As regards climate, according to a wide range of bibliographical sources on paleoclimatic analyses, in Central Asia and Hindu Kush-Himalayan

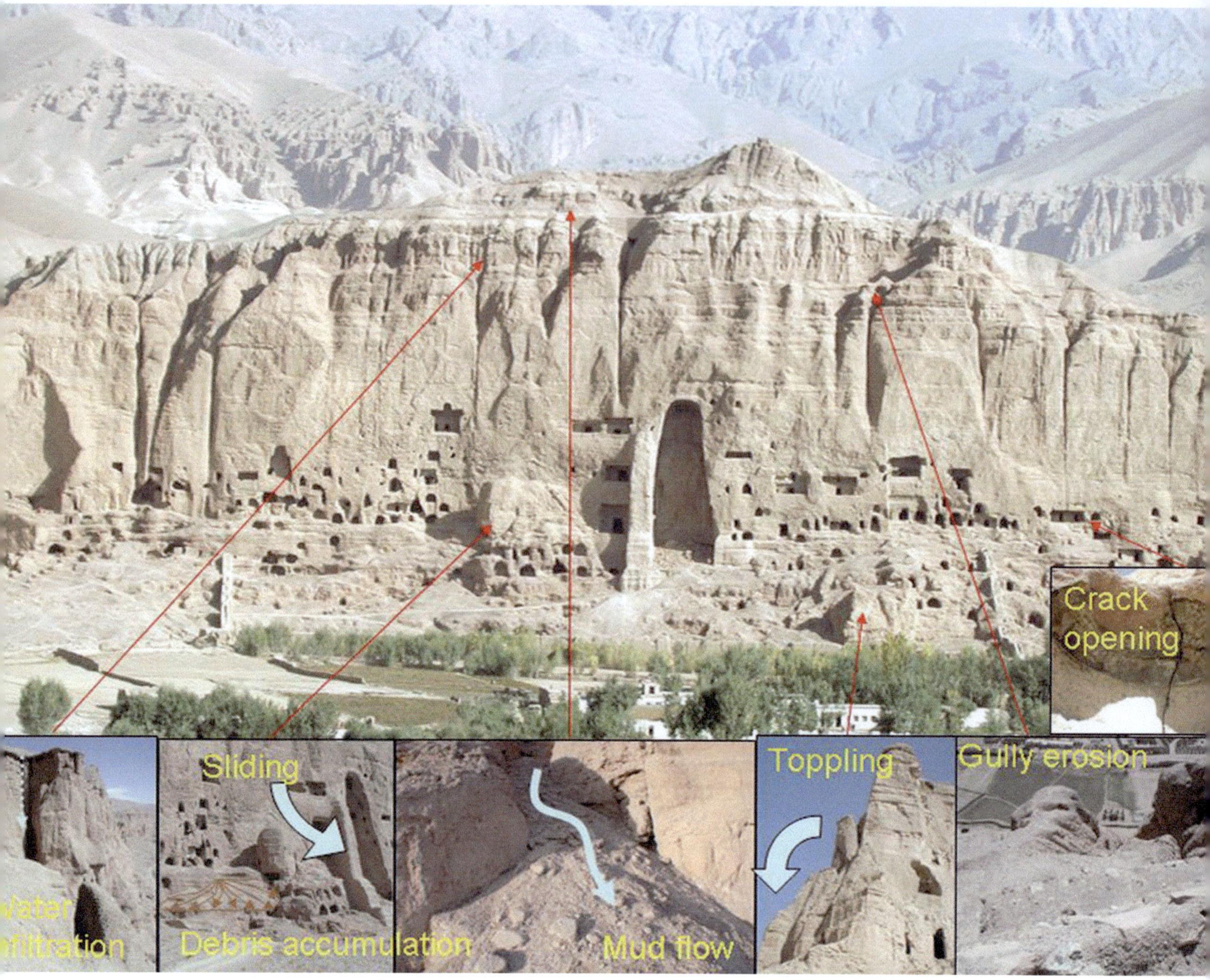

Fig. 5.11 Major geomorphological features occurring in the Bamiyan cliff

mountain chains, these areas have experienced the same climatic fluctuations as other areas of the Northern Hemisphere at least from the Late Pleistocene era up to the present time, as drawn from geological and environmental proxy data (see, e.g., Esper et al. 2002; Kamp et al. 2004; Bush 2005). It means that after the Wurmian glaciation (20,000–13,500 years BP) long-term and medium short-term cold and warm periods have occurred in the same way as in Europe or North America. As regards the latter, tree ring analysis and core sample analysis of soils and ice undertaken in Central Asia and Himalaya have recorded, in the last millennium, the Medieval Warm Epoch (eleventh to thirteenth century A.D.) and the so called Little Ice Age (1450–1830 A.D.).

Considering the main geomorphological features, briefly described above, and the long-term evolution of the cliff with respect to climate and tectonic activity, the potential evolution of the cliff may be the following (Fig. 5.14).

Stage 1

At the end of the last maximum glacial period, around 13.5 ky BP, the valley of Bamiyan was,

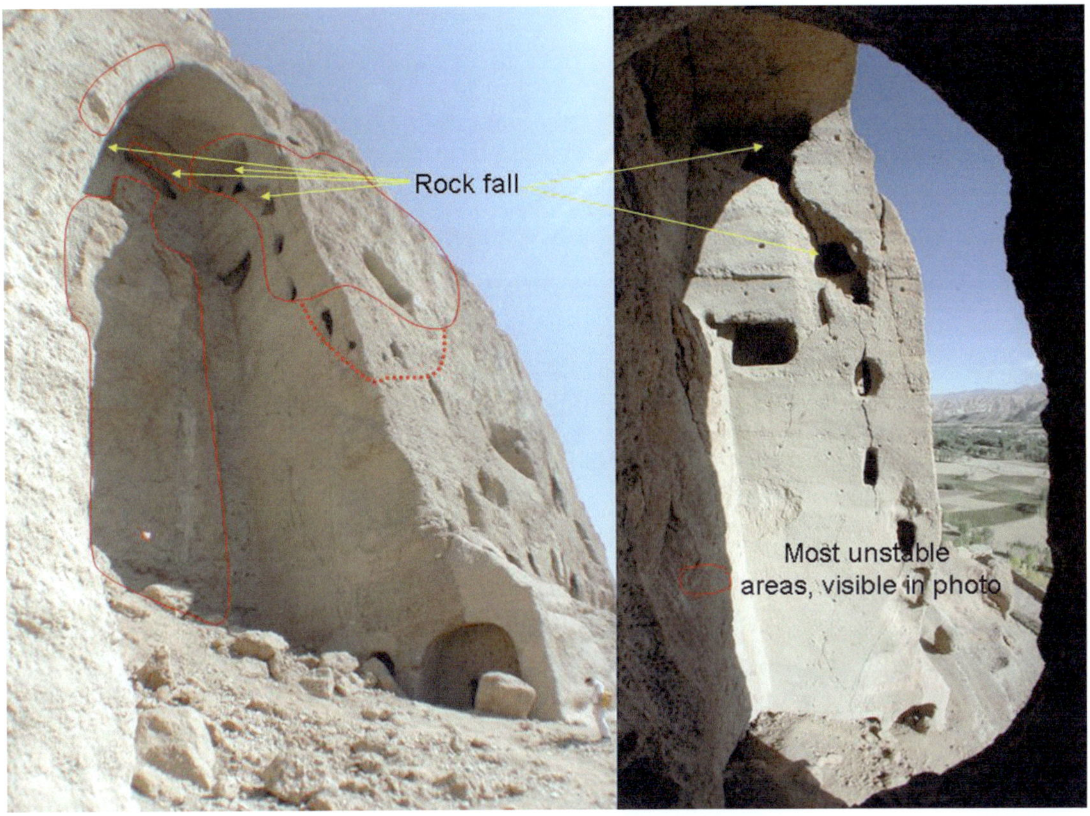

Fig. 5.12 Eastern Giant Buddha niche: effects of explosion and major unstable areas

presumably, deeper than at the present time. In the absence of any direct detailed geological information, it is not possible to quantitatively assess such a depth.

Siltstone and conglomerate rocks, because of the high energy of the relief and lateral unloading derived from river erosion in the flood plain, have suffered heavy tensional stress mainly concentrated at the toe of the cliff. This resulted in straining of rocks and development of a system of parallel cracks and joints, oriented roughly E–W, very deeply until the bedrock (Fig. 5.15).

This main system of cracks, developed in the upper part of the cliff, encountered another secondary system originated by tectonic stress and located mainly in the conglomerate and siltstone layers placed at the base of the cliff. This system is clearly visible in the caves located in the farthest eastern side of the cliff, at lowest elevation, where this part of the Buddha formation is visible. This system has a general S-quadrant dipping orientation with slope angles between 30 and 50° (Fig. 5.16).

This situation has caused, especially in the past, rock sliding phenomena at the base of the cliff, still clearly visible along all the border of the slope, mostly in an inactive or quiescent state of activity (Fig. 5.17). The presence of quite stable caves with horizontal floor surface (Fig. 5.17, top) in the majority of accumulation material, demonstrates that this type of mass movement is ancient and occurred, mainly, prior to the anthropisation of the cliff. Nevertheless there are some caves, 2 of 13, located in the two most evident rock slides west of the Eastern Giant Buddha and east of the Western Giant one (see geomorphological map of Fig. 5.23), which clearly exhibit cave floors tilted or displaced by the cliff instability (Fig. 5.18). They demonstrate that some rock slides occurred also after the fifth to sixth century A.D., when the anthropisation of the valley begun.

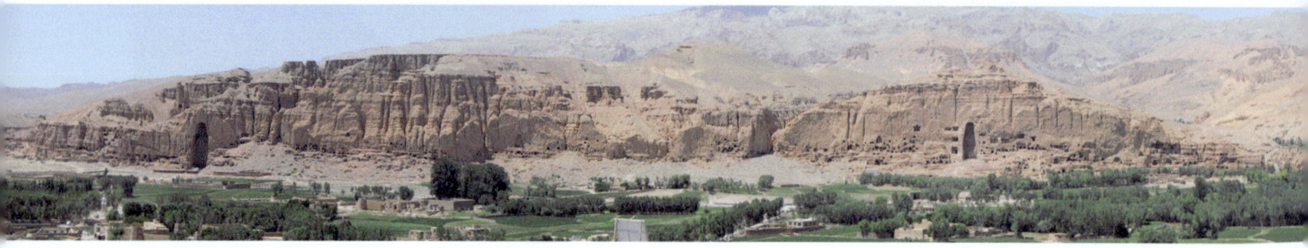

Fig. 5.13 Front view of the Bamiyan valley

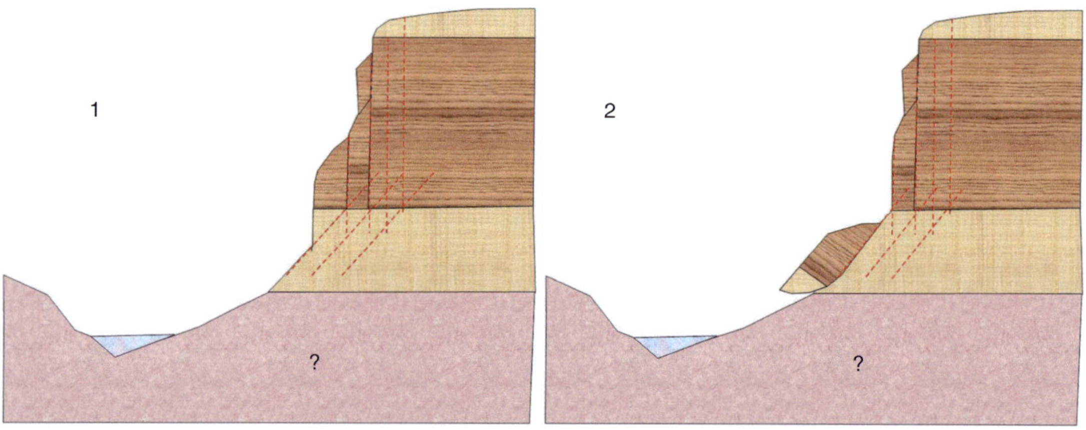

Fig. 5.14 Evolution of the Bamiyan valley, Stage 1, with development of vertical cracks (sketch 1) and rock sliding phenomena (sketch 2)

Stage 2

After the cold peak occurred in the Late Pleistocene–Early Holocene period, warmer and more humid climatic conditions as well as a rising of the sea level promoted a phase of large deposition of debris and alluvial soils in the valley until the present state (Fig. 5.19). The reduction of the potential energy in the slope and a consequent decrease of stress conditions that have caused large and diffuse rock sliding events changed the general trend of kinematic phenomena in the Bamiyan cliff. Now the main landslide process acting in the slope is toppling and rock fall that affect the medium-high sectors of the cliff, also involving the cultural heritage developed since the Buddhist period in Bamiyan (Fig. 5.19).

Stage 3

The so called Little Ice Age (fifteenth to nineteenth centuries A.D.) with cooler and more humid conditions than the present time may have promoted an increase of erosion and debris production from the upper catchments (Fig. 5.20) especially in the western sector of the cliff. In the middle of the slope, where the two segments of the cliff with different orientation converge, is the most active area of debris production, evidenced by the coalescent debris cones that have diverted the flowing of the Bamiyan river through SSE.

Nevertheless, slope debris widely outcrops at the toe of the cliff (Fig. 5.21) with a tendency toward erosion, caused by present climatic conditions characterized by aridity, low yearly total precipitation, and concentrated rainfall events.

Erosion also affects the slope face of the cliff and the upper part of the slopes, where a network of gullies are well developed, mostly concentrated in small catchments located in the vicinity of the cliff border. The deepening

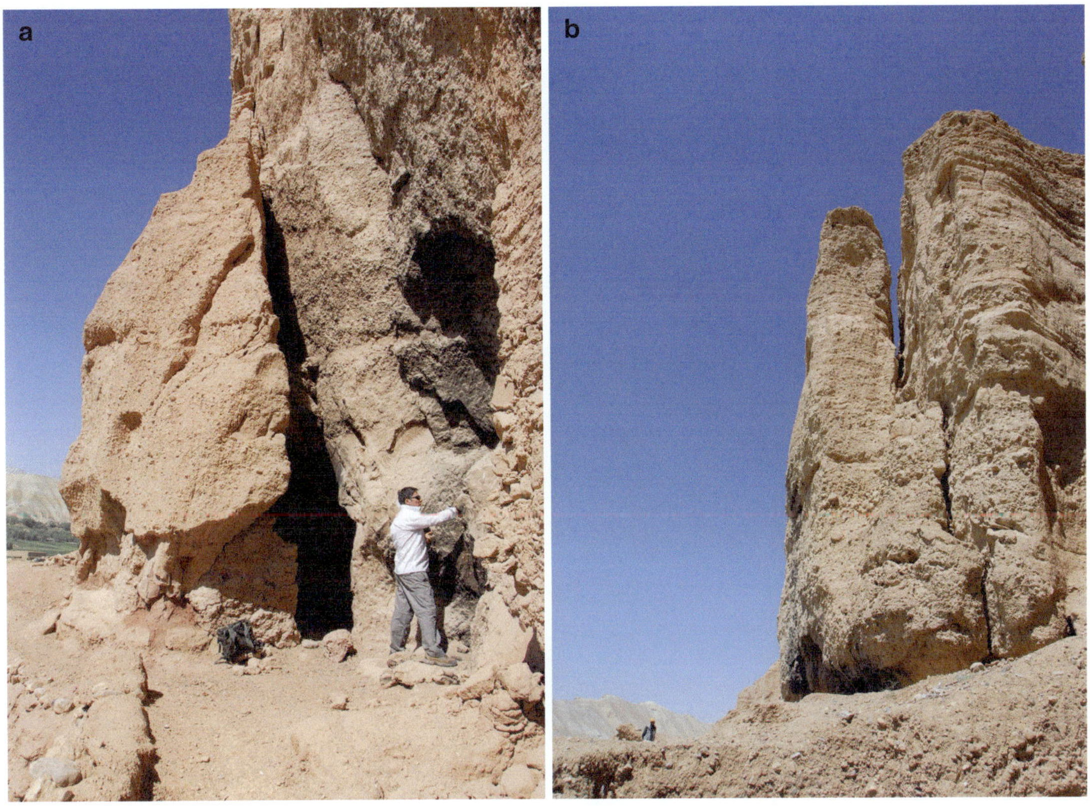

Fig. 5.15 Example of joints parallel to the cliff main orientation

of the gully system is promoted by the weakly cemented materials, especially siltstone, as well as by the development of straining in those areas (Fig. 5.22).

The water infiltration inside the rock-forming slope now represents one of the most severe problems for the general stability of the Buddhas' cliff; as a matter of fact, the circulation of superficial water along gullies placed at the correspondence of a vertical joint (see the photo at left in Fig. 5.22) and its infiltration inside the rock mass can contribute to a decrease in cohesion along the joint, isolate a block, and accelerate the process of toppling/falling.

All the geomorphological features of Bamiyan Valley are reported in the following map (Fig. 5.23). From this it is possible to highlight the major past rock slides and the present day erosional phenomena.

Kinematic Analysis

Introduction

This section reports the studies carried out to investigate the Bamiyan cliff stability conditions. As already reported, the Bamiyan Valley represents a quite unique condition in the whole Afghan territory in terms of slope geomorphological setting. This may lead to an assumption that the present slope conditions (extremely steep slope angle) could be caused by human intervention (slope cutting) in order to excavate the giant Buddha statues and niches, also promoted by the presence of soft and jointed rocks. The field mission undertaken in June 2007 has the main scope to analyze the geomorphological evidence, especially with respect to the structural setting and instability failure modes of the slope-forming

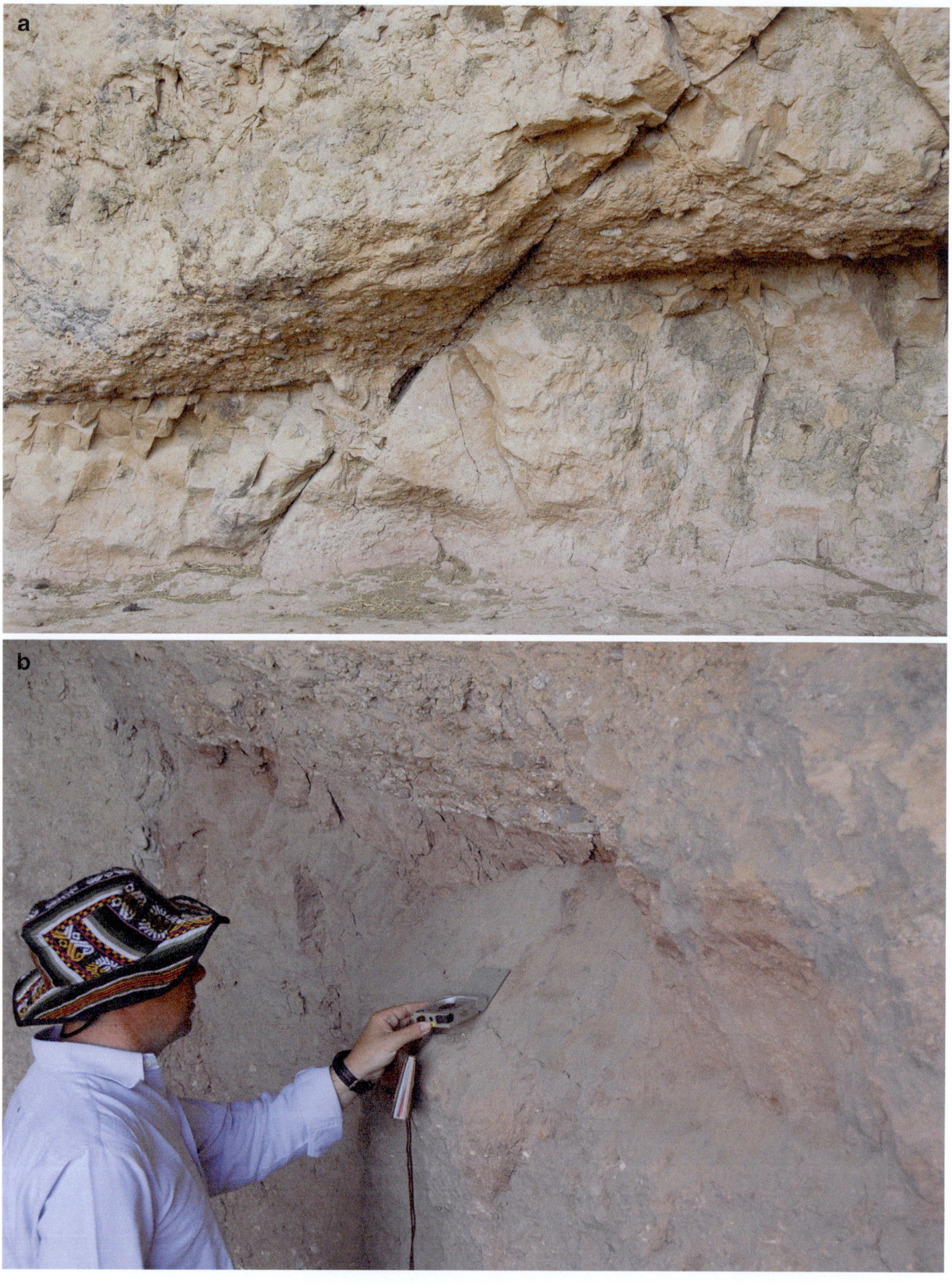

Fig. 5.16 View of medium-low angle faults affecting the outcropping base rocks in the eastern side of the cliff

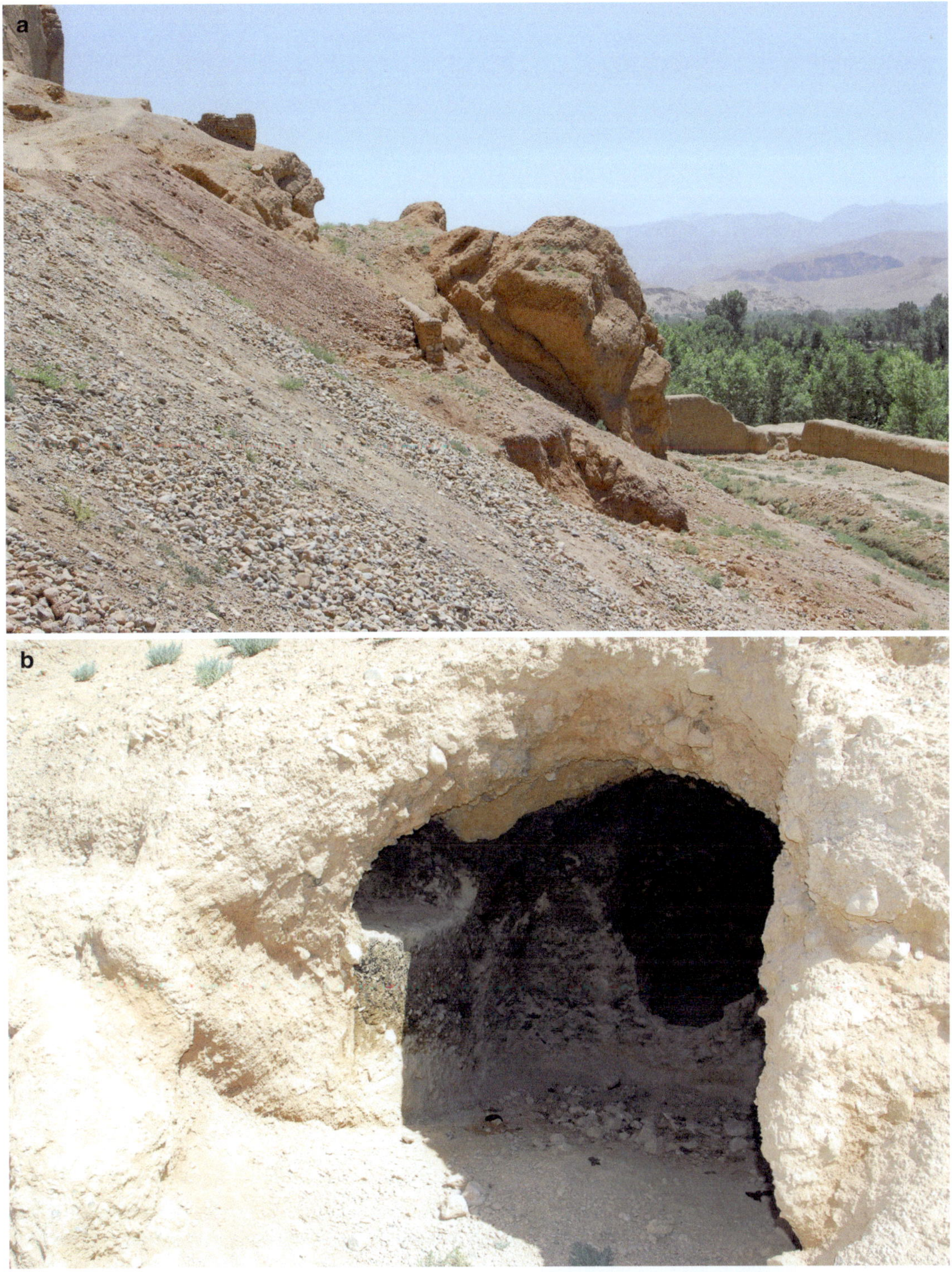

Fig. 5.17 Remains of a block involved by sliding (*top*) and a cave built inside a slid block (*bottom*)

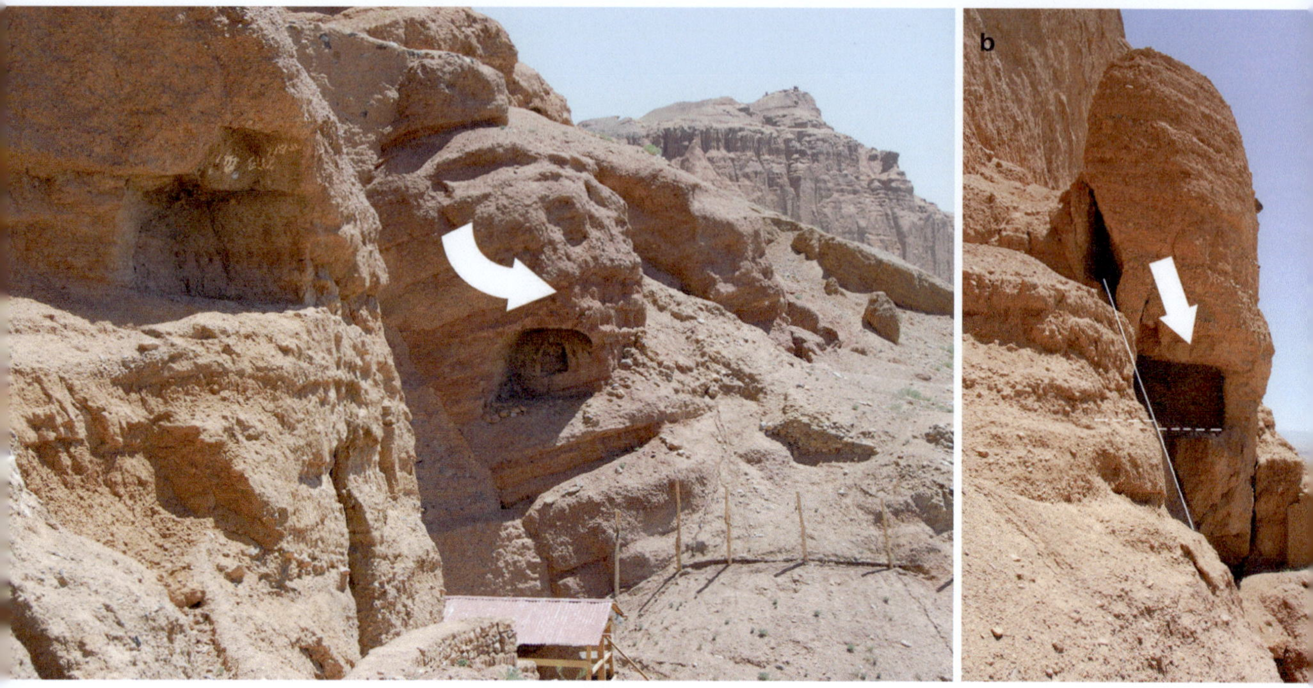

Fig. 5.18 Cave's ground floor tilted after a rock slide east of Western Giant Buddha (*left*) and a cave displaced by a rock slide's plane west of Eastern Giant Buddha (*right*)

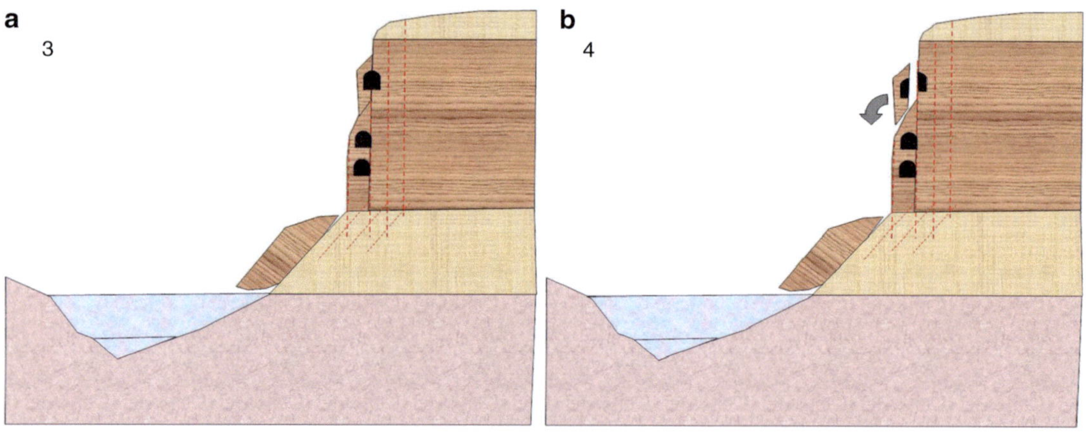

Fig. 5.19 Structural slope conditions at the beginning of the anthropisation of the cliff, with toppling/rock fall landslide types as major phenomena

rocks, providing some ideas on potential morphological evolution of the Bamiyan cliff.

The angle for most of the rock face is approximately 80–88°. The outcropping soft rocks present prominent discontinuity sets whose origin, especially the joint system parallel to the slope face, can be associated with the geomorphological evolution of the valley as well as to tectonic setting (Ambrasey and Bilham 2003). This situation has caused apparent slope instability phenomena, somewhat aggravated by the explosions during the destruction of the Buddha statues in 2001 around the niches areas. To analyze the assessment of the geomorphological conditions of the cliff, a total of 17 structural stations were selected by visual inspections in

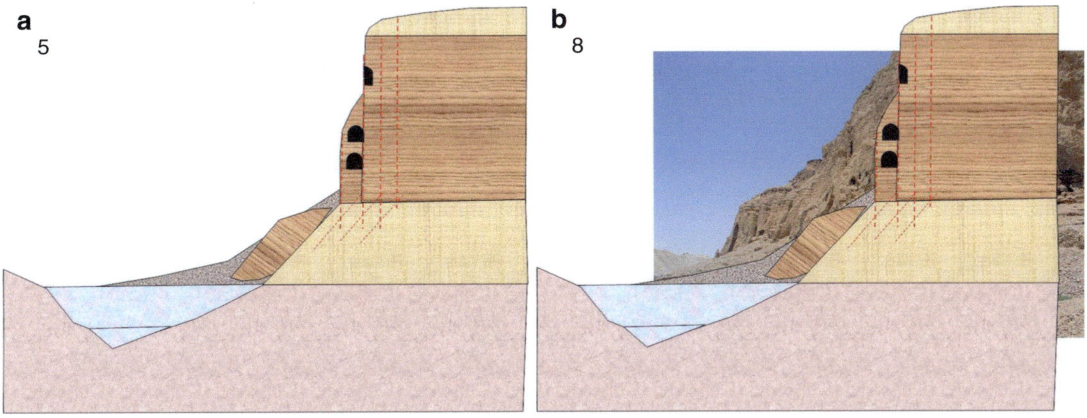

Fig. 5.20 Present conditions of the Bamiyan cliff

the areas of the cliff where historical structures (e.g., Buddha niches, external and underground caves) present prominent or potential instability conditions (Table 5.1).

The joint mapping has been performed reporting the collection of joint geometry data integrated with in situ mechanical tests such as Schmidt hammer rebound values and joint roughness analysis in the selected test sites.

Structural Analysis

The main joint orientation data in each observation point were represented with the Schmidt equal angle stereonet (Fig. 5.24, stations 1–17), whereas total observations have been represented also as rose diagrams (Fig. 5.24, a and b diagrams). To find the dense points and the resulting joint sets, the Terzaghi cluster algorithm has been applied. The majority of stations exhibit a high variability of joints orientation that is partly reflected by the low amount of data available for orientation analysis. Nevertheless, the rose diagram for totals shows that most of the surveyed discontinuities are E–W oriented, whereas a secondary system has a NW–SE orientation.

Kinematic analysis is fundamental to detect and investigate possible failure modes of rock masses affected by joints, as the Bamiyan cliff.

Failure involving movement of rock blocks on joints combine one or more of the three basic modes: plane sliding, wedge sliding, and toppling. For the selected stations, kinematic analyses have been implemented to estimate the potential aforementioned failure modes that may develop along the slope.

The slope has been divided into two main sections: respectively at W, where the statue of the Big Buddha is located (stations 1–10), and the eastern side that includes the area of the Eastern Giant Buddha (stations 11–17). This diversification of the analysis has been decided considering that the two sides show a different general orientation of the slope (dip direction 155° for the west side; dip direction 185° for the eastern part), probably because of tectonic effects. The western side is larger than the eastern one, showing as well a general higher elevation. The eastern side has a lower base level with respect to the western one so that some layers of the Buddha geological formation, prevalently siltstone and subordinately conglomerate, are clearly visible. This part of the formation has been more intensively excavated during the construction of the cave system; in addition, the upper slope in this side of the cliff is less developed and presents a smaller catchment area than the west side. This has caused lesser production of debris that, conversely, is more abundant in the western sector of the cliff.

For these reasons, it has been decided to operate with such differentiation of the cliff analysis because the aforementioned factors may have promoted some difference in jointing and, as a consequence, in potential failure modes.

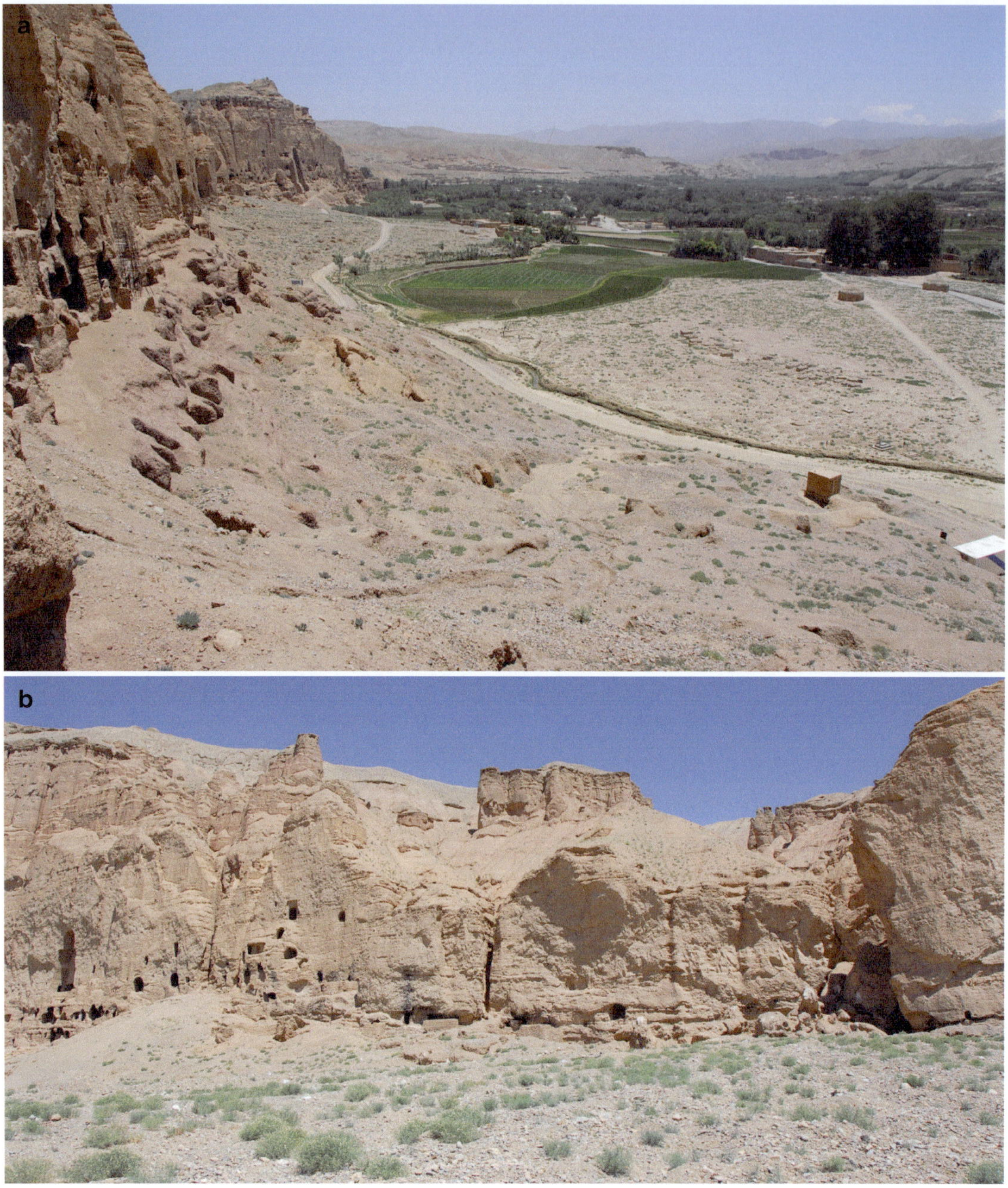

Fig. 5.21 Debris outcropping along the slope toe (*top*) and view of coalescent debris cones located in the middle of the cliff (*bottom*)

Figure 5.25 reports the analysis of 38 main joints orientation in the western sector, represented through an equal angle stereonet and a rose diagram. The analysis of diagrams shows three main orientations for joints: one set is almost parallel to the slope and two other sets are mutually orthogonal and respectively oriented NW–SE and ENE–WSW. Many other minor sets of joints are present and oriented in a scattered way.

The eastern sector joints orientation is represented in Fig. 5.26. The 34 main joints surveyed

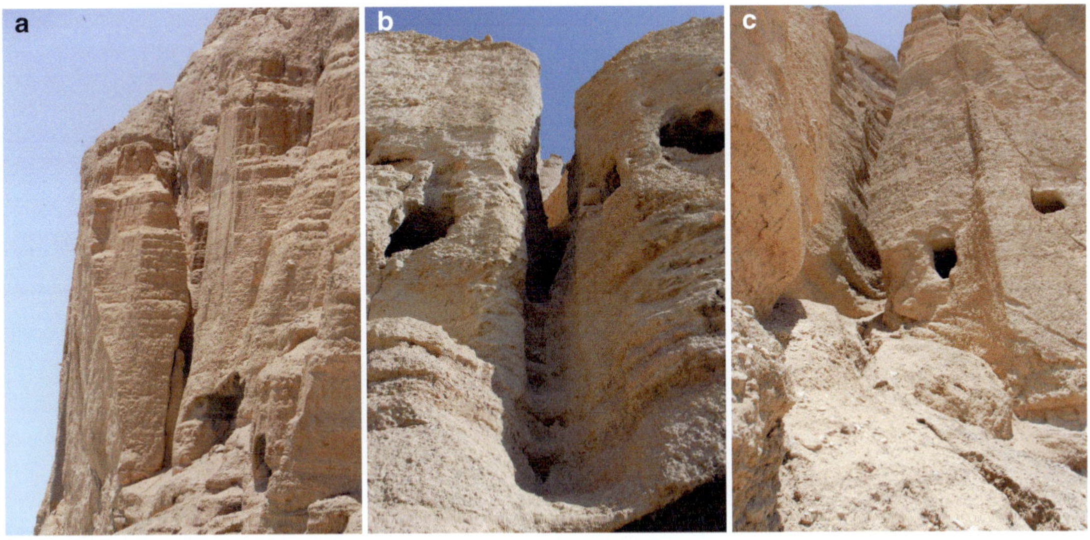

Fig. 5.22 Examples of gullies developed along the face of the cliff

display, also visually, a different general orientation trend with respect to the west side. The main joints families are almost parallel to the slope (E–W orientation) and there is a more prominent scattering of discontinuities in this part of the cliff. The way in which this may contribute to a distinctive difference between the two sectors will be thoroughly described in the following chapters.

It is interesting to notice also that the two Buddha niches reflect the same pattern of the cliff, with some disturbance in the Eastern Giant Buddha niche (Fig. 5.27).

Toppling

Toppling failures most commonly occur in rock masses that are subdivided into a series of slabs or columns formed by a set of fractures that strike approximately parallel to the slope face. In a toppling failure the rock column or slab rotates about a fixed point (pivot) at or near the base of the slope at the same time that slippage occurs between the layers. In order for toppling to occur, the center of gravity of the column or slab must fall outside the dimension of its base.

The necessary conditions for toppling failure are the following:

- The strike of the layers must be approximately parallel to the slope face. Differences in these orientations of 15–30° have been considered in literature. In our analysis, a value of 30° has been considered.
- The dip of the layers must be into the slope face.
- As suggested by Goodman (1980), in order for interlayer slip to occur, the normal toppling plane must have a plunge less than the inclination of the slope face, less the friction angle of the surface. This condition can be formulated as follows:

$$(90^\circ - \Psi_p) \leq (\Psi_f - \Phi_p)$$

where

- Ψ_p = dip of geologic layers (planes)
- Ψ_f = dip of slope face
- Φ_p = friction angle along planes

According to the above statements, the toppling analysis has provided the following results (Fig. 5.28), considering a friction angle of 30°, to be on the conservative side, calculated for siltstone materials that are the weaker lithotypes where higher stress condition can develop.

For both areas rock topple (or rock fall) is probable. The west side of the cliff exhibits a wider potential area for toppling phenomena as also surveyed in the field during the mission.

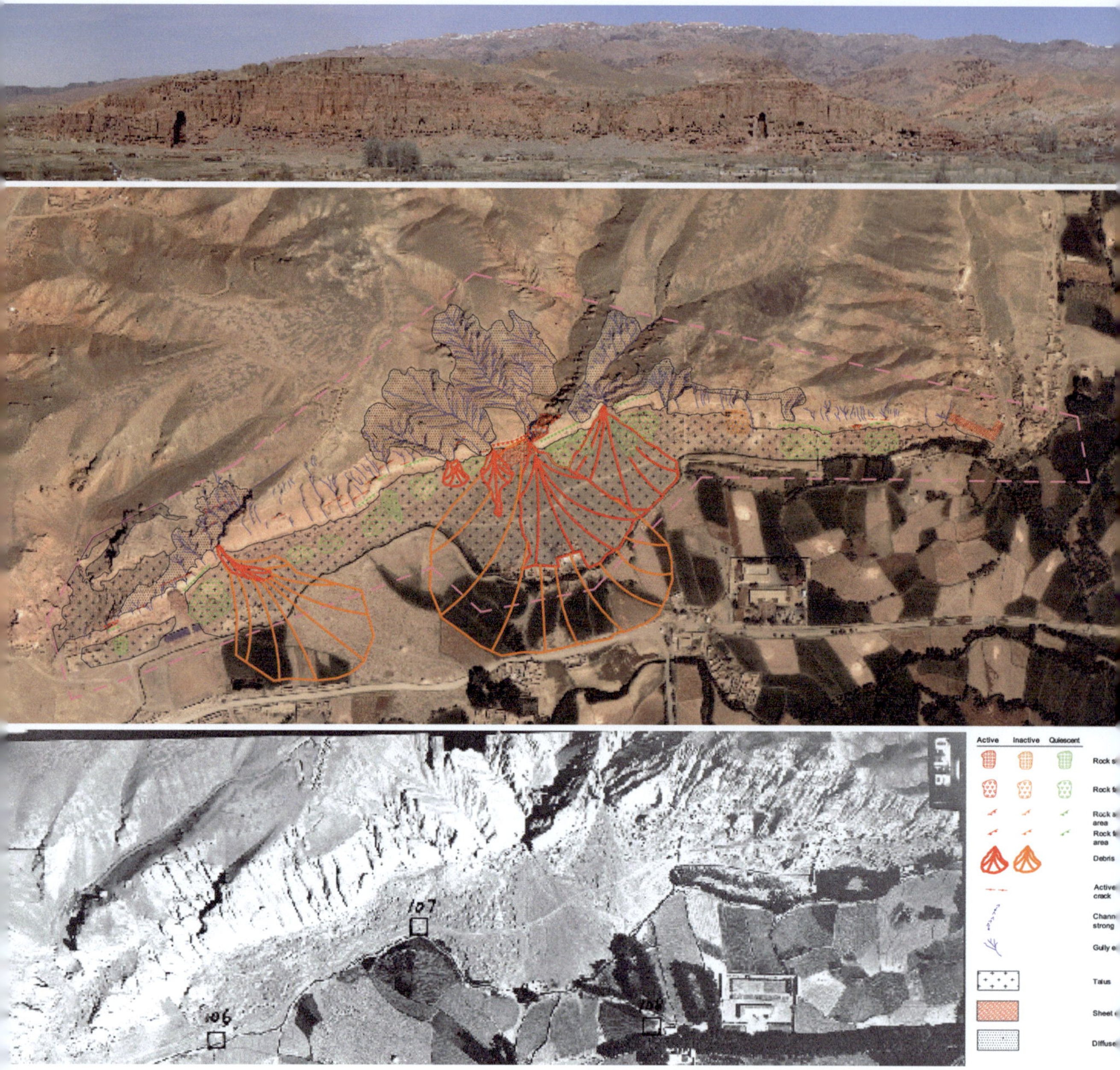

Fig. 5.23 Geomorphological map of Bamiyan cliff area. At the *top* is the present day frontal view of the cliff; at the *bottom*, the only available aerial photograph from late '60s is reported

Figure 5.29 shows some example of toppling failure mode surveyed in the cliff of Bamiyan. The photo on the left represent a potential block sliding detected in the western side of the cliff, developed along a deep discontinuity parallel to the slope, previously affected by gully erosion that promoted a deep infiltration of waters in the rock mass. The photo on the right depicts an old fallen block caused by rock fall/toppling in the vicinity of the Eastern Giant Buddha.

In general, the toppling potential seems to be higher in the western part of the cliff, especially

Table 5.1 List of stations with orientation and global location

Site ID	Dip direction°	Dip°	Location
Station 1	141	86	N34°49′44,1″ E067°49′02.2″
Station 2	141	86	N34°49′47,3″ E067°49′07.6″
Station 3	158	82	N34°49′47,4″ E067°49′09.1″
Station 4	155	82	N34°49′47,5″ E067°49′10.7″
Station 5	164	81	N34°49′47,7″ E067°49′11.0″
Station 6	162	80	N34°49′49,1″ E067°49′14.3″
Station 7	162	85	N34°49′51,0″ E067°49′17.6″
Station 8	153	88	N34°49′51,4″ E067°49′21.7″
Station 9	153	85	N34°49′51,8″ E067°49′23.5″
Station 10	157	84	N34°49′54,8″ E067°49′29.8″
Station 11	189	88	N34°49′54,5″ E067°49′31.9″
Station 12	178	85	N34°49′54,1″ E067°49′33.3″
Station 13	185	83	N34°49′53,3″ E067°49′38.5″
Station 14	167	86	N34°49′52,9″ E067°49′40.3″
Station 15	164	85	N34°49′53,3″ E067°49′42.2″
Station 16	192	82	N34°49′53,2″ E067°49′46.2″
Station 17	198	88	N34°49′52,9″ E067°49′47.9″

in terms of rock volumes potentially involved. In the eastern side, a potential toppling failure mode has been detected but with minor volumes involved because of a higher density of fractures in the jointed mass that presents, as well, a higher number of orientation sets.

Planar Failure

Planar failures are those in which movement occurs by sliding on a single discrete surface that approximates a plane.

The four necessary structural conditions for planar failures can be summarized as follows:

- The dip direction of the planar discontinuity must be within 20° of the dip direction of the slope face.
- The dip of the planar discontinuity must be less than the dip of the slope face and thereby must "daylight" in the slope face.
- The dip of the planar discontinuity must be greater than the angle of friction of the surface.
- The lateral extent of the potential failure mass must be defined either by lateral release surfaces that do not contribute to the stability of the mass or by the presence of a convex slope shape that is intersected by the planar discontinuity.

The stereographic analysis for planar failure has provided the following results (Fig. 5.30).

The analysis suggests that this kind of failure mode is highly potential in both sides of the cliff, also as a secondary movement connected with toppling failure, that, factually, determines the sliding of vertical blocks previously deformed following a typical toppling evolution. This occurs especially when the "pivot" of the block is located inside a siltstone layer where the major stress is concentrated. In that case the evolution of failure is that typical of sliding, sometimes with the development of circular-shaped rupture surface in cohesive materials (Mohr-Coulomb behavior of weak siltstone).

Along the slope many rock slid blocks affecting the base of the cliff are visible. Most of these events have occurred in the past (see section "The geomorphological survey") although some limited reactivation of blocks is possible because of deformation of the bedrock, especially in areas where blocks are lying over debris. In that case, erosion of debris may cause some small readjustment of the displaced block as attested to by Fig. 5.31, where the cave has been clearly displaced by further movement along the old failure surface.

Presently, especially in the eastern side, some pervasive and well developed joints with medium-low angle (30–40°) have been detected, especially inside the caves, that are particularly dangerous for the development of large rock slides that are affecting many parts of the cultural heritage in that part of the slope.

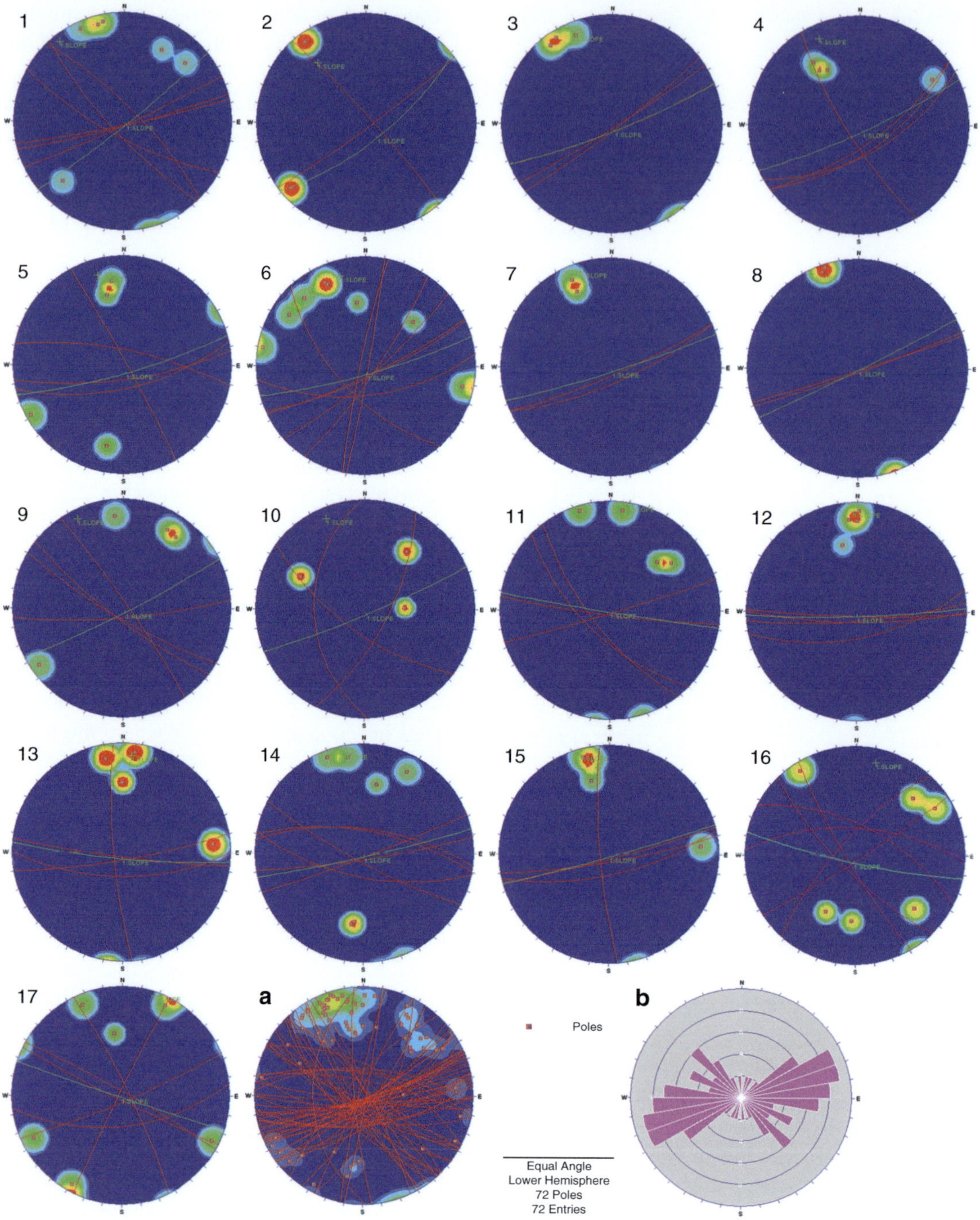

Fig. 5.24 Joints orientation sets of stations *1–17*. Stereonet (*a*) shows all joints surveyed, represented in (*b*) as rose diagram

This kind of joint can be correlated also to the old tectonic activity that has acted on this area, evidenced also by shear stress (slickenside striae) surveyed along the surface of exposed joints (Fig. 5.32).

Wedge Failure

Wedge failures result when rock masses slide along two intersecting discontinuities, both of which dip out of the cut slope at an oblique angle to the cut face, forming a wedge-shaped block.

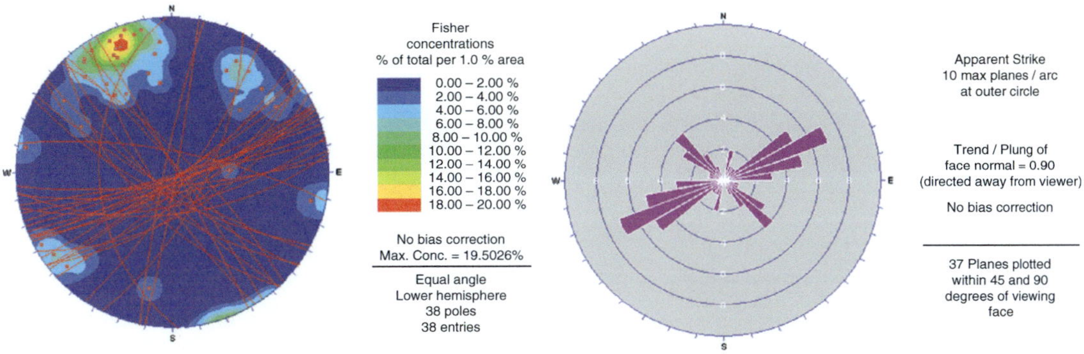

Fig. 5.25 Joints orientation sets of stations *1–10* represented through stereonet (*left*) and rose diagram (*right*)

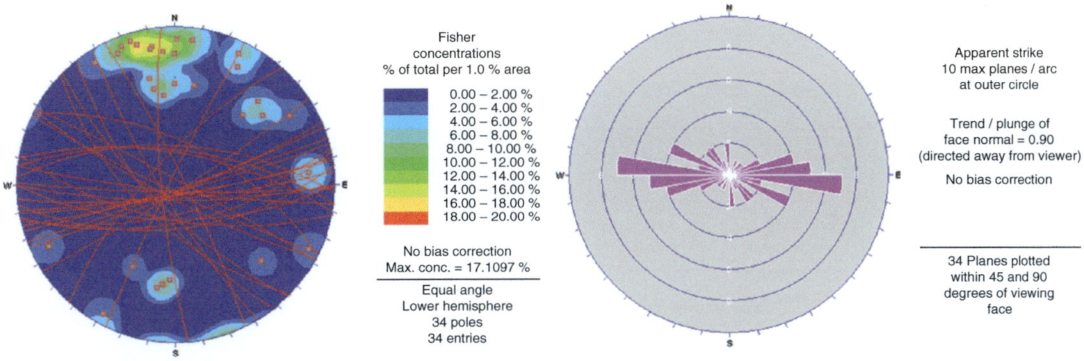

Fig. 5.26 Joints orientation sets of stations *11–17* represented through stereonet (*left*) and rose diagram (*right*)

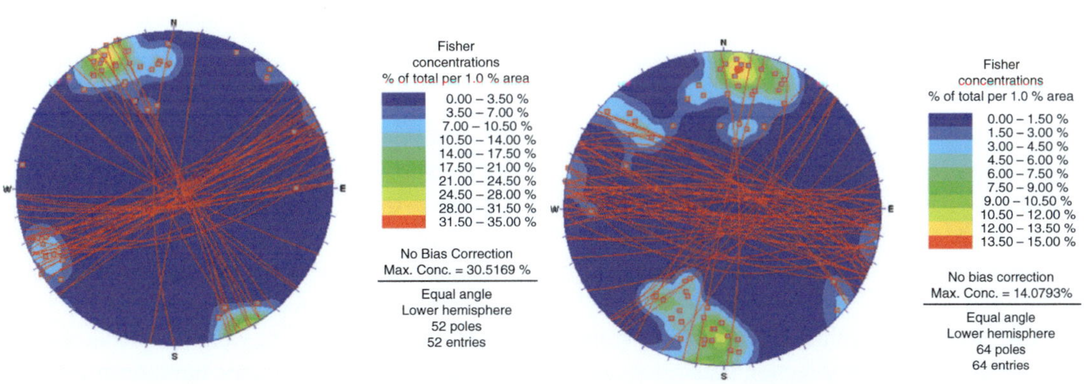

Fig. 5.27 Western and Eastern Giant Buddha niches with discontinuities exhibiting the same pattern of the related cliff (respectively, *left* and *right*). Only in the Eastern niche some disturbance caused by the explosion can be noticed

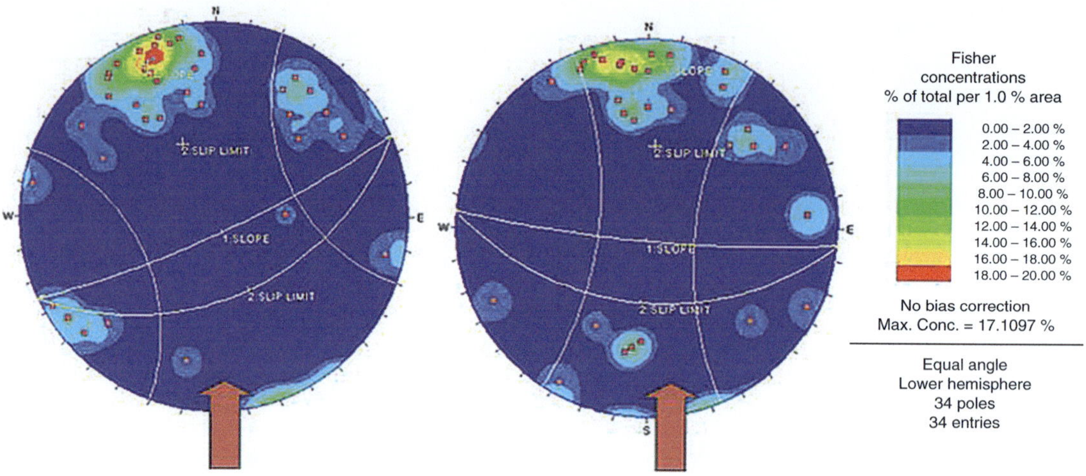

Fig. 5.28 Results of kinematic analysis for toppling for the western side (*left*) and eastern side of the cliff (*right*); the arrows show the region where toppling is possible (between slip limit and lower stereonet border)

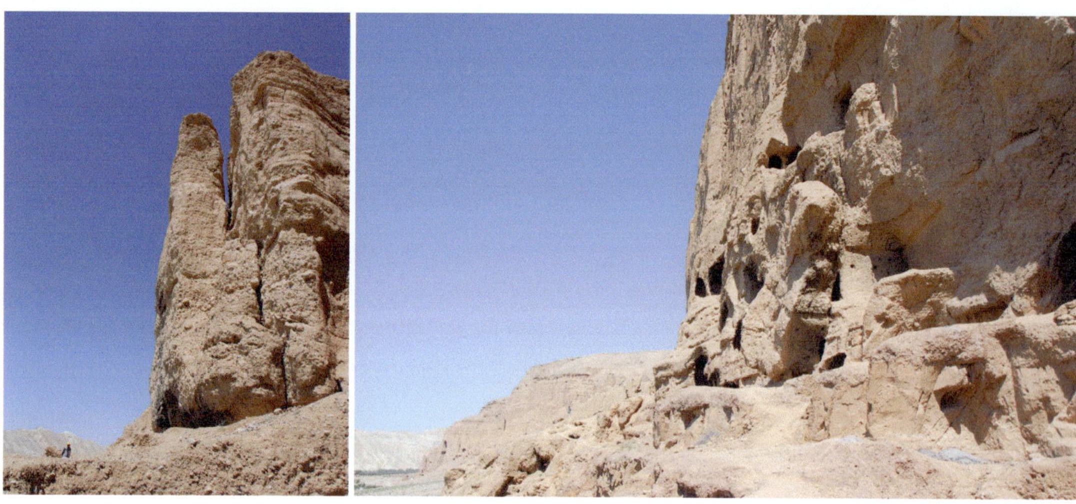

Fig. 5.29 Potential rock toppling in the western side (*left*) and fallen block caused by toppling in the eastern side (*right*)

The movement can occur either along both planes simultaneously or along the steeper of the two planes in the direction of maximum dip.

The necessary structural conditions for wedge failure are:

- The trend of the line of intersection must approximate the dip direction of the slope face.
- The plunge of the line of intersection must be less than the dip of the slope face. Under this condition, the line of intersection is said to daylight on the slope.
- The plunge of the line of intersection must be greater than the angle of friction of the surface.

Kinematic analysis for wedge failure has provided the following results (Fig. 5.33).

Through the Terzaghi weighted mean statistical technique, major planes have been selected for kinematic analysis.

In the western side, wedge failure is possible in rock blocks delimitated by joints 1–2 (oriented respectively 158°/76° and 220°/72°) and 1–3

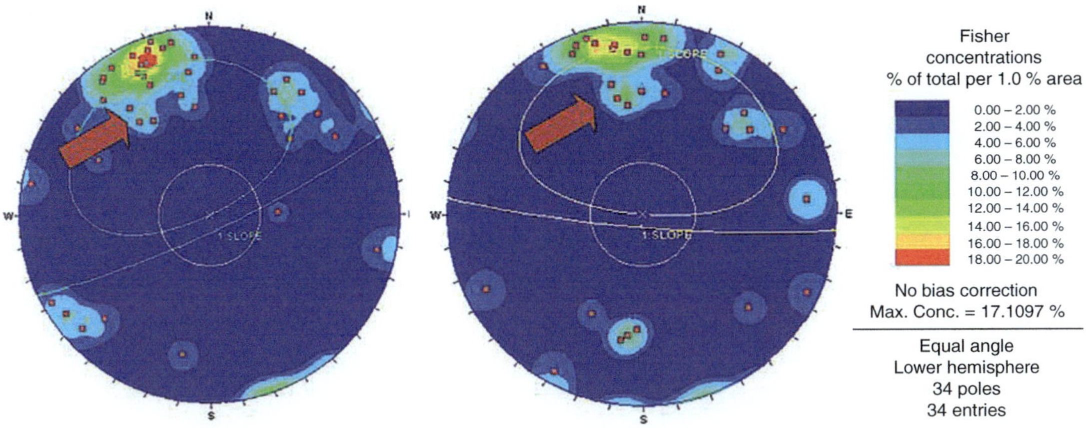

Fig. 5.30 Results of kinematic analysis for planar sliding for the western side (*left*) and eastern side of the cliff; the arrows show the region where toppling is possible (inside the slip limit area and the upper friction cone line)

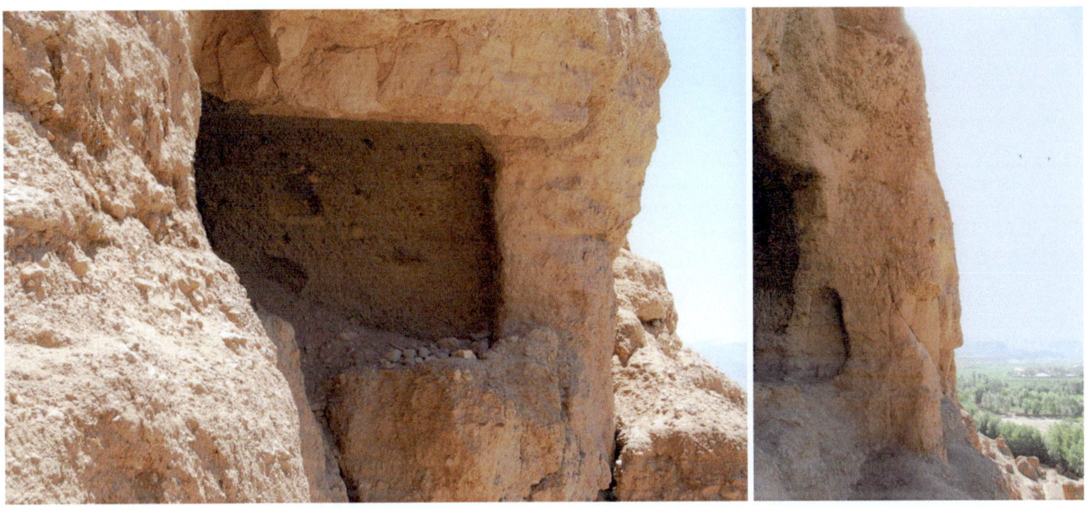

Fig. 5.31 Old rock-block slide reactivated (*left*) and potential rock sliding

(158°/76° and 052°/83°). In the eastern side, wedge failure can be promoted by joints 1–2 (oriented 172°/78° and 207°/82°) and joints 1–3 (172°/78° and 227°/69°).

It can be affirmed that in the western portion, because the most important system is the discontinuity oriented parallel to the slope, this kind of failure mode is very difficult to occur, because this system primarily produces rock falls and toppling phenomena. As a matter of fact, no special evidence of wedge potential, although theoretically possible, has been surveyed in this area. On the contrary, the eastern side has shown wide sectors of the slope where wedge failure has been detected, especially in the lower parts of the slope where siltstone is prevalent, although this kind of failure mode can mobilize a small volume of rocks owing to high frequency of discontinuities (Fig. 5.34).

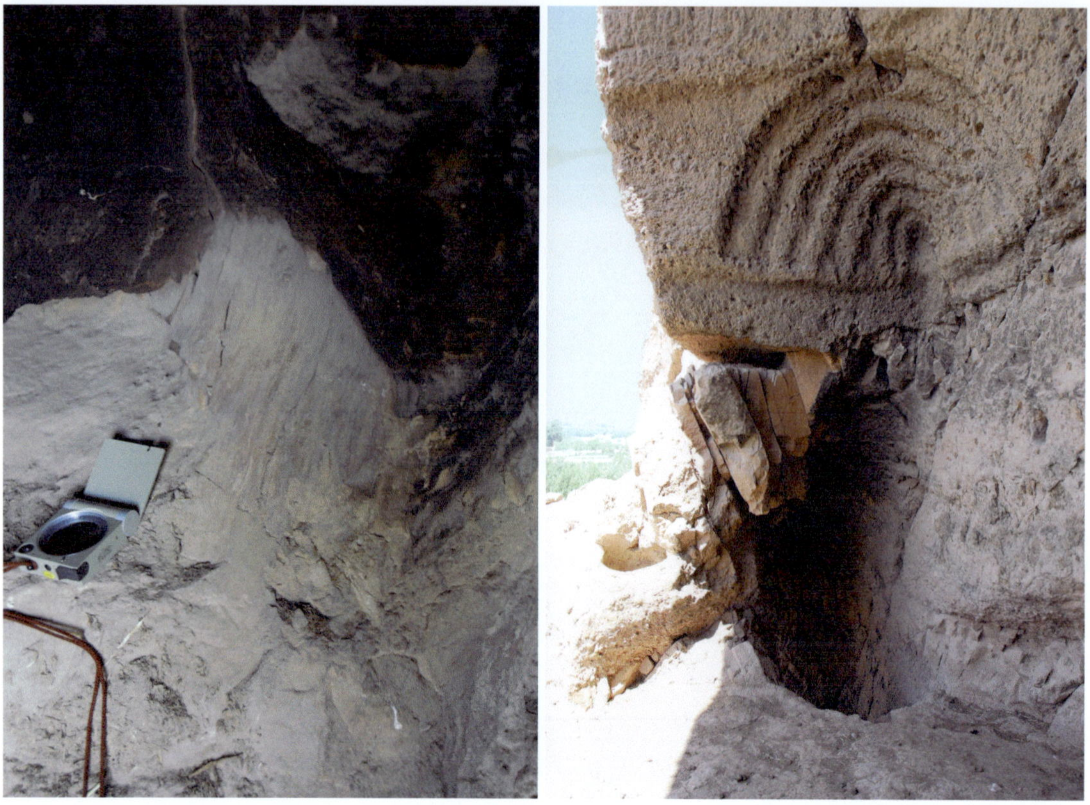

Fig. 5.32 Fault surface with evidence of stress (*left*) and persistent discontinuity where potential planar sliding can develop (*right*)

Conclusions

In the study area, the analysis of results from the kinematic analysis undertaken in the structural stations along the cliff of Bamiyan has resulted in a potentiality of toppling, planar sliding, and wedge failure, although with distinct perspectives.

Planar sliding is the most diffuse failure potential on both sides of the cliff, although most of the movements have occurred in the past. This failure type can be considered as the secondary movement type after toppling evolution of unstable blocks, especially when the failure surface is located inside siltstone layers.

Toppling of rock blocks is equally diffuse and may be considered the most hazardous landslide type for all of the cliff, even if on the western side a higher magnitude of events can be expected with respect to the eastern part of the cliff.

Wedge sliding is potentially developing in both parts of the cliff. Nevertheless, the structural conditions suggest that this type of movement is more probable in the eastern sector, characterized by high potential frequency and low magnitude of events, as also surveyed during the field mission of June 2007.

Multitemporal Investigation

Multitemporal analysis is becoming one of the most powerful tools in applied geomorphology. This is because in many part of the world it is possible to collect historical documents and depictions, photos, and satellite images suitable to highlight recent evolution in a given area.

Not too much material is available in Afghanistan and, as a major source of information, historical

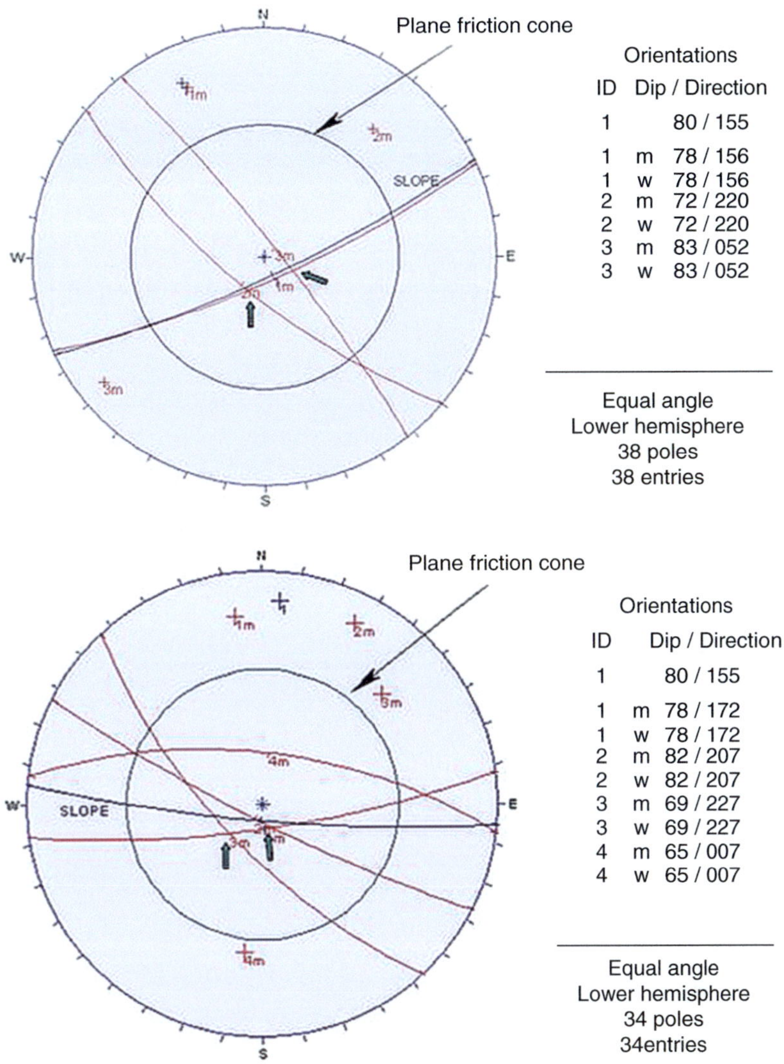

Fig. 5.33 Results of kinematic analysis for wedge failure for the western side (*left*) and eastern side of the cliff; the *arrows* show the intersections of planes that may cause sliding inside the potential area (crescent shaped region between slope limit and the lower friction cone line)

pictures can give some general impression on the evolution of the cliff. All the collected materials are presented in Fig. 5.35, reporting the same cliff over time.

The collected photos were mainly used to evaluate the potential evolution of debris accumulation in time and, secondarily, to identify potential large mass movement(s) occurring in the last century. The obtained information is not very accurate and only a rough estimation is possible. This is because of camera resolution and different positions.

As a major conclusion it can be mentioned that the investigation did not revealing major changes of morphology. The main feature of the cliff remained almost unaltered in time, with the only exception being the area below the Eastern Giant Buddha. In this site, the large construction work performed by the French archaeological mission in about 1968–1972 by the Indian Archaeological Survey, produced a large amount of debris, clearly visible when comparing the image of 1910 with 2004 (Fig. 5.36).

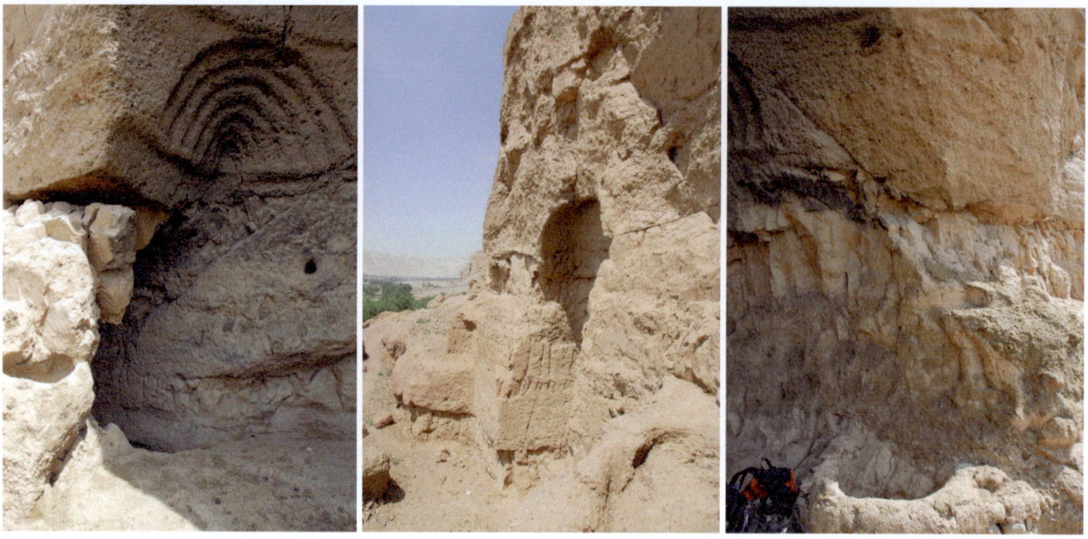

Fig. 5.34 Example of wedge failure potential area in the east side of the cliff

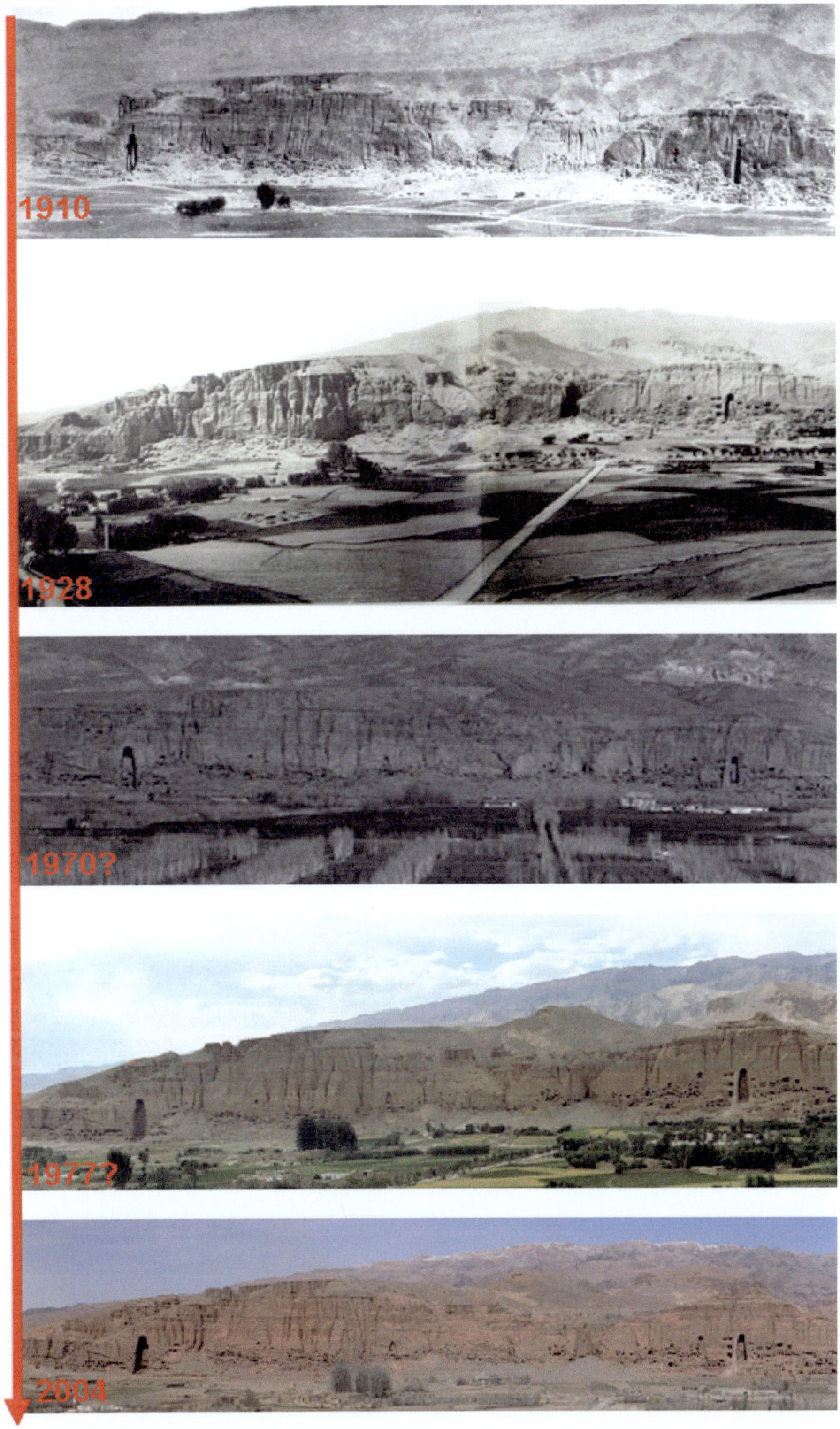

Fig. 5.35 Multitemporal view of the Bamiyan cliff, from 1910 until today

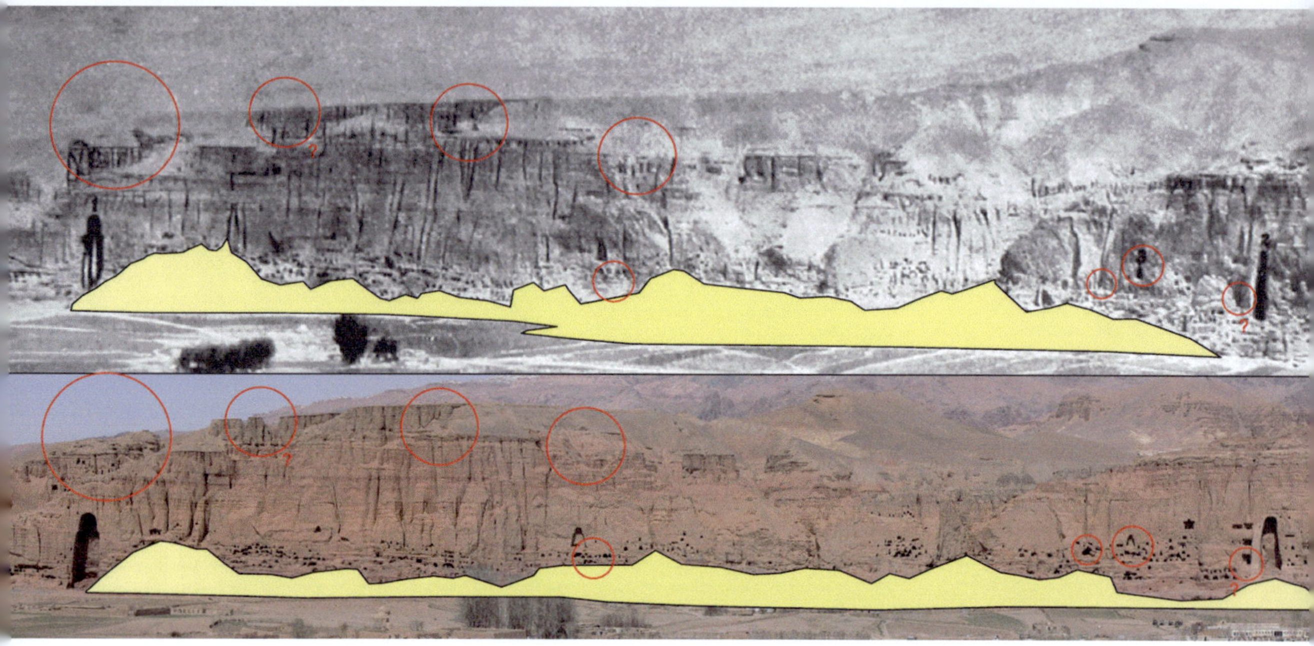

Fig. 5.36 Comparison in debris accumulation at the toe of the Bamiyan cliff between 1910 and 2004 and areas exhibiting some changes in the morphology (*red circle*). The different view, and resolution of adopted equipment do not allow a detailed estimation; nevertheless the debris below the site of the Eastern Giant Buddha produced by the consolidation works developed in late 1950s and 1960s, is quite evident, as well as some changes in the morphology of the top

References

Ambrasey N, Bilham R (2003) Earthquakcs in Afghanistan. Seismological Res Lett 74:107–123

Bush ABG (2005) CO2/H2O and orbitally driven climate variability over central Asia through the Holocene. Quaternary Int 136:15–23

Delmonaco G, Margottini C (2007) Geomorphological features of the Bamiyan valley. Report on the UNESCO mission to Bamiyan (Afghanistan), 18–29 June 2007. UNESCO, Paris, 26 pp

Esper J, Schweingruber FH, Winiger M (2002) 1300 years of climatic history for Western Central Asia inferred from tree-rings. Holocene 12(3):267–277

Goodman RE (1980) Introduction to rock mechanics. Wiley, Toronto

Gruen A, Remondino F, Zhang L (2002) Reconstruction of the Great Buddha of Bamiyan, Afghanistan. International Archives of Photogrammetry and Remote Sensing 34(5):363–368, Corfu (Greece)

Gruen A, Remondino F, Zhang L (2004) Photogrammetric reconstruction of the Great Buddha of Bamiyan, Afghanistan. The Photogrammetric Record 19(107): 177–199

Higuchi T (ed) (1984–1985) Bamiyan. vol 4. Dohosha, Kyoto (re-printed edition, Dohosha Media plan, Kyoto, 2001)

IRI (2003) Central and South-West Asia drought: climate analysis and monitoring

Kamp U Jr, Haserod K, Shroder F, Jr JF (2004) Quaternary landscapc cvolution in the eastern Hindu Kuslı, Pakistan. Geomorphology 57(1–2):1–27

Lang J (1968) La serie neogene de Bamian (Afghanistan). C R Acad Sci 266(26):S.2383–S.2384, Paris

Lang HD (1971) Über das Jungtertiär und Quartär in Süd-Afghanistan. Beih Geol Jb 96 (Hannover):167–208

Lang J (1972) Bassins intramontagneux néogènes de l'Afghanistan Central. Rev Geogr Phys Geol Dynam, 14, 4(2):415–427, Paris

NCDC (2003) Climate of Afghanistan – an overview. http://lwf.ncdc.noaa.gov/oa/climate/research/afghan/overview.html

Operational Climatic Data Summary (2002) https://www2.afccc.af.mil/ocds_mil/products

PASCO (2003a) Bamiyan mapping project. UNESCO report

PASCO (2003b) Buddhas laser scanner project. UNESCO report

Reineke T (2006) Environmental assessment of the Bamiyan Valley in the Central Highlands of Afghanistan. In: University of Aachen, Bamiyan Masterplan Campaign 2005. Aachen

Santana Quintero M (2002) Preliminary report Bamiyan (great – small Buddha). ICOMOS-UNESCO mission to Afghanistan July 2002

6 The Tectonic Setting of Bamiyan and Seismicity in and Near Afghanistan for the Past Twelve Centuries

Nicholas Ambraseys[†] and Roger Bilham

Overview

The ancient city of Bamiyan is located on the Herat fault, a 1,200-km-long, east–west suture through central Afghanistan, that trends northward into the Hindu Kush mountains north of Kabul at its eastern end. The Bamiyan region is located in the transition zone between the intense seismic activity that characterises the Indo-Asian plate boundary in eastern Afghanistan, and the largely inactive central part of Afghanistan. We describe 52 earthquakes that occurred in the period A.D. 734–2004. Data for some centuries are unavailable, and only after the mid-nineteenth century does the earthquake record become more complete. Historically, the western Herat fault has remained largely inactive; however, a significant earthquake occurred near Bamiyan in 1956. The causal fault that slipped in this $M=7.4$ earthquake is not known, although it appears to have occurred in the region bounded by the Herat fault and the Andarabad fault, 80 km to the north of Bamiyan.

N. Ambraseys (✉)
Department of Civil Engineering,
Imperial College London,
South Kensington Campus,
London SW7 2AZ, UK
e-mail: n.ambraseys@imperial.ac.uk

R. Bilham
Department of Geological Sciences,
University of Colorado,
2200 Colorado Avenue, Boulder,
CO 80309-0399, USA
e-mail: roger.bilham@colorado.edu

Introduction

We discuss earthquakes in the region surrounding the Bamiyan region of Afghanistan (Figs. 6.1 and 6.2). For completeness we include earthquakes in an area bounded by 29–38°N and 58–73°E, including the whole of Afghanistan, the eastern part of Iran, southernmost Turkmenistan, Uzbekistan, and Tajikistan, and northwestern Pakistan (Ambraseys and Bilham 2003a, b).

Written records of historical earthquakes in Afghanistan are sparse. Even in the nineteenth and early twentieth century, communications have remained poor because of the skeletal development of roads, phone lines, and government infrastructures, resulting in few published notices about earthquake locations and damage. Newspapers were unavailable until the first quarter of the twentieth century and contain news mostly from the Kabul area. Travelers' accounts and the narratives of early explorers contain macroseismic information often of more utility than that available for the last half century.

Despite the antiquity of Bamiyan, we find no information in Arabic and Persian sources regarding earthquakes in its vicinity (34.83°N, 67.83°E). The reason for this is attributable to the fact that the site as a population center ceased to exist almost 800 years ago. The city of Bamiyan was the capital of a great district of the same name,

[†]deceased

C. Margottini (ed.), *After the Destruction of Giant Buddha Statues in Bamiyan (Afghanistan) in 2001*, Natural Science in Archaeology, DOI 10.1007/978-3-642-30051-6_6, © Springer-Verlag Berlin Heidelberg 2014

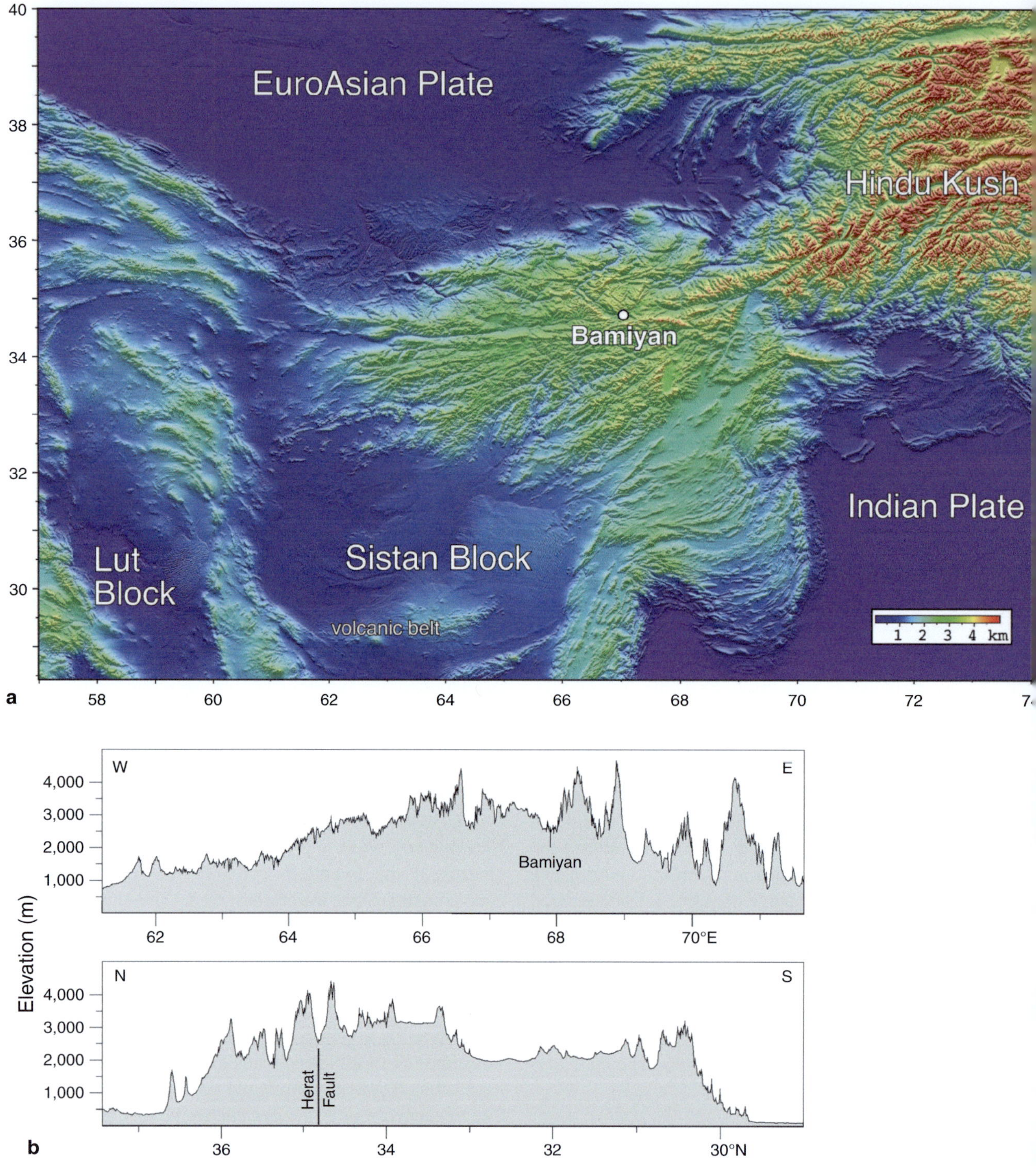

Fig. 6.1 (**a**) Bamiyan lies at the center of the mountains of central Afghanistan on an ancient trade route linking east to west through the valley of the great Herat fault. (**b**) North–south and east–west topographic sections through Bamiyan show its location relative to surrounding mountains. The east–west approaches to Bamiyan along the Herat fault are impeded by 3-km-high passes that isolate the 2,500-km-high Bamiyan Valley. The Shekari River drains the valley to the north-east toward Baghlan

and as its very ancient remains show, was a great Buddhist center long before the days of Islam. In the tenth century it was the trade center of Khorasan and its territory included many large cities, such as Basghurfand, Sakiwand, and Lakhrab, the sites of which are now completely lost.

Arab writers of the beginning of the thirteenth century describe in some detail the great sculptured statues of Buddha. These were high on the mountainside, in a chamber supported on columns, and on its walls were sculptured every species of bird. Outside the chamber entrance were two enormous statues in rock of the hillside, from base to summit, known as the Surkh (red) Bud and the Khing (gray) Bud.

The abandonment of Bamiyan and all its provinces was caused by the wrath of Genghis Khan, whose grandson was killed at the siege of Bamiyan. The Mongol troops were ordered to level to the ground the town walls and all houses, and Genghis forbade any to build or live here ever again, the name of Bamiyan being changed to Marv Balik. Since that time, Bamiyan has been an uninhabited waste (for early illustrations of the great sculptures see Talbot and Maitland 1886). Thus, news from Bamiyan terminates abruptly in the early part of the thirteenth century,

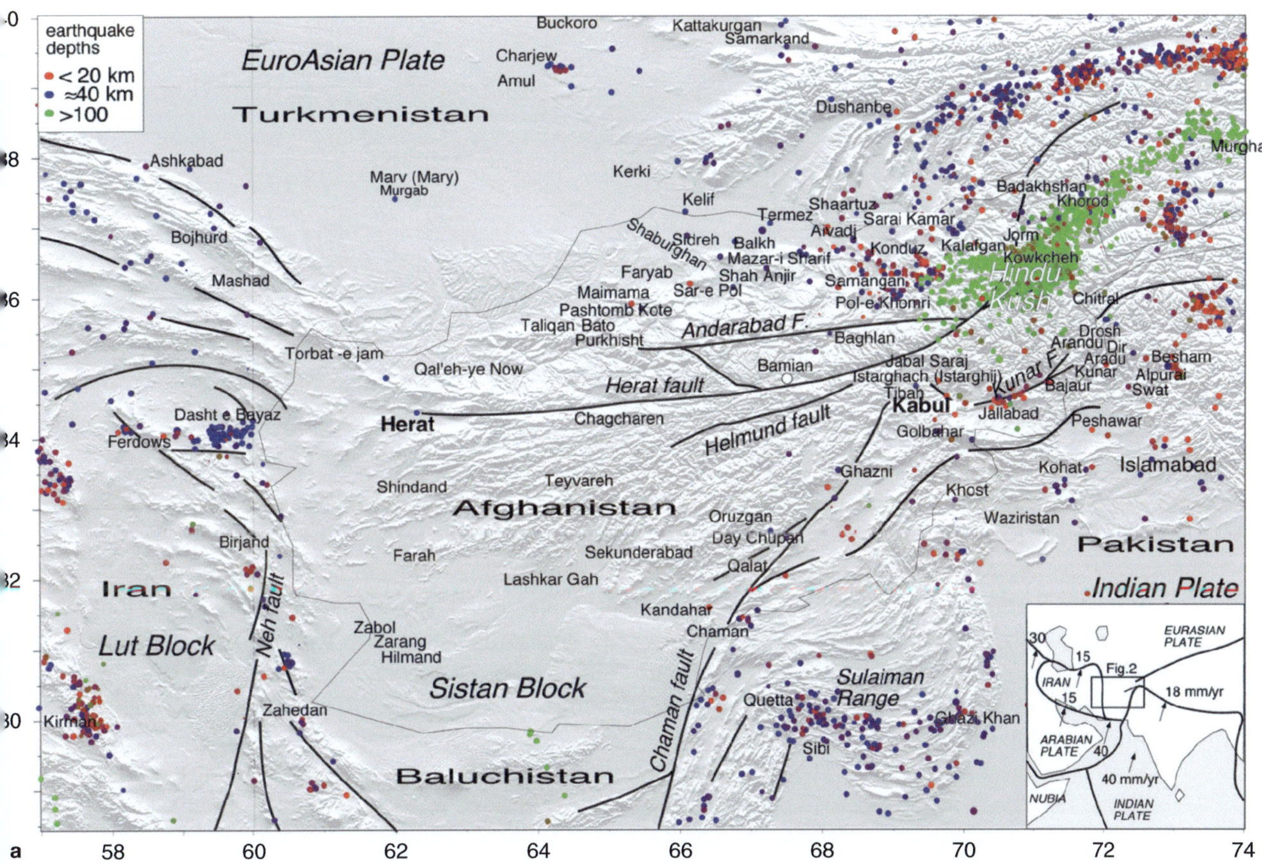

Fig. 6.2 (a) (*Top*) A map of Afghanistan showing prominent faults and major earthquakes described in the chapter. In 1956 a significant earthquake occurred near Bamiyan in the block bounded by the Andarabad fault and the Herat fault. (b) (*Bottom*) Principal stress directions from earthquakes reveal largely oblique compression on the eastern and western edges of Afghanistan, a 1,200 km southward promontory of the Asian plate (stippled) (Data from World Stress Map)

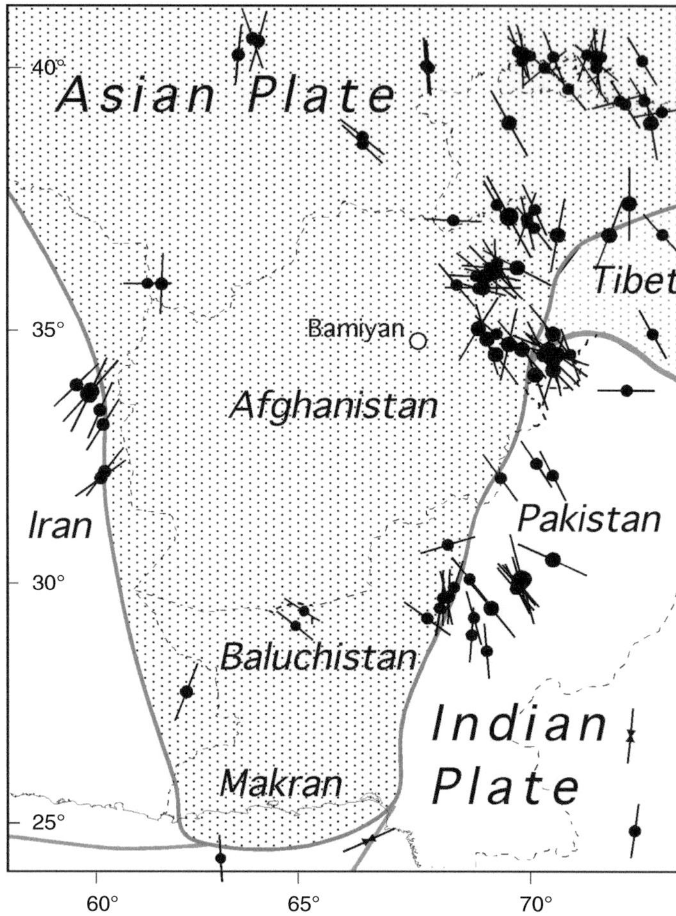

Fig. 6.2 (continued) **b**

and with it reliable information regarding earthquakes in this part of Afghanistan.

Tectonic Setting

Afghanistan lies on a promontory of the Eurasian plate, which converges with the Arabian Plate to the south and the Indian plate to the southeast at rates of approximately 31 mm/year, and with the Lut Block to the west at approximately 15 mm/year (see Fig. 6.2). Minor shortening may be occurring across the mountains of NE Afghanistan, causing a slow counterclockwise rotation of the Sistan block, an old continental fragment that collided with Asia before the continent–continent collision of India with southern Tibet. Afghanistan has been assembled over the past 100 million years by the accretion of small continental masses from the south, the Herat fault representing a suture between two of these former continental terrains.

A GPS measurement on the eastern edge of the Sistan Block southeast of Farah indicates that southeastern Afghanistan behaves as part of the Euro-Asian plate (Tatar et al. 2002). Seismicity in the region is not distributed uniformly. Within the wide deforming belt are several large areas, such as western and central Afghanistan, that appear to have relatively little seismicity during the twentieth century and to behave as effectively rigid blocks (see Figs. 6.2a and 6.3).

Bamiyan lies on the northern edge of the valley that follows one of the largest and most prominent faults in Afghanistan, the Herat fault. This fault, like others with similar strike in the northwest corner of the Indo-Asian collision zone, has slipped in a dextral sense, but unlike the

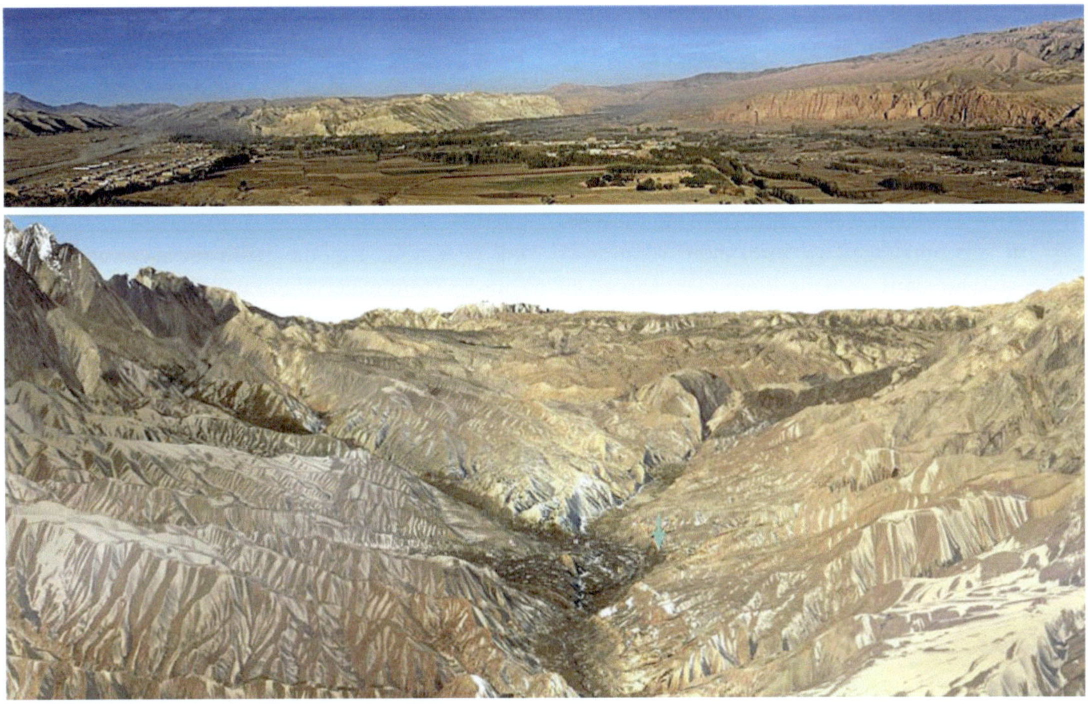

Fig. 6.3 Panoramic views of Bamiyan looking west (Photo credit: Carl Montgomery, www.carlmontgomery.com) (*Lower panel*) © enhanced view of topography from GoogleEarth. Bamiyan lies within the 2-km-wide valley at the confluence of two valleys: a minor valley trending to the NW, and a valley trending SW (the ancient trade route) that for a short distance follows the 600-km-long Herat Fault. The valley floor lies 2.5 km above sea level and the mountains visible to the south exceed 4.5 km. The Bamiyan statues were carved into a north-facing conglomerate cliff (*green arrow*)

Talas Ferghana fault to its north, the Herat fault is relatively aseismic at present. A measure of its structural importance is the continuity and width of the 600-km-long east–west valley whose path it controls. The fault can be traced from west of 61°E in the plains near Herat, to meet the Chaman fault system at 69°E, north of Kabul and 200 km to the east of Bamiyan. The fault then curves to the northeast for a further 400 km continuing the approximate curve of the Chaman fault system through the Hindu Kush mountains.

One reason for the apparent absence of present-day seismic inactivity on the western half of the Herat fault may be its approximately east–west orientation, a direction that is unfavorable for accommodating the approximately north–south convergence between the Arabian plate and the Asian plate. We are unaware of published calculations for stress directions in central Afghanistan but, assuming central Afghanistan is reasonably homogeneous, and averaging stress directions to its sides, would also suggest that stresses here are directed in a north–south direction, normal to the Herat fault. However, geological evidence suggests that the sub-parallel Andarabad fault 80 km north of Bamiyan is currently active.

Earthquakes with focal depths >100 km are associated with subduction of Arabian Sea floor in the Makran in the south, and with a descending slab beneath the Hindu Kush in the north (Pegler and Das 1998). Seismicity throughout the remainder of Afghanistan, including the Bamiyan area, is restricted to the upper 30 km of the continental crust (Maggi et al. 2000a, b).

Faults in western Afghanistan along the Iran border permit the southward motion of the Sistan block through a series of faults that slip in a right-lateral sense. Faults in eastern Afghanistan along the Pakistan border slip in left-lateral sense at 19–24 mm/year (Lawrence et al. 1992; Bernard et al. 2000). Whereas both borders exhibit a component of convergence, thrust faulting is well developed in Baluchistan (Ambraseys and Bilham 2003a) and especially in northern Pakistan, where the Chaman fault veers northwest into the collisional tectonics of the Himalaya and Karakorum. The plate boundary is less well defined along the Iran border than on the Pakistan border, and in the absence of a through-going fault, right lateral shear is partly accommodated by "bookshelf" rotation of the east–west fault systems there (Jackson et al. 1995; Ambraseys and Bilham 2003b).

To the north, the seismicity of eastern Iran merges with that of the Kopet Dagh and eastern Alborz, and the mountains north of Afghanistan merge with the Tadjik basin, overridden by the Hindu Kush to the south and the Pamir to the north (Burtman and Molnar 1993). The northeastern corner of Afghanistan merges with the western end of the Hindu Kush intermediate-depth seismic zone, with many earthquakes in the depth range 70–300 km, forming a contorted slab dipping steeply north in the Hindu Kush and steeply south in the Pamir (Pegler and Das 1998). This represents a relict ocean basin consumed by subduction within the last 10–15 Ma.

Historical Macroseismic Data

For the early period our main sources of macroseismic information are Persian documents, whereas for the later period, British and French consular reports are available that occasionally refer to earthquakes outside the Kabul region. Figure 6.3 illustrates the location of historic earthquakes in different regions of Afghanistan. Numerous shocks have been reported from the capital, Kabul, but although these events have caused general alarm they are typically associated with little damage (Furon 1925a). A few documents written by European residents in the capital (Niedermayer 1936; Stenz 1945), and expedition reports (Danby et al. 1972) add minor additional data, and demonstrate the difficulty of retrieving reliable macroseismic information outside Kabul. An important source of data in the twentieth century is the little-known work of Heuckroth and Karim (1970), who retrieved reports from the Kabul press for the period 1928–1969. Secondary publications and catalogs are available that contain no important new data (e.g., Samizay 1998), NEIS Catalogue of Significant Earthquakes, and the US National Earthquake Information Center's Earthquake Database. The GSHAP catalog, in particular, is uncritical and occasionally misleading Giardini (1999).

An annotated summary of case histories for shallow earthquakes in Afghanistan is listed in the Appendix, together with the more important sources from which these data have been derived. Notices of felt earthquakes at single locations, such as in Kabul, and at a few other urban centers are numerous but because they add little information to the overall seismicity of the region they have been excluded.

It is clear from the limited number of events described in the Appendix that the historic record for Afghanistan is far from complete compared with contiguous surrounding regions. Historical earthquakes in Pakistan, Tajikistan, Uzbekistan, and Turkmenistan are described in Kondorskaya and Shebalin (1997), earthquakes in eastern Iran are described in Ambraseys and Melville (1982) and Moinfar et al. (1994), and earthquakes in Baluchistan are described by Ambraseys and Bilham (2003a, b). Data discussed in these sources are not repeated in the present chapter.

Recent Instrumental Data

An abrupt increase in information about Afghan earthquakes occurred near the start of the twentieth century (Fig. 6.4). Instrumental data are available from station bulletins worldwide since the end of the nineteenth century, particularly from Russian and Indian stations, that are useful for the assessment of magnitude before the advent of the magnitude scale in the mid-1950s. Earthquake locations of diverse quality are given by various agencies and authors: BAAS, ISS, ISC (see abbreviation list in references) by the Russian network and Gutenberg and Richter (1949). Some events in the region have been relocated by Nowroozi (1971), Quittmeyer and Jacob (1979), and Engdahl et al. (1998).

Early instrumental epicenters in Afghanistan, like elsewhere, are unreliable and are frequently based on a poor distribution of global seismic stations. For this reason it is important to correlate

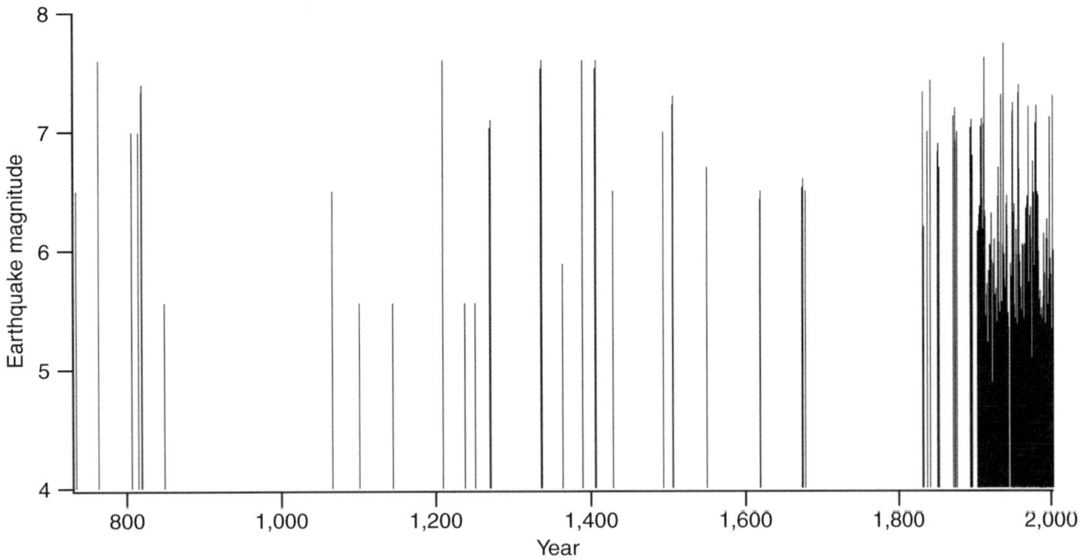

Fig. 6.4 Earthquake history in the complete catalog illustrating the sparsity of data throughout most of the past 1,200 years compared with the increase in accounts from explorers in the mid-nineteenth century and the abundant twentieth-century instrumental record

early epicentral determinations with felt reports where these are available. A special difficulty attends this type of epicentral verification, especially in northwestern Afghanistan where earthquakes occur in both the crust at shallow depths and at subcrustal depths (>70 km). Shallow earthquakes cause heavy local damage and loss of life, but if they occur in a remote part of the country they may fail to be reported. In contrast, large deep earthquakes that cause little or no damage are felt over a large area, and are therefore reported from multiple urban centers. This may have the effect of skewing the perceived location of instrumentally determined earthquakes toward populated areas, and almost certainly will bias the historic record. Thus, deep events are unlikely to have escaped notice historically, but shallow earthquakes were recorded only when they occurred near trade routes and literate population centers.

The case of the recent pair of earthquakes of March 3 and 25, 2002 in the Hindu Kush is illustrative of the problem. The first of these damaging events occurred at a depth of 250 km with a magnitude of 7.4, and the second occurred at a depth of less than 10 km with a magnitude of 6.1. The deeper and much larger event was felt over a large area, and caused widespread but relatively minor damage, including about 150 fatalities. In contrast the second, shallow shock, 120 km southwest of the first, caused great damage within a relatively small area, killing about 1,200 people, but it was felt within a radius of only 200 km. If this pair of earthquakes had occurred in the preinstrumental period it is very likely that their effects could have been conflated into a single event of large epicentral intensity and large radius of perceptibility, to which one could assign a shallow depth and a large magnitude.

Few earthquakes before the mid-1970s have hitherto been assigned a surface wave magnitude. Gutenberg and Richter (1949), Abe (1981), and Abe and Noguchi (1983a, b) have calculated M_s or m_b values for fewer than 8 % of the 147 post-1892 instrumentally recorded events discussed in this chapter.

Our study commenced by reviewing each event in the parametric catalogs described in the preceding to remove double entries, obvious errors, and spurious events. For 1,410 remaining events we merged a large body of macroseismic information derived from diverse primary sources, both published and unpublished.

Early earthquakes are far less well located and it is often difficult to ascertain even their true epicentral area, although for most of them there is little ambiguity about their general location. Although it is certain that many small to medium magnitude events are absent in the record, it can reasonably be assumed that those few for which damage details survive were important events.

The final dataset consists of 1,312 shallow earthquakes of all magnitudes, 98 fewer than in recent parametric catalogs. Discarded entries include spurious events, repeated entries, field explosions in Uzbekistan, and underground detonations in Turkmenistan in the 1960s and 1970s. Macroseismic epicentral locations before the 1970s, if well defined, were selected in preference to instrumental epicentral locations, and constitute 12 % of the total number of entries. Otherwise instrumental locations and focal depths were adopted, in a descending order of preference, 9 % from Engdahl et al. (1998), for well-determined earthquakes (DEQ); and 6 % from Quittmeyer and Jacob (1979) after checking. Locations in northern Afghanistan reported by the Russian network (15 %) and the ISC were also used (48 %), whereas a number of events before 1966 of particular interest were relocated in this study using the ISC procedure (4 %). Fewer than 4 % of the entries in the dataset were adopted from Nowroozi (1971), BCIS, and USGS.

For all the earthquakes identified in the region after 1896 we examined station bulletins for associated surface-wave readings of amplitudes and periods, and for events in which such data were found, surface wave magnitudes were calculated from the Prague formula (Willmore 1979) without restricting its validity to the arbitrary chosen narrow period and distance range employed by NEIS and ISC. The calculation of station magnitudes with station corrections required the compilation of a magnitude database that required extracting about 9,000 surface wave amplitude/period readings from station bulletins, which were then used to estimate M_s event values for 545 earthquakes.

Having calculated M_s for the most important events of the last century we proceeded to assess semi-empirically their moment magnitude, M_W. Seismic moments M_0 of earthquakes after 1977 were taken from the Harvard Moment Tensor Solution Catalog (CMT). To these we added a few seismic moments for post-1969 events that were calculated from special studies of body-wave modeling.

First, we considered the global average log $(M_0) - M_s$ relations of Ekström and Dziewonski (1988). However, as these authors point out, there is regional bias in M_0 and such global relationships may be inappropriate for the estimation of tectonic motion in continental regions. We therefore calculated a regional M_s – log (M_0) relation that is based on the 94 M_s – log (M_0) pairs from crustal events (h<40 km), binned into units of 0.2 in M_s and 0.2 in log M_0. We used a bilinear relation with slope 1.0 at M_s<6.2 and slope 1.5 at M_s>6.2, for which there is theoretical justification (Ambraseys and Douglas 2000), and obtained

$$\log M_o = 19.09 + M_s \quad \text{for} \quad M_s \leq 6.2 \quad (6.1)$$

and

$$\log M_o = 15.94 + 1.5\, M_s \quad \text{for} \quad M_s > 6.2 \quad (6.2)$$

with M_0 in dyn-cm. These relations fit the data well (Fig. 6.5). In this figure the two small events (<4.5) that fall well above the fitted relation are probably subcrustal events whose M_s have not been corrected appropriately for depth. The data fit equally the relation obtained in the preceding if the regression coefficients are determined from individual, rather than binned, data points. These relations differ little from (6.1 and 6.2) and for a given M_s yield smaller M_0 values than the global average relation of

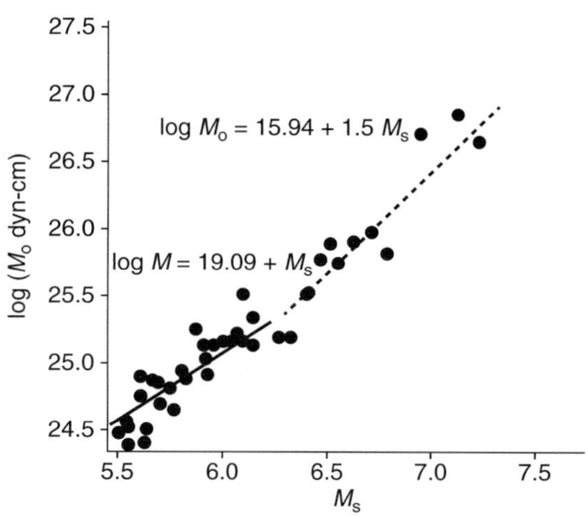

Fig. 6.5 A plot of M_s vs log (M_0) for recent well-located events reveals a change in scaling at approximately M_s = 6.2. We use these relations to derive the moment magnitudes for other earthquakes in the catalog for which we have obtained surface-wave magnitudes

Ekström and Dziewonski (1988) by an average factor of 0.68 for $M_s \leq 6.2$, and by 0.62 for $M_s > 6.2$.

A similar overestimation of M_o from global relations was found from a much larger number of data for the Eastern Mediterranean region, which confirms that in these areas global M_s-M_o relations yield M_o values that are too high, whereas regional relations yield values closer to those calculated directly (Ambraseys 2001). From the derived M_s to M_o relations (6.1 and 6.2) we estimated seismic moments for those events for which only M_s was previously available.

Table 6.1 lists all 52 earthquakes before 1891 for which we could find information from contemporary sources. Table 6.2 presents a list of 148 shallow $M_s \geq 5.5$ earthquakes from 1891 to 2002. Appendix summarizes the macroseismic effects of the more important earthquakes in these tables. For shocks in surrounding countries, this information is given in the publications quoted. Table 6.3

Table 6.1 Significant historic Afghanistan earthquakes, 1734–1891

Y	M	D	OT	N	E	R	M_s	r	log M_o	q	M_w	Location
734	0	0	0	31.6	60.5	3	6.5	5	25.7	2	6.4	Sistan-I
763	0	0	0	33.3	59.3	3	7.6	5	27.3	2	7.5	Qayin-I
805	12	2	0	29.5	60.5	3	7	5	26.4	2	6.9	Sistan-I
815	0	0	0	29.5	60.5	3	7	5	26.4	2	6.9	Sistan-I
819	6	0	0	36.4	65.4	3	7.4	5	27	2	7.3	Balkh
849	0	0	0	34.3	62.2	3	5.3	5	24.4	2	5.6	
1066	5	0	0	33.9	59.2	3	6.5	5	25.7	2	6.4	Kuhistan-I
1102	2	28	1,800	34.4	62.2	3	5.3	5	24.4	2	5.6	
1145	0	0	0	36.2	58.8	3	5.3	5	24.4	2	5.6	Nishapur-I
1209	0	0	1,200	36.4	58.7	3	7.6	5	27.3	2	7.5	Nishapur-I
1238	0	0	0	34.3	58.7	3	5.3	5	24.4	2	5.6	Gonabad-I
1251	0	0	0	36.2	58.8	3	5.3	5	24.4	2	5.6	Nishapur-I
1270	10	7	1,200	36.2	58.8	3	7.1	5	26.6	2	7	Nishapur-I
1336	10	21	600	34.7	59.7	3	7.6	5	27.3	2	7.5	Kwaf-I
1364	2	10	0	34.9	61.7	3	5.8	5	24.9	2	5.9	
1389	2	0	0	36.2	58.8	3	7.6	5	27.3	2	7.5	Niishapur-I
1405	11	23	0	36.2	58.8	3	7.6	5	27.3	2	7.5	Nishapur-I
1410	0	0	0	36.7	66.8	3	0	0	0	0	0	Balkh
1428	0	0	0	35.9	63.8	3	6.5	5	25.7	2	6.4	Taliqan
1493	1	10	600	33	59.8	3	7	5	26.4	2	6.9	Muminabad-I
1505	7	6	0	34.5	69.1	3	7.3	5	26.9	2	7.2	Kabul
1519	1	3	0	35	71.5	3	0	5	0	0	0	Bajaur
1549	2	15	2,400	33.7	60	3	6.7	5	26	2	6.6	EQayin-I
1619	5	0	0	35.1	58.9	3	6.5	5	25.7	2	6.4	Dughabad-I
1673	7	30	0	36.3	59.3	3	6.6	5	25.8	2	6.5	Mashhad-I
1678	0	0	2,400	34.3	58.7	3	6.5	5	25.7	2	6.4	Gonabad-I
1687	4	0	0	36.3	59.6	3	0	0	0	0	0	Mashhad-I
1827	3	17	0	36.3	59.6	0	0	0	0	0	0	Kabul
1829	0	0	0	36.3	59.6	0	0	0	0	0	0	Kabul
1832	1	22	0	36.5	71	3	7.4	5	27	2	7.3	Badakhshan
1833	0	0	0	37.3	58.1	3	6.2	0	25.3	2	6.2	Shirvan-I
1833	4	19	0	36.3	59.6	0	0	0	0	0	0	Kabul
1836	0	0	0	36.3	59.6	0	0	0	0	0	0	Kabul
1837	12	14	0	36.3	59.6	0	0	0	0	0	0	Kabul
1838	0	0	0	29.6	59.9	3	7	0	26.4	2	6.9	Nasratabad-I

(continued)

Table 6.1 (continued)

Y	M	D	OT	N	E	R	M_s	r	log M_o	q	M_w	Location
1838	1	7	0	35.4	66.8	3	0	0	0	0	0	Jurm
1840	1	26	0	36.3	59.6	0	0	0	0	0	0	Kabul
1842	2	19	1,140	35	71	3	7.5	5	27.2	2	7.4	Kunar
1851	6	0	0	36.8	58.4	3	6.9	5	26.3	2	6.8	Quchan-I
1852	1	24	345	29.3	68.6	3	6.7	5	26	2	6.6	Kahun
1857	6	0	1,230	31.6	65.7	3	0	0	0	0	0	Kandahar
1858	8	25	0	28.3	68.4	3	0	0	0	0	0	Jacobabad-P
1871	12	23	1,800	37.4	58.4	3	7.2	0	26.7	2	7.1	Quchan-I
1872	12	15	0	29.6	67.9	0	0	0	0	0	0	Sibi-P
1874	10	18	0	35.1	69.2	3	7	5	26.4	2	6.9	Kohistan
1880	2	8	0	33.9	70	3	0	0	0	0	0	Khurum
1888	12	28	0	0	0	0	0	0	0	0	0	Quetta-P
1889	4	0	0	36.3	59.6	0	0	0	0	0	0	Kabul
1889	10	0	0	36.7	67.1	3	0	0	0	0	0	Mazar
1890	9	0	2,100	36.3	59.6	0	0	0	0	0	0	Kabul
1891	6	0	0	34.5	68.5	3	0	0	0	0	0	Paghman

Fifty-one earthquakes are listed, of which 11 are described in narrative accounts

Y, *M*, *D*, and *OT* correspond to year, month, day, and origin time in hours and minutes

E and *N* corresponds to longitude and latitude in degrees, respectively

R Epicentral locations were adopted from the following sources:

0. Locations in northern Afghanistan taken from reports of the Soviet network
1. Epicenters computed by BAAS/ISS/ISC
2. Rough macroseismic locations or epicenters adopted by BAAS/ISS with no calculation, calculated from the Prague formula with station corrections
3. Macroseismic epicenters from well-defined macroseismic observation
4. Relocated positions using ISC procedures
5. For early earthquakes, locations calculated by Ambraseys and Melville for western Afghanistan
6. Quittmeyer and Jacob (1979) relocated events in the time period 1914 through 1965, using a computer program similar to the one described by Bolt (1960)
7. From Engdahl et al. (1998). We adopted only recomputed locations that have a well determined depth (DEQ in their columns 2–4)
8. Few locations were taken from BCIS for the period 1953 through 1965 during which ISS did not report small events
9. In the absence of other locations, we accepted some data marked LEQ and FEQ from Engdahl et al. (1998)
10. Nowroozi recalculated epicentral locations of earthquakes in the region between 1950 and 1965, using the program by Sykes and Landisman (1964) and Nowroozi (1971)
11. PDE locations

n Number of seismographic stations used by ISS/ISC to calculate epicentral position

p Actual number of stations greater than that given by ISS/ISC

M_s Surface wave magnitude

r 1: calculated from the Prague formula with station corrections
2: calculated from the Prague formula from few stations with station corrections
3: ISC
4: Moscow
5: calculated from macroseismic data

h Focal depth from sources in *R*

log (M_o) dyn cm

q 0: M_o not available
1: CMT Harvard
2: converted from M_s from equations described in text
3: from other sources (Bernard et al. 2000)

M_w Moment magnitude

m_b (or m_B) Body wave magnitude from ISC

M Surface wave magnitude from ISC

Location P: Pakistan
I: India

Table 6.2 Significant Afghanistan earthquakes, 1892–2002

Y	M	D	OT	N	E	RR	*n*	M_s	*R*	*h*<40	log M_o	b_q	M_w	m_B	M_b	Location
1892	12	20	20	30.9	66.5	3	3	6.5	1	0	25.8	2	6.5	0	0	Chaman-P
1893	2	13	400	30.7	67.4	3	3	5.9	2	0	25	2	6	0	0	Chaman-P
1893	11	17	1,500	37	58.4	3	3	7.1	0	0	26.6	2	7	0	0	Quchan-I
1895	1	17	1,100	37.1	58.4	3	0	6.8	0	0	26.1	2	6.7	0	0	Quchan-I
1901	10	17	557	31	68.4	3	0	6.1	1	0	25.2	2	6.1	0	0	Loralai-P
1902	9	20	632	38.5	67	0	0	6.2	4	40	25.3	2	6.2	0	0	
1903	1	20	824	37	71	0	0	5.5	4	0	24.6	2	5.7	0	0	
1903	3	22	1,435	33.2	59.7	3	0	6.2	1	0	25.3	2	6.2	0	0	Durukhsh
1903	9	25	120	35.2	58.2	3	0	5.9	2	0	25	2	6	0	5.3	I
1903	12	23	300	29.5	67.5	3	0	5.9	2	0	25	2	6	0	0	Bolan
1904	7	27	520	33	72	1	0	5.7	1	0	24.8	2	5.9	0	0	
1904	11	9	328	36.9	59.8	3	0	6.4	1	0	25.5	2	6.3	0	0	Tedzhen-Tu
1905	6	19	127	29.9	60	3	0	6	1	0	25.1	2	6	0	6.8	Nasradabad
1905	9	26	128	30.3	69.9	3	0	6.4	1	0	25.6	2	6.4	0	0	Sulaiman
1906	10	4	652	37.2	67.3	0	0	5.7	4	20	24.8	2	5.8	0	0	
1906	10	24	1,539	36.5	68	0	0	7.1	1	32	26.6	2	7	0	6.8	Ayvadzh
1907	10	23	2,025	37.7	65.4	3	0	6.1	1	0	25.1	2	6.1	0	6.1	Kersk
1908	1	12	1,019	30.2	67.7	3	0	5.6	2	0	24.7	2	5.8	0	0	
1908	3	5	220	30.2	67.7	3	0	6.4	1	0	25.6	2	6.4	0	0	Harnai-P
1908	6	3	1,556	28	67	1	0	6.2	1	0	25.3	2	6.2	0	0	Quetta-P
1909	9	7	1,528	33	70	1	0	6	1	0	25.1	2	6	0	0	
1909	10	20	2,341	28.9	68.3	3	0	7.1	1	0	26.5	2	7	0	0	Kachhi-P
1910	8	17	1,158	28.4	67	4	0	6.3	1	0	25.4	2	6.3	0	0	Sind-P
1911	1	1	1,018	36.5	66.5	0	0	6.9	1	40	26.4	2	6.9	0	6.6	Mazar-i
1911	1	1	1,459	36.5	66	3	0	6.5	1	20	25.7	2	6.4	0	6.3	Shuburghan
1911	2	18	1,841	38	72.8	0	0	7.7	1	26	27.5	2	7.6	0	7.4	Serez
1912	8	23	1,402	35	71.5	3	0	6.4	1	0	25.5	2	6.3	0	0	Kunar
1913	3	27	913	29.5	67.5	3	16	5.6	1	0	24.7	2	5.8	0	0	Dhadar-P
1914	5	21	826	32	69.5	1	33	5.7	1	0	24.8	2	5.8	0	0	
1914	11	4	1,106	32	70	3	22	5.7	1	0	24.8	2	5.8	0	0	
1917	8	29	1,300	37.4	58.1	3	0	5.7	2	0	24.8	2	5.8	0	0	
1917	11	28	1,442	36.5	59.1	0	0	5.7	4	0	24.8	2	5.8	0	0	
1917	12	1	947	30	71	1	13	5.6	1	0	24.6	2	5.7	0	0	
1918	3	24	2,314	34.9	60.7	3	23	6	1	0	25.1	2	6.1	6.3	0	Torbat-I
1918	11	29	1,041	32.7	67.7	3	24	6.2	1	0	25.3	2	6.2	0	0	Ghazni
1919	5	23	610	32.5	68	3	35	6.3	1	0	25.4	2	6.2	0	6.48	Zarghun
1919	6	1	1,246	30	71	2	8	5.5	1	0	24.6	2	5.7	0	5.81	
1919	6	15	1,849	30	71	2	15	5.5	1	0	24.6	2	5.7	0	0	
1920	2	27	351	35	69	1	25	5.9	1	0	25	2	6	0	0	
1923	5	25	2,221	35.2	59.1	3	39	5.8	2	0	24.9	2	5.9	0	0	I
1923	7	16	1,323	37.5	70.5	1	14	5.5	2	15	24.6	2	5.7	0	5.6	
1923	9	14	810	29	59.3	4	14	5.6	2	0	24.7	2	5.8	0	0	
1923	10	1	816	29	67.5	4	41	6.1	1	0	25.2	2	6.1	0	0	Kachhi-P
1928	3	8	1,814	31.5	60.1	3	26	5.5	1	0	24.6	2	5.7	0	0	I
1928	8	21	1,902	32.3	58.7	3	25	5.6	1	0	24.6	2	5.7	0	5.3	
1928	9	1	609	28.8	69.6	6	88	6.5	1	33	25.6	2	6.4	0	0	?

(continued)

Table 6.2 (continued)

Y	M	D	OT	N	E	RR	n	M_s	R	$h<40$	$\log M_o$	b_q	M_w	m_B	M_b	Location
1928	10	15	1,419	28.5	67.4	4	106	6.7	1	33	26	2	6.6	0	0	Katra-P
1928	12	14	28	28.8	68.1	1	35	5.6	1	33	24.7	2	5.8	0	0	
1929	6	4	704	37.3	66.5	0	33	5.7	2	14	24.8	2	5.8	0	5.7	
1929	7	13	736	37.2	58.2	3	64	6.1	1	0	25.2	2	6.1	0	0	Quchan
1931	8	24	2,135	30.1	67.6	3	136	6.8	1	40	26.1	2	6.7	0	0	Sharigh-P
1931	8	27	1,527	29.9	67.2	4	155	7.3	1	33	26.9	2	7.2	0	0	Mach-P
1931	8	28	1,940	28.8	67.4	6	22	5.6	1	33	24.6	2	5.7	0	0	
1932	9	8	725	31.6	58.2	3	45	5.6	1	0	24.7	2	5.7	0	0	I
1933	10	16	435	32.7	66.5	3	45	5.6	1	0	24.7	2	5.7	0	0	Uruzgan
1935	5	15	201	28.4	67.5	3	68	6	1	33	25.1	2	6	0	0	Kotra-P
1935	5	30	2,133	28.9	66.4	6	231	7.7	1	33	27.6	2	7.7	0	0	Quetta-P
1935	6	2	916	30.1	66.9	6	104	6	1	13	25.1	2	6.1	0	6	Quetta-P
1935	10	28	1,205	31.3	69.3	0	52	5.6	1	33	24.7	2	5.8	0	0	
1936	6	30	1,926	33.6	60	3	78	6	1	0	25.1	2	6	6.2	0	Bamrud
1937	11	11	1,711	37.5	72.2	0	−1	5.5	4	0	24.6	2	5.7	0	0	
1937	11	13	1,150	38	69.5	0	31	5.6	4	10	24.7	2	5.8	0	5.6	
1938	12	19	1,856	36.6	58.5	3	0	5.6	2	0	24.7	2	5.8	0	0	Khorasan
1940	5	4	2,101	35.8	58.5	3	106	6.5	1	0	25.6	2	6.4	6.2	0	Khorasan
1940	7	17	636	36.7	72.2	6	35	5.6	2	33	24.7	2	5.8	0	5.7	
1940	7	17	1,144	36.8	70.7	0	43	5.6	4	15	24.7	2	5.8	0	5.6	
1940	8	1	1,945	38	72.5	0	28	5.5	4	20	24.6	2	5.7	0	0	
1941	2	16	1,639	33.4	58.9	3	74	6.2	1	0	25.3	2	6.2	6.4	0	Muhammadabad
1945	10	1	516	29	67.3	6	49	5.9	1	33	25	2	6	0	0	Nagau
1946	6	20	37	29.5	66	1	31	5.8	1	0	24.9	2	5.9	0	0	
1947	9	23	1,228	33.7	58.7	3	126	6.9	1	0	26.3	2	6.8	6.4	0	Dustabad
1947	9	26	304	33.8	58.6	3	57	6	1	0	25.1	2	6.1	0	0	Ferdows
1948	1	28	1,551	36.4	67.7	6	80	6.5	1	33	25.8	2	6.5	6.3	6.9	Samangan
1948	6	18	1,844	37.5	58	3	54	5.5	1	0	24.6	2	5.7	0	0	
1948	10	5	2,012	38	58.3	0	175	7.2	1	0	26.8	2	7.2	0	7.3	Ashkhabad
1949	2	24	2,302	30.1	69	6	55	5.9	1	33	24.9	2	5.9	0	0	
1950	5	9	1,116	37.9	58.4	3	120	6.4	1	0	25.5	2	6.3	6.1	0	I
1950	9	24	2,256	34.5	60.7	3	91	5.6	1	0	24.6	2	5.7	5.6	0	
1951	5	14	407	30.2	70.1	6	76	5.7	2	33	24.8	2	5.8	0	0	
1952	9	15	1,128	30.6	70	6	50	5.5	1	33	24.6	2	5.7	5.8	0	
1952	10	10	1,847	30.4	69.3	3	147	6.2	1	2	25.3	2	6.1	0	0	Sulaiman-P
1952	12	25	2,222	29.3	70	4	124	6	1	33	25.1	2	6	0	6	Sulaiman-P
1954	1	23	1,606	37.4	72.5	0	67	5.8	4	9	24.9	2	5.9	0	5.8	
1955	2	18	2,248	30.2	67	3	85	5.7	1	33	24.8	2	5.8	0	0	Chaman-P
1956	1	11	2,216	30	69.6	6	−1	5.5	4	24	24.6	2	5.7	0	0	
1956	5	13	750	30	69.9	3	90	5.9	1	33	25	2	6	0	0	Sulaiman-P
1956	6	9	2,313	35.1	67.5	6	334	7.4	1	33	27	2	7.3	0	7.2	Bamyan
1956	6	10	101	34.9	67.7	6	−1	5.6	1	20	24.6	2	5.7	0	5.5	
1956	9	16	837	33.8	69.6	3	250	6.7	1	11	26	2	6.6	0	0	Jaji
1956	9	24	1,020	34	69.6	6	114	5.6	1	1	24.7	2	5.8	0	0	Kabul
1958	3	22	1,107	35.3	67.4	6	120	5.9	1	20	25	2	6	0	5.9	
1960	8	14	2,237	36.3	69.9	0	90	5.5	4	15	24.6	2	5.7	0	5	

Table 6.2 (continued)

Y	M	D	OT	N	E	RR	*n*	M_s	*R*	*h*<40	log M_o	b_q	M_w	m_B	M_b	Location
1962	9	12	2,057	36.4	69	4	147	6	1	35	25.1	2	6.1	0	6.3	
1965	2	2	1,556	37.4	73	1	152	6	1	15	25.1	2	6.1	5.3	6	
1966	2	7	426	29.8	69.5	3	245	6.7	1	17	26	2	6.6	5.8	0	Sulaiman-P
1966	2	7	2,306	30.2	69.8	7	230	6.5	1	17	25.6	2	6.4	5.6	0	Loralai-P
1966	8	1	1,909	29.9	68.7	3	229	6.1	1	19	25.4	1	6.2	5.5	6	Duki-P
1966	8	1	2,030	29.9	68.6	7	179	5.8	4	10	24.9	2	5.9	5.4	0	
1966	8	1	2,102	30.2	68.8	3	286	6.9	1	9	26.3	2	6.8	5.7	5.7	Duki-P
1966	10	25	1,007	29.9	68.8	7	122	5.8	2	19	24.9	2	5.9	5.2	0	
1968	8	31	1,047	34	59	3	301	7.2	1	33	26.8	2	7.1	6	0	Dasht
1968	9	1	727	34.1	58.2	1	262	6.5	1	14	25.6	2	6.4	5.9	6.5	Ferdows
1968	9	11	1,917	34	59.5	3	162	6	1	33	25.1	2	6.1	5.7	0	Gonabad
1968	11	15	625	38	58.1	3	127	5.9	1	18	25	2	6	5.1	5.6	
1971	5	26	241	35.5	58.3	3	185	5.6	1	25	24.6	2	5.7	5.4	5.6	
1971	9	8	1,253	29	60.2	3	151	5.6	4	34	24.7	2	5.8	5.3	0	
1972	6	24	1,529	36.3	69.7	7	336	6.4	1	28	25.5	2	6.3	5.9	0	Takhar
1973	1	20	1,246	29.4	68.6	7	98	5.6	4	17	24.7	2	5.8	5	0	
1974	12	28	1,211	35	72.9	7	358	6.3	1	14	25.2	1	6.1	5.9	0	
1975	10	3	514	30.2	66.3	7	281	6.8	1	1	26.1	2	6.7	5.4	0	Spin-Tezha-P
1975	10	3	1,731	30.4	66.4	7	293	6.5	1	2	25.6	2	6.4	5.5	0	Spin-Tezha-P
1976	3	19	1,303	36.6	67.8	7	266	5.5	4	5	24.6	2	5.7	5.9	5.7	Khulm
1976	11	7	400	33.8	59.2	3	321	6.5	1	0	25.7	2	6.4	5.8	6.5	Vandik
1977	7	13	809	29.9	67.3	7	196	5.6	1	10	24.4	1	5.6	5.1	0	
1978	3	16	159	29.9	66.2	7	239	5.9	1	12	25.3	1	6.1	5.3	5.8	Nushki-P
1978	5	6	1,116	29.8	66.1	7	291	5.8	1	17	24.9	1	5.9	5.4	5.8	
1979	1	16	950	33.8	59.5	3	384	6.8	1	11	25.8	1	6.5	5.9	6.7	
1979	11	14	221	34	59.8	4	399	6.6	1	10	25.9	1	6.6	6	6.6	Khwaf
1979	11	27	1,710	34.1	59.8	1	411	7.2	1	8	26.7	1	7.1	6.2	7.3	Buniabad
1979	12	7	923	34.1	59.9	1	319	6	1	10	25.2	1	6.1	5.9	6.1	
1982	12	16	40	36.1	69	7	530	6.5	1	36	25.8	1	6.5	6.1	6.7	
1984	2	1	1,422	34.6	70.5	7	485	5.9	3	9	25	1	6	5.9	5.9	
1985	5	6	304	30.9	70.3	7	436	5.7	1	17	24.9	1	5.9	5.6	5.6	
1990	3	4	1,946	29	66.4	4	515	6.1	1	20	25.2	1	6.1	5.8	6.1	Kalat-P
1992	5	10	404	37.2	72.9	7	508	5.6	3	38	24.8	1	5.8	5.5	5.8	
1992	5	20	1,220	33.3	71.3	7	607	6	1	20	25.2	1	6.1	5.9	6	Kohat
1992	8	28	50	29.2	66.8	7	436	5.6	1	5	24.4	1	5.6	5.6	5.6	
1993	11	16	1,552	30.8	67.2	7	379	5.6	1	29	24.5	1	5.6	5.4	5.6	Pishin
1994	2	23	802	30.8	60.5	7	650	6.1	1	12	25.2	1	6.1	6	6.1	Sistan
1994	2	24	11	30.8	60.5	7	640	6.1	1	10	25.5	1	6.3	6	6	Sistan
1994	2	26	231	30.8	60.5	7	541	5.9	3	7	25.2	1	6.1	5.7	5.9	
1994	2	28	1,135	30.6	60.6	5	400	5.5	3	30	24.5	1	5.6	5.5	5.5	
1994	5	1	1,200	36.9	67.1	7	630	6.3	1	23	25.2	1	6.1	5.9	6.3	
1995	6	11	2,155	32.6	69.7	7	353	5.6	1	27	24.5	1	5.7	5.1	5.5	
1997	2	27	2,108	30	68.2	7	678	6.9	1	21	26.7	1	7.1	6.1	6.9	Sibi-P
1997	2	27	2,130	30	68	7	498	6.4	3	25	25.5	2	6.3	5.8	6.4	
1997	3	4	1,303	29.4	68.8	7	476	5.8	1	24	24.7	1	5.7	5.3	5.8	
1997	3	20	850	30.1	68	7	441	5.6	3	27	24.9	1	5.9	5.4	5.6	
1997	5	10	757	33.9	59.8	1	725	7.1	1	7	26.9	1	7.2	6.2	7	Zirkuh

(continued)

Table 6.2 (continued)

Y	M	D	OT	N	E	RR	n	M_s	R	h<40	log M_o	b_q	M_w	m_B	M_b	Location
1997	6	25	1,938	33.9	59.5	1	541	5.7	1	11	24.9	1	5.9	5.4	5.8	
1998	2	4	1,433	37.1	70.1	7	552	5.9	1	29	24.9	1	5.9	5.5	6	
1998	4	10	1,500	32.5	60.1	1	458	5.7	3	26	24.7	1	5.8	5.2	5.7	Birjand
1998	5	30	622	37.1	70.1	7	760	6.5	1	29	25.9	1	6.6	5.7	6.7	
1999	2	11	1,409	34.3	69.2	1	432	5.8	3	32	25	1	5.9	5.3	5.8	
1999	7	12	342	30	69.5	1	565	5.5	1	20	24.6	1	5.7	5.5	5.6	
2002	3	25	1,456	36	69.4	P	0	6	0	8	25.2	1	6.1	6	6.1	
2002	4	12	400	35.9	69.2	P	0	5.9	P	10	25	2	6	0	5.9	

One hundred and forty-seven are listed and 36 are described in narrative accounts
Fifty-one earthquakes are listed, of which 11 are described in narrative accounts.
Y, *M*, *D*, and *OT* correspond to year, month, day, and origin time in hours and minutes
E and *N* corresponds to longitude and latitude in degrees, respectively
R Epicentral locations were adopted from the following sources:

0. Locations in northern Afghanistan taken from reports of the Soviet network
1. Epicenters computed by BAAS/ISS/ISC
2. Rough macroseismic locations or epicenters adopted by BAAS/ISS with no calculation, calculated from the Prague formula with station corrections
3. Macroseismic epicenters from well-defined macroseismic observation
4. Relocated positions using ISC procedures
5. For early earthquakes, locations calculated by Ambraseys and Melville for western Afghanistan
6. Quittmeyer and Jacob (1979) relocated events in the time period 1914 through 1965, using a computer program similar to the one described by Bolt (1960)
7. From Engdahl et al. (1998). We adopted only recomputed locations that have a well determined depth (DEQ in their columns 2–4)
8. Few locations were taken from BCIS for the period 1953 through 1965 during which ISS did not report small events
9. In the absence of other locations, we accepted some data marked LEQ and FEQ from Engdahl et al. (1998)
10. Nowroozi recalculated epicentral locations of earthquakes in the region between 1950 and 1965, using the program by Sykes and Landisman (1964) and Nowroozi (1971)
11. PDE locations

n Number of seismographic stations used by ISS/ISC to calculate epicentral position
p Actual number of stations greater than that given by ISS/ISC
M_s Surface wave magnitude
r 1: calculated from the Prague formula with station corrections
2: calculated from the Prague formula from few stations with station corrections
3: ISC
4: Moscow
5: calculated from macroseismic data
h Focal depth from sources in *R*
log (M_o) dyn cm
q 0: M_o not available
1: CMT Harvard
2: converted from M_s from equations described in text
3: from other sources (Bernard et al. 2000)
M_w Moment magnitude
m_b (or m_B) Body wave magnitude from ISC
M Surface wave magnitude from ISC
Location P: Pakistan
I: India

Table 6.3 1,312 Earthquakes in and near Afghanistan, A.D. 734–2002

Y	M	D	OT	N	E	R	n	M_s	r	h<40	log M_0	q	M_w	m_b	M	Location
734	0	0	0	31.6	60.5	3	0	6.5	5	0	25.69	2	6.43	0	0	Sistan-I
763	0	0	0	33.3	59.3	3	0	7.6	5	0	27.34	2	7.53	0	0	Qayin-I
805	12	2	0	29.5	60.5	3	0	7	5	0	26.44	2	6.93	0	0	Sistan-I
815	0	0	0	29.5	60.5	3	0	7	5	0	26.44	2	6.93	0	0	Sistan-I
819	6	0	0	36.4	65.4	3	0	7.4	5	0	27.04	2	7.33	0	0	Balkh
849	0	0	0	34.3	62.2	3	0	0	5	0	24.39	2	5.56	0	0	Herat
1066	5	0	0	33.9	59.2	3	0	6.5	5	0	25.69	2	6.43	0	0	Kuhistan-I
1102	2	28	1,800	34.4	62.2	3	0	0	5	0	24.39	2	5.56	0	0	Herat
1145	0	0	0	36.2	58.8	3	0	5.3	5	0	24.39	2	5.56	0	0	Nishapur-I
1209	0	0	1,200	36.4	58.7	3	0	7.6	5	0	27.34	2	7.53	0	0	Nishapur-I
1238	0	0	0	34.3	58.7	3	0	5.3	5	0	24.39	2	5.56	0	0	Gonabad-I
1251	0	0	0	36.2	58.8	3	0	5.3	5	0	24.39	2	5.56	0	0	Nishapur-I
1270	10	7	1,200	36.2	58.8	3	0	7.1	5	0	26.59	2	7.03	0	0	Nishapur-I
1336	10	21	600	34.7	59.7	3	0	7.6	5	0	27.34	2	7.53	0	0	Kwaf-I
1364	2	10	0	34.9	61.7	3	0	0	5	0	24.89	2	5.89	0	0	Herat
1389	2	0	0	36.2	58.8	3	0	7.6	5	0	27.34	2	7.53	0	0	Niishapur-I
1405	11	23	0	36.2	58.8	3	0	7.6	5	0	27.34	2	7.53	0	0	Nishapur-I
1410	0	0	0	36.7	66.8	3	0	0	0	0	0	0	0	0	0	Balkh
1428	0	0	0	35.8	64.2	3	0	6.5	5	0	25.69	2	6.43	0	0	Taliqan
1493	1	10	600	33	59.8	3	0	7	5	0	26.44	2	6.93	0	0	Muminabad-I
1505	7	6	0	34.5	69.1	3	0	7.3	5	0	26.89	2	7.23	0	0	Kabul
1519	1	3	0	35	71.5	3	0	0	5	0	0	0	0	0	0	Bajaur
1549	2	15	2,400	33.7	60	3	0	6.7	5	0	25.99	2	6.63	0	0	EQayin-I
1619	5	0	0	35.1	58.9	3	0	6.5	5	0	25.69	2	6.43	0	0	Dughabad-I
1673	7	30	0	36.3	59.3	3	0	6.6	5	0	25.84	2	6.53	0	0	Mashhad-I
1678	0	0	2,400	34.3	58.7	3	0	6.5	5	0	25.69	2	6.43	0	0	Gonabad-I
1687	4	0	0	36.3	59.6	3	0	0	0	0	0	0	0	0	0	Mashhad-I
1827	3	17	0	0	0	0	0	0	0	0	0	0	0	0	0	Kabul
1829	0	0	0	0	0	0	0	0	0	0	0	0	0	0	0	Kabul
1832	1	22	0	36.5	71	3	0	7.4	5	0	27.04	2	7.33	0	0	Badakhshan
1833	0	0	0	37.3	58.1	3	0	6.2	0	0	25.29	2	6.16	0	0	Shirvan-I
1833	4	19	0	0	0	0	0	0	0	0	0	0	0	0	0	Kabul
1836	0	0	0	0	0	0	0	0	0	0	0	0	0	0	0	Kabul
1837	12	14	0	0	0	0	0	0	0	0	0	0	0	0	0	Kabul
1838	0	0	0	29.6	59.9	3	0	7	0	0	26.44	2	6.93	0	0	Nasratabad-I
1838	1	7	0	35.4	66.8	3	0	0	0	0	0	0	0	0	0	Jurm
1840	1	26	0	0	0	0	0	0	0	0	0	0	0	0	0	Kabul
1842	2	19	1,140	35	71	3	0	7.5	5	0	27.19	2	7.43	0	0	Kunar
1851	6	0	0	36.8	58.4	3	0	6.9	5	0	26.29	2	6.83	0	0	Quchan-I
1852	1	24	345	29.3	68.6	3	0	6.7	5	0	25.99	2	6.63	0	0	Kahun
1857	6	0	1,230	31.6	65.7	3	0	0	0	0	0	0	0	0	0	Kandahar
1858	8	25	0	28.3	68.4	3	0	0	0	0	0	0	0	0	0	Jacobabad-P
1871	12	23	1,800	37.4	58.4	3	0	7.2	0	0	26.74	2	7.13	0	0	Quchan-I
1872	12	15	0	29.6	67.9	0	0	0	0	0	0	0	0	0	0	Sibi-P
1874	10	18	0	35.1	69.2	3	0	7	5	0	26.44	2	6.93	0	0	Kohistan
1880	2	8	0	33.9	70	3	0	0	0	0	0	0	0	0	0	Khurum
1888	12	28	0	0	0	0	0	0	0	0	0	0	0	0	0	Quetta-P
1889	4	0	0	0	0	0	0	0	0	0	0	0	0	0	0	Kabul

(continued)

Table 6.3 (continued)

Y	M	D	OT	N	E	*R*	*n*	M_s	*r*	*h*<40	log M_o	*q*	M_w	m_b	*M*	Location
1889	10	0	0	36.7	67.1	3	0	0	0	0	0	0	0	0	0	Mazar
1890	9	0	2,100	0	0	0	0	0	0	0	0	0	0	0	0	Kabul
1891	6	0	0	34.5	68.5	3	0	0	0	0	0	0	0	0	0	Paghman
1892	12	20	20	30.9	66.5	3	3	6.54	1	0	25.75	2	6.47	0	0	Chaman-P
1893	2	13	400	30.7	67.4	3	3	5.9	2	0	24.99	2	5.96	0	0	Chaman-P
1893	11	17	1,500	37	58.4	3	3	7.1	0	0	26.59	2	7.03	0	0	Quchan-I
1895	1	17	1,100	37.1	58.4	3	0	6.8	0	0	26.14	2	6.73	0	0	Quchan-I
1899	12	31	0	0	0	0	0	0	0	0	0	0	0	0	0	Kabul
1900	0	0	0	30.9	67.3	3	0	0	0	0	0	0	0	0	0	
1901	10	17	557	31	68.4	3	0	6.14	1	0	25.23	2	6.12	0	0	Loralai-P
1902	0	0	0	30.63	66.76	3	0	0	0	0	0	0	0	0	0	Pishin-P
1902	7	20	0	35.33	59.15	5	0	0	0	0	0	0	0	0	0	I
1902	9	20	632	38.5	67	0	0	6.2	4	40	25.29	2	6.16	0	0	
1903	1	10	1,317	35.3	59.5	5	0	4.3	5	0	23.39	2	4.89	0	0	I
1903	1	20	824	37	71	0	0	5.5	4	0	24.59	2	5.69	0	0	
1903	3	22	1,435	33.16	59.71	3	0	6.19	1	0	25.28	2	6.15	0	0	Durukhsh
1903	6	19	0	35.3	59.2	5	0	4	5	0	23.09	2	4.69	0	0	I
1903	6	22	2,148	35.3	59.2	5	0	0	0	0	0	0	0	0	0	I
1903	9	25	120	35.2	58.2	3	0	5.9	2	0	24.99	2	5.96	0	5.3	I
1903	9	30	1,900	35.2	58.5	5	0	0	0	0	0	0	0	0	0	I
1903	12	23	300	29.5	67.5	3	0	5.9	2	0	24.99	2	5.96	0	0	Bolan
1904	7	27	520	33	72	1	0	5.74	1	0	24.83	2	5.85	0	0	
1904	11	9	328	36.94	59.77	3	0	6.37	1	0	25.5	2	6.3	0	0	Tedzhen-Tu
1904	11	9	930	37.01	59.91	3	0	5.3	2	0	24.39	2	5.56	0	0	
1905	2	25	1,035	37	70.5	0	0	5.4	4	30	24.49	2	5.63	0	0	
1905	6	19	127	29.89	59.98	3	0	5.96	1	0	25.05	2	6	0	6.8	Nasradabad
1905	9	26	128	30.3	69.9	3	0	6.44	1	0	25.6	2	6.37	0	0	Sulaiman
1906	10	4	652	37.2	67.3	0	0	5.7	4	20	24.79	2	5.83	0	0	
1906	10	24	1,539	36.5	68	0	0	7.11	1	32	26.61	2	7.04	0	6.8	Ayvadzh
1907	4	13	710	36.5	70.5	0	0	5.3	2	30	24.39	2	5.56	0	0	
1907	9	14	1,919	37.59	61.85	3	0	4.6	2	0	23.69	2	5.09	0	0	I
1907	10	23	2,025	37.65	65.4	3	0	6.05	1	0	25.14	2	6.06	0	6.1	Kersk
1907	10	24	510	38	68.8	0	0	5.2	4	18	24.29	2	5.49	0	0	
1908	1	12	1,019	30.2	67.7	3	0	5.58	2	0	24.67	2	5.75	0	0	
1908	3	5	220	30.2	67.7	3	0	6.43	1	0	25.59	2	6.36	0	0	Harnai-P
1908	6	3	1,556	28	67	1	0	6.18	1	0	25.27	2	6.15	0	0	Quetta-P
1908	7	26	332	34.59	61.49	3	0	5.3	2	0	24.39	2	5.56	0	0	Herat
1909	9	7	1,528	33	70	1	0	5.99	1	0	25.08	2	6.02	0	0	
1909	10	20	2,341	28.9	68.3	3	0	7.06	1	0	26.53	2	6.99	0	0	Kachhi-P
1909	11	13	2,400	34.35	62.16	3	0	4	2	0	23.09	2	4.69	0	0	Herat
1910	8	17	1,158	28.4	67	4	0	6.33	1	0	25.44	2	6.26	0	0	Sind-P
1911	1	1	1,018	36.5	66.5	0	0	6.94	1	40	26.35	2	6.87	0	6.6	Mazar-i
1911	1	1	1,459	36.5	66	3	0	6.47	1	20	25.65	2	6.4	0	6.3	Shuburghan
1911	2	18	1,841	38	72.8	0	0	7.69	1	26	27.48	2	7.62	0	7.4	Serez
1912	2	24	1,436	37	59.5	0	0	5.3	2	0	24.39	2	5.56	0	0	
1912	8	23	1,402	35	71.5	3	0	6.35	1	0	25.47	2	6.28	0	0	Kunar
1912	11	1	1,900	29	67	3	0	5.45	1	0	24.54	2	5.66	0	0	Bagh-P
1913	3	27	913	29.5	67.5	3	16	5.63	1	0	24.72	2	5.78	0	0	Dhadar-P

Table 6.3 (continued)

Y	M	D	OT	N	E	R	n	M_s	r	$h<40$	$\log M_o$	q	M_w	m_b	M	Location
1913	5	14	850	34.5	69.2	3	0	5.07	1	0	24.16	2	5.41	0	0	Kabul
1913	8	7	0	30.38	67	3	0	0	0	0	0	0	0	0	0	Quetta-P
1913	9	9	1,843	37.1	68.1	3	0	5	2	18	24.09	2	5.36	0	0	Ayvadzh-T
1914	3	8	0	29.5	67.6	3	0	0	0	0	0	0	0	0	0	Dhadar-P
1914	5	4	1,800	30.4	67.7	3	0	0	0	0	0	0	0	0	0	Ziarat-P
1914	5	21	826	32	69.5	1	33	5.69	1	0	24.78	2	5.82	0	0	
1914	11	4	1,106	32	70	3	22	5.72	1	0	24.81	2	5.84	0	0	
1915	3	3	145	32	73	3	23	5.19	1	0	24.28	2	5.49	0	0	
1915	10	19	0	0	0	3	0	0	0	0	0	0	0	0	0	Kabul
1916	2	11	2,053	36	59.5	0	0	4.8	5	0	23.89	2	5.23	0	0	
1917	8	29	1,300	37.37	58.05	3	0	5.7	2	0	24.79	2	5.83	0	0	
1917	9	30	1,830	37.3	71.5	0	0	5	4	0	24.09	2	5.36	0	0	
1917	11	28	1,442	36.5	59.1	0	0	5.7	4	0	24.79	2	5.83	0	0	
1917	11	28	1,743	36.5	58.4	0	0	5.2	4	0	24.29	2	5.49	0	0	
1917	12	1	947	30	71	1	13	5.55	1	0	24.64	2	5.73	0	0	
1918	3	24	2,314	34.9	60.7	3	23	6.04	1	0	25.13	2	6.05	6.3	0	Torbat-I
1918	4	24	930	30.4	67.3	3	0	0	0	0	0	0	0	0	0	Quetta-P
1918	11	29	1,041	32.7	67.7	3	24	6.21	1	0	25.3	2	6.17	0	0	Ghazni
1919	5	23	610	32.5	68	3	35	6.31	1	0	25.41	2	6.24	0	6.48	Zarghun
1919	6	1	1,246	30	71	2	8	5.51	1	0	24.6	2	5.7	0	5.81	
1919	6	15	1,849	30	71	2	15	5.53	1	0	24.62	2	5.71	0	0	
1919	8	24	830	32.6	59.2	3	0	4.6	5	0	23.69	2	5.09	0	0	I
1920	2	27	351	35	69	1	25	5.92	1	0	25.01	2	5.97	0	0	
1920	4	26	1,000	29.8	67.9	3	0	0	0	0	0	0	0	0	0	Sibi-P
1920	12	15	1,245	30.1	67.9	3	0	0	0	0	0	0	0	0	0	Harnai-P
1922	8	23	900	37	59.8	3	0	4.3	5	0	23.39	2	4.89	0	0	I
1923	5	25	2,221	35.19	59.1	3	39	5.8	2	0	24.89	2	5.89	0	0	I
1923	7	16	1,323	37.5	70.5	1	14	5.5	2	15	24.59	2	5.69	0	5.6	
1923	9	14	810	28.97	59.33	4	14	5.61	2	0	24.7	2	5.77	0	0	
1923	10	1	816	29	67.5	4	41	6.09	1	0	25.18	2	6.09	0	0	Kachhi-P
1923	10	2	1,122	31.2	70.3	2	8	5.17	2	0	24.26	2	5.47	0	0	
1923	10	15	349	31.2	70.3	2	8	4.96	2	0	24.05	2	5.33	0	0	
1923	11	29	336	33.6	59.4	3	16	5.29	1	0	24.38	2	5.55	0	0	I
1924	9	12	9	33.2	71.4	1	5	4.98	2	0	24.07	2	5.35	0	0	
1924	9	12	859	33.2	71.4	2	6	5.06	2	0	24.15	2	5.4	0	0	
1925	5	2	257	35	58.1	5	5	4.7	5	0	23.79	2	5.16	0	0	I
1925	5	14	710	35.83	70.63	6	18	0	0	0	0	0	0	0	0	
1925	8	30	1,315	38	69.5	1	14	5.4	4	10	24.49	2	5.63	0	5.4	
1926	3	16	1,438	37.7	66.1	0	2	4.4	4	15	23.49	2	4.96	0	0	
1926	3	18	631	35	69	2	8	5.1	2	0	24.19	2	5.43	0	0	
1926	3	22	1,624	36	70	0	29	5.4	4	15	24.49	2	5.63	0	0	
1926	9	13	0	29.9	67.3	3	0	0	0	0	0	0	0	0	0	Mach-P
1927	1	30	854	30.18	70.32	6	23	5.49	1	33	24.58	2	5.69	0	0	
1927	3	25	1,814	37.7	58.5	0	2	4.5	4	0	23.59	2	5.03	0	0	
1927	3	29	1,805	37.7	58.5	5	7	4.3	1	0	23.39	2	4.89	0	0	I
1927	4	24	1,120	36.4	70.5	2	12	5.1	2	15	24.19	2	5.43	0	5.1	
1927	5	4	0	30.6	67	3	0	0	0	0	0	0	0	0	0	Pishin-P

(continued)

Table 6.3 (continued)

Y	M	D	OT	N	E	R	n	M_s	r	h<40	log M_o	q	M_w	m_b	M	Location
1927	5	21	804	29.5	68	3	16	5.01	1	0	24.1	2	5.37	0	0	Shahpur-P
1927	6	27	812	34	73	1	7	0	0	0	0	0	0	0	0	
1927	6	29	2,136	38	61.6	0	2	5	4	0	24.09	2	5.36	0	0	
1927	7	11	1,612	36	67	0	5	4.92	2	12	24.01	2	5.31	0	5.2	
1927	7	15	346	35	69	2	29	5.26	1	40	24.35	2	5.53	0	5.4	
1927	9	5	2,010	34.2	72	1	11	5.02	2	0	24.11	2	5.37	0	0	
1927	9	30	1,840	34.2	72	2	11	5.48	2	0	24.57	2	5.68	0	5.7	
1927	10	2	307	35	69	2	8	5.18	2	40	24.27	2	5.48	0	0	
1927	10	7	2,134	36	69.5	2	21	5.4	1	15	24.49	2	5.63	0	5.2	
1928	1	21	1,504	36.5	70.5	2	9	0	0	0	0	0	0	0	0	
1928	3	8	1,814	31.5	60.1	3	26	5.5	1	0	24.59	2	5.69	0	0	I
1928	5	1	1,558	37.9	58.4	5	7	4.8	5	0	23.89	2	5.23	0	0	I
1928	5	2	1,403	34.2	72	1	5	4.87	2	0	23.96	2	5.27	0	0	
1928	8	19	354	30	64	2	15	5.2	2	0	24.29	2	5.49	0	0	
1928	8	21	1,902	32.32	58.69	3	25	5.55	1	0	24.64	2	5.73	0	5.3	
1928	9	1	609	28.77	69.57	6	88	6.45	1	33	25.62	2	6.38	0	0	?
1928	10	15	1,419	28.5	67.4	4	106	6.7	1	33	25.99	2	6.63	0	0	Katra-P
1928	12	12	125	29.68	69.12	1	13	5.29	2	17	24.38	2	5.55	0	0	
1928	12	14	28	28.81	68.08	1	35	5.64	1	33	24.73	2	5.79	0	0	
1929	3	26	1,400	28	62	2	16	5.1	2	0	24.19	2	5.43	0	0	
1929	5	1	2,242	37.3	58.2	0	11	4.8	4	0	23.89	2	5.23	0	0	
1929	5	3	1,620	37.7	58.2	0	32	5.1	4	0	24.19	2	5.43	0	0	
1929	5	4	631	37.6	58.2	0	21	4.9	4	0	23.99	2	5.29	0	0	
1929	6	4	704	37.3	66.5	0	33	5.7	2	14	24.79	2	5.83	0	5.7	
1929	7	13	736	37.24	58.22	3	64	6.07	1	0	25.16	2	6.07	0	0	Quchan
1930	3	12	0	29.8	69.5	3	0	0	0	0	0	0	0	0	0	Loralai-P
1930	9	23	1,015	37.03	70.05	1	12	0	0	33	0	0	0	0	0	Khozdar-P
1930	10	10	0	29.8	66.9	3	0	0	0	0	0	0	0	0	0	Mastung-P
1931	6	2	1,736	32.9	69.3	0	24	5.48	1	0	24.57	2	5.68	0	0	
1931	6	9	29	32.9	69.3	2	11	5.05	1	0	24.14	2	5.39	0	0	
1931	8	7	1,049	37.93	58.37	3	14	5.1	2	15	24.19	2	5.43	0	5.2	I
1931	8	8	854	37	59	0	32	5.3	4	12	24.39	2	5.56	0	0	
1931	8	24	2,135	30.1	67.6	3	136	6.79	1	40	26.13	2	6.72	0	0	Sharigh-P
1931	8	25	31	30.2	67.7	1	38	5.27	1	33	24.36	2	5.54	0	0	
1931	8	25	306	31.49	66.52	2	18	5.44	1	0	24.53	2	5.65	0	0	
1931	8	25	1,545	30.16	67.79	2	10	4.95	1	0	24.04	2	5.33	0	0	
1931	8	25	1,853	30.42	67.82	6	10	5.18	1	33	24.27	2	5.48	0	0	
1931	8	27	1,527	29.9	67.2	4	155	7.31	1	33	26.91	2	7.24	0	0	Mach-P
1931	8	28	42	29.25	68.42	6	32	5.17	1	40	24.26	2	5.47	0	0	
1931	8	28	318	30.2	67.7	2	12	5.01	1	0	24.1	2	5.37	0	0	
1931	8	28	1,940	28.82	67.39	6	22	5.55	1	33	24.64	2	5.73	0	0	
1931	9	3	1,711	29.8	67.3	2	12	5.13	1	0	24.22	2	5.45	0	0	
1931	9	6	1,433	29.72	67.85	6	18	5.43	1	33	24.52	2	5.65	0	0	
1931	9	8	1,612	29.8	67.3	2	10	4.94	2	0	24.03	2	5.32	0	0	
1931	9	10	2,227	34.4	62.2	3	11	5	2	0	24.09	2	5.36	0	0	Herat
1931	10	3	1,035	29.8	67.3	2	15	5.44	2	0	24.53	2	5.65	0	0	
1931	11	4	1,521	36.8	69.5	2	13	0	0	33	0	0	0	0	0	

Table 6.3 (continued)

Y	M	D	OT	N	E	R	n	M_s	r	h<40	log M_o	q	M_w	m_b	M	Location
1932	9	8	725	31.59	58.15	3	45	5.56	1	0	24.65	2	5.73	0	0	I
1933	8	11	1,349	31.2	70.3	2	4	0	0	0	0	0	0	0	0	
1933	10	16	435	32.7	66.5	3	45	5.57	1	0	24.66	2	5.74	0	0	Uruzgan
1933	10	26	15	31.6	59.9	5	0	3.9	5	0	22.99	2	4.63	0	0	I
1933	12	2	215	36.34	69.6	0	15	5.24	2	15	24.33	2	5.52	0	5.3	
1933	12	9	752	37.01	69.26	6	21	5.42	2	33	24.51	2	5.64	0	5.4	
1934	3	30	0	35.8	64.8	3	0	0	0	0	0	0	0	0	0	Maimana
1934	9	26	7	37.1	69.9	0	24	5	4	15	24.09	2	5.36	0	5	
1934	11	15	2,314	36.6	71.4	0	31	5	4	0	24.09	2	5.36	0	0	
1935	5	12	520	37.1	71.37	6	34	5.3	2	33	24.39	2	5.56	0	0	
1935	5	15	201	28.4	67.5	3	68	5.97	1	33	25.06	2	6.01	0	0	Kotra-P
1935	5	16	1,724	37.14	68.88	6	50	0	0	33	0	0	0	0	0	
1935	5	30	2,133	28.87	66.4	6	231	7.74	1	33	27.55	2	7.67	0	0	Quetta-P
1935	5	31	1,712	30	67.19	6	35	5.46	1	35	24.55	2	5.67	0	0	
1935	6	2	916	30.14	66.93	6	104	6.04	1	13	25.13	2	6.05	0	6	Quetta-P
1935	7	5	1,752	37.98	67.3	6	128	0	0	8	0	0	0	0	0	
1935	10	4	1,451	29.26	66.34	6	28	5.4	2	33	24.49	2	5.63	0	0	
1935	10	28	1,205	31.3	69.3	0	52	5.59	1	33	24.68	2	5.75	0	0	
1936	1	17	740	30	66.8	3	0	0	0	0	0	0	0	0	0	Quetta-P
1936	5	7	153	36.53	69.45	6	24	5.35	2	33	24.44	2	5.59	0	5.4	
1936	6	30	1,926	33.6	60	3	78	5.97	1	0	25.06	2	6.01	6.2	0	Bamrud
1936	7	22	856	28.55	68.11	6	24	5.18	2	33	24.27	2	5.48	0	0	
1936	9	7	852	29	61	2	16	0	0	0	0	0	0	0	0	
1937	6	29	1,804	37.3	70	0	10	0	0	33	0	0	0	0	0	
1937	11	11	1,711	37.5	72.2	0	−1	5.5	4	0	24.59	2	5.69	0	0	
1937	11	13	1,150	38	69.5	0	31	5.6	4	10	24.69	2	5.76	0	5.6	
1937	11	16	1,349	35.46	70.34	6	14	0	0	33	0	0	0	0	0	
1938	7	19	1,945	34.92	59.35	3	21	5.2	2	0	24.29	2	5.49	0	4.9	
1938	8	5	1,417	36.8	71	0	23	5.22	2	15	24.31	2	5.51	0	5.2	
1938	8	8	0	31.6	66.5	3	0	0	0	0	0	0	0	0	0	Arghestan
1938	11	1	47	35.87	71.33	6	14	0	0	33	0	0	0	0	0	
1938	12	1	1,809	30.43	68.53	6	0	5.43	1	0	24.52	2	5.65	0	0	
1938	12	19	1,856	36.64	58.5	3	0	5.6	2	0	24.69	2	5.76	0	0	Khorasan
1939	6	6	124	36.96	70.14	6	22	0	0	0	0	0	0	0	0	
1939	6	19	42	37.02	71.27	6	32	5.18	2	33	24.27	2	5.48	0	5.3	
1939	8	28	2,135	36.14	58.58	3	17	5.2	2	0	24.29	2	5.49	0	0	
1939	9	6	1,127	36.6	66.5	0	35	5.38	2	30	24.47	2	5.61	0	5.2	
1939	10	10	2,042	36.3	69	0	18	5	4	15	24.09	2	5.36	0	5	
1939	11	8	1,721	36.32	58	4	34	5.49	1	0	24.58	2	5.69	0	0	
1940	2	2	1,953	37	70.5	2	12	0	0	33	0	0	0	0	0	
1940	5	4	1,704	35.76	58.53	3	19	4.9	2	0	23.99	2	5.29	0	0	I
1940	5	4	2,101	35.76	58.53	3	106	6.46	1	0	25.63	2	6.39	6.2	0	Khorasan
1940	7	17	636	36.66	72.21	6	35	5.64	2	33	24.73	2	5.79	0	5.7	
1940	7	17	1,144	36.8	70.7	0	43	5.6	4	15	24.69	2	5.76	0	5.6	
1940	8	1	1,945	38	72.5	0	28	5.5	4	20	24.59	2	5.69	0	0	
1940	9	1	1,846	36.1	69	0	29	5.4	4	15	24.49	2	5.63	0	5.4	
1940	10	5	1,444	37.2	69	0	15	5.2	2	0	24.29	2	5.49	0	5.8	

(continued)

Table 6.3 (continued)

Y	M	D	OT	N	E	*R*	*n*	M_s	*r*	*h*<40	log M_o	*q*	M_w	m_b	*M*	Location
1941	2	16	1,639	33.41	58.87	3	74	6.21	1	0	25.3	2	6.17	6.4	0	Muhammadabad
1941	5	16	846	36.88	70.71	6	16	0	0	33	0	0	0	0	0	
1941	9	29	232	30.19	67.09	6	26	5.47	1	33	24.56	2	5.67	0	0	Spin-Baldak
1941	10	29	743	28.3	63.5	1	22	5.3	2	0	24.39	2	5.56	0	0	
1942	5	23	2,001	29.56	68.62	6	23	5.35	1	33	24.44	2	5.59	0	0	
1943	11	27	844	31.95	70.09	6	13	5.3	2	33	24.39	2	5.56	0	0	
1943	11	27	954	31.8	70.2	2	10	5.09	1	0	24.18	2	5.42	0	0	
1945	10	1	516	28.95	67.28	6	49	5.89	1	33	24.98	2	5.95	0	0	Nagau
1946	2	10	1,313	32.52	59.32	3	13	5.2	2	0	24.29	2	5.49	0	0	I
1946	6	20	37	29.5	66	1	31	5.78	1	0	24.87	2	5.88	0	0	
1947	6	30	336	36.3	59.6	3	7	4.1	5	0	23.19	2	4.76	0	0	I
1947	9	20	1,856	36.2	69.2	0	27	5.2	4	15	24.29	2	5.49	0	5.2	
1947	9	23	1,228	33.7	58.7	3	126	6.89	1	0	26.28	2	6.82	6.4	0	Dustabad
1947	9	23	2,223	33.7	58.7	5	8	4.8	2	0	23.89	2	5.23	0	0	I
1947	9	24	450	33.7	58.7	5	4	3.9	4	0	22.99	2	4.63	0	0	I
1947	9	25	1,825	33.7	58.7	5	8	4.7	4	0	23.79	2	5.16	0	0	I
1947	9	26	304	33.8	58.56	3	57	6.03	1	0	25.12	2	6.05	0	0	Ferdows
1947	9	26	1,216	36.2	58.9	5	2	4.5	2	0	23.59	2	5.03	0	0	I
1947	9	27	816	33.56	58.92	3	12	5.1	2	0	24.19	2	5.43	0	0	I
1947	10	6	1,518	33.54	58.85	3	33	5.4	2	0	24.49	2	5.63	0	0	I
1947	11	8	1,626	36.9	68.25	4	30	5.18	2	40	24.27	2	5.48	0	5	I
1947	11	13	326	33.52	58.78	3	32	5.2	2	0	24.29	2	5.49	0	0	I
1948	1	28	1,551	36.36	67.73	6	80	6.54	1	33	25.75	2	6.47	6.3	6.9	Samangan
1948	3	14	2,118	33.08	58.9	3	31	5.1	2	0	24.19	2	5.43	0	0	
1948	4	9	1,459	29.88	68.78	6	22	5.19	2	33	24.28	2	5.49	0	0	
1948	6	18	1,844	37.53	58	3	54	5.54	1	0	24.63	2	5.72	0	0	
1948	9	8	1,118	34.02	59.34	3	11	4.9	2	0	23.99	2	5.29	0	0	
1948	10	5	2,012	37.95	58.32	0	175	7.24	1	0	26.8	2	7.17	0	7.3	Ashkhabad
1948	10	6	124	37.4	58.8	0	60	5.27	1	6	24.36	2	5.54	0	0	
1948	10	17	616	37.36	58.46	3	12	4.5	2	0	23.59	2	5.03	0	0	
1948	10	24	2,306	37.7	58.5	0	8	4.3	4	0	23.39	2	4.89	0	0	
1948	10	29	221	37.85	58.5	0	8	4.8	4	0	23.89	2	5.23	0	0	
1948	11	2	1,524	38.2	58.5	0	18	4.6	4	0	23.69	2	5.09	0	0	
1948	11	4	1,642	37.6	58.6	0	3	4.4	2	0	23.49	2	4.96	0	0	
1948	11	6	1,425	37.6	58	0	−1	4.3	4	0	23.39	2	4.89	0	0	
1948	11	30	2,240	37.8	58.2	0	2	4.3	2	0	23.39	2	4.89	0	0	
1948	12	18	1,302	37.7	58.4	0	1	4.3	4	0	23.39	2	4.89	0	0	
1948	12	19	458	37.6	58.8	0	5	4.8	4	0	23.89	2	5.23	0	0	
1948	12	24	1,442	37.6	58.4	3	2	4.6	4	0	23.69	2	5.09	0	0	
1949	1	11	2,007	37.8	58	0	8	4.6	4	0	23.69	2	5.09	0	0	
1949	2	24	2,302	30.09	69	6	55	5.85	1	33	24.94	2	5.93	0	0	
1949	4	23	555	36.2	58.8	0	0	4	4	0	23.09	2	4.69	0	0	I
1949	7	2	227	30.38	69.17	6	15	0	0	33	0	0	0	0	4.6	
1949	12	14	343	36	59	0	x	5	4	0	24.09	2	5.36	0	0	
1949	12	18	1,913	35.83	69.25	2	14	0	0	33	0	0	0	0	0	
1949	12	19	458	37.6	58.6	0	x	4.8	4	0	23.89	2	5.23	0	0	
1950	2	19	352	30.1	68.8	2	11	4.94	1	0	24.03	2	5.32	0	4.4	

Table 6.3 (continued)

Y	M	D	OT	N	E	*R*	*n*	M_s	*r*	*h*<40	log M_o	*q*	M_w	m_b	*M*	Location
1950	3	12	206	30.12	70.42	4	30	0	0	0	0	0	0	0	0	
1950	3	23	24	29.98	68.3	6	23	4.78	1	33	23.87	2	5.21	0	0	
1950	4	6	243	37.92	58.37	3	58	5.3	2	0	24.39	2	5.56	0	0	I
1950	5	2	1,643	37.93	58.37	3	38	5.3	2	0	24.39	2	5.56	0	0	
1950	5	9	1,116	37.93	58.37	3	120	6.39	1	0	25.53	2	6.32	6.1	0	I
1950	8	8	529	37.9	58.4	3	29	4.9	2	0	23.99	2	5.29	0	0	I
1950	9	24	2,256	34.5	60.7	3	91	5.55	1	0	24.64	2	5.73	5.6	0	Herat
1951	3	14	2,302	37.4	58.5	5	9	4.6	2	0	23.69	2	5.09	0	0	I
1951	5	14	407	30.15	70.13	6	76	5.7	2	33	24.79	2	5.83	0	0	
1951	10	29	2,116	30.31	68.43	6	22	5.09	2	33	24.18	2	5.42	0	0	
1951	11	1	1,849	37.04	69.49	6	19	0	0	33	0	0	0	0	0	
1951	11	1	2,019	37.02	69.58	6	24	0	0	33	0	0	0	0	0	
1952	1	7	2,143	37.56	69.96	N	13	0	0	0	0	0	0	0	0	
1952	1	30	2,333	37.6	68.5	0	x	4.3	4	0	23.39	2	4.89	0	0	
1952	2	2	109	37.22	69.23	6	14	0	0	33	0	0	0	0	0	
1952	2	4	1,145	37.1	72.5	1	19	0	0	33	0	0	0	0	0	
1952	2	6	937	36.7	70.5	2	16	0	0	33	0	0	0	0	0	
1952	3	30	129	29.32	67.7	6	16	0	0	0	0	0	0	0	4.4	
1952	4	26	2,013	37.93	69.04	N	13	0	0	33	0	0	0	0	0	
1952	6	13	515	37.38	72.98	6	16	0	0	33	0	0	0	0	0	
1952	8	2	2,258	37.17	71.58	N	−1	0	0	14	0	0	0	0	0	
1952	8	13	314	37.44	69.99	6	40	4.92	2	10	24.01	2	5.31	0	4.9	
1952	8	13	1,406	36.6	70.4	N	14	0	0	0	0	0	0	0	0	
1952	9	15	431	37.85	58.3	3	−1	5.1	1	0	24.19	2	5.43	0	0	
1952	9	15	1,128	30.58	70.01	6	50	5.53	1	33	24.62	2	5.71	5.8	0	
1952	9	19	308	37.93	72.37	6	14	0	0	0	0	0	0	0	0	
1952	10	10	1,847	30.37	69.34	3	147	6.17	1	2	25.26	2	6.14	0	0	Sulaiman-P
1952	10	22	2,108	37.63	69.66	6	13	0	0	0	0	0	0	0	0	
1952	10	26	823	29.34	67.87	6	41	5.07	2	7	24.16	2	5.41	6.4	0	
1952	10	29	320	37.95	69.15	6	13	0	0	0	0	0	0	0	0	
1952	11	7	224	36.5	71	2	19	0	0	33	0	0	0	0	0	
1952	12	4	452	36.15	71.74	N	13	0	0	0	0	0	0	0	0	
1952	12	5	1,748	36.86	71.38	6	30	0	0	33	0	0	0	0	0	
1952	12	17	1,208	36.6	70.1	2	15	0	0	33	0	0	0	0	0	
1952	12	25	137	37.14	70.61	6	15	0	0	0	0	0	0	0	0	
1952	12	25	2,222	29.25	70	4	124	5.96	1	33	25.05	2	6	0	6	Sulaiman-P
1953	5	1	2,118	33.54	72.67	6	54	5.14	1	39	24.23	2	5.45	0	0	
1953	6	6	2	35.3	59.2	0	−1	4.71	2	0	23.8	2	5.17	0	0	
1953	6	25	2,122	37	58	0	−1	4.39	2	0	23.48	2	4.95	0	0	
1953	7	12	53	36.5	71	0	−1	5	4	0	24.09	2	5.36	0	0	
1954	1	23	1,606	37.4	72.5	0	67	5.8	4	9	24.89	2	5.89	0	5.8	
1954	1	23	1,711	37.4	72.5	0	66	5.2	4	15	24.29	2	5.49	0	5.2	
1954	7	23	1,445	34.04	69.56	6	59	5.38	2	9	24.47	2	5.61	0	0	
1954	10	13	2,211	37.4	69.3	0	52	5.02	2	11	24.11	2	5.37	0	5.1	
1954	10	25	1,319	37.62	58.42	3	−1	4.9	4	0	23.99	2	5.29	0	0	
1955	2	18	2,248	30.2	67	3	85	5.71	1	33	24.8	2	5.83	0	0	Chaman-P
1955	6	18	2,021	34	58.5	0	−1	4.81	2	0	23.9	2	5.23	0	0	

(continued)

Table 6.3 (continued)

Y	M	D	OT	N	E	R	n	M_s	r	h<40	log M_o	q	M_w	m_b	M	Location
1955	7	28	306	36.7	66.7	0	−1	5	2	40	24.09	2	5.36	0	5	
1956	1	11	2,216	29.99	69.55	6	−1	5.5	4	24	24.59	2	5.69	0	0	
1956	5	13	750	30	69.9	3	90	5.94	1	33	25.03	2	5.99	0	0	Sulaiman-P
1956	6	8	407	35.17	67.49	6	92	5.43	1	19	24.52	2	5.65	0	5.5	
1956	6	9	2,313	35.13	67.48	6	334	7.39	1	33	27.03	2	7.32	0	7.2	Bamyan
1956	6	9	2,353	34.99	67.59	6	19	0	0	33	0	0	0	0	0	
1956	6	10	101	34.9	67.66	6	−1	5.55	1	20	24.64	2	5.73	0	5.5	
1956	6	10	2,343	34.11	68.21	6	−1	0	0	33	0	0	0	0	0	
1956	6	25	1,252	31.89	60.38	N	23	5.02	2	0	24.11	2	5.37	5.7	0	
1956	8	8	2,302	31.68	67.16	N	−1	5.17	1	29	24.26	2	5.47	0	0	
1956	9	16	837	33.8	69.6	3	250	6.68	1	11	25.96	2	6.61	0	0	Jaji
1956	9	24	1,020	33.98	69.57	6	114	5.58	1	1	24.67	2	5.75	0	0	Kabul
1956	10	2	826	34	66	5	−1	0	0	0	0	0	0	0	0	
1956	10	13	821	36.3	71.2	0	−1	5.1	4	0	24.19	2	5.43	0	0	
1957	2	17	723	37.6	65.3	3	−1	4.7	2	0	23.79	2	5.16	0	0	
1957	4	4	1,136	36.1	69.9	0	−1	5.3	4	15	24.39	2	5.56	0	5.3	
1957	4	25	46	37.3	58.4	0	−1	4.7	4	0	23.79	2	5.16	0	0	
1957	6	10	446	31.23	69.57	6	−1	4.5	2	35	23.59	2	5.03	0	0	
1957	6	14	1,136	31.71	67.1	4	50	5	4	10	24.09	2	5.36	0	0	
1957	8	7	1,528	34.33	70.52	N	−1	0	0	0	0	0	0	0	0	
1957	11	26	41	37.22	72.22	6	−1	0	0	33	0	0	0	0	0	
1957	12	13	907	36.75	70	8	−1	0	0	33	0	0	0	0	0	
1957	12	18	950	37.51	70.68	8	−1	0	0	33	0	0	0	0	0	
1958	1	28	1,714	35.65	58.38	N	−1	4.84	2	0	23.93	2	5.25	0	0	
1958	2	27	355	35.3	58.9	0	−1	4.8	4	0	23.89	2	5.23	0	0	
1958	3	22	1,107	35.31	67.44	6	120	5.9	1	20	24.99	2	5.96	0	5.9	
1958	4	8	959	33.04	68.21	6	80	5.17	2	27	24.26	2	5.47	0	0	
1958	5	7	1,328	30.98	69.61	5	−1	0	0	0	0	0	0	0	0	
1958	6	13	733	37.2	66.5	0	−1	5.28	1	20	24.37	2	5.55	0	5.3	
1958	8	13	1,732	29.42	67.69	6	−1	4.96	2	32	24.05	2	5.33	0	0	
1958	9	16	1,422	34.37	59.52	N	35	5.01	1	0	24.1	2	5.37	0	0	
1958	12	8	1,405	29.06	69.64	6	−1	0	0	33	0	0	0	0	0	
1959	2	22	853	31.44	69.78	6	−1	0	0	26	0	0	0	0	0	
1959	3	8	20	36.2	58	0	−1	4.4	4	10	23.49	2	4.96	0	0	
1959	3	25	603	30.24	69.89	6	−1	0	0	33	0	0	0	0	0	
1959	5	19	1,517	33.04	68.16	6	119	5.46	1	37	24.55	2	5.67	0	0	
1959	5	26	635	37	69.9	0	124	5.3	4	30	24.39	2	5.56	0	5.3	
1959	7	12	807	36.2	60	0	−1	4.2	4	0	23.29	2	4.83	0	0	
1959	8	23	319	29.07	69.55	6	29	4.85	2	40	23.94	2	5.26	0	0	
1959	8	30	2,256	36.5	68.28	6	85	5.3	4	1	24.39	2	5.56	0	5.3	
1959	10	14	729	33.5	59.5	0	−1	4.5	4	0	23.59	2	5.03	0	0	
1959	10	23	1,654	33.34	58.84	1	29	4.5	4	0	23.59	2	5.03	5	0	
1959	11	4	954	32.5	66	5	−1	0	0	0	0	0	0	0	0	
1959	12	7	1,903	29.92	66.79	6	−1	0	0	31	0	0	0	0	0	
1959	12	15	1,047	36.3	69.8	0	72	5.1	2	20	24.19	2	5.43	0	5	
1960	1	18	2,201	29.68	67.66	6	−1	0	0	33	0	0	0	0	0	
1960	6	2	722	33.21	60.41	6	83	5.16	2	13	24.25	2	5.47	0	5.2	

Table 6.3 (continued)

Y	M	D	OT	N	E	R	n	M_s	r	$h<40$	$\log M_o$	q	M_w	m_b	M	Location
1960	7	7	2,237	36.49	68.21	N	−1	0	0	37	0	0	0	0	0	
1960	7	17	514	36.48	68.25	6	96	5.28	3	25	24.37	2	5.55	0	5.3	
1960	7	29	1,433	31.78	67.26	6	91	5.2	2	6	24.29	2	5.49	0	0	
1960	8	14	2,237	36.3	69.9	0	90	5.5	4	15	24.59	2	5.69	0	5	
1960	8	16	1,001	37.95	58.45	0	x	4	4	9	23.09	2	4.69	0	0	
1960	8	23	858	29.02	60	N	78	5.28	2	40	24.37	2	5.55	5.5	0	
1960	10	3	49	29.76	68.25	6	74	0	0	25	0	0	0	0	0	
1960	10	3	49	29.76	68.25	6	74	5.4	2	25	24.49	2	5.63	0	0	
1960	11	19	1,312	30.18	69.71	6	−1	0	0	33	0	0	0	0	0	
1961	5	1	137	35.24	58.9	N	−1	4.51	2	0	23.6	2	5.03	0	0	
1961	7	18	129	37.71	58.6	3	−1	4.2	4	0	23.29	2	4.83	0	0	
1961	10	16	1,812	35.2	58.45	3	−1	4.7	4	9	23.79	2	5.16	0	0	
1961	11	2	659	33.99	69.52	6	−1	0	0	33	0	0	0	0	5.5	
1962	4	1	45	33.21	58.87	3	111	5.49	1	0	24.58	2	5.69	5.5	0	Birjand
1962	4	21	107	36.68	69	N	−1	0	0	17	0	0	0	0	0	
1962	5	9	1,212	36.6	68.5	0	−1	5	4	15	24.09	2	5.36	0	5	
1962	6	4	1,521	30.06	69.97	6	−1	4.2	2	5	23.29	2	4.83	0	4.2	
1962	7	11	104	31.88	67.32	6	93	0	0	29	0	0	0	0	0	
1962	7	21	603	32.03	67.01	N	−1	0	0	0	0	0	0	0	0	
1962	9	12	2,057	36.4	69	4	147	6.04	1	35	25.13	2	6.05	0	6.3	
1962	9	20	1,310	35.5	58.6	0	−1	4.5	4	0	23.59	2	5.03	0	0	
1962	10	5	2,000	35.09	58.72	N	90	5.03	2	0	24.12	2	5.38	0	0	
1962	10	29	157	35.98	70.48	N	−1	0	0	0	0	0	0	0	0	
1962	11	1	1,346	37.84	70.08	0	−1	5	4	7	24.09	2	5.36	0	5	
1962	11	1	1,526	37.78	70.14	N	−1	0	0	40	0	0	0	0	0	
1962	11	2	1,322	34.91	71.85	8	−1	0	0	33	0	0	0	0	4	
1963	1	1	1,927	35.23	58.84	3	−1	4.55	2	0	23.64	2	5.06	0	0	Kashmar
1963	3	1	319	35.88	59.85	N	−1	4.51	2	0	23.6	2	5.03	4.8	0	
1963	4	1	923	35.4	69.6	0	−1	5.3	4	0	24.39	2	5.56	0	0	
1963	6	9	745	35	58	0	−1	4	4	0	23.09	2	4.69	0	0	
1963	6	11	325	37.08	70.06	6	128	4.95	1	21	24.04	2	5.33	0	5.1	
1963	7	22	745	30.03	68.11	N	−1	0	0	0	0	0	0	0	0	
1963	8	17	2,158	36.82	59.76	N	−1	4.62	2	3	23.71	2	5.11	4.8	0	
1963	9	19	1,631	31.05	66.7	6	−1	4.5	2	30	23.59	2	5.03	0	4.5	
1963	10	31	2,010	37.53	69.93	N	−1	0	0	36	0	0	0	0	0	
1964	1	13	1,834	32	65	1	3	0	0	0	0	0	0	0	0	
1964	2	17	1,558	36.99	68.71	N	19	0	0	40	0	0	0	4.7	0	
1964	2	21	104	34.25	58.35	3	35	4.32	2	0	23.41	2	4.91	4.8	0	
1964	5	13	950	37.8	58.3	0	x	4	4	0	23.09	2	4.69	0	0	
1964	5	15	2,231	36.3	58.85	1	28	4.5	4	16	23.59	2	5.03	0	0	
1964	6	18	418	29.8	71.7	1	5	0	0	0	0	0	0	0	0	
1964	9	24	1,002	31	67.2	1	3	0	0	0	0	0	0	0	0	
1964	10	4	701	28.29	69.32	1	42	5.43	1	39	24.52	2	5.65	4.8	0	
1964	10	28	1,935	36.58	71.29	N	62	0	0	39	0	0	0	5.3	0	
1964	11	1	317	34.3	63	5	−1	0	0	0	0	0	0	0	0	
1964	11	3	605	31.56	66.44	6	45	0	0	35	0	0	0	5	0	
1964	11	3	614	32.71	59.22	3	10	4.2	4	0	23.29	2	4.83	0	0	

(continued)

Table 6.3 (continued)

Y	M	D	OT	N	E	R	n	M_s	r	h<40	log M_o	q	M_w	m_b	M	Location
1965	2	2	1,556	37.42	73	1	152	6.04	1	15	25.13	2	6.05	5.3	6	
1965	2	26	137	35.19	58.46	3	63	5.1	2	0	24.19	2	5.43	5.1	0	
1965	4	2	2,226	36.8	66.8	7	85	5.3	2	15	24.39	2	5.56	5	5.3	
1965	4	24	2,002	35.87	65.38	N	27	4.95	2	37	24.04	2	5.33	0	0	
1965	6	14	459	37.34	68.83	6	11	0	0	33	0	0	0	0	0	
1965	8	3	702	36.33	69.75	N	42	0	0	28	0	0	0	4.3	0	
1965	12	3	2,117	36.29	69.33	7	143	5.32	2	19	24.41	2	5.57	5.3	0	
1966	1	24	215	32.63	67.44	1	80	0	0	17	0	0	0	4.9	0	
1966	1	24	1,532	29.93	69.69	7	86	5.3	0	19	24.39	2	5.56	5	0	
1966	2	7	521	30.11	69.91	7	106	0	0	13	0	0	0	5.2	0	
1966	2	7	530	29.95	69.82	7	104	0	0	23	0	0	0	5.2	0	
1966	2	7	657	30.1	70.08	1	26	0	0	40	0	0	0	4.8	0	
1966	2	7	838	30.1	69.89	1	44	0	0	6	0	0	0	4.7	0	
1966	2	8	3	30.5	70	1	7	0	0	34	0	0	0	0	0	
1966	2	8	11	30	69.94	1	12	0	0	25	0	0	0	4	0	
1966	2	8	25	29.91	69.92	1	10	0	0	15	0	0	0	0	0	
1966	2	8	553	30.15	70.04	1	21	0	0	14	0	0	0	0	0	
1966	2	8	1,208	30.1	69.6	1	6	0	0	0	0	0	0	0	0	
1966	2	9	822	29.95	69.83	1	66	0	0	16	0	0	0	5	0	
1966	2	9	2,226	30.3	69.95	1	12	4.24	1	35	23.33	2	4.85	4.4	0	
1966	2	13	1,909	29.95	69.67	1	115	5.25	1	9	24.34	2	5.53	4.9	0	
1966	2	14	541	29.94	69.69	1	52	0	0	31	0	0	0	4.7	0	
1966	2	16	944	29.88	69.75	1	11	0	0	34	0	0	0	4.5	0	
1966	2	17	1,826	30.13	69.88	1	26	0	0	40	0	0	0	4.5	0	
1966	2	19	943	37.4	70.45	1	19	0	0	39	0	0	0	0	0	
1966	3	4	601	30.01	69.94	1	40	0	0	2	0	0	0	4.5	0	
1966	3	15	914	30	69.91	1	21	0	0	19	0	0	0	4.4	0	
1966	3	16	1,740	30.06	69.82	1	23	0	0	34	0	0	0	4.7	0	
1966	4	15	2,031	31.95	68.74	1	16	0	0	39	0	0	0	0	0	
1966	4	25	1,811	30.06	69.93	1	29	4.82	1	6	23.91	2	5.24	4.5	0	
1966	5	13	2,312	29.97	69.95	1	30	0	0	26	0	0	0	4.3	0	
1966	5	22	1,223	32.1	68.7	1	9	0	0	35	0	0	0	0	0	
1966	5	24	435	31.1	70	1	7	0	0	26	0	0	0	4.7	0	
1966	7	24	507	30.24	70.06	8	26	0	0	6	0	0	0	4.3	0	
1966	8	1	2,135	29.73	69.2	1	26	0	0	33	0	0	0	4.8	0	
1966	8	1	2,230	29.93	68.77	7	106	0	0	20	0	0	0	5.1	0	
1966	8	2	541	29.97	68.68	1	66	0	0	2	0	0	0	5	0	
1966	8	2	919	29.91	69.18	1	78	0	0	37	0	0	0	5	0	
1966	8	4	2,229	29.88	68.64	1	40	0	0	15	0	0	0	4.8	0	

Y, M, D, OT correspond to year, month, day, and origin time in hours and minutes
E and *N* correspond to longitude and latitude in degrees, respectively
R Epicentral locations were adopted from the following sources:

0. Locations in northern Afghanistan taken from reports of the Soviet network
1. Epicenters computed by BAAS/ISS/ISC
2. Rough macroseismic locations or epicenters adopted by BAAS/ISS with no calculation, calculated from the Prague formula with station corrections
3. Macroseismic epicenters from well-defined macroseismic observation
4. Relocated positions using ISC procedures
5. For early earthquakes, locations calculated by Ambraseys and Melville for western Afghanistan

Table 6.3 (continued)

6. Quittmeyer and Jacob (1979) relocated events in the time period 1914 through 1965 using a computer program similar to the one described by Bolt (1960)
7. From Engdahl et al. (1998). We adopted only recomputed locations that have a well determined depth (DEQ in their columns 2–4)
8. Few locations were taken from BCIS for the period 1953 through 1965, during which ISS did not report small events
9. In the absence of other locations we accepted some data marked LEQ and FEQ from Engdahl et al. (1998)
10. Nowroozi recalculated epicentral locations of earthquakes in the region between 1950 and 1965, using the program by Sykes and Landisman (1964) Nowroozi (1971)
11. PDE locations

n Number of seismographic stations used by ISS/ISC to calculate epicentral position
p Actual number of stations greater than that given by ISS/ISC
M_s Surface wave magnitude
r 1: calculated from the Prague formula with station corrections
2: calculated from the Prague formula from few stations with station corrections
3: ISC
4: Moscow
5: calculated from macroseismic data
h Focal depth from sources in *R*
$log\ (M_0)$ dyn cm
q 0: M_0 not available
1: CMT Harvard
2: converted from M_s from equations described in text
3: from other sources (Bernard et al. 2000)
M_w Moment magnitude
m_b (or m_B) Body wave magnitude from ISC
M Surface wave magnitude from ISC
Location P: Pakistan
I: India

presents a list of 1,313 earthquakes from earliest times to the present. The temporal density of known historic earthquakes as a function of time is illustrated in Fig. 6.5.

Discussion

Most pre-nineteenth century earthquakes were reported along trade routes (Fig. 6.6) that have remained unchanged since pre-Hellenistic times, governed as they are by the physiography of Afghanistan and the availability of fresh water. This, and the propensity for established and persistent settlements to develop along and at the confluence of trade routes, brings with it the danger that the surviving historic record provides a somewhat biased geographic view of Afghanistan's seismicity. Yet the absence of recent earthquakes in the large desert depressions of Afghanistan suggests that these regions are empty of both roads and earthquakes. The one exception to this observation is the cluster of deep subduction earthquakes beneath the sparsely populated Chagai Mountains along the southern border of Afghanistan with Baluchistan.

Deformation along the eastern and western margins of Afghanistan is highlighted by seismicity throughout historical and recent time. Afghanistan appears as a promontory of the Eurasian plate penetrating southward between Iran and India, with relatively minor seismic evidence for significant convergence in the northern Afghan mountains, as suggested by recent GPS data from southwest of Farah. The total cumulative seismic moment release in the entire catalog (1.6×10^{28} dyn cm) is roughly 50 % higher in eastern Afghanistan than in western Afghanistan (1.1×10^{28} dyn cm). If the Hindu Kush and Kopeh Dag regions north of 35°N are ignored (thereby removing the contribution from thrust and deep earthquakes from the

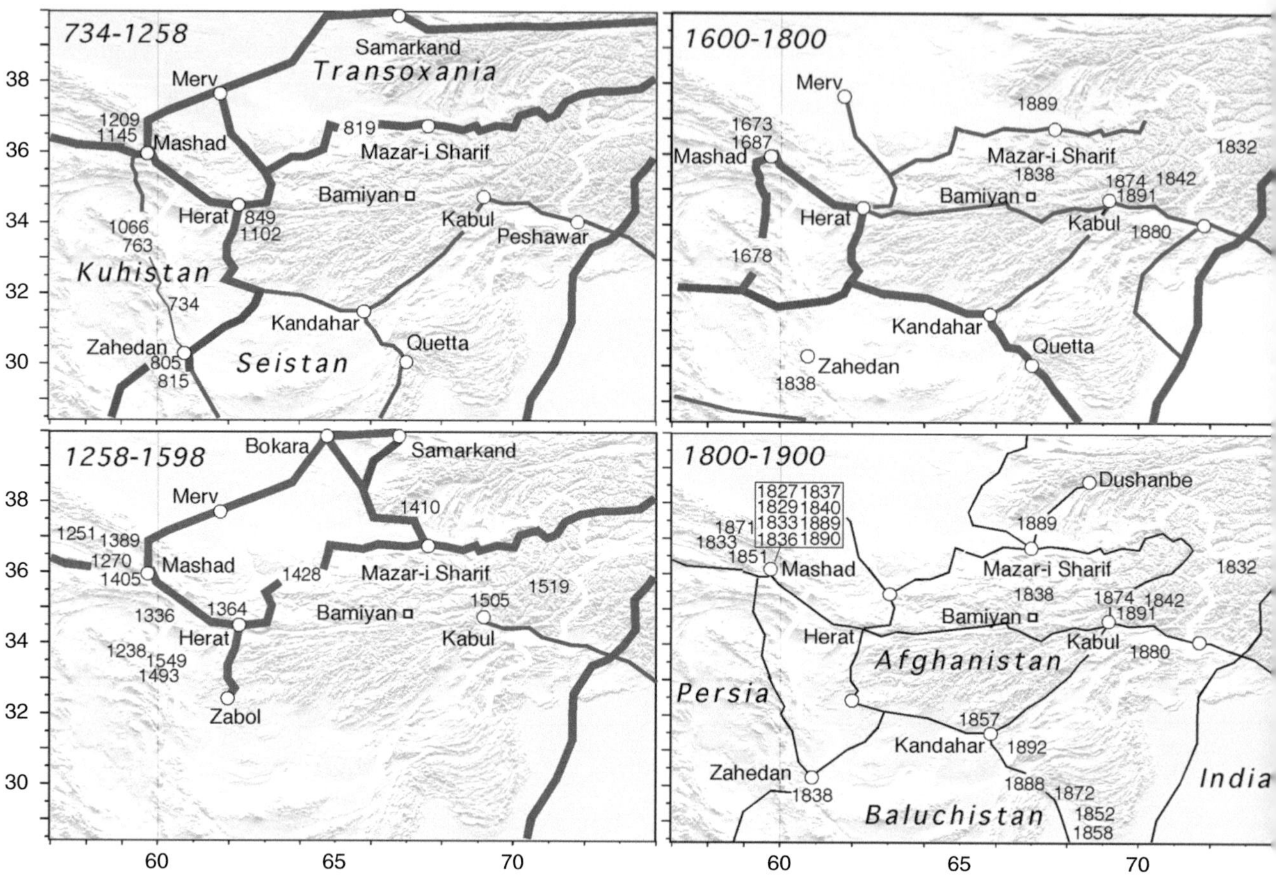

Fig. 6.6 Evolution of Afghan trade routes in the past 1,300 years showing the dates of historical earthquakes discussed in the text. Pre-1800 major routes are indicated by thick lines, minor routes by thin lines

summation), the rates are comparable (≈8×10^{27} dyn cm). Despite this similarity, the seismic record is considered too incomplete to estimate a meaningful slip rate on Afghanistan's eastern and western borders.

The historic record in eastern Afghanistan within the Chaman fault system on Afghanistan's eastern boundary with the Indian plate is much sparser than in western Afghanistan, with merely three significant earthquakes in the period 800–1,800. Moreover, different plate boundary processes prevail.

An example of strain partitioning occurred in the 1930s (Ambraseys and Bilham 2003a, b). The Mach 1931 ($M_s = 7.3$) earthquake absorbed approximately 1 m of east–west convergence, followed in 1935 by the Quetta ($M_s = 7.7$) earthquake with more than 3 m of inferred sinistral slip. Distributed seismicity in the 400-km-wide Sulaiman lobe of western Pakistan represents significant shallow thrusting in this region, which also does not contribute directly to plate boundary slip. A scalar summation of moment release along this eastern boundary thus results in a misleading estimate of plate boundary slip rates along Afghanistan's eastern boundary.

Silent both in the historic record and in the record of recent seismicity is a 300- to 400-km segment of the Chaman fault system on the

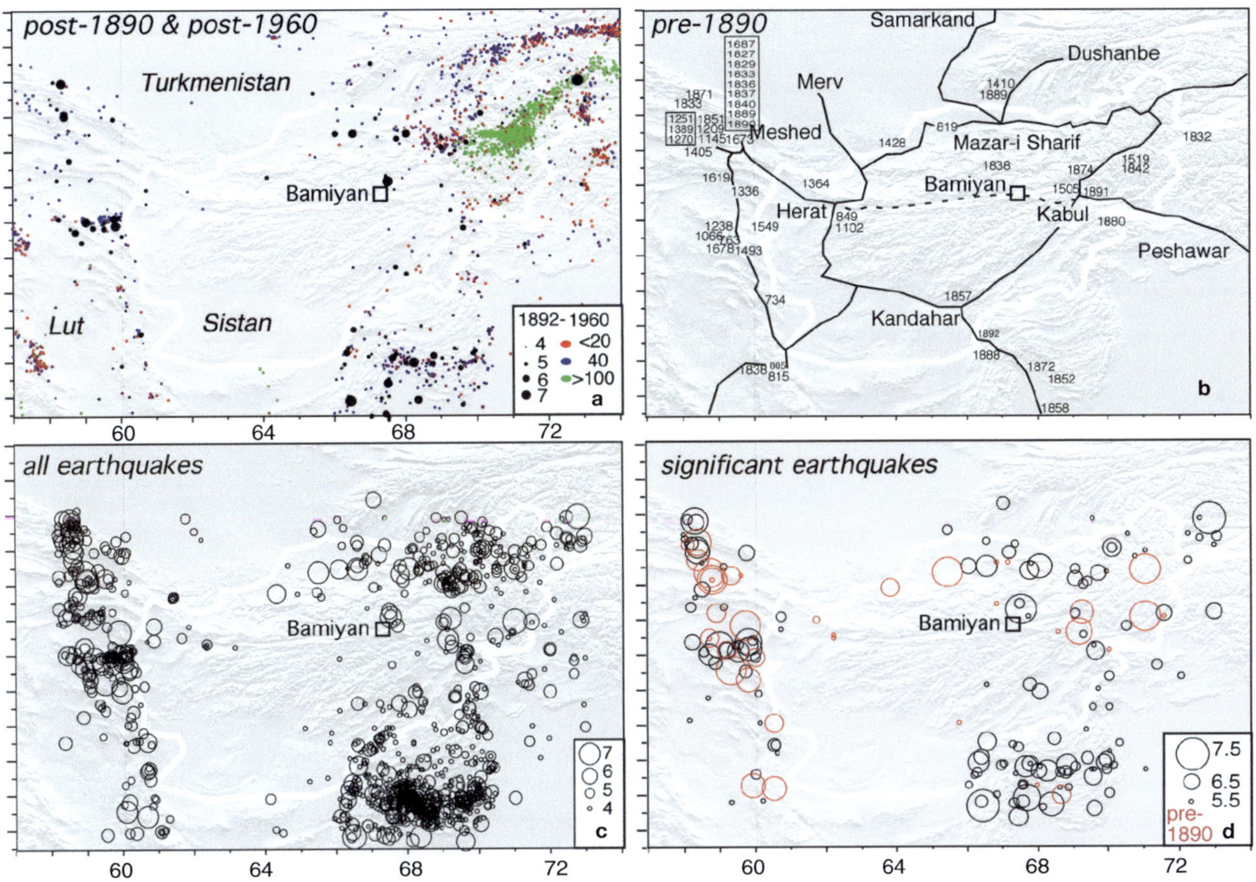

Fig. 6.7 Seismicity of Afghanistan. (**a**) Historical earthquakes 73–1890 from Table 6.1. (**b**) Significant events since 1890 from Table 6.2. (**c**) The entire catalog (Tables 6.1, 6.2, and 6.3). (**d**) Historical and recent significant earthquakes from (a, b)

Afghan/Baluchistan border between 31° and 33.5°N. The absence of historical earthquakes here, with the exception of the 1892 $M=6.8$ Chaman earthquake at the southern end of this segment, may represent a gap in the historical record, but it is unlikely that events since 1890 are missing.

The locations of the 1857 Kandahar, and the 1505 and 1891 Kabul earthquakes appear to be too far from the plate boundary to represent activity on the Chaman fault. The plate boundary velocity at this location is estimated to be 2–4 cm/year suggesting that $M>7$ events could occur at less than 200-year intervals.

The historical record reveals damaging events along the northern mountains fronting the Turam depression and in Turkmenistan. This suggests that some convergence is occurring between Asia and the Sistan block. The absence of recent earthquakes in the western half of this region and their occurrence intermittently in the past 1200 years suggests that future earthquakes may not be unexpected in this region (Figs. 6.6 and 6.7). Recent earthquakes along this zone have occurred with decreasing frequency and magnitude toward the west suggesting that convergence between the Sistan block and Asia may result in minor counterclockwise rotation of the Sistan block

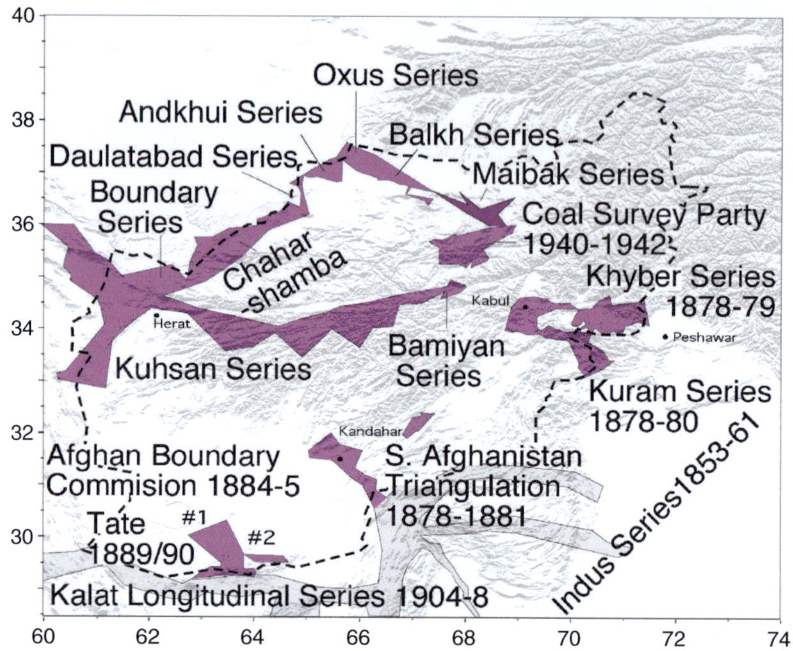

Fig. 6.8 Violet-shaded regions outline pre-1950 triangulation surveys first measured in 1884–1886 except where indicated (Survey of India 1948). To our knowledge none of these surveys, with the exception of points overlapping the south Afghan border region (Burrard 1909), have been remeasured. The re-measurement of the 1940 Survey, and the 1884 Kuhsan and Bamiyan Series would be of value in quantifying deformation during the 1956 Bamiyan earthquake

relative to Asia. This would have the effect of reducing both the left-lateral slip rate and the convergence rate on the Chaman transform fault system.

In 1956 a crustal $M_w = 7.4$ earthquake occurred near and north of Bamiyan. The trend of maximum damage strikes to the north. A $M_w = 7.4$ earthquake is typically associated with 1–2 m of slip on a fault rupture 50–100 km long, but there is no mapped fault between the Herat and Andarabad faults on which this could have occurred, and there are no reports of surface rupture on any of the faults in the region (see Appendix).

The Herat fault is a suture between earlier continental margins, similar to the Indus-Tsangpo suture north of the Himalaya. Like that suture, the western Herat fault may be a tectonically inert lithological boundary. Few earthquakes occurred there during the medieval period in Herat, and the fault west of Bamiyan lies in a region where no teleseimically recorded earthquakes have been reported. However, the apparent absence of large earthquakes on the Herat fault for the past 1,200 years should not be taken to imply that these events cannot occur. Currently the most seismicity-free segments of the San Andreas system are those two segments that sustained the largest ruptures: the 1906 San Francisco and the 1857 Fort Tejon earthquakes.

In principle, additional insight into tectonic activity near the Bamiyan region is to be obtained from the resurvey of triangulation series first measured more than 120 years ago in the region (Fig. 6.8). Accurate survey measurements along the Herat Fault from Herat to beyond the Bamiyan region were undertaken in 1884–1886 in the measurement of the Kuhsan and Bamiyan Triangulation Series (Survey of India 1948). These surveys were typically accurate to 10 ppm in scale and 7 μ radian in angle. Early triangulation surveys in the south, east, and west were linked to Survey of India survey networks in Baluchistan, Iran, and India (Burrard 1909; Ambraseys and Bilham 2003a, b). To our knowledge none of these surveys have been remeasured.

Conclusions

The historical record in Afghanistan for the past 1,200 years delineates a tectonic view of the country ploughing southward into the Arabian and Indian plates as a promontory of the Eurasian plate. Both its western margin with Iran and its eastern boundary with Baluchistan have a long history of damaging earthquakes, with the exception of an ≈300-km segment of the Chaman fault system. This segment represents a real seismic gap in significant earthquakes in the past century, and a possible gap in historical knowledge in earlier centuries, because no evidence for creep on the faults in the region between 31°N and 33°N has been reported. We note that a large earthquake on the northern Chaman fault would result in significant damage in the Kabul region.

Deep earthquakes are associated with subduction of the Arabian plate beneath the Makran coast along its southern border, and with a descending slab beneath NE Afghanistan. Because of incomplete spatial reporting, these deep earthquakes are difficult to distinguish in the historical record from shallow earthquakes that occur at similar times. This may have the effect of biasing upward the estimated magnitude of an earthquake sequence if unwittingly assessed as a single event. A pair of earthquakes in March 2002 provide a recent example of such a sequence: the March 4 $M=7.2$ earthquake at greater than 50 km depth causing widespread low-intensity damage, followed on March 25–27 by a sequence of shallow $M_s \leq 6$ earthquakes with severe surface intensities.

The greater part of the interior of Afghanistan is seismically inactive, but the more heavily populated north and east experience significant seismicity. In particular, northeastern Afghanistan, near and north of the capital, Kabul, has a long history of damaging deep and shallow earthquakes. Historical earthquakes along the north-facing frontal ranges in western Afghanistan have no recent instrumental counterparts, suggesting that infrequent future damaging earthquakes will recur in this region.

Bamiyan lies 230 km to the east-northeast of Kabul in a transition region between the aseismic interior of Afghanistan and the eastern plate boundary. Although a $M_w=7.4$ earthquake occurred near Bamiyan on June 9, 1956 there is no historical evidence for previous earthquakes in this region. The strike of the 1956 earthquake appears to have been to the north or northeast, and may have occurred on a conjugate fault linking the Herat and Andarabad fault. Historical geodesy may provide additional insight into the tectonics of the Bamiyan region, because several surveys were measured in 1880–1940. Their remeasurement may reveal surface deformation associated with the tectonics of the region. However, as yet none of these early surveys have been re-measured.

Acknowledgments We thank James Jackson for discussions concerning the active tectonics of the region and for his critical reading of the manuscript, and Ruben Tatevosian and George Hallows for helping with the retrieval of information from the Russian and Pakistani press, respectively. RB was supported by National Science Foundation grant number NSF EAR000399.

Appendix: Narrative Descriptions of Earthquakes in Afghanistan 819–2000

819 Jun (7.4) The earliest earthquake in Afghanistan for which we have information occurred in Dhu'l-Hijja 20 aH = June 819, in the region between modern Meymaneh, Andkhvoy, and Mazar-i Sharif (Abu'l Fida: ii.26, Ibn al-Shihna: vii.59. Ibn al-Athir). It affected a large area in which many houses were destroyed, with heavy casualties severely affecting other urban centers many tens of kilometers apart, including Faryab (36.42N, 64.91), Taliqan (35.78N, 64.16E), and others that are not named in the districts of Juzjan in the west and Tukharistan in the east of these places. The shock destroyed one-fourth of the city of Balkh (36.75N, 66.90E) and ruined the masjid-i jami there (Ibn al Jauzi quoted by al-Suyuti:24).

As a result of the earthquake, the desert of Sidreh (36.75N, 66.22E), which lies between Shaburghan (36.67N, 65.74E) and Balkh, was flooded by an excessive rise of the water table, which turned the country into a fertile area in some places. Some of this flooding seems to have been permanent, resulting in the creation of new oases (Qudama: 210; al-Masudi: pp. 43–45). The shock was felt at Marv (37.56N, 63.32E) and probably also at Amul (39.11N, 63.57E) and Transoxania (Ibn al-Athir: vi.252). Aftershocks lasted for a month or so. That this earthquake is mentioned by several sources suggests its importance, and from the radius of the felt area we assess a magnitude of $M_s \approx 7.4$.

Damage in both Transoxania and the towns of Faryab and Taliqan to the west is considered unlikely, and it is possible that Ma vara'al-nahr (Transoxania) mentioned in some early sources should stand for Marv al-Rud, or Lesser Marv; that is, Bala Murghab, which is confirmed by Qudama, who says that the regions of Marv and Tukharistan were affected, and probably also Amol (Gorshkov 1947). An earthquake of this magnitude would certainly have been felt across the Oxus, but probably without damage.

849 A series of earthquakes was reported from heart. Information from other areas is lacking. The first damaging shock in Herat in 234 aH = 849 caused some houses to collapse [Ibn al-Jauzi 38v; al Soyuti 26]. It is most likely the reference is to Herat in Afghanistan, but it should be borne in mind that there was also a town of this name in Fars, near Istakhr (Yaqut iv.959).

1102 Feb 28 This second Herat earthquake destroyed a number of houses and other buildings, including heavy damage to the masjid-i jami, with some casualties in the city. According to Isfizari (ii.55), on the night of Friday 8 (Jumada I, 495), there was a calamity (nazileh) in heart. The western side of the mosque and most of the northern and southern sides were destroyed. A rather different version specifies that a strong shock, with a north–south motion, caused the collapse of many buildings in Herat, the masjid-i jami suffering heavy damage. The closeness of nazileh to zalzaleh (earthquake), and the apparent support of this reading in a variant text, confirms that an earthquake was responsible for the damage (Fahmi in Barbier 1860).

1364 Feb 19 On February 19, 1364 (6 Jumada I, 765), a third destructive earthquake occurred in Herat (Hafiz-i Abru: xxxvii; Hafiz-i Abru, Manjmu: 48; Fasih: iii.96). Most of the buildings in the city were ruined, particularly the tall structures. The shock caused the battlements to fall from the ramparts and several meters fell from the top of the Falak al-Din minaret. The masjid-i jami in the city was again damaged; the main arch collapsed, although its two supporting pillars remained intact.

The information about this earthquake suggests that the shock originated some distance from Herat, possibly in the Gulran district, with damage in Herat caused by long-period ground movements.

It is important to note that in spite of the prominence on Herat during the Middle Ages, no information regarding serious damage has been found in the voluminous source material.

1410 Earthquakes were felt in Balkh (36.75N, 66.90E) and Bukhara (39.77N, 64.42E). Landslides in the mountains dammed streams to form a deep lake at a place that is not given (al-Umari: fol.153v). It is not possible to assess the effects of the earthquake on the cities mentioned. Balkh and Bukhara were both brought under the authority of Shah Rukh this year, but there is no reference to the earthquake in any of the Persian and Timourid sources, so al-Umari's authority remains obscure. Damage to Balkh cannot have been serious or one would expect it to have been more widely described in historic materials. The earthquake is comparable to the event of A.D. 819 in the extent of the area affected and the number of shocks experienced. The mention of mountains in this region suggests an epicentral region east of Bukhara (Melville 1978).

1428 All that is known about this event is that in the year 831 aH = 1,428, there was an earthquake in Taliqan, with shocks lasting 10 days, in which many people were killed (al-Umari: fol.159v). Even the location can be questioned, because Taliqan of Qazvin or Talikan of Merv may have been shaken, both areas being almost equally seismically active. Earthquakes in Merv are more likely to have been recorded than those further west at this period, and because al-Umari's source may have been the same as for the 1,410 event, a tentative location of this event in northern Afghanistan may be preferred (Melville 1978).

1505 Jul 6 (7.3) For this earthquake (3rd Safar 911 aH, or July 6, 1505) we have an eyewitness account. In Kabul (34.53N, 69.13E) the shocks ruined the ramparts of the fort, and even the walls of gardens. Paghman (34.58N, 68.95E) was particularly badly affected, all houses there being destroyed and 70 or 80 people were killed, with numerous casualties in nearby towns and villages. Most of the houses in Tipa (Tibah, 34.68N, 69.01E) were leveled to the ground.

Between Istarghach (Istarghij, 34.91N, 69.07E) and the plain (maidan) for about 6 or 8 farsakhs (31–42 km), in some places the ground rose as high as an elephant, in others, sank as deep. It is not clear from the text whether maidan here refers to the plain or the town of Maidan (shahr), which is at the southern end of the Paghman range, west–southwest of Kabul. Villages and groves slipped from their place and many rising grounds were leveled and dust rose from the tops of the mountains. Between Paghaman and Begtut, the valley just north of Paghaman, there was a landslide in which water springs emerged to the surface. Thirty-three aftershocks are reported on the day of the earthquake with two or three a day for the next month.

A greatly abbreviated notice of this earthquake in Kabul is given also by al-Asafi, who puts the event in 912 aH (1506) and says that destruction was general among citadels (qal'at) and houses in which many people perished al-Asafi (p. 934).

At the time of the earthquake, Babur was outside Kabul, preparing for his campaign against Kandahar. It took him about a month of hard work to repair the fort (Bala Hissar) at Kabul (Babur: fols. 157r-158r, pp. 247–248). We have no information of the effects of the earthquake from other places and it is not known how far the shock was felt. Babur's memoirs do not mention any damage or repairs in other places through which he traveled during his campaign against Kandahar.

Baird-Smith (1843b) assumes that the effects of the shock extended hundreds of kilometers away to Kandahar, Ghazni, and Jalalabad, for which we could find no evidence, whereas other modern writers confuse the word *qal'at* mentioned by al-Asafi with the town of *Qalat* northeast of Kandahar. The available information clearly suggests that the earthquake was associated with at least a 40-km-long surface rupture of the Paghman fault, 20 km northwest of Kabul, which struck at N20°E. Much of the destruction was reported from Istargij, Tibeh, to Paghman, localities that lie along the fault. The vertical offset on the fault appears to have been approximately 3 m. An unknown amount of strike-slip faulting also may have occurred.

Some authors have conflated this earthquake with the M>8.2 earthquake in Kumaon and western Nepal a month later that damaged monasteries in Tibet and resulted in much damage in Agra and other northern Indian cities.

1519 Jan 3 There is little macroseismic information about this earthquake in northeast Afghanistan. It is mentioned in the sources that describe Babur's campaigns to Swat. They say that the earthquake occurred on Monday, 1st Muharram 925 (Jan 3, 1519) in the direction of the region adjoining Bajaur (c. 34.9N, 71.4E), and that it lasted half an hour (sic). After the earthquake Babur conquered the citadel of Bajaur (Abu 'l-Fadl, al-Asafi, sub.ann.).

The region meant here should be along the middle course of the river Kunar, northeast of Jalalabad (34.43N, 70.45E) near Arandu (35.32–71.56). The survival of this notice in the sources suggests that the earthquake, which occurred in this sparsely populated and mountainous area, was a relatively large event.

Wilson (1930) commits a double error with this earthquake. He reads Bujnurd for Bajaur, places the event in Iran, and confuses this earthquake with that of Kabul in 1505. The quality of the data allows only an estimate of the general location of the event, which may have occurred on the Kunar fault, but it does not permit an accurate assessment of its presumably sizable magnitude.

1832 Jan 22 (7.4) This was a large earthquake with an apparent epicentral area in the district of Badakhshan in northeastern Afghanistan. It occurred at 11 p.m. local time on January 22, 1832, and together with its damaging aftershock of February 21 destroyed most of the villages in the district, allegedly killing thousands of people.

In Kalafgan (36.77–69.93) all forts and houses were destroyed and many lost their lives. In Jorm (36.84–70.7) 35 km east of Kalafgan, houses collapsed and 12 people out of 25 were killed. From a total population of 310 in three nearby villages, 156 were killed. In the valley of Kowkcheh (36.60–70.85), the shock triggered numerous rockfalls, and in the neighboring valley of Varaduj (36.68–71.13), a fort and houses were destroyed and a whole mountainside fell into the valley damming the river for 8 days before the dam was breached. In the Sargulam valley, 72 out of a population of 155 perished.

The earthquake was felt strongly in Kabul (34.53–69.14), caused some panic in Lahore (31.56–74.35), and was reported from Srinagar (34.08N, 74.8E), Kokand (40.52–70.95), and Bukhara (39.78–64.43) (J. Asiat. Soc. Bengal, vol. 1, pp. 34, 146, 1832; Burnes 1835 i.17, ii.203; Masson 1844; Wood 1872; Trans. Geol. Soc. London, vol. 3, p. 492).

The earthquake was felt over an area of 450 km radius, an indication that, on the authority of Musketof and Orloff (1893, 242) and Kondorskaya and Shebalin (1977, 202), is interpreted as a subcrustal event of 180 km focal depth, consistent with depths of recent large events in the region. They assign to it an epicentral intensity of IX and a magnitude of m_b of 7.4. The earthquake was followed by a long series of aftershocks, one of which (Feb 21) caused rockfalls that blocked valleys in Badakhshan, adding to the damage, a rather unusual characteristic of deep earthquakes.

1842 Feb 19 (7.5) This earthquake in northeast Afghanistan occurred during the first Afghan War on February 19, 1842. A famous account of the earthquake and its aftershocks is related by Lady Sale (Plate 6.1), who was held hostage with her colleagues during the several months following the earthquake (Sale 1843). On January 5, British forces had surrendered Kabul and began their retreat to Jalalabad, 120 km west of Kabul, which was still garrisoned by the British, who had hastily restored the dilapidated defenses damaged by the earthquake. Much of what has been written about this earthquake is based on contemporary accounts and press reports (Englishman 1842; Eyre 1843; Gleig 1846; Sale 1846) and on less useful later works and earthquake catalogs (Stenz 1945).

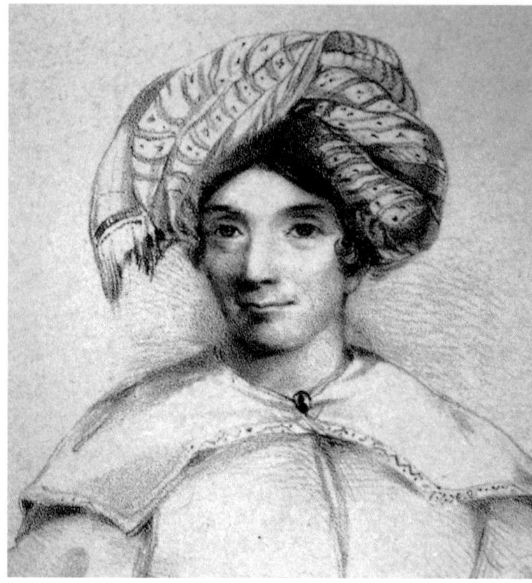

Plate 6.1 Lady Sale sporting a makeshift Afghan turban, sketched by fellow captive Capt. Eyre, while both were held hostage shortly after the outbreak of the Second Afghan War. Their journals, and those of field officers nearby, detail the main shock damage and aftershocks of the 1842 Kunar earthquake

The earthquake occurred at 11 h·40 m local time. At Jalalabad (34.43N, 70.45E) situated on the right bank of the Kabul river, ground movements were very strong; not a man could keep his legs; everyone was prostrated, and nausea affected all. The defenses of Jalalabad itself, which had just been repaired, were seriously affected. A substantial (2.4-km) length of the newly built parapets were damaged without loss of life. Several breaches were made in the old adobe bastions and in the curtains in the Peshawar (east) face. The Kabul gate was reduced to a shapeless mass of ruins. However, in a matter of days the damage was repaired (IO 1843; Sale 1846; Abbott 1879).

Within the walls a third of the local houses were destroyed, the collapse of tall houses choked the streets, leaving no room for escape. Some people were injured, but few of the inhabitants were killed. In the British garrison only four men were killed, and many were injured, although none seriously. On the whole, loss of life was small compared with the nature and extent of

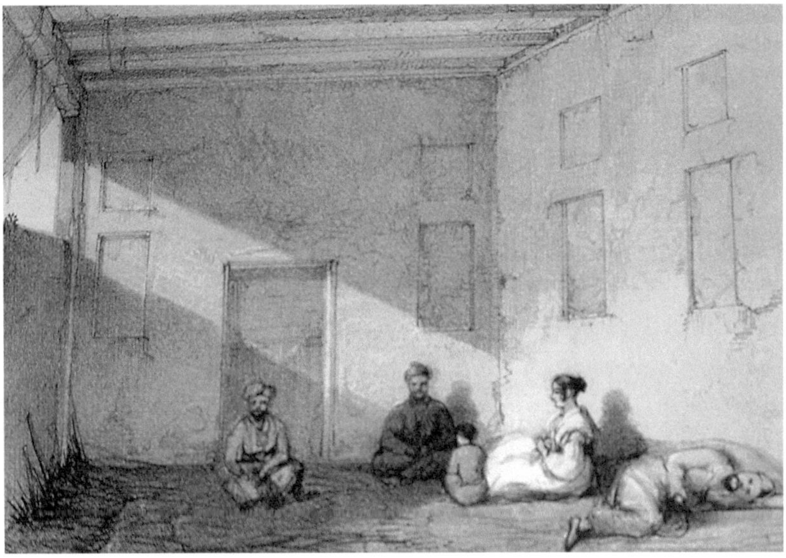

Plate 6.2 Lady Sale and companions in captivity drawn by Captain Eyre. Lady Sale and her companions were forced to march from town to town during the 1842 aftershock sequence, numerous shocks of which were recorded by Lady Sale in her published diary (Sale 1843)

damage. A report written after the earthquake contains a complete set of sketches indicating the extent of damage to the walls of Jalalabad (IO 1942; Heuckroth and Karim 1970). Outside of the walls, presumably along the river face, the ground opened in several places and water appeared on the surface. According to Baird-Smith (1842–1844), the waters of the Kabul river were twice thrown from their bed (Plate 6.2).

North of Jalalabad the shock caused considerable damage to settlements on a portion of the Suffid Kuh range of mountains (Baird-Smith 1843b). In the regions of Laghan and Kunar, villages were ruined, killing dozens of children and women. The large settlements of Chaharbag and Tigri in the Alingar valley suffered severely, scarcely a house being left standing, and several hundreds of people were killed (Eyre 1843).

Further to the north, damage was considerable in the fort of Budeeabad, where several English prisoners (including Eyre and Lady Sale) were confined. The motion of the ground was so severe that people could only maintain their balance with difficulty. Walls, gateways, and corner towers, were all much shaken or thrown down. Almost all the houses in the fort were damaged and a few collapsed, but all hostages entirely escaped injury. Eyewitness accounts indicated that the fort of Budeeabad suffered less than the other 40 forts in the valley. In one fort a tower fell, killing five people; in others not a wall remained (Sale 1843).

Along the valley from Buddeabad to Tigri, none of the forts escaped damage, with few inhabitable, and mostly masses of ruins (Eyre 1843). In the Kunar valley, where the shock triggered landslides and rockfalls from the hills, the forts of Shewa and Pashat were totally destroyed (IO 1842) (Plate 6.3).

Outside this region, damage was far less severe but widespread. To the west, some parts of the fort of Tezeen is said to have been destroyed (Heuckroth and Karim 1970), whereas at Kabul the shock caused little more than general panic. The walls of the European ward were badly shaken and came down a few days later (Baird-Smith 1843b). There is no evidence that Kabul and its fort needed any repairs after it was re-taken by the British forces on Sep 15, 1842 (IO 1842). Nor are there damage reports from Argandeh, Bala Maidan, and Jalriz, west of Kabul, where the shock caused some concern, but details are lacking (Sale 1843) (Plate 6.4).

It appears that the effects of the earthquake extended more to the east of the Laghan-Kunar region. At the camp of Kawulsur, 13 km from Peshawar, the shock was violent. Many of the camels that were carrying the baggage of the troops were thrown down, people were obliged to

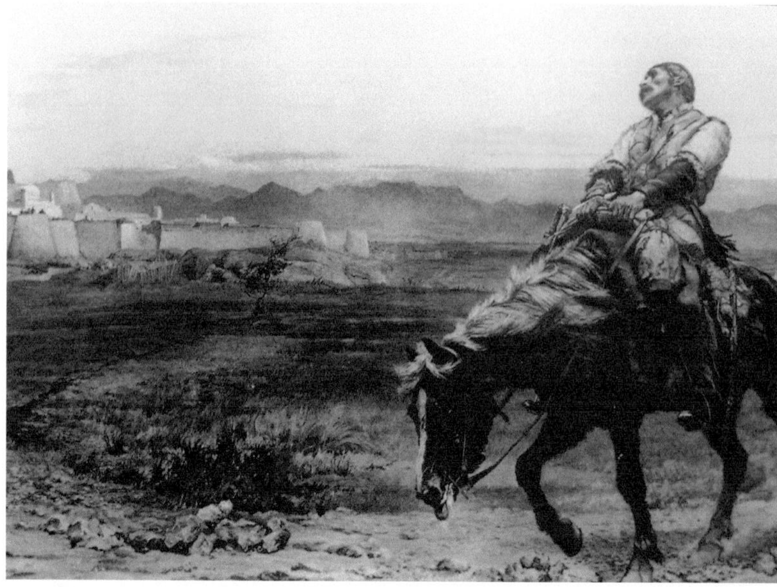

Plate 6.3 Dr. Bryden, the surgeon from the British army in Kabul was the only person permitted to return following the massacre of the British army in their retreat from Kabul to Jalalabad by the Afghans in 1842. The fort in Jalalabad (seen in the distance) was damaged during the 1842 earthquake and hastily repaired by the British after the earthquake. Painting in the National Gallery, London

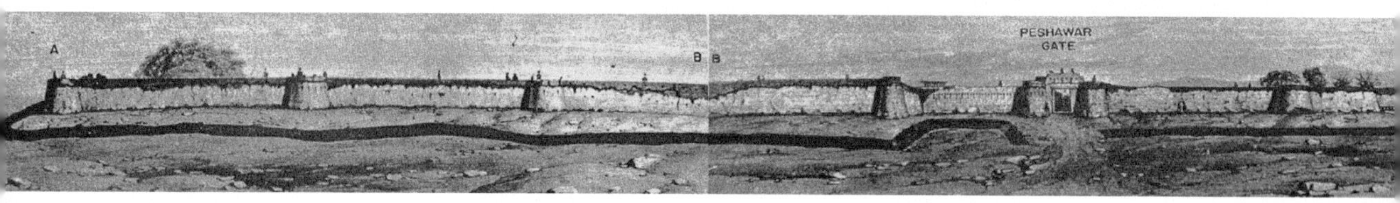

Plate 6.4 First two panels of a 14-segment panorama of the outer walls of Jalalabad Fort (from a view in the British Embassy, Kabul)

support themselves, and many suffered severe nausea (Delhi Gazette).

According to press reports from Peshawar, the largest urban center east of the apparent epicentral region, the shock lasted intermittently for almost 2 min and destroyed one tenth of its adobe houses, killing 40–50 people (Delhi Gazette; Sale 1846). There is some evidence that some small damage extended to Kalabagh, particularly in the part of the town on the right bank of the Indus, but it is not certain whether this was a result of the earthquake, the great flood of the Indus in 1842, or both (GM Binnie, personal communication, 1995).

Interestingly, several reports have survived for this earthquake recorded at remote distances from the epicenter, implying that the magnitude of the earthquake was considerable. At Shalkur, in Little Tibet near Skardu, at an epicentral distance of 410 km, the shock was strong enough to be mentioned in official dispatches (Baird-Smith 1843a). Although we have no information about damage between Kalabagh and Ferozopore, at Ferozopore (560 km) the shock was widely felt and was rather strong (Baird-Smith 1843a). At Ludiana (650 km) it lasted more than 90 s. Eastward along the Himalayan front it was felt at Lansdour and Dehra (Baird-Smith 1843a, b). At Simla (720 km) the earthquake was barely perceptible; however, it disturbed the records in the magnetic observatory (Baird-Smith 1843b; Lloyd 1846; Mallet 1853). Mussoorie (820 km) was the most easterly limit of the earthquake where the shock was perceptible (Baird-Smith 1843b).

In the plains fronting the Himalaya the shock was barely felt, but at Poojna and along the Doab Canal at Kulsea (860 km) the water in the canal was unusually muddy and was disturbed by a high swell (Baird-Smith 1843a). The shock was felt at Saharanpur (860 km), but attracted no particular attention (Baird-Smith 1843a). In Delhi (910 km), the shock was generally felt (Baird-Smith 1843b). About 30 km southwest of Delhi, at Sonub (930 km), the shock was not felt but allegedly was responsible for a change in the flow rate of a hot spring (Baird-Smith 1843b).

To the south and southeast of the epicenter the shock was felt slightly at Quetta (650 km), and was scarcely felt in the Sind (1,100 km).

Although it is not clear as to precisely where the epicentral area was located, some reasonable deductions can be made from the available evidence. We know that maximum damage occurred over a large area in the eastern Laghan and Kunar regions between Budeeabad, Jalalabad, and Pashtat, which could have well extended toward the northeast into an area from which we have no information. It is unlikely that the epicentral area extended to the southeast along the supply route of the British forces, and most certainly did not extend much to the west. Significant damage apparently did not extend to Tezeen and beyond Kabul.

These observations and the fact that all known aftershocks were reported from the Jalalabad-Budeeabad region, very few from Peshawar to the eastward, and none to the westward (Baird-Smith 1843b) suggest that the location of the epicentral region must be sought in eastern Laghan in the Kunar district, possibly associated with oblique thrust faulting along the southern section of the Kunar fault system about 35.0N, 71.0E (Chmyriov and Mirzad 1972). Aftershocks continued to the beginning of the following year, most of them being strong to damaging chiefly in the region of Jalalabad (Baird-Smith 1844). The large number of aftershocks and the prolonged duration of the aftershock sequence that extended for months suggest a crustal event. From the area within which the earthquake was felt, we estimate the surface wave magnitude of this event to be $M_s \approx 7.5$.

1874 Oct 18 (7.0) Little is known about this earthquake, which occurred on October 18, 1874 and caused serious damage in the region north of Kabul. According to local sources (Furon 1925a, b; Stenz 1945, 3) it occurred on an afternoon during Ramadan about the middle of Sher Ali Khan's reign. This is consistent with the date of an earthquake in Kabul on October 18, 1874 reported in the press (The Times 1/12/1874, p. 5).

The densely populated region of Kohestan (35.12N, 69.30E) and the nearby villages of Golbahar (35.14N, 69.30E) and Jabal Saraj (35.13N, 69.24E), 70 km north of Kabul were almost totally destroyed with many casualties. The ground opened up, presumably because of liquefaction, in the vicinity of Jabal Saraj.

Press reports say that in Kabul (34.53N, 69.13E) more than 1,000 houses were destroyed and many people were killed (Fuchs 1886, 485). However, no corroborating evidence has been found to support this, and this statement may refer to the losses sustained in the Kohestan.

Press reports add that the shock was felt at Sekunderabad (32.42N, 65.05E) (Times of India: 7.11.76) and Kandahar (31.61N, 65.70E) (Ballore 1905, 207) about 490 km south of Kohestan. However, it is not clear which of the many Sekunderabads in Afghanistan and India is meant here. The shock was perceptible 530 km north of Kohestan, in Samarqand (39.66N, 66.95E) at 13 h 30 m lasting for 1–2 min (Musketof and Orloff 1893, 460). We can find no mention in consular dispatches from Mashhad of an earthquake having been felt in the city (FO. Persia.60.361.1874).

The earthquake occurred near the north terminus of the Pagham fault, from where ground deformation was reported. Its magnitude may be assessed roughly from the area of perceptibility, which suggests M_s 7.0 (Plate 6.5).

1892 Dec 20 M=6.5 (*Chaman*) The event occurred about 90 km northwest of Quetta, near the Pakistan–Afghanistan border. Aftershocks continued to be felt until the end of February 1893.

Old Chaman, the only large settlement and the head of the North-Western Railway line, which was at the time under construction, was almost totally destroyed without loss of life. The stations of Sanzal and Shalabagh, situated on either side of the 3.8-km-long Khojak tunnel and about 5 and 8 km from Old Chaman, were damaged. However, no locomotives and carriages at these places were overturned or derailed by the shock, and no water supply tanks along the track were destroyed. The tunnel itself was undamaged, but workmen engaged on the roofing were thrown from their scaffolds and the upper crenulations of the tunnel's block-house were cracked. At Shalabagh and its vicinity, the shock was strong enough to throw down several local houses, making many of them unsafe.

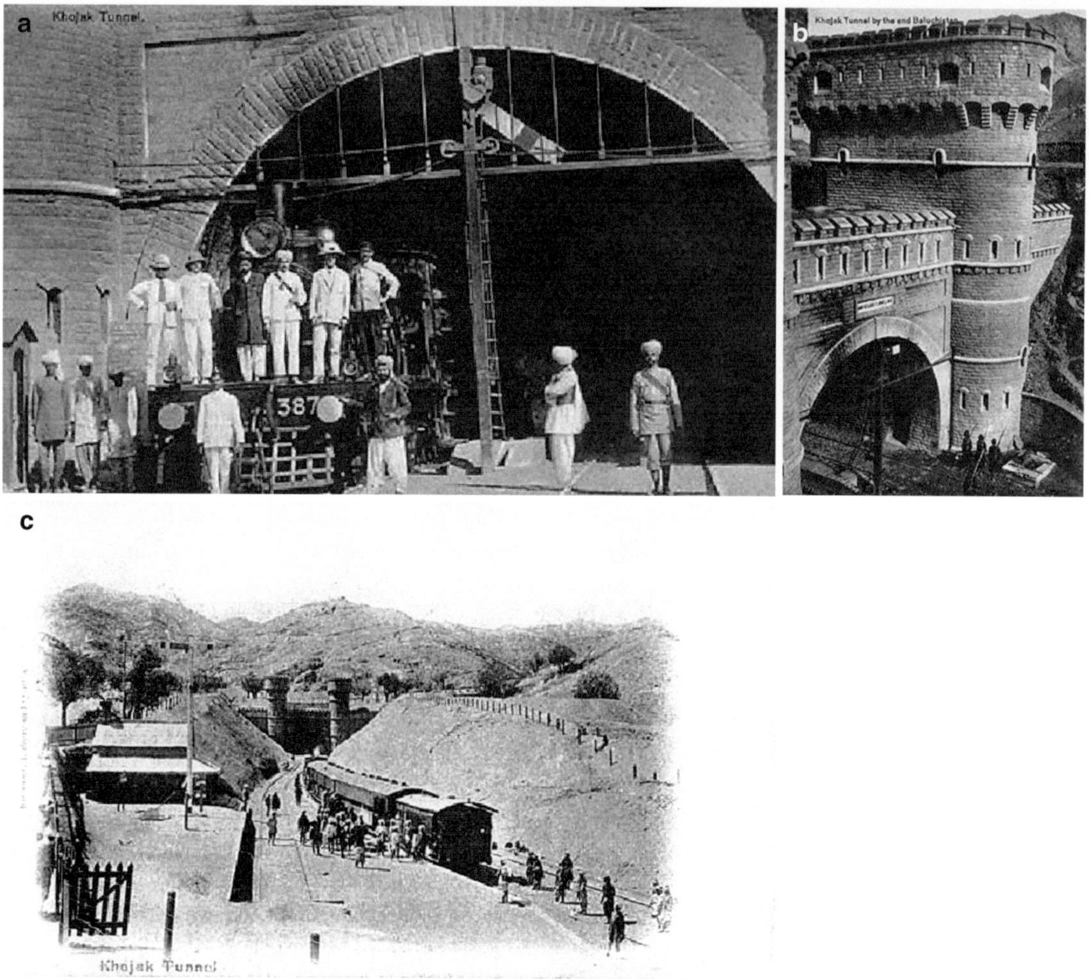

Plate 6.5 Three views of the Khojak tunnel before the 1892 earthquake. The tesselated portal was damaged by the earthquake

Spin Baldak, a small fortified station in Afghanistan, about 20 km northwest of Chaman, was probably damaged. The country to the East of Old Chaman was very sparsely inhabited, hence the limits of the damage area must probably remain unknown. After the earthquake Chaman was relocated to its present position about 10 km further to the northwest of its old site.

The shock was not felt over great distances. It was rather strong at Quetta, 90 km to the southeast, and probably was also felt at Kandahar in Afghanistan, 115 km northwest of Old Chaman. We could find no evidence that the shock was felt along the Shalabagh section of the North-Western Railway, at Sibi, Jacobabad, and Sukkur.

The earthquake was associated with surface faulting, with Old Chaman located directly on the active trace of the Chaman Fault. At a place outside of the Khojak tunnel, near the Chaman end, where the line emerges on the plain of Kandahar, the fault break crossed the railway track at 25° and followed the Khojak Range (NNE–SSW), about 4.8 km to the west of the southward continuation of the railway line. Where they crossed the fault, both the tracks together with their sleepers were buckled in the same manner, suggesting an oblique thrust with 60–75 cm left-lateral motion combined with 20–30 cm downthrow of the west block (Griesbach 1893; Davison 1893). From the length of the rail removed and the obliquity of the fault trace, (Griesbach 1893) estimated 75 cm of left-lateral slip, although from the numerical shortening of the rails and the obliquity in the photograph it is possible to derive a fault slip of just over 80 cm. The two nearest bench marks on either side of the fault-break were checked by (Egerton 1893), the location of which are not given, show a difference of 4.3 m more than before the earthquake, indicating that either the western side was downthrown or the eastern side was raised, but it is not stated which side is relatively the higher. Conflicting with this observation, Davison points out that a revision of these levels shows the actual difference was not more than 5 cm (Davison 1893).

Egerton traced the fault-break for several miles in both directions from its crossing with the railway lines, but it was not possible to trace the break to its end in either direction. To the north it passed from British territory into Afghanistan, and to the south it was lost in the snow of the Khowaja Amran peak, a distance that measured on the map that accompanies his paper is approximately 16 km (Egerton 1893). However, Griesbach states that the fault-break was 32 km long, with displacement continuing out of sight to both north and south (Griesbach 1893). He does not indicate on what he based this estimate.

Attempts in recent years to identify the 1892 rupture in the field are inconclusive. The fault in Afghanistan north of Chaman, mapped in the early 1970s by (Denikaev and Kafarski 1973), is shown on copies of the geological map of eastern Afghanistan. These authors could find no evidence that this segment of the fault was activated in 1892 or the recent past. A minimum surface rupture length of 60 km is inferred by Lawrence et al. (1992). This they deduced from a single very young scarp of low relief and great continuity, which includes Egerton's reported rupture (Plate 6.6).

This is the earliest event for which we have scant but reasonably good instrumental data. We used Abe's (1994) method to calculate its magnitude from trace amplitudes recorded by undamped instruments other than Milne recorders; that is, by Rebeur-Paschwitz undamped seismographs at Nikolaiev and Strasbourg (von Rebeur-Paschwitz 1895, 1893). Using Abe's method we calculate a magnitude of 6.7, which is considerably larger than the value of 6.2 assessed by Abe for this earthquake.

Although not confirmed by contemporary corroborating evidence, the following estimates of rupture parameters are available: Montessus De Ballore (1906) $L=20$ km $H=80$ cm,

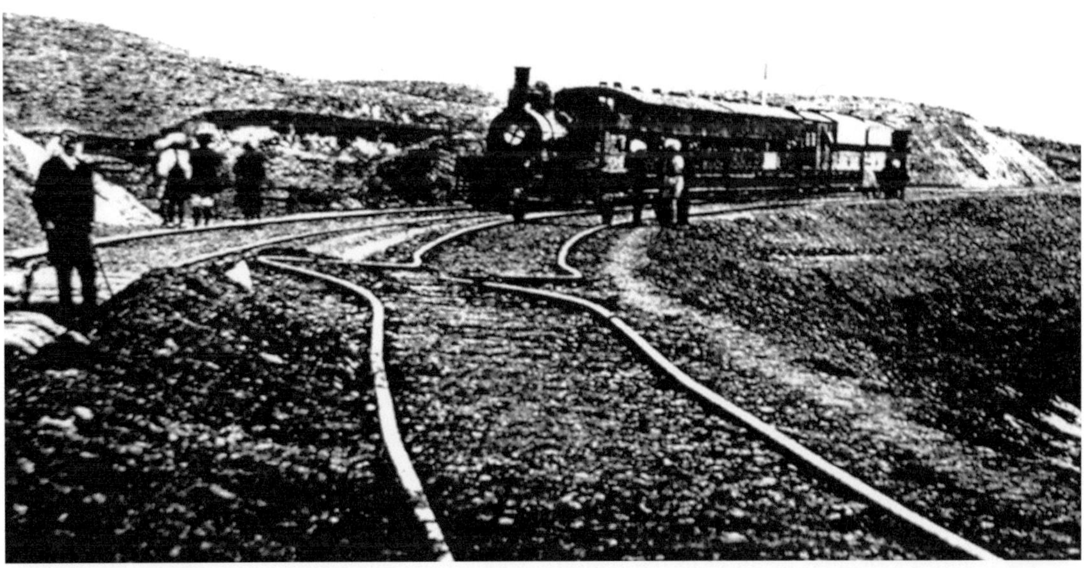

Plate 6.6 Contraction of the railway line where it crossed the 1892 Chaman fault (From Griesbach 1893). The sinistral fault slip is estimated here to have exceeded 70 cm

V=20–30 cm; Montessus de Ballore (1924) L=24 km H=60–75 cm; Quittmeyer and Jacob (1979) L>30 km H=75 cm; Lawrence et al. (1992) and Yates et al. (1997) L=60 km H=60–75 cm, V=20–30 cm. As a check on the instrumental estimate of magnitude we calculated M_s using the empirical formula of Ambraseys and Jackson (1998), using the maximum length (60 km) and slip (75 cm) inferred for this event. The derived value of M_s 6.8 is the close to the instrumental value. This magnitude is smaller than earlier estimates by previous investigators but is consistent with relatively small area from which the shock was reportedly felt (Ambraseys and Bilham 2003b).

1893 Feb 13 M =5.9 (*Chaman*) This earthquake caused some concern at Chaman but no damage. In Quetta it was much stronger than the shock of December 20, 1892, which implies that this was not an aftershock but a separate event not far from the town. It may be the shock that is said to have caused damage at Pishin and Baghihindu sometime in early 1893, NT(47.470; 48.348-349). The earthquake was recorded at Strasbourg (von Rebeur-Paschwitz 1895, 1893) and if it be assumed that it occurred somewhere near Quetta, its magnitude should be about 5.9.

1899 Dec 31 An earthquake in Kabul on December 31, 1899 at about 10 pm, brought down many adobe houses and several people were killed. It was apparently preceded by foreshocks in October and November (Martin 1907).

1906 Oct 24 (7.1) This was a large, globally recorded earthquake, with an epicenter somewhere on the border of Uzbekistan with Afghanistan. The shock was felt with low intensities within a radius of about 380 km at Tsardzhui, Bukhara, Katta Kurgan, Samarkand, Tashkent, Khodzent (Leninabad 40.29N, 69.63E), Khorog, Kelif, and Kerki, as far as Ufra (40.02N, 53.04E), but it was only at Aivadz (36.95°N–68.95°E) and Termez (37.22N, 67.27E) that the shock was strong enough to cause damage. We could find no macroseismic information from Afghanistan.

Macroseismic information is insufficient to define an epicentral area. Instrumental data merely confirm a general location in the region north of Termez (Seismische Monatsber. Physikal. Observ. Tiflis, 1906, p. 9, Levitski 1908; Spesivtseva 1933; Mitshkevich 1937 Tbilisi; Gubin 1960; Izvest. Post. Tsentr. Seism. Komis vol. 3, no. 2).

1907 Oct 23 (6.1) Little is known about the effects of this earthquake in the region of Kerki in Turkmenistan, except that it caused some damage at Kizil Ayak on the Amu Darya. This event occurred at the time of an earthquake in Sicily and instrumental readings are confused (Scheu and Lais 1912).

1908 Jul 26 (5.3) An earthquake in northwest Afghanistan caused damage to a few villages in the Injil area (34.6N 61.5E) northeast of Herat, but details are lacking. The shock was strong at Herat, nearby Karizak, and everywhere in the Herat valley, where it is said to have lasted for 1 min. The shock was felt at Torbat-i Jam but there is no evidence that it was felt in Mashhad, 320 km northwest of Herat (PRO.FO.Mashad 248/939 [01.08, 15.08, 01.09.1908]; Patterson 1908).

1909 Jul 7 (≈7.5) This earthquake, probably consisting of two events, one shallow and the other deep, was widely recorded 1 min apart by the global seismographic network of the time. The solution by Gutenberg gives a rough location (±3°) at 36.5°N, 70.5°E, which places a single event somewhere in Badakhshan in northeastern Afghanistan, at a focal depth of 230 ± 30 km, to which Gutenberg assigns M=7.1(±0.5) and m_B=7.6 (±0.4) (Gutenberg 1947), a location and depth that were adopted by later authors who assign to the earthquake a larger magnitude of 8.0, presumably by adding to 7.1 a depth correction (Duda 1965; Kondorskaya and Shebalin 1997).

Until recently, published macroseismic information exclusively derived from Russian territory north of the Amu Darya river, favors a deep location, not only because of the large area over which the earthquake was felt, but also because in spite of its large magnitude the shock did not cause damage anywhere in the territory. Along the alluvial valley of the Amu Darya river a few local houses suffered minor damaged in Patta Kesran (37.1, 67.2), Aivadj (36.98, 68.03), and Sarai Kamar (near Kirovabad 37.24, 69.09) but not in nearby Termez (37.2, 67.3), where the shock was rather strong but caused no damage. To the north of these places the earthquake was felt chiefly in alluvial valleys: Khorog (37.49, 71.55), Samarkand (39.66, 66.95), Khodzent

(40.29, 69.63), and Kokand (40.53, 70.93), which at great epicentral distances tend to enhance ground motions in large earthquakes. It was perceptible at Kerki (37.83, 65.20), Katta Kurgan (39.90, 66.25), Chust (41.00, 71.23), Andijan (40.79, 72.34), and Murgab (38.17, 73.95).

This information has led to a macroseismic location near the large urban settlement of Aivadj on the Amu Darya, after which the Russian catalog names the event and assigns to it a magnitude m_b of 7.7, and a radius of perceptibility of 700 km (Kondorskaya and Shebalin 1997). However, unpublished reports of the political agents in India, Iran and the Hindu Kush, as well as the Indian press show that damage was serious south of the Amu Darya in Badakhshan (37.6, 70.8), in Afghanistan and in the region of the northwest frontier. Maximum effects were reported along the Kunar valley, from Asmar (35.03, 71.36), Aradu (35.32, 71.31), and Drosh (35.55, 71.80) as far north as Chitral (36.02, 71.75), and from the levy posts between these places.

Unspecified damage, probably caused by landslides and rockfalls, was also reported in Alpurai (34.90, 72.65), Karori (34.88, 72.76), and Besham (34.93, 72.87). The Lady Minto hospital in Swat (34.90, 72.49) was damaged and in villages between this place and Dir (35.20, 71.88) houses collapsed without fatalities.

In the Northwest Frontier Province rockfalls destroyed levy posts, disrupted telephone lines, and blocked many passes into Badakhshan, where summer camps were destroyed. We have no information from Jalalabad in Afghanistan, but in Kabul (34.53, 69.14), several houses collapsed and about 10 people were killed in addition to many cattle.

We have no details from Khost (33.3, 69.9) and Waziristan (32.9, 70.6) south of Jalalabad, except that the shock was felt in these regions. Because of hostilities access to these regions was restricted. Some houses were damaged in Kohat (33.60, 71.44) to the south, and in Gilgit to the northeast (35.92, 74.29). To the west, in Lahore (31.56, 74.35) the shock was strong enough to awake people and cause some panic. The earthquake was felt strongly at Mussoorie (30.45, 78.08), it was reported from Srinagar (34.08, 74.81), and it was barely perceptible in Tashkent (41.31, 69.29).

The earthquake was followed by many aftershocks that were reported from the upper Kunar region, the largest of which on September 7 caused some additional damage in Swat.

We interpret this sequence to have been a double event consisting of a shallow earthquake that caused damage and aftershocks in the Kunar area, followed by a deep earthquake centered further west. Instrumental readings are too few and too inconsistent to attempt a revised instrumental location, but the data suggest two consecutive shocks each of $M_s \approx 7.5$, separated by an interval of little more than 1 min. The second event appears to have occurred west of the first and its focal depth may have been subcrustal (O.L.P/S.7.230.1102,1114,1146 [Swat] 08,10, 15.07.1909; IO.L.P/S 7.230., 1114,1171 [Kabul] 17.07.1909; IO.L.P/S 7.230.1275 [Gilgit] 07.1909; IO.L.P/S 7.231.1341.1401 [Mashhad] 31.07.1909; IO.L.P/S 7.231.1466 [Dir] 15.09.1906; IO.L/P&S Turkestan no.1401, Gubin 1960; Paisa Akhbar.10-14.07.1909, Spesivtseva 1933; Kondorskaya and Shebalin 1997; Biullet. Postr. Tsentr. Seism. Komm. for 1909, BCIS Strasbourg Archives Files:1909. no. 11).

1911 Jan 1 (7.1) This was a relatively large earthquake in northern Afghanistan, followed 4 h later by an aftershock of $M_s = 6.5$. Macroseismic information is sparse, some of which is confused with the effects of the large ($M_s = 8.3$) Kebin earthquake 2 days later.

Maximum damage in the January 1 earthquake sequence occurred between Feyzabad (37.12N, 70.56E), Khnabad (36.68N, 69.11E), and Kalan (35.19N, 69.23E). In the region of Kalan, 60 houses collapsed, killing 240 people. In Khanabad, 70 houses were destroyed and two people were killed. In the region of Feyzabad, houses were ruined with fatalities, although numbers are not given. Damage extended to the region of Kabul (34.53N, 69.13), where about 300 houses collapsed killing 460 people.

The shock was rather strong at Shuburghan (36.67N, 65.74E), Termez (37.22N, 67.28E), and Mazar-I Sharif (36.70N, 67.10), and less so at Kerki (37.82N, 65.20E) and Peshawar (34.01N, 71.54E). At Takhtebazar (35.96N, 62.91E), lamps were set swinging and plaster fell off ceilings. At nearby Kushka (35.31N, 62.41E), the earthquake was hardly felt but it was noticed because of the swinging of lamps. The shock was generally noticed in Bukhara (39.77N, 64.42E) and Kattakurgan (39.90N, 66.25E), but it was hardly perceptible in Tashkent (41.31N, 69.30), Samarqand (39.66N, 66.96E), and Kushka (Zakaspiskoe Obozrenie: 21–23.12.10; Turkestanskie Vedomosti: 21.12.10; Shpil'kov 1914; Gorshkov 1941). The India Office consular correspondence gives no evidence that the shock was felt at Herat (34.14N, 62.17E) or Mashhad (36.16N, 59.51E). It is important to point out that far-field information comes chiefly from sites on alluvial basins that could have enhanced ground motions at large distances.

The epicentral area of the earthquake that can be assessed from macroseismic information, and therefore must be sought near 36.0N, 70.0E, but it is not clear whether it belongs to the first or second shock. The main shock and its largest aftershock were widely recorded instrumentally. An early crude instrumental location places the main shock in the Altai (43.4N, 74.0E), which is 880 km to the NE of the macroseismic epicenter (Nikiforov 1912), whereas later relocations bring it closer to the adopted epicenter; that is, to 38.0N, 66.0E given by Gutenberg and Richter (1949) and 36.5N, 66.5E by Kondorskaya and Shebalin (1997).

Instrumental data are sufficient to allow calculation of the surface-wave magnitude of the main shock from 29 stations using the Prague formula, which gives M_s=7.1(±0.3), compared with m_B=7.2 in Gutenberg and Richter. For the foreshock at 14 h·59 m, from seven stations we find M_s=6.5(±0.2). In these calculations we have assumed normal depth. Regarding the depth of these events, Gutenberg and Richter estimate 50 km for the main shock and 20 km for the foreshock. The pattern of intensity distribution does suggest somewhat greater depth than normal for the main shock, but the small difference between body and surface magnitudes does not. We have no means that could help decide on the actual depth.

1913 Sep 9 (5.0) A strong earthquake at Ayvadz (36.98, 68.03) on September 9, 1913 cause the collapse of adobe and damage of stone masonry houses. The shock caused liquefaction

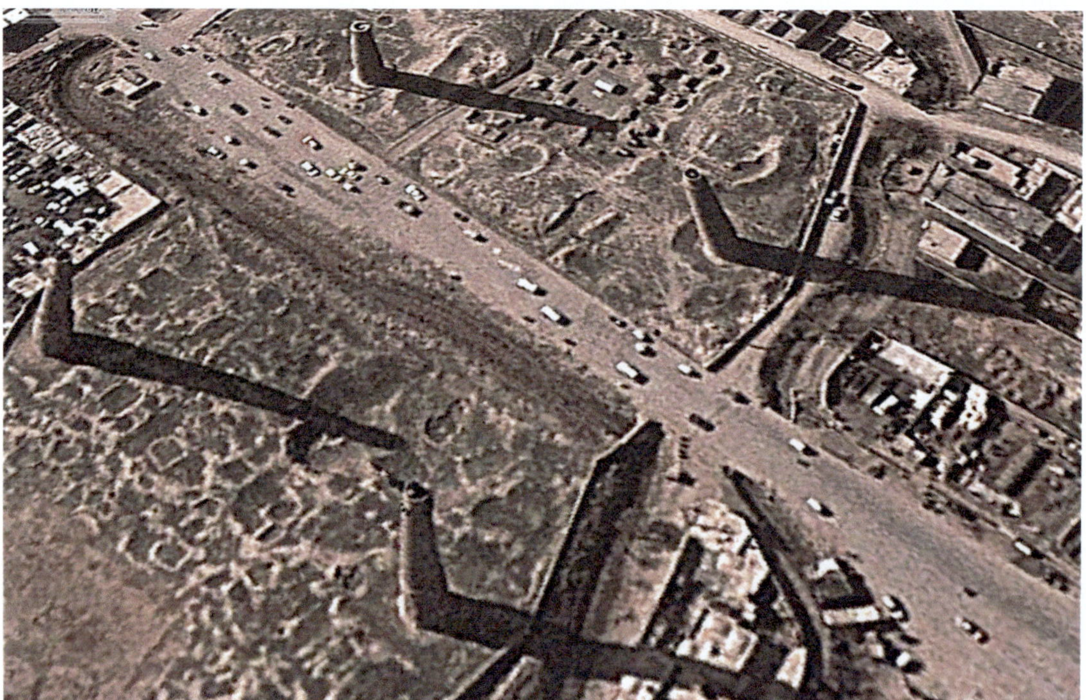

View from space of four of the five remaining minarets in Herat in 2012 (© GoogleEarth). The minarets (~ 55 m high) lean perilously far from the vertical and are the subject of UNESCO restoration

of the ground in the Shaartuz area (37.27, 68.13) (Gorshkov 1941). See Russian Intensity Catalog for Ayvadz 1906–1913 (Plate 6.7).

1918 Nov 29 (6.2) A violent earthquake took place in Afghanistan, between Kalat-i Ghilzai (32.11N, 66.90E) and Ghazni (33.56N, 68.42E), causing many deaths because of the collapse of houses. It is also said to have increased the flow of water in springs and qanats, and that even some dry qanats began to yield water (IO L/P&S Baluchistan/10/814). The shock was strong at Jabal Saraj (35.13N, 69.24E), where some walls were thrown down (Heuckroth and Karim 1970).

1931 Sep 10 (5.0) In 1855 a group of nine 36- to 45-m high minarets existed in the complex of the musalla of Gauhar Shad in Herat (34.34, 62.19), two of which collapsed in the earthquake of September 10, 1931. The earthquake shock was felt at Murgab (37.49, 61.98) (Byron 1937, 99; Gorshkov 1941; Blunt 1957).

1933 Oct 16 (5.6) A damaging earthquake on 16 October 1933 caused the collapse of three forts in the Oruzgan (32.93, 66.63) region. Slides and rockfalls were reported from this area and from Day Chupan (32.63, 66.77), where the shock caused great concern (Heuckroth and Karim 1970).

1934 Mar 30 A damaging earthquake occurred on March 30, 1934 in the hills south of Maimana (35.92, 64.78). The villages of Pashtunb Kote (35.90, 64.78), Bato (35.72, 64.80), and Purkhisht (35.72, 64.87) were totally destroyed, and people fled the region and took refuge in the hills. The shock triggered slides and caused cracks in the ground. There is no evidence that damage extended to Maimana. Aftershocks continued for 10 days. No teleseismic data have been found for this event (Stenz 1945; Furon 1925a, b; Huckriede et al. 1962).

1935 Jul 5 This earthquake occurred on the borders of Afghanistan and Uzbekistan and affected the region on both sides of the Amu Darya river. Information is too poor to allow location of its epicentral region, which must be sought between Baisun and Shirabad in Tadzhikistan, where some damage was done to unnamed localities and the irrigation system of Talishkan and Tazhdiuli.

The shock was strongly felt to the south at Shibarghan in Afghanistan at Sari-i Pul, Aq Chah, Balkh (Wazirabad), Mazar-i Sharif, and Tashkurgan, where it caused insignificant damage. The earthquake was reported from a large area disproportionate for its magnitude. For example it is alleged that it was perceptible at Srinagar, at an epicentral distance of more than 700 km (India Weather Review for 1935, Simla; Gorshkov DATE, Spesivtseva 1933; Heuckroth and Karim 1970).

1935 May 30 $M = 7.7$ (Quetta) This earthquake occurred ≈100 km outside the borders of Afghanistan, but is the largest in the catalog and is therefore included in this account (Ambraseys and Bilham 2003b). Although Baluchistan in 1935 had one of the lowest population densities on the subcontinent of India, the earthquake occurred where most of the population lived. For this reason, and despite the initiation of earthquake-resistant designs triggered by the recent 1931 sequence, the 1935 Quetta earthquake resulted in the largest number of fatalities of any earthquake in the Indian continent or its boundaries before the Sumatra/Andaman earthquake of 2004 and the Kashmir earthquake of 2005.

Quetta was ceded to the British in 1877. Before then Quetta was a station that controlled the trade route from India to Kandahar, as well as the Bolan Pass to the south, as well as the road to Kalat and Persian Baluchistan. Although a sketch of the fort at this time looks quite imposing, an 1839 report describes Quetta as "a most miserable mud town with a small castle on a mound having one small gun on a rickety carriage" (Spate 1954). Photographs of the town just before and after the 1935 earthquake reveal a busy frontier town that was severely damaged by the event.

Detailed information survives for the Quetta earthquake (Situation Reports 31.05-13.08.1935, Files no. 1 to 73 in the Archives of the Government of India Bureau in Simla), as well as from (Skrine 1936; West 1936; Pinhey 1938; Jackson 1960). Most of the damage to rural and urban houses, chiefly of mud brick construction, was enclosed within a narrow zone, with Baleli and Quetta in the north extending in a southwesterly direction

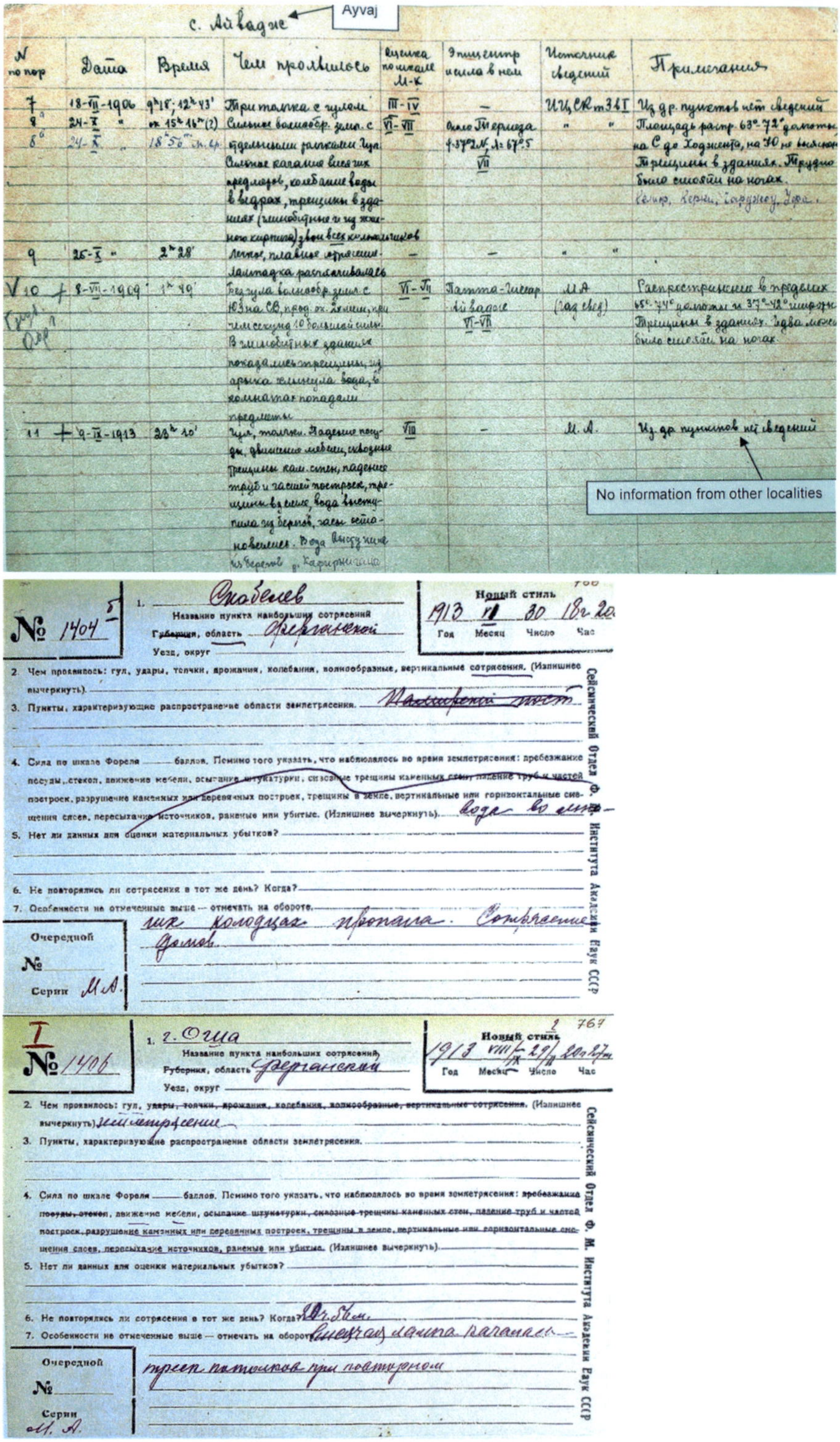

Plate 6.7 Two 1913 cards from the Russian "Card Catalog" of earthquakes. Data for September 9, 1913 Ayvadz earthquake listed in the Russian Catalog

into the Harboi Hills, about 160 km long and 25 km broad. A great deal of the land in the zone is unproductive, containing only a few large villages in which water could be found in underground irrigation and water supply conduits (qariz, qanat). In addition to the towns of Quetta and Mastung, at least 100 villages in Quetta subdivision and Kalat State were totally destroyed.

Starting from the north, Baleli was totally destroyed, 108 people were killed, and 23 were injured. At Kuchlagh all houses were ruined and the railway depot collapsed with the loss of eight lives and nine injured. Further south, Sheik Manda was razed to the ground, and in nearby Nauhissar, 77 people were killed and 28 injured.

Quetta, a military garrison town with a population of about 40,000 (summer population 65,000), is built on a slope crossed by two nullahs (watercourses). The Habib Nullah separated the Civil Lines and town from the cantonment, and the Durani Nullah, a kilometer further north, ran parallel to it for most of its length. Both nullahs were crossed by bridges at several points. The Civil Lines was the more densely populated part of the town. It occupied an area of about 4 km^2 and it was located south of Habib Nullah, an area with a high water table in the spring. This low-lying part of Quetta was utterly destroyed, and about 15,000 people lost their lives (Pinhey 1938). The Police Lines, the Durbar Hall, the Civil and Mission Hospitals, and the Club were ruined, and the Residency was damaged. The only buildings that survived the earthquake with minor damage were the few reinforced concrete structures and the new railway quarters, constructed after the 1931 earthquake using earthquake-resistant principles and situated in the most damaged part of the Civil Station.

North of the Habib Nulla, on higher ground, the cantonment was much less affected and only a few houses collapsed. The garrison church and the British and Indian military hospitals were undamaged. The only serious damage done to the cantonments was a belt about 1 km wide immediately adjoining the Durani watercourse and the Civil Lines, damage decreasing rapidly toward the northeast. Here a good deal of damage was done to the fort and some of its buildings collapsed.

The airfield, with its modern hangars and barrack blocks, stood to the northwest, apart from the city and cantonment. In the RAF lines the hangars were left standing but little else. Every aircraft was so damaged that it was unsafe to fly. The earthquake caused no serious damage to the piped water supply or the power stations, which continued to work on restricted load.

South of Quetta, Kansi was totally destroyed: 1,010 people were killed and 370 were injured. Also, Sariab was razed to the ground with the loss of 1,206 lives and 641 injuries. At Durani, 101 people were killed and 114 were injured. At Spezand, Dingar, and Mand-i Haji, local houses were flattened but the railway station at Spezand was not destroyed. Tiri was utterly destroyed, with the loss of 710 lives and 275 people injured. Mastung, 65 km south of Kuchlagh, was flattened by the shock together with the Khan's palace, killing 1,736 people and injuring 716. Southwest of Mastung, small settlements in the Shirinab valley, which extends from Kuhnak to Manguchar along a distance of 90 km by road, were destroyed. This region was within the thinly populated tribal territory belonging to the Khan of Kalat, and damage details are lacking. Pringabad, the only large settlement in this region, was destroyed: 369 people were killed and 234 were injured. Manguchar, 100 km south of Kuchlagh, was also destroyed: 185 people were killed and 185 were injured.

Destruction extended south into the State of Kalat, 155 km from Kuchlag. Kalat itself was ruined, with 120 people killed and 50 injured. In Kalat State, out of a population of 10,000, 2,900 were killed and 5,000 injured. It was estimated that all villages between Quetta and Kalat were destroyed, with 70 % of the population either dead or injured.

Outside the epicentral region, damage was widespread to dilapidated rural houses, particularly in many places in the Indus valley and in the Spin Baldak (Qla-i Jadid) and Kandahar regions of Afghanistan (Stenz 1945).

Liquefaction of the ground and mud volcanoes were reported in the valley northwest of Quetta.

Some 20 km south of Kalat on the main road to Surab, about 5 km east off the road near the village of Thok (28.333°N, 66.517°E), large quantities of liquid mud were observed coming out from the top of an old mud volcano at the time of the earthquake, the eruption lasting for 9 h. None of the local inhabitants could recall any similar event in the past. The new flows had spread out beyond the limits of the old, occupying a radius of 140 m.

Official figures for the loss of life in the earthquake are no more than estimates (Pinhey 1935). In Quetta, about 26,000 people were killed, of which a few thousand bodies were left buried in the ruins of the town. Outside Quetta, numbers are even more uncertain, particularly in the Kalat tribal area, where more than 8,410 deaths were recorded. Altogether, the earthquake could have killed about 35,000 people, but reliable figures are lacking.

The telegraph lines from Kalat and Quetta to Chaman and Jacobabad were broken, but communication with the government of India at Simla was established by radio. The railway and road communications, including the section through the Bolan Pass, were not badly damaged. Several small road bridges suffered slumping of their abutments and five segments of the Quetta–Nushki rail track had to be replaced at its crossing with the zone of ground fissures (Thomson 1936).

Administration became difficult owing to the fact that nearly all the subordinate civil officers and police had been killed. However, the fact that the troops escaped with few casualties allowed a quick rescue and evacuation of survivors, the disposing of thousands of dead by burial or burning, sealing the town to prevent looting and the outbreak of epidemic disease, protecting and salvaging property, and controlling the rehabilitation of the region. Two battalions of the 7th Gurka Rifles who were posted in Zhob and Chaman respectively at the time of the earthquake felt the earthquake and returned to find their regimental institute damaged beyond repair (Mackey 1962).

Following the earthquake, new laws were enacted for regulation of the distribution to relatives of property salvaged in the earthquake, settlement of property claims, and the compulsory application of earthquake-resistant design for all new public buildings and engineering structures (Robertson 1948; Khan 1956). Priority was given to the repair and reconstruction of destroyed qanats (underground irrigation channels) throughout the affected area to secure the next harvest.

Contrary to what has been said by a number of authors, that the shock was felt within a radius of only 280 km, Urdu and Hindi press reports confirm that the shock was felt over a large area, as far as Amritsar, Sultanpur, and Simla to the east (1,000 km), to Jatti (610 km) on the mouth of the Indus to the south, Dera Ismail Khan (500 km) to the north, and Chagai (250 km) to the east. In the last two directions information is lacking beyond these points into Afghanistan.

The earthquake was followed by a long sequence of relatively small-magnitude earthquakes, the largest of which in the south part of the epicentral area did not exceed a magnitude of M_s 6.0. Shocks continued to be felt until the beginning of October. No shocks were reported before the earthquake, but a bright orange glow was seen over Quetta to the west, and further south, near Kalat, flashes of light were reported along the flanks of the mountains on both sides of the valley.

Following the earthquake, a survey of the northwest dipping thrust faults southwest of Quetta, which runs for a few kilometers to the southwest along the northern flanks of Chiltan, showed no signs of any movement, although 20 cm of uplift was detected on bedrock benchmarks near the Quetta brewery that lies 5 km west of the town on the hanging wall of the fault. Ground deformations that extended discontinuously for about 105 km followed the south side of the Chiltan Range toward Kalat, striking N15E. Over the greater part of this distance they took the form of strands of open cracks, 2–20 cm wide, mostly in alluvium. About 8 km west of Mastung, the ground of the west side of these cracks was downthrown on average by about 80 cm, although a little further south the sense of vertical movement was reversed. In some places, instead of a throw or open cracks, the ground had been heaved up, the uplifted portion being 30 cm or more high and several meters wide.

Where the locus of ground deformation crossed the railway that runs from Spezand to Nushki, about 3 km west of Mastung Road Station, the track had been uplifted and the rails buckled. Where bedrock intervened along these zones of fissures in alluvium, the cracks died out, with rock-alls and shattered rock taking the place of fissuring along the same line. This was well seen to the northwest of Mastung Road, on the southern flanks of the Chiltan Range. The zone of fissures extended further south, past Kalat. A few kilometers of southwest of the village, ground cracks passed beneath an adobe house, displacing its walls but leaving them standing. It appears, therefore, that the earthquake was associated with the zone of faults that lies along the east edge of the Chiltan range and that this zone extends to the south passing near the towns of Mastung and Kalat. Leveling data confirm abrupt uplift of the foot of the range, but triangulation data that would reveal the extent and sense of the strike-slip motion have yet to be released. The first-order North Baluchistan Series of the Survey of India measured in 1909 (Burrard 1912), passes southward through Quetta and along the Chiltan range. Re-measurements of some of these stations were in progress during the earthquake.

Our instrumental position, a relocation using p-waves from 231 stations and present ISC location procedures, is 160 km south of Quetta, about 40 km west of Kalat, and close to the location computed by Engdahl et al. (1998) and Ramanathan and Mukherji (1938). The formal errors are about 15 km, but because most of the stations are grouped to the northwest, the true position could be further to the northwest. The surface wave magnitude calculated from 25 station magnitudes is 7.7, which corresponds to a seismic moment of 17.0 × 1,027 dyn cm, estimated by Singh and Gupta (1980).

Lawrence et al. (1992) and Yeats et al. (1997) associate these ruptures with the Ghazaband fault zone, one of a series of large north–south left-lateral strike-slip systems that accommodate plate boundary shear. A temporary seismic network operated in the region in 1978 suggests that microseismicity was at that time concentrated near the ends of the 1935 rupture (Armbruster et al. 1980). Fault plane solutions in the area confirm left-lateral faulting on this trend, and in spite of the lack of reported observations of strike-slip offsets at the time, this is the most likely mechanism for the 1935 earthquake.

1948 Jan 28 (6.5) A damaging earthquake occurred in the region of Balkh in northern Afghanistan over a large area between Shah Anjir (36.34, 67.22) and Mazari Sharif (36.70, 67.11). Much damage occurred near the Dulan Pass, at Shah Injir, Yakatal, and Quduk Mulla, where a number of people and domestic animals were killed. At Mazari Sharif (36.70, 67.11), 45 km from the pass, old houses and shops were destroyed and the dome and towers of the shrine building fell down. At Samangan, 90 km away, a few houses were ruined and the sugar storage building was destroyed. The shock was strong at Maimana (35.92, 64.76), but it did not cause significant damage. The earthquake was widely felt, as far as Kabul (34.53, 69.12), Dushanbe (38.57, 68.77), and Samarkand (370 km), but not at Herat (Ittilaat 13.11.1326) (Heuckroth and Karim 1970).

1950 Sep 24 (5.6) An earthquake occurred on September 24, 1950 with an instrumental location on the Iran–Afghanistan border, just south of Taiabad. There are no macroseismic data for this event except for the unconfirmed information that about this time a shock caused the collapse of a sixteenth-century minaret in the Musalla of Gawrshad in Herat, 130 km away, which was already in ruins, leaving seven minarets leaning in different directions (Byron 1937, 99; Blunt 1957; Dupree 1971).

1952 An earthquake at noon on a day at the end of autumn, probably in 1952, occurred in the Schkurigal-Bashgal region, at Ozzuk (35.80, 71.27), Appsigh (35.80, 71.25), and the upper Bashgal basin and Deywanababa (35.90, 71.30), but not in Lulook (35.77, 71.22), where damage was caused by falling rocks. Cracks in the ground and snowfields were reported from Deywanababa and Barge Matal (35.67, 71.33). In the Munjam valley (36.02, 70.76), the shock was strong but did not destroy any houses, even though falling rocks were numerous (Danby et al. 1972).

1955 Aug 29 This was one of the many shocks that preceded and followed the August 29th event over a period of 1 month, ruining many houses at Gulran in northwest Afghanistan. The shock was felt strongly 100 km away, across the border in Iran, at Torbat Sheykh Jam, where it caused no damage. No association of this event can be made with the shocks recorded by the Soviet and Indian networks during the period of August 28–30 (Bozorgnia 1962; Heuckroth and Karim 1970).

1956 June 9 (7.4) A large magnitude earthquake occurred in the Bamiyan district of Afghanistan. Much of the destruction was in the valley between the Kahmard and Saighan mountains, a sparsely populated region with no large villages, at an altitude over 3,000 m. Little is known about damage to individual settlements, except that the few small mountain villages around Kahmard (35.33N, 67.50E) and Saighan (35.17N, 67.70E) were totally destroyed, and those in the district of Yakwalang (34.73N, 66.97E) were heavily damaged with loss of life. Contrary to information in the press the earthquake caused surprisingly little loss of life, fewer than 70 people.

The epicentral region may be defined roughly within an area about 120 km long and 40 km wide, running from Yakwalang in the south to Kahmard and Doab (35.55N, 67.81E) in the northeast. Within this zone the earthquake triggered rockfalls and landslides, which caused additional destruction. The largest slide occurred at Kami Kharqushaq (35.36N, 67.53E), about 15 km northwest of the village of Kahmard. An estimated 100,000 cubic meters of limestones and marls slid down, damming the upper valley of the Kamar river and holding back eight million cubic meters of water for 4 days. The dam gave way on June 14, and the flood swept away the settlements in the valley, drowning about 350 people, and leaving behind a small lake that marks the site of the slide. Another large landslide at Darra-i Shikari (34.88N, 67.78E), about 25 km northeast of Bamiyan (34.83N, 67.83E), blocked roads and disrupted communications, killing a number of domestic animals. Also at Yakwalang, slides blocked roads and killed herds of animals.

Outside this region, around Pul-I Khuni (35.95N, 68.71E), Baghlan (36.14N, 68.70E), Kunduz (36.73N, 68.86E), Seh Kundi (33.35N, 68.40E), and Kabul (34.53N, 69.13E), the shock caused some damage and great panic, but no loss of life.

The earthquake was felt as far as Termez (37.22N, 67.27E), Motovabad (37.34N, 68.67E), and Chitral (36.02N, 71.75E), and was perceptible at Stalinabad (38.57N, 68.78E) but not in Peshawar (Furrer 1956; Heuckroth and Karim 1970; Wesson 1972).

1956 Sep 16 (6.7) A damaging earthquake occurred in the Lohgar district (33.98N, 68.99E) in northeast Afghanistan near the border with Pakistan. Little is known in detail about the damage in the epicentral area, which includes the villages of Said Karani (33.68N, 69.37E), Laza, and Jaji (33.78N, 69.67E), where a number of houses collapsed and a few people were killed. In the Hindu Kush the earthquake caused landslides and snow-avalanches depending on local conditions. Some villages suffered slight damage, whereas others suffered none. Deywanababa (35.90N, 71.30E) was outstanding in being the scene of an avalanche that killed a few people. The shock triggered rockfalls from the Mangal Mountain that killed a number of animals. Also in Nuristan, the shock caused snow-avalanches from area above the snow-line.

The earthquake was felt in Afghanistan and Pakistan in an area disproportionately large for its magnitude. Felt reports are available from the districts of Hazarajat, Parwan (35.1N, 69.2E), Ghazni (33.63N, 68.95E), Kohat (33.59N, 71.44E), Parachinar (33.90N, 70.10E), Behsud (34.38N, 67.53E), Rawalpindi (33.69N, 73.04E), and Srinagar (34.08N, 74.81E), within a radius of more than 300 km.

A large aftershock on September 16 was felt in the Lohgar district in Kabul, in the region of Jalalabad, and across the border at Parachinar (Seism. Bull. Sep. 1956, p. 5,10, Met. Dept. India, Delhi; Heuckroth and Karim 1970; Danby et al. 1972; Wesson 1972).

1962 Sep 12 (6.0) An earthquake occurred in the Thakar province of northeast Afghanistan. The available macroseismic information is insufficient to define an epicentral region. In Afghanistan the shock was strong in the districts of Baghlan and Qatghan, and was felt up to 250 km away at Jalalabad, Kabul, and Mazar-i Sherif. In the north, in Tadjikistan and Uzbekistan, the shock was equally strong at Karshi, Samarkand, Leninabad, and Khorg. The shock was felt within a radius of 450 km. The large felt area and small magnitude suggests that this was a lower crust event (Zemletriasenia v SSSR v 1962, 1964, pp. 70, 82, Izd. Nauka, Moscow; Heuckroth and Karim 1970).

1968 Sep 3 (5.0) This relatively small earthquake has been relocated by Engdahl et al. (1998) to northern Afghanistan between Nahrin and Khanabad, at a focal depth of 40 km. Although we can find no macroseismic information from Afghanistan, the event was widely felt in the plains of Amu Darya and reported from 37 localities in Tajikistan, up to epicentral distances of 250 km. It was felt strongly at Dangana (37.60, 69.78), Dangara (38.09, 69.33), Iol (37.77, 70.18), Khorog (37.49, 71.54), Kulyab (37.91, 69.78), Moskovski (37.61, 69.70), Obi Garm (38.72, 69.70), and Pyandzh (37.23, 69.09), and was perceptible at Dushanbe (38.57, 68.78) and Regar (38.52, 68.18) (Zakharova et al. 1972).

1972 Jun 24 (6.4) A damaging earthquake occurred in the district of Takhar in northern Afghanistan on June 24, 1972. The large villages of Khost Fering, Nahrin (36.07N, 69.13E), Ishkimish (36.38N, 69.32E), and smaller settlements within a radius of about 25 km were ruined. A few hundred houses collapsed and approximately 20 people were killed. Damage to local houses was reported from Baghlan, Pul-i Khumri, and Warsaj at epicentral distances of 60 km. The shock was felt in Kabul, Peshawar, and Rawalpindi in Pakistan, and as far as Kulyab (37.91, 69.78), Khorog (37.49N, 71.54), Obi Garm (38.72N, 69.70E), on the Nurek dam site (38.39N, 69.32E), and Samarkand (39.66N, 66.95E) in Soviet Central Asia; that is, within a radius of 370 km. It was perceptible at Dushanbe (38.57, 68.78) (Seismological Bulletin for 1972, Seism. Center Kabul Univ., Kabul; Danby et al. 1972).

1975 Oct 3 (6.8) An earthquake occurred on the Chaman fault zone, close to the border with Afghanistan at Spin Tezha, followed 12 h later by an aftershock of 6.5. Little is known about the effects of these events in the epicentral region, a sparsely inhabited area of the Afghan border, except that they caused minor damage at Quetta. The shock was felt strongly at Quetta and in northwest Baluchistan.

The trace, probably of only a small segment, of a north–south-trending discontinuous surface fault break was found in alluvium that could be followed for about 5 km south of Spin Tezha along the Chaman fault zone. It showed an average left-lateral displacement of about 4 cm, with minor dip-slip up to the west consistent with the earthquake focal mechanism (Farah 1976; Lawrence and Yeats 1979).

1976 Mar 19 (5.5) A relatively small earthquake occurred on March 19, 1976 but caused considerable damage in Samangan province, killing about 50 people and ruining more than 1,000 local houses. Maximum damage was reported from the areas of Khulm (36.69, 67.69) and the nearby Tashkurgan gorges. Rockfalls and slides in the eastern section of the Khulm Gorge buried several vehicles, adding to the loss of life. The shock was strong at Kornilovka; it was felt at Termez, Denay, Khorog, Dushanbe, and Kulyab, and was perceptible at Samarkand (Zemltr. v CCP1976(280) p. 160).

1981 Jun 13 (5.4) A damaging earthquake occurred in northern Afghanistan. Little is known about its effects in the epicentral area, which was somewhere between Samangan and Jozjan, where a number of people were killed. The shock was felt along the Amu Darya from Termez to Parkhar and further north in Uzbekistan at Kurgan Tyube and Dushanbe. It was perceptible at Samarkand at an epicentral distance of 390 km.

1982 Dec 16 (6.5) A destructive earthquake occurred in north Afghanistan. Little is known about the effects of this event apart from the fact that it destroyed approximately 7,000 houses, killing 450 and injuring 3,000 people in the Baghlan district. The shock caused serious damage and loss of life in the coal mines in the district. The earthquake was felt as far as Tashkent, Murgab, and Tarbela, over an area of a radius of 520 km.

1983 Dec 7 (4.6) A small earthquake occurred in Afghanistan with an instrumental epicenter in the region of Takhar. The shock was felt at Kuliab and Khorog at an epicentral distance of about 200 km.

1984 Jul 3 (5.2) We have no macroseismic data from Afghanistan for this earthquake in the Takhar district. The shock was felt at Khorog, Kuliab, Nurek, Kabodien, and was perceptible at Dushambe, Samarkand, and Tashkent (Barinova et al. 1987; Kondorskaya and Shebalin 1997).

1986 Jan 12 (5.5) An earthquake occurred in northeast Afghanistan with an instrumental location near Kabul. No macroseismic information is available with the exception that the shock was felt strongly in the Kabul area and at Peshawar in Pakistan.

1998 Feb 4 (5.9) A destructive earthquake occurred in the Rustaq area (37.12, 69.82) of northeast Afghanistan that resulted in estimated losses of 2,300 people killed, 800 injured, and 8,100 houses destroyed, the shock making 8,000 homeless. The earthquake triggered extensive landslides, which added to the damage, killing more than 6,000 livestock. The shock was felt at Dushanbe in Tadjikistan and was perceptible in Tashkent.

1998 May 30 (6.5) A destructive earthquake occurred in northeast Afghanistan. The shock killed approximately 4,000 people, injuring many thousands in the districts of Badakhshan and Takhar.

The shock was strong at Mazar-i Sharif (36.70, 67.11). It was felt in Kabul (34.53, 69.13), and in northern Pakistan at Peshawar (34.00, 71.54), Rawalpindi, and Islamabad (33.72, 73.06), as well as at Dushanbe (38.57, 68.77) in Tajikistan and Andijan (40.79, 72.34), and was perceptible at Samarkand (39.66, 66.95) and Tashkent (41.31, 69.30).

1999 Feb 11 (5.8) An earthquake in the Lowgar (Logar) and Vardak provinces of Afghanistan destroyed approximately 7,000 houses, killing 70, injuring about 500 people, and making at least 14,000 homeless. The damage extended to Kabul, where several people were injured. The shock was felt in Pakistan at Peshawar and Islamabad.

2002 Jan 3 13:40 (6.3) An earthquale occurred in the Takhar province of the Hindu Kush. No damage or casualties have been reported. Buildings shook in several major cities, including Kabul, Islamabad, Delhi, Peshawar, and Mazar-e-Sharif in northern Afghanistan.

2002 March 3 (7.2) The epicenter of the earthquake was in the Hindu Kush region with a depth of 195 km. The earthquake was felt in Kabul, Jalalabad, Faizabad, Mazar-i-Sharif, and Bamiyan. In Kabul, 32 houses were destroyed, 20 persons injured, and 6 persons killed in the city. Except for cracked walls, there were no reports of destruction or injuries in rural areas adjacent to Kabul. A landslide 20 km southeast of Aybak, on the road to the Ruy-e-Du Ab district in the Surkenda Valley, created a transient lake. Damage was extensive in the villages of Souchi Bala Payan, Targane, and Yawan, as well as two other settlements. Although 340 houses were destroyed, the absence of reports of significant injuries or deaths is probably owing to the fact that the earthquake took place during the day when most people were not at home. The reports from the Baharak district, located east of Faizabad, indicate that a limited number of houses were destroyed in villages around the district capital and one death was registered in Malang Ab. In Jurm, there are reports of two houses destroyed in Dashteq, another 8 in Abad, 50 in Souch, and 6 in Zardeh (UN Office for the Coordination of Humanitarian Affairs [OCHA] OCHA/GVA – 2002/0052 4 Mar 2002. Afghanistan-Earthquake OCHA Situation Report No. 1).

2002 March 25. (M_s=6.0, 5.1, 5.8) A series of shallow earthquakes hit the Nahrin district of Baghlan province on the evening of March 25, the early morning of March 26, and the afternoon of March27. The epicenter of the earthquakes was located southeast of the Nahrin district in Baghlan province. The affected area included a radius of 12–15 km around Nahrin. Other affected areas included Burkha, Panshjiri, Lakankhel, and Toli. In all, 78 villages were affected by the earthquakes, resulting in approximately 1,200 deaths (OCHA Situation Reports No. 1–7 Afghanistan–Earthquake. OCHA/GVA – 2002/0076, 30 March 2002).

References

Newspapers With very few exceptions, references to press reports are omitted

ZAK Zakaspiskoe Obozrenie. Ashkhabad, 21–23 Dec 10

TUR Turkestanskie Vedomosti. Tashkent, 21 Dec 10

Abbreviations

BAAS British Association for the Advancement of Science; Seismological Committee Reports (1890–1917)

BCIS Bureau Central de l'Association International de Sismologie. Catalogues & Reports. Strasbourg, 1902–1914

BPTS Biullet Postr Tsentr Seism Komm. 1905–1911

DLG Delhi Gazette, 19–20 Feb 1842

ENG Englishman (Calcutta) 16 May 1842 et seq

FO Public Records Office, Foreign Office, London

GSH Global Seismic Hazard Assessment Program (GSHAP see Annali di Geofisica 1999)

JASB J Asiat Soc Bengal, vol 1, p 34, 146, 1832; ii 439 & 564; xii.1049

IO India Office, Political proceedings of the Government of India, P/S various, London

ISC International Seismological Centre, Regional Catalogue of Earthquakes 1964–2000

ISS International Seismological Summary, 1918–1963

ITT Ittilaat (Tehran)

NAT Nature (London)

PAI Paisa (Lahore)

SMT Seism Monatschr Tiflis Seism Observatory

SZ Sredneaziatskaya Zhizn 1906:222, 262

TGS Trans Geol Soc. London, vol 3, p 492

TIM The Times (London)

TIN The Times of India (Delhi)

TV Turkestanskie Vedomosti (Tashkent) 1906.161, 1910. 282, 286

USG U.S. Geological Survey Earthquake Data Base, National Earthquake Information Center, (Updated 2001)

ZEM Zemletriasenia v CCCP, Institut Fiziki Zemli, Izd. Nauka, Moscow, 1960–1989

ZO Zakaspiyskoe Obozrenie (Ashkhabad) 1910.280–281

Historical Sources

Abu 'I-Fadl 'Allami, Akbar-namah, i.236, Bibl. Ind. 1881, [1569]

Abu 'I-Fida Ismail b. Ali (1869) Al-mukhtasir fi akhbar al-bashar, vol 2. Istanbul, p 26

al-Asafi, Ulughkhani (d.1605) Zafar al-walih bi-muzaffar wa alih, ed. E. Ross, London 1928

Babur, Zahir al-Din (d.1530) Vaqi'at-i Baburi, fasc. A.Beveridge, Gibb Memorial Ser. 1905, trans. Beveridge, The memoirs of Babur, London 1921

Barbier de Meynard C (1860) Extraits de la chronique persane de. J Asiat 16:461

Fasih Khwafi, Mujmil-i Fasihi ed. M. Farrukh, 3 vols, Mashhad, 1340/1961

Hafiz-i Abru, Nur al-Din Majmu' al tawarikh, in F. Tauer, Cinq opuscules de Hafizi Abru concernant l'histoire de l'Iran au temps du Tamerlan, Prague 1959

Hafiz-i Abru, Nur al-Din, Jughrafiya: qismat-i rub'-i Khurasan, ed. M. Naravi, Tehran 1349/1970

Ibn al-Athir, 'Izzal-Din, al-Kamil fi 'l-tarikh, (13th c.) ed. C.J. Tornberg, Leiden 1851–1876

Ibn al-Jauzi, 'Abd al-Rahman, Kitab al-muntazam fi tarikh al-muluk wa 'l-uman, ed. Hyderabad, 1938–41 [1181]

Ibn al-shihna, Muhammad, Raudat al munazir fi akhbar al-awa'il wa'l-awakhir, Marginal work in vols 7–9 of Ibn al-Athir (vol 7, p 59)

Isfizari, Mu'in al-Din, Raudat al jannat fi ausaf madinat Harat, ed. M. Kazim Imam, 2 vols, Tehran 1338/1959

Qudama b. Jafar, Kitab al kharaj, excerpts ed. J.M. de Goeje, Bibliotheca Geograph Arab, vol 4, pp 210–211

al-Suyuti, Jelal al-Din, kashf al-salsala 'an wasf al-zalzala, ed. A.L.Sa'dani, Fez. 1971

al-Umari, Yasin al-Khatib, Al-athar al-jaliya fi 'l-hawadith al-ardiya

Yaqut, Abdallah, Mu'jam al-buldan, ed. Wüstenfeld, Leipzig, 1866–1873

Journal Articles

Abbott A (1879) The Afghan War, 1838–1842, a journal and correspondence of the late Major-General Agustus Abbott, ed. C.R. Low, Bentley, London

Abe K (1981) Magnitudes of large shallow earthquakes from 1904 to 1980. Phys Earth Plant Inter 27:72–93

Abe K (1994) Instrumental magnitudes of historical earthquakes, 1892 to 1898. Bull Seism Soc Am 84:415–425

Abe K, Noguchi N (1983a) Determination of magnitude for large shallow earthquakes 1898 to 1917. Phys Earth Plant Inter 32:45–59

Abe K, Noguchi S (1983b) Revision of magnitudes of large shallow earthquakes 1897–1912. Phys Earth Plan Inter 33:11

Ambraseys N, Melville C (1982) A history of Persian earthquakes. Cambridge University Press, New York, pp 219

Ambraseys N (2001) Reassessment of earthquakes, 1900–1999, in the Eastern Mediterranean and the Middle East. Geophys J Int 145:471–485

Ambraseys N, Douglas J (2000) Reappraisal of surface-wave magnitudes in the Eastern Mediterranean and Middle East. Geophys J Int 141:357–373

Ambraseys N, Bilham R (2003a) Earthquakes and crustal deformation in northern Baluchistan. Bull Seism Soc Am 93(4):1573–160
Ambraseys N, Bilham R (2003b) Earthquakes in Afghanistan. Seism Res Lett 74(2):107–123
Ambraseys N, Jackson D (2003) Early earthquakes in northern India and Tibet. Curr Sci 84(4):570–581
Ambraseys N, Jackson J (1998) Faulting associated with historical and recent earthquakes in the Eastern Mediterranean region. Geophys J Int 133:390–406
Armbruster J, Seeber L, Quittmeyer R, Farrah A (1980) Seismic network data from Quetta Pakistan: the Chaman fault and the fault related to the 30 May 1935 earthquake. Geol Bull Univ Peshawar 13:129–142
Baird-Smith R (1842) Notes on the recent earthquakes on the North-West Frontier. J Asiat Soc Bengal 11: 242–275
Baird-Smith R (1843a) Memoir on Indian earthquakes, part I. J Asiat Soc Bengal 11:1046–1054, Calcutta
Baird-Smith R (1843b) Memoir on Indian earthquakes, part II. J Asiat Soc Bengal 12(2), New series, 136:1029–1056
Baird-Smith R (1844) Memoir on Indian earthquakes, part III. J Asiat Soc Bengal 156:964–983
de Ballore M (1905) La géographie séismologigue. A. Colin, Paris
de Ballore M (1924) La géologie séismologigue. A. Colin, Paris
Barinova A, Shakirzhadova G et al (1987) Zemletriasenia Srednei Azii i Kazakhstan, *Zemletrisenia v CCCP v 1987 g*, Moscow, pp 46–59
Bellew HW (1862) A political mission to Afghanistan in 1857. Smith Elder and co., London
Berberian M, Jackson JA, Qorashi M, Khatib MM, Priestley K, Talebian M, Ghafuri-Ashtiani M (1999) The 1997 May 10 Zirkuh (Qa'cnat) earthquake (MW 7.2): faulting along the Sistan suture zone of eastern Iran. Geophys J Int 136(3):671–694
Bernard M, Shen-Tu B, Holt WE, Davis DM (2000) Kinematics of active deformation in the Sulaiman Lobe and Range, Pakistan. J Geophys Res 105:13253–13253
Blunt W (1957) A Persian spring. Barrie, London
Bolt B (1960) The revision of earthquake epicenters, focal depths and origin-times using a high-speed computer. Geophys J 3:433–440
Bozorgnia M (1962) Newspaper earthquake reports in Iran for the period 1944 to 1962. Report Natl Iranian Oil Co., Tehran
Burnes A (1835) Travels into Bokhara, vol 2. J. Murrey, London, p 15
Burnes A (1842) Cabool; personal narrative of a journey to, and residence in that city. J. Murray, London
Burrard SG (1909) Extracted from the Narrative Report of Captain C. M. Browne, D.S.O., R.E., in charge of No. 24 Party (Triangulation) for Season 190607. 126- In Extracts form narrative Reports of Officers of the Survey of India for the Season 190405. Government Printing Office, Calcutta
Burrard SG (1912) Annual reports 1909–1910. Records of the Survey of India, vol 1. Superintendent Government Printing, Calcutta, pp 24–25
Burtman VS, Molnar P (1993) Geological and geophysical evidence for deep subduction of continental crust beneath the Pamir, vol 281, Geological society America special paper. Geological Society of America, Boulder
Byron R (1937) The road to oxiana. Black, London
Chmyriov VM, Mirzad SH (1972) Tectonic map of Afghanistan. Ministry Mines and Industries, Kabul
Danby CJ, Hore M, MacDermott K, Pelly RC, Taylor GC, Wesson ES (1972) Report of the Cambridge Hindu Kush expedition. St. John's College, Cambridge, pp 66–83
Davison C (1893) Note on the Quetta earthquake of December 20th, 1892. Geol Mag 350:356–361
De Mets C, Gordon RG, Argus DF, Stein S (1990) Current plate motions. Geophys J Int (101) 425–478
Denikaev S, Kafarski A (1973) Recent movements of the Chaman fault in Afghanistan. Report Tekhnoimport, Department of Geology, Kabul
Dupree NH (1971) A historical guide to Afghanistan. Afghan Air Authority, Kabul
Duda S (1965) Secular seismic energy release in the circum-Pacific belt. Tecto-nophysics 2:409–452
Egerton RW (1893) Effects of earthquakes on North-Western railway, India. Engineering (May 19): 698–699.
Ekström G, Dziewonski A (1988) Evidence of bias in estimation of earthquake size. Nature 332:319–323
Engdahl ER, Van der Hildt RD, Buland RP (1998) Global teleseismic earthquake relocation with improved travel times and procedures for depth determination. Bull Seism Soc Am 88:722–743
Eyre V (1843) The military operations at Cabul. J. Murray, London
Farah A (1976) A study of recent seismotectonics in Pakistan presented at the CENTO Working Group on Recent Tectonics, Instanbul (manuscript)
Fuchs CW (1886) Statistik der Erdbebe von 1865–1885, Sitzungsb. k. Akad. Wissen. Math.-Natur. Cl.92(3): 215–625, Vienna
Furrer KH (1956) Annual geological report UNESCO mission to Kabul, (dated Geol.Service Kabul 28.08.1956) Dept. Earth Sci., UNESCO, Paris
Furon R (1925a) Les séismes de la region de Caboul. Compt Rend Acad Sci, Paris
Furon R (1925b) L'Afghanistan, Paris
Giardini D (1999) The Global Seismic Hazard Assessment Program (GSHAP)–1992/1999. Annali de Geofisica 42:957–974
Gleig NA (1846) Sale's brigade in Afghanistan. J. Murray, London
Gorshkov GP (1941) Katalog zemletriasenii na territorii CCCP. Trudi Seismolog Inst 95:28, Moscow
Gorshkov GP (1947) Zemletriasenii Turkmenii, Trudi Seism., no.122, Akad. Nauk, Moscow

Griesbach CL (1893) Notes on the earthquake of Baluchistan on the 20th December 1892. Rec Geol Surv India 26:57–64

Gubin IE (1960) Zakonomernosti seismicheskih proiavlenii na territorii Tadzhikistana. Izd Akad Nauk, Moscow

Gutenberg B, Richter C (1949) Seismicity of the earth and associated phenomena. Princeton University Press, Hafner, New York, pp 267

Gutenberg B. Unpublished worksheerts, Millikan Memorial Library, Cal. Inst. Tech

Hall LA (1979) Archival resources for Afghan history in Afghanistan and Pakistan, vol 2. Afghan Studies, London, pp 75–77

Heuckroth LE, Karim RA (1970) Earthquake history, seismicity and tectonics of the regions of Afghanistan. Publ. Seism. Center, Kabul University, Kabul

Huckriede R, Kursten M, Venzlaff H (1962) Zur Geologie des Gebietes zwischen Kerman und Sagand (Iran). Beihefte zum Geologischen Jahrbuch 51:197

IO (1842) India Office, Political Proceedings of the Government of India. British Library, London

IO (1843) India Office, Political Proceedings of the Government of India. British Library, London

Jackson D (2002) The great western-Himalayan earthquake of 1505: a rupture of the central Himalayan gap? Tibetan Studies I. Brill's Tibetan Studies Library, Leiden, pp 147–159

Jackson J, Haines AJ, Holt WE (1995) Accommodation of Arabia-Eurasia plate convergence in Iran. J Geophys Res 100(15):205–15, 219

Jackson J, Priestley K, Allen M, Berberian M (2002) Active tectonics of the South Caspian Sea. Geophys J Int 148:214–245

Jackson R (1960) Thirty seconds at Quetta. Evans, London, p 248

Khan AQ (1956) Earthquake and aseismic design in Pakistan. In: Proceedings of the 1st world conference on Earthquake Engineering, Berkeley, vol. 19, pp 1–9

Kondorskaya N, Shebalin N (1997) Nov'i katalog sil'nih zemletriasenii na territorii SSSR. Izd. Nauka, Moscow

Levitski G (1908) Blulletin postoiannoi tsentralnoi seismicheskoi komissii 1902–1907, Imperat. Akad. Nauk, vols 1–3, St. Petersburg (continued to 1912 by P. Nikiforov)

Lawrence R, Yeats R (1979) Geological reconnaissance of the Chaman fault in Pakistan. In: Farah A, DeJong KA (eds) Geodynamic of Pakistan, Quetta, Elite Publishers, Karachi, pp 351–357

Lawrence RD, Khan SH, Nakata T (1992) Chaman Fault, Pakistan-Afghanistan. In: Buckham RC, Hancock P (eds) Major active faults of the world. Results of IGCP Project 206, Annales Tectonicae, Special Issue suppl. vol 6, pp 196–223

Lloyd H (1846) Extracts of a letter from Dr. Lloyd to Sir J. Herschel, Trinity College Dublin 1845. Reports on the State of Science. In: Report of the fifteenth meeting of the British Association for the Advancement of Science, 1845, Cambridge. John Murray, London, pp 3–5

Mackey JN (1962) Regimental history of the 7th Duke of Edinburgh Gurkha Rifles. W. Blackwood, London

Maggi A, Jackson J, McKenzie D, Pristley K (2000a) Earthquake focal depths, effective elastic thickness and the strength of the continental lithosphere. Geology 28:495–498

Maggi A, Jackson J, Priestley K, Baker C (2000b) A re-assessment of focal depth distribution in southern Iran the Tien Shan and northern India: do earthquakes really occur in the continental mantle? Geophys J Int 142:629–661

Mallet R (1853) Third report on the facts of earthquake phenomena. In: Report of the twenty-first meeting of the British Association for the Advancement of Science, Belfast 1852, John Murray, London, pp 1–176

Martin FA (1907) Under the absolute amir. Harper, London, pp 255–258

Masson C (1844) Narrative of various journeys in Baluchistan, Afghanistan, the Punjab and Kalat, vol 2. R. Bentley, London

Melville C (1978) Arabic and Persian source material on the historical seismicity of Iran. PhD thesis, University of Cambridge, Cambridge

Minshull TA, White RS, Barton PJ, Collier JS (1992) Deformation at plate boundaries around the Gulf of Oman. Mar Geol 104:265–277

Mitshkevich A (1937) Epitsentralnaia zona zemletriasenii Samarkandskogo raiona. *Trudi* no.2, Samark Seism Stan, Tashkent

Moinfar A, Mahdavian A, Maleki E (1994) Historical and instrumental earthquake data collection of Iran. Publ. Iranian Cultur. Affairs Inst, Tehran, pp 446

Molnar P, Tapponnier P (1975) Cenozoic tectonics of Asia: effects of a continental collision. Science 189:419–426

Musketof N, Orloff A (1893) Katalog zemletriasenii Possiskoi imperii. Zapis Imper Puss Geograf Ob, vol 26, Sanktpeterbug, p 242

Niedermayer R (1936) Der Wüstenzug der deutschen Expedition nach Persien und Afghanistan Hamburg, p 108

Nikiforov PM (1912) Byullet Postoiannoi Tsentral Seism Comm, 1911, St. Petersburg, p 3

Nowroozi A (1971) Seismotectonics of the Persian plateau, Eastern Turkey, Caucasus and Hindu Kush regions. Bull Seism Soc Am 61:317–342

Patterson J (1908) Seismic disturbances; Simla observatory, Monthly Weather Review, July issue, Calcutta, p 71

Pegler G, Das S (1998) An enhanced image of the Pamir-Hindu Kush seismic zone form relocated earthquake hypocenters. Geophys J Int 134:573–595

Pinhey LA (1938) The Quetta Earthquake of 31 May 1935. Govt of India, New Delhi

Quittmeyer R, Jacob K (1979) Historical and modern seismicity of Pakistan, Afghanistan, northwest India and southeastern Iran. Bull Seism Soc Am 69:777

Ramanathan K, Mukherji S (1938) A seismological study of the Baluchistan, Quetta, earthquake of May 31, 1935. Rec Geol Surv India 73:483–513

Robertson RG (1948) Earthquake-resistant structures, the seismic factor and the use of reinforced brickwork in Quetta civil construction, Paper no. 5381. J Inst Civil Eng 171–187, London

von Rebeur-Paschwitz E (1893) Ueber die Aufzeichnung der Fernewirkungen von Erdbeben. Petermann Mitteil 9:201–212

von Rebeur-Paschwitz E (1895) Horizontalpendel-Beobachtungen auf der Kaiserlichen Univesitäts Sternwaarte zu Strassburg 1892–1894. Beiträge z Geoph 2:211–534, Leipzig

Sale F (1843) A journal of the disasters in Afghanistan 1841–1842, vol 2. Galognani, Paris

Sale R (1846) The defense of Jalalabad. Hogarth, London

Samizay M (1998) Earthquake history and hazard zones in Afghanistan, Afghan Mosaic Magazine, July issue

Scheu E, Lais R (1912) Catalogue régional des tremblements de terre 1907. Publ. Bureau Centr. Assoc. Intern. Seism, Strasbourg

Sella GF, Dixon TH, Mao A (2002) REVEL: a model for recent plate velocities from space geodesy. J Geophys Res 107(4). doi:10.1029/2000JB000033

Skrine CP (1936) The Quetta earthquake. Geogr J 88:414–430

Spate OHK (1954) India and Pakistan. Methuen, London

Spesivtseva VP (1933) Istoriko-statsticheskie materiali po seismichnosti Sredi Azii. Trudi Seismol. Instit, Material. po Seismotekt. Akad. Nauk, no.34, Leningrad

Spil'kov GA (1914) Zemletriasenie 1911 g. Pamirah i ego posledstviya. Izvest Imper Russ Geograf Ob-va 50: 69–97, Petrograd

Stein S, Sella G, Okal E (2002) The January 26, 2001 Bhuj earthquake and the diffuse western boundary of the Indian plate. In: Stein S, Freymueller J (eds) Plate boundary zones, vol 30, American geophysical union geodynamics series. American Geophysical Union, Washington, D.C., pp 243–254

Stenz E (1945) Strong earthquakes in Afghanistan. Q Bull Polish Inst Arts Sci (January 1945)

Sykes LR, Landisman M (1964) The seismicity of east Africa and Gulf of Aden and the Arabian Seas. Bull Seism Soc Am 54:1927–1940

Survey of India (1948) Technical report 1947 part III geodetic work. Dehra Dun, pp 139

Talbot and Maitland (1886) J Royal Asiat Soc, 1886, p 323, and Encyclopedia of Islam, item Bamiyan)

Tatar M, Hatzfeld D, Martinod J, Walpersdorf A, Ghafori-Ashtiany M. Chéry J (2002) The present-day deformation of the central Zagros from GPS measurements. 29(19):33-1–33-4. doi:10.1029/2002GL015427

Thomson ROC (1936) Notes on the Quetta earthquake. Technical paper 307, Railway Board, Lucknow

Willmore PL (1979) Manual of seismological observatory practice, World Data Center A for Solid Earth Geophysics, Rep. SE-20, Natl. Oceanic and Atmosph. Admin., Denver

Wesson PS (1972) The 1972 Cambridge Hindu Kush expedition. St. John's Coll, Cambridge, pp 66–82

West WD (1936) Preliminary geological report on the Baluchistan (Quetta) earthquake of May 31, 1935. Rec Geol Surv India 69:204–241

Wilson AT (1930) Earthquakes in Persia. Bull School Orient Afr Stud 6:103

Wood J (1872) A journey to the sources of the river Oxus. J. Murrey, London

Yeats R, Sieh K, Allen C (1997) The geology of earthquakes. Oxford University Press, Oxford, p 483

Yule H (1858) A narrative of the mission sent by the Govt.-General of India to the court of Ava in 1855, with notices of the country. London

Zakharova A, Ibragimov R, Iodko V et al (1972) Zemletriasenia Sredni Azii. Zemletriasenia v CCCP v 1968 godu, Izd. Nauka, Moscow, pp 65–72

Properties of Local Materials 7

Claudio Margottini

Mineralogical and Petrographical Composition of Local Materials

Mechanical properties of rocks cannot be investigated without a clear identification of the mineralogical and petrographical composition of materials. On the samples coming from the Bamiyan cliff, the following investigations have been performed:

- X-ray diffractometry
- Mineralogical phases
- Petrographical composition and sedimentological aspects

Introduction

The preceding three items are strictly related and point out the pertinent information about genesis, chemical composition, and stability in time as a consequence of meteorological conditions (weathering) and/or external contamination, as in potential chemical grouting.

X-ray diffractometry has been developed in six different samples: conglomerate matrix, conglomerate matrix with grain diameter less than 0.1 mm, yellow siltstone (two samples), and red siltstone (two samples). All these samples are from the lower medium part of the cliff and are not related to the statues.

The conglomerate is composed of quartz, calcite, mica, feldspar group minerals, clay minerals, and heavy minerals, in order of importance (x-ray diffractometry). Quartz is associated with individual grains in which calcite is present as well as secondary deposition in terms of carbonuric cement. Matrix is composed mainly of clay minerals (Fig. 7.1).

Upon thin-section microscope analysis, the conglomerate results were shown to be composed of mono-mineral grains and lithic fragments, moderately classed, and cemented by carbonates (Fig. 7.2). Mono-mineral grains comprise scarcely rounded mono-crystalline quartz and poli-crystalline aggregation of moderately rounded quartz. Less frequently are observed poli-crystalline aggregation of moderately rounded carbonates and rarely mono-crystals or aggregation of slightly rounded feldspar. Dimensions of mono-crystals may vary from tens of ~tm to about 1.5 mm. The dimensions of aggregations can reach 3 mm.

Lithic fragments exhibit dimensions from 1 to 15 mm. The majority of them can be grouped into three main families:

- Fragments of rocks constituted by small grains (sandstone?) composed mainly of quartz, calcite, and clay minerals.
- Fragments of moderately rounded claystone-siltstone. This family seems to be similar to the siltstone formation surveyed on the cliff.
- Fragments of probably volcanic/subvolcanic origin highly altered in clay minerals, oxides and/or hydroxides, and rarely epidote. In some

C. Margottini
Department of Geological Survey of Italy,
ISPRA - Italian Institute for Protection and Environmental Research,
Via Brancati, 60, 00143 Rome, Italy
e-mail: claudio.margottini@isprambiente.it

C. Margottini (ed.), *After the Destruction of Giant Buddha Statues in Bamiyan (Afghanistan) in 2001*,
Natural Science in Archaeology, DOI 10.1007/978-3-642-30051-6_7, © Springer-Verlag Berlin Heidelberg 2014

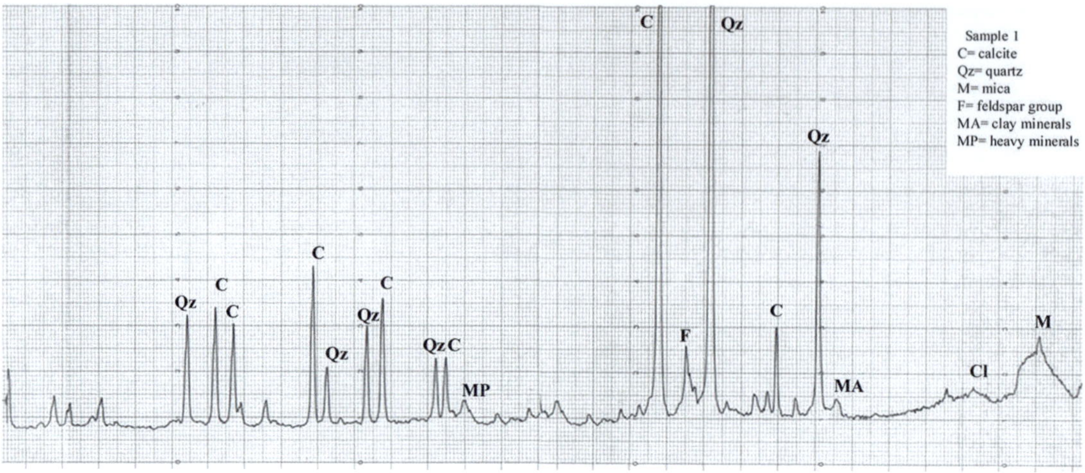

Fig. 7.1 X-ray diffractogram of conglomerate matrix

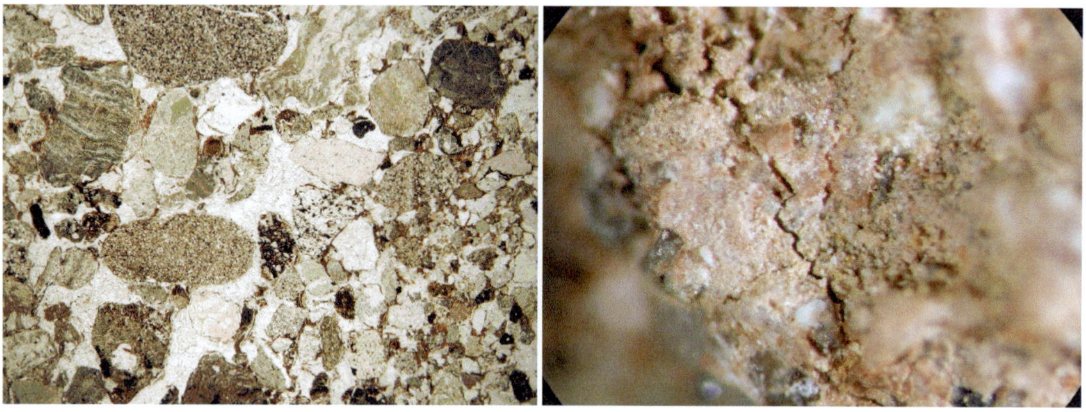

Fig. 7.2 Thin section (*left*) and 200× microscope view (*right*) of conglomerate

fragments, scarcely altered feldspar fenocrystals can still be seen and femic minerals altered in oxides and hydroxides, in a microcrystalline background paste. Rarely rounded quartz crystal can be detected. More often they are fragments of altered background paste.

In some cases lithic fragments are totally transformed into a dark/brown color phase, probably iron hydroxides. Iron hydroxides and/or iron sulfate are frequently near lithic fragments and grains (between carbonate cements and clasts), within the cement, and in some fractures, affecting the fragments.

According to the preceding considerations, the recorded field uniaxial compression strength mainly depends on the carbonuric cement, developed secondarily during the diagenetic phase (see later in this chapter).

The siltsone has been surveyed on the lower part of the cliff (yellow component) and in some strata in the middle-upper part of the cliff. It is composed of quartz, calcite, mica, clay minerals, and heavy minerals, in order of importance (x-ray diffractometry). Quartz can be approximately estimated as 70 %, calcite 20 %, and all others about 10 %. According to Brown's (1972) methodology, the following clay minerals have been detected: illite, chlorite, and smectite.

X-ray diffractometry of yellow and red material did not reveal any mineralogical difference (Figs. 7.3, 7.4, and 7.5).

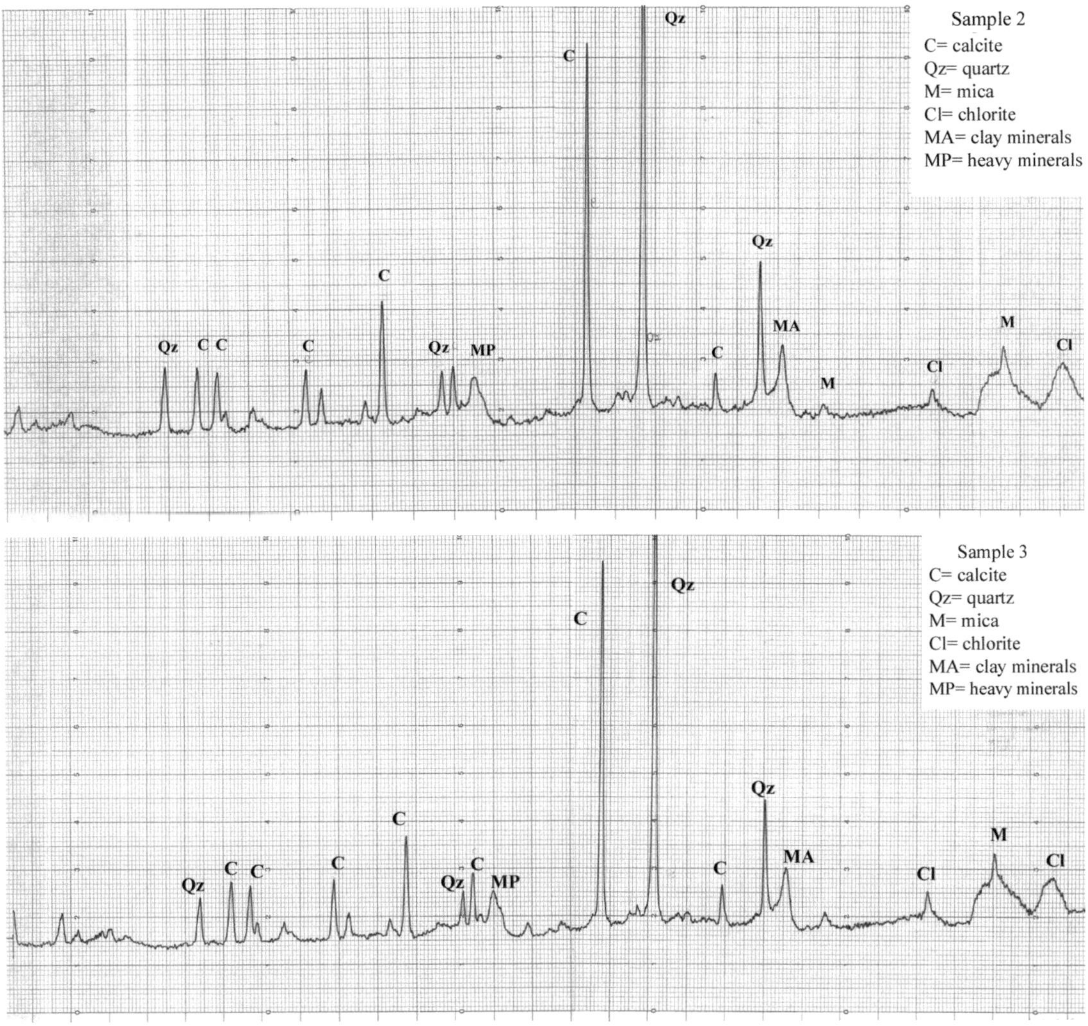

Fig. 7.3 X-ray diffractogram for yellow siltstone (sample 2) and red siltstone (sample 3)

In microscopic thin section (Fig. 7.6), the rock can be recognized as an aggregation of clay minerals crossed by small irregular veins of carbonates. Clay minerals seem to be associated with iron hydroxides and/or sulfates. Within the aggregate quartz grains, scarcely rounded with dimensions ranging from 0.01 to 1 mm as well as fragments, moderately rounded crypto-crystalline carbonuric rocks with dimensions less than 2 mm are evident.

No cement has been identified on microscopic thin section, as confirmed by the total loss of mechanical cohesion when the sample is weathered. Only red siltstone flocculates in water, testifying to the presence of salts, which are residuals of an evaporite depositional environment. In conclusion, siltstone can be considered desiccated mud, with an apparent cohesion depending on dryness, at least in part.

In both conglomerate and siltstone (yellow), x-ray diffractometry does not identify soluble salts (e.g., gypsum, alite). This is in contrast to the flocculation of red siltstone, but it testifies to an alternation of environmental deposits in time.

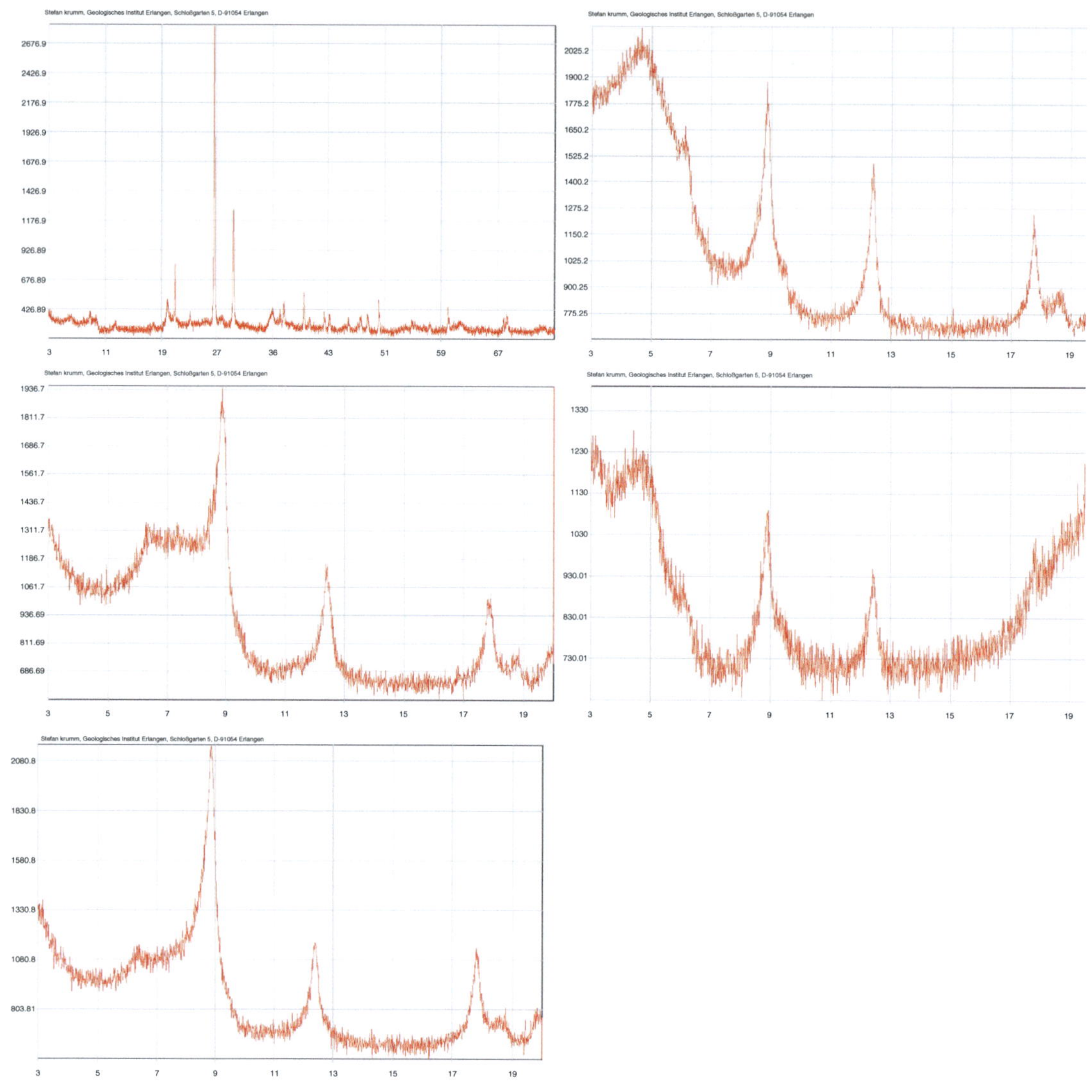

Fig. 7.4 X-ray diffractometry of yellow siltstone. (1) A tout venant sample. (2) *Mg* sample processed with magnesium salt. (3) *K* sample processed with potassium salt. (4) *Mg glic* sample processed with salt and then with glycerine. (5) *K* 100 sample processed with salt and then heated to 100 °C

Petro-geophysical Properties of Local Materials

*V*p has been measured in the laboratory for both conglomerate and siltstone (Fig. 7.7). Conglomerate exhibits values of about 2.2–2.5 km/s and a Poisson ratio of about 0.300–0.400 (samples perpendicular and transverse to deposition flow). The value of *V*p is approximately 1.5–1.7 km/s and a Poisson ratio of approximately 0.100–0.200 has been measured in the direction of depositional flow.

The siltstone exhibits a *V*p of approximately 1.7–2.0 km/s and a Poisson ratio of approximately 0.150–0.200 without any geometrical anisotropy (Table 7.1).

The lower seismic velocity in conglomerate is parallel to the sedimentary flux (stratification plan). This implies that, in this direction, it is likely that seismic waves exhibit lower values

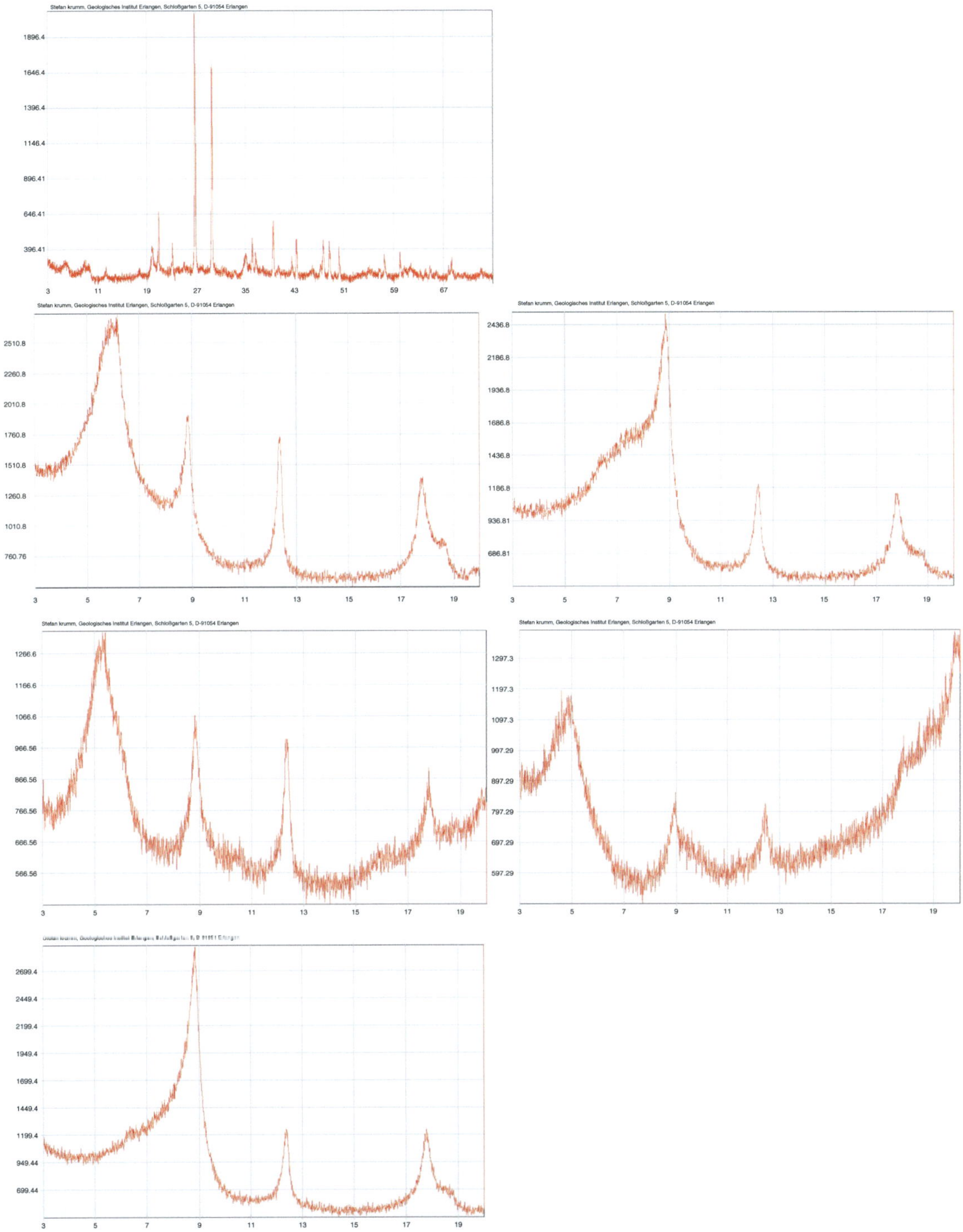

Fig. 7.5 X-ray diffractometry of yellow siltstone. (1) A tout venant sample. (2) *Mg* sample processed with magnesium salt, (3) *K* sample processed with potassium salt, (4) *Mg GL* sample processed with salt and then glycolated, (5) *Mg glic* sample processed with salt and then with glycerine. (6) *K* 100 sample processed with salt and then heated to 100 °C

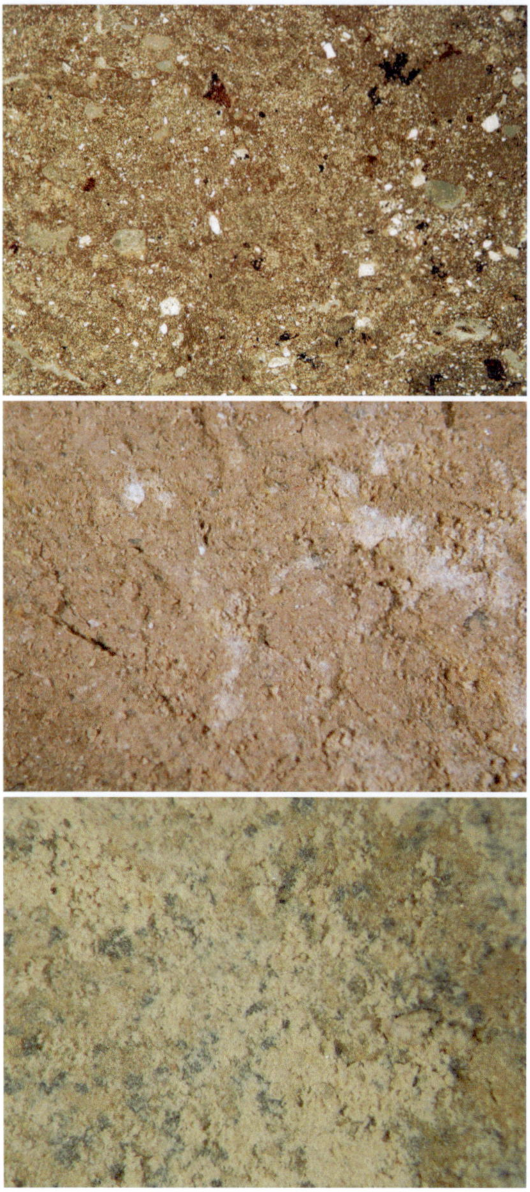

Fig. 7.6 Thin section and 200× microscope view of yellow and red siltstone

because they cross the matrix mainly composed of cemented silt material (axe 2, Fig. 7.8). The velocity is higher when the same seismic waves cross the samples perpendicular and transverse to the sedimentary flux/stratification plan (axes 1 and 3, Fig. 7.8), probably because it intersects greater thickness of rock. Finally, the values obtained in the sedimentary flux/stratification plan of conglomerate, when the matrix is predominant, are closer to the values measured in siltstone, indirectly confirming a similar origin and composition.

Investigations of Siltstone Under Scanning Electronic Microscope

The siltstone samples collected in Bamiyan exhibited very severe slaking (Fig. 7.9), completely disaggregating the material when submerged in water, after just a few minutes (3–10 min, according to different samples).

The swelling behavior of fine-grained sedimentary rocks (e.g., claystones, mudstones, shales) is a complex phenomenon (Franklin and Chandra 1972; Crosta 1998). When in contact with water, these materials can present with considerable volume change, resulting in swelling or breakdown and sometimes mobilizing high swelling pressures. The behavior of swelling rock is controlled by numerous factors acting jointly and resulting in an effect whose main agent is frequently hard to identify. The last two decades have been marked by a growing concern about weak rock behavior and there has been an increased number of studies. This type of rock generally behaves in a way that is intermediate between rock and soil, which complicates the sampling and testing required in order to forecast its geotechnical behavior. Some weak rocks are more sensitive to wetting and drying cycles than others, resulting in an almost total breakdown after a number of cycles or, like the siltstone from the Bamyian cliff, just after watering. Identification of this behavior is crucial in planning work in areas in which weak rocks occur (Pejon and Zuquette 2002).

Sample preparation and microscopic observations were carried out following the recommendations of Goldstein et al. (1992) and Le Roux (1971). Obtained by simple rupture, all the observed samples were approximately 15–20 mm long, 5–10 mm wide, and 10 mm high. They were dried by a cryosublimation process (Le Roux 1971; Shi et al. 1999) before being observed using the microscope. This procedure is intended to preserve rock microtexture and original

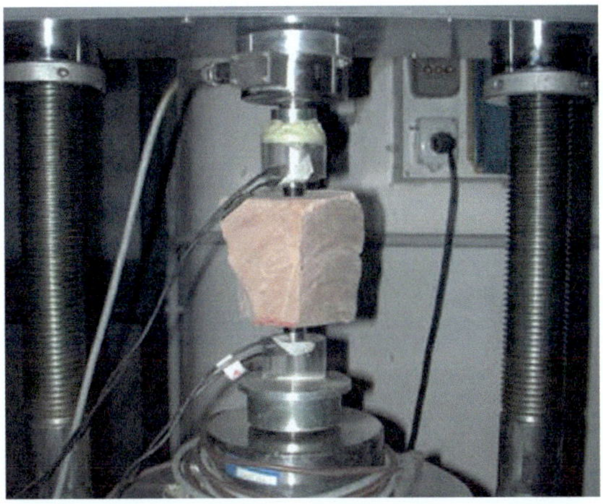

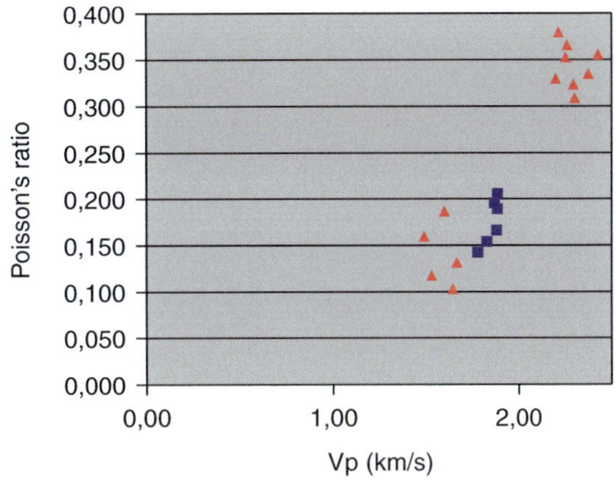

Fig. 7.7 Vp velocity and Poisson ratio in conglomerate (*red*) and siltstone (*blue*)

Table 7.1 Petro-geophysical properties of siltstone and conglomerate

Sample	High mm	t_P Micro s	t_S Micro s	VP km/s	VS km/s	VP/VS	pois_dyn	Notes
S1_1	108.80	59.8	93.4	1.82	1.16	1.56	0.153	
S1_2	104.20	58.7	90.8	1.78	1.15	1.55	0.141	
S2_1	74.10	39.4	63.6	1.88	1.17	1.61	0.189	
S2_2	65.00	34.7	54.8	1.87	1.19	1.58	0.165	
S1_3	103.20	55.2	89.6	1.87	1.15	1.62	0.194	
S2_3	70.00	37.3	61.2	1.88	1.14	1.64	0.205	
C1_2	114.40	75.1	114.1	1.52	1.00	1.52	0.118	Vs uncertain
C1_2a	111.20	67.8	101.9	1.64	1.09	1.50	0.103	Vs uncertain
C1_1	104.90	44.2	88.7	2.37	1.18	2.01	0.335	
C1_1a	102.90	44.8	87.7	2.30	1.17	1.96	0.323	
C1_2b	114.60	68.9	105.7	1.66	1.08	1.53	0.131	Vs good
C2_1	93.50	42.2	95.7	2.22	0.98	2.27	0.379	Vs good
C2_1a	98.10	42.6	81.0	2.30	1.21	1.90	0.309	
C2_2	115.40	77.9	122.4	1.48	0.94	1.57	0.160	
C2_2a	115.20	72.3	116.4	1.59	0.99	1.61	0.186	
C2_3	109.20	49.7	98.7	2.20	1.11	1.99	0.330	
C2_3a	107.80	47.6	103.0	2.26	1.05	2.16	0.364	Vs uncertain
C1_3	120.20	53.3	111.4	2.26	1.08	2.09	0.352	
C1_3a	120.20	49.5	104.7	2.43	1.15	2.12	0.356	

The legend of sample S1_2: S1 stands for siltstone (C conglomerate) sample n.1; the second number represents the axes and then the seismic wave direction; 3 is perpendicular to stratification plane; a and b stand for repetition of measurements.

structure and allow observation of modifications caused by the swelling process. Surfaces perpendicular to the rock lamination were preferred for observation with scanning electronic microscope (SEM), although some surfaces parallel to the bedding plane were also examined, mainly when generated during swelling tests. All the samples were observed with the electron microscope both before and after the swelling tests, to analyze textural, structural, or mineralogical modifications caused by the swelling process. The following figures report SEM observations for 300× and 1k×, undisturbed and after weathering and breakdown of the structure (Figs. 7.10 and 7.11).

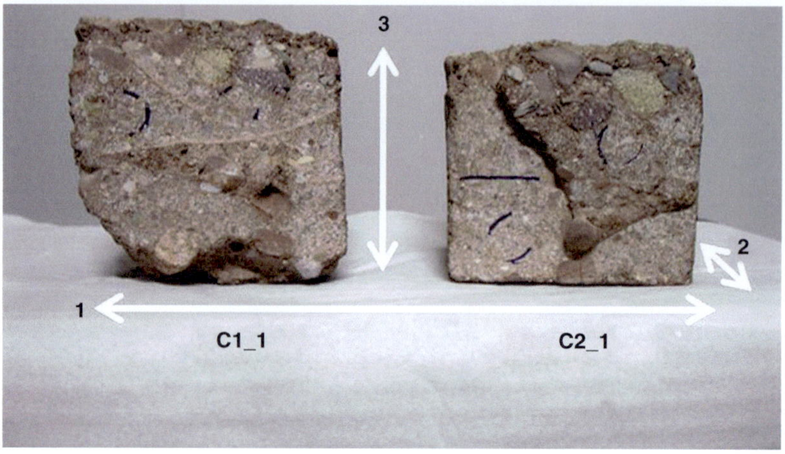

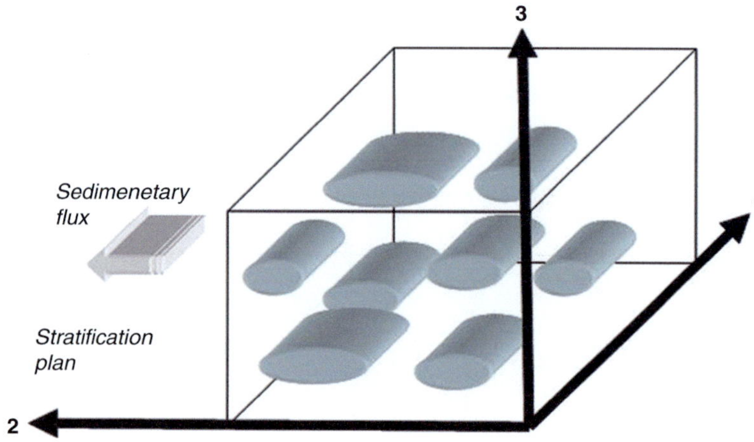

Fig. 7.8 Axis used in seismic wave measurement (*left*) and sedimentary flux/stratification plan of lithotypes (*right*)

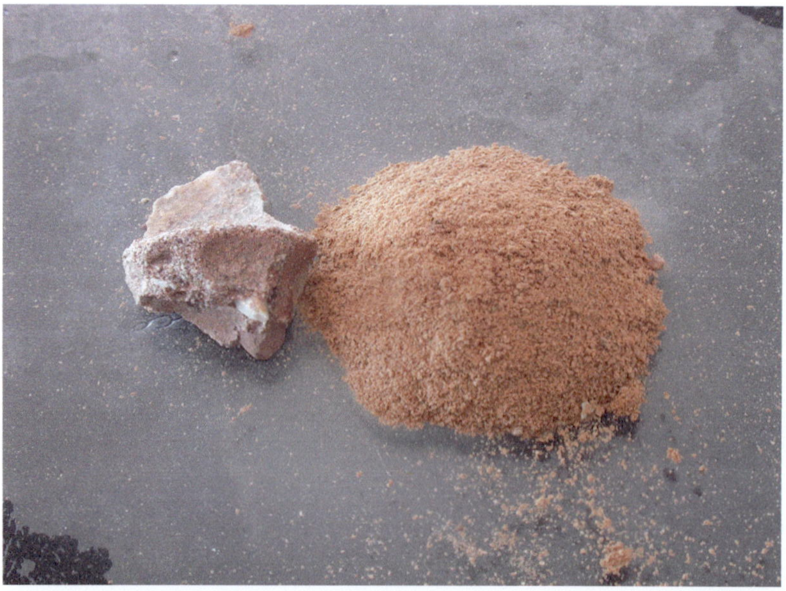

Fig. 7.9 Stability of conglomerate (*left*) and degradation of siltstone (*right*) after only a few minutes of immersion in water

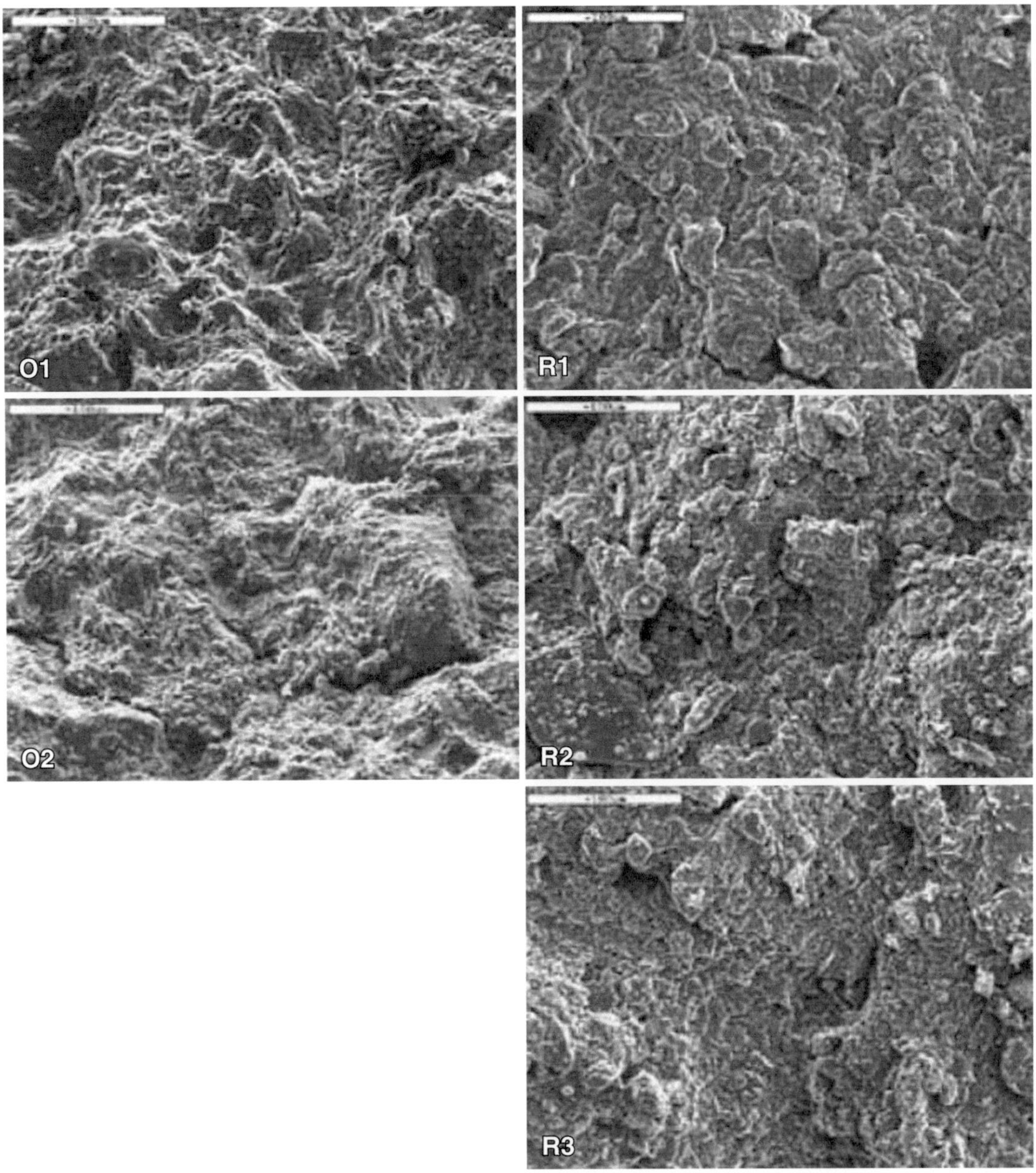

Fig. 7.10 Observation under scanning electronic microscope at 300×. O1 and O2 are and undisturbed sample, whereas R1, R2, and R3 are from material weathered, broken down, and then dried to obtain a feature similar to natural conditions

In spite of the low concentration of swelling clay mineral in the studied materials, breakdown was intense, and occurred in a few minutes when the samples were submitted to wetting.

Observation of the rocks under SEM demonstrated that the presence of a compact and laminar structure (left side of the figures) no longer exists after breakdown (right side of the figures) and drying. The laminar structure could be the consequence of the clay minerals' orientation, completely filling the voids of the siltstone, and producing a high-density material. Wetted

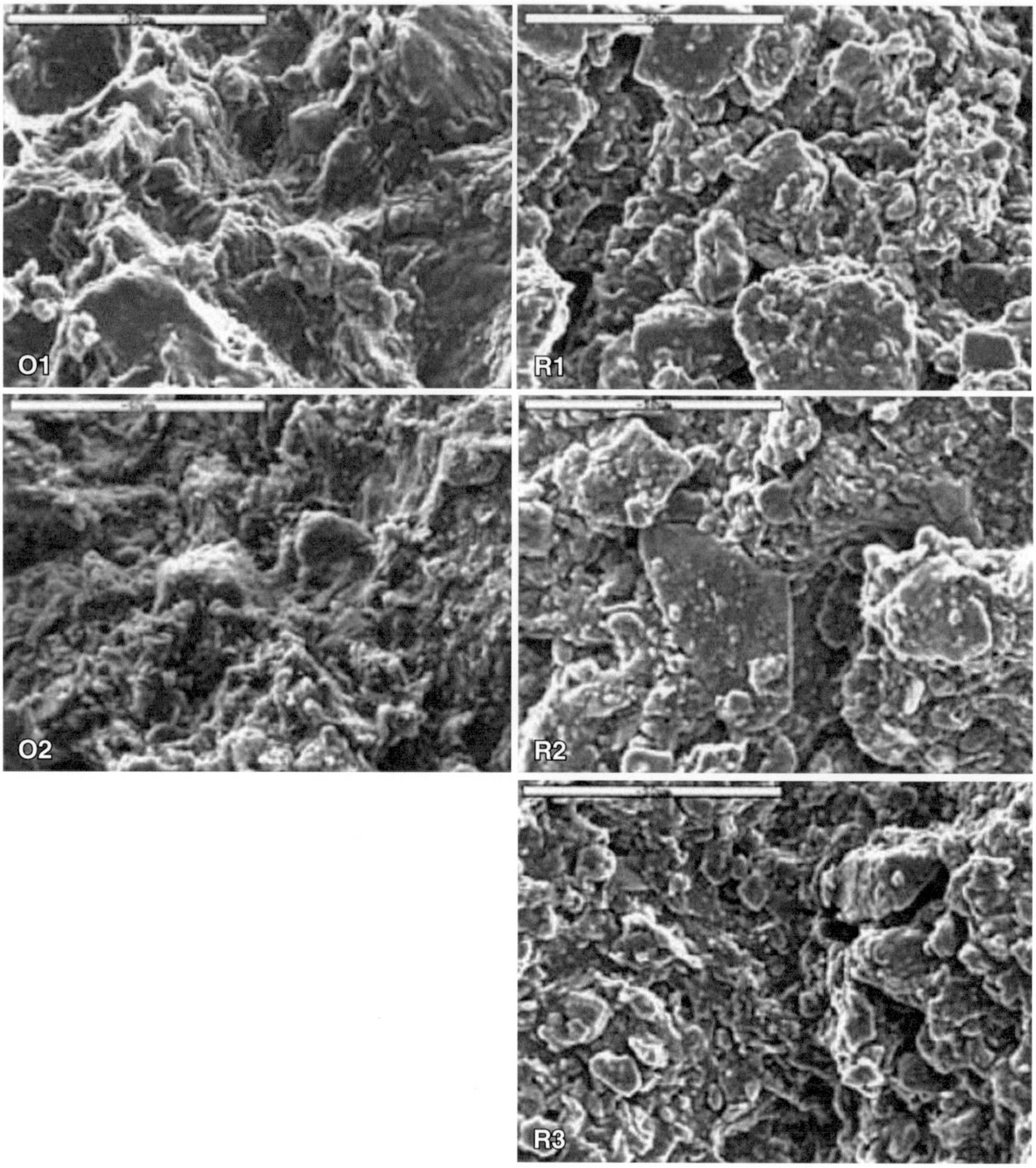

Fig. 7.11 Observation under scanning electronic microscope at 1,000×. O1 and O2 are an undisturbed sample, whereas R1, R2, and R3 are from material weathered, broken down, and then dried to obtain a feature similar to natural conditions

and reconstituted material no longer presents the typical structure of undisturbed samples; it clearly presents an evident grain composition, some laminar, probably clay minerals. The number of voids is clearly higher with respect to undisturbed siltstone, so the original high density is not reached after a wetting and drying cycle on the same material.

Possibly the small amounts of swelling clay (e.g., smectite) in certain positions in the rock texture could promote their swelling and breakdown, as observed by Pejon and Zuquette (2002), in

French mudrocks with similar granulomatic and mineralogical composition.

The compression of entrapped air was shown to also be an important factor in the swelling and breakage of mudrocks, also observed by Taylor and Spears (1970).

Physical and Mechanical Properties of Materials

A field survey as well as laboratory tests were conducted to collect as much data as possible. The laboratory test comprised physical properties (Atterberg limits and grain size distribution) for siltstone and a point load test as an indicator of uniaxial compressive strength for both the conglomerate and siltstone. The field investigation included in situ density for the conglomerate, rock classification (RMR and Q system), and uniaxial compressive strength by means of a Schmidt hammer for both siltstone and conglomerate.

Soil and rock mechanic tests on both materials were necessary because the two lithotypes exhibited completely different behavior when weathered. As described earlier, the conglomerate exhibited permanent cohesion because of the carbonate cement in the matrix, whereas the siltstone, a desiccated mud, easily lost its apparent cohesion when saturated. From a physical and mechanical point of view, it is extremely important to define the major parameters characterizing the two lithotypes as well as define their respective behavior to the possible consolidation techniques or, stated another way, define the consolidation measurements according to the real behavior and property of local materials.

The grain size distribution of siltstone, determined with an electronic sedigraph, was discovered to be 7 % for clay, 17 % for sand, and 76 % for silt. This distribution identifies "sandy silt" in the Shepard (1964) diagram (Fig. 7.12). According to Atkinson (1997), this grain size distribution is typical for a river estuary environment, contributing to the hypothesis of deposition in a small lagoon or lake.

The grain size distribution of a second sample of siltstone, investigated with a standard hydrometer, revealed 28 % for clay, 66 % for silt, and 6 % for sand.

Plasticity characters reflect the low clay content, pointing out a liquid limit of 39.80 % and a plastic limit of 16.96 %. The material is characterized as CL, according to the Unified Soil Classification System reported in the American Society for Testing of Materials (ASTM) (1977), which fits in the category of "inorganic clays of low to medium plasticity, gravelly clays, sandy clays, silty clays, lean clays" (Fig. 7.13).

Siltstone also exhibits a high natural density (2.28 g/cm^3), despite a dry density of 2.23 as displayed in the phase diagram of the figure. The values of porosity and void index are quite low—14.93 and 0.176 %, respectively—because of the high compaction of siltstone. These data are from samples taken from the site and may not exactly reflect the actual conditions. Nevertheless, the high natural density is a matter of fact (Fig. 7.14).

A major consideration from Fig. 7.14 is that the siltstone is almost in a state of critical natural density (maximum allowed). Any breakdown and re-consolidation may occur only with increasing volume, as testified by SEM.

Table 7.2 provides a summary of the major physical properties of siltstone.

A general survey of the mechanical properties of the entire cliff was executed by means of a field Schmidt hammer. This type of test, even if very rough, allows us to understand the possible vertical and lateral variability of rock mechanic properties better than does the absolute value of uniaxial compressive strength (UCS), which is obviously more precise, but is tested on few samples and then represents only the point measured.

The uniaxial compressive strength, evaluated by a field Schmidt hammer, highlights the general satisfactory condition of the materials, at least in the dry season. Lateral variability can be also numerically evaluated by standard deviation, which assumes a consistency of approximately 13 % from the mean for both the conglomerate and siltstone. These numbers and the data reported in Fig. 7.15 do not suggest important variations in the pattern of mechanical properties.

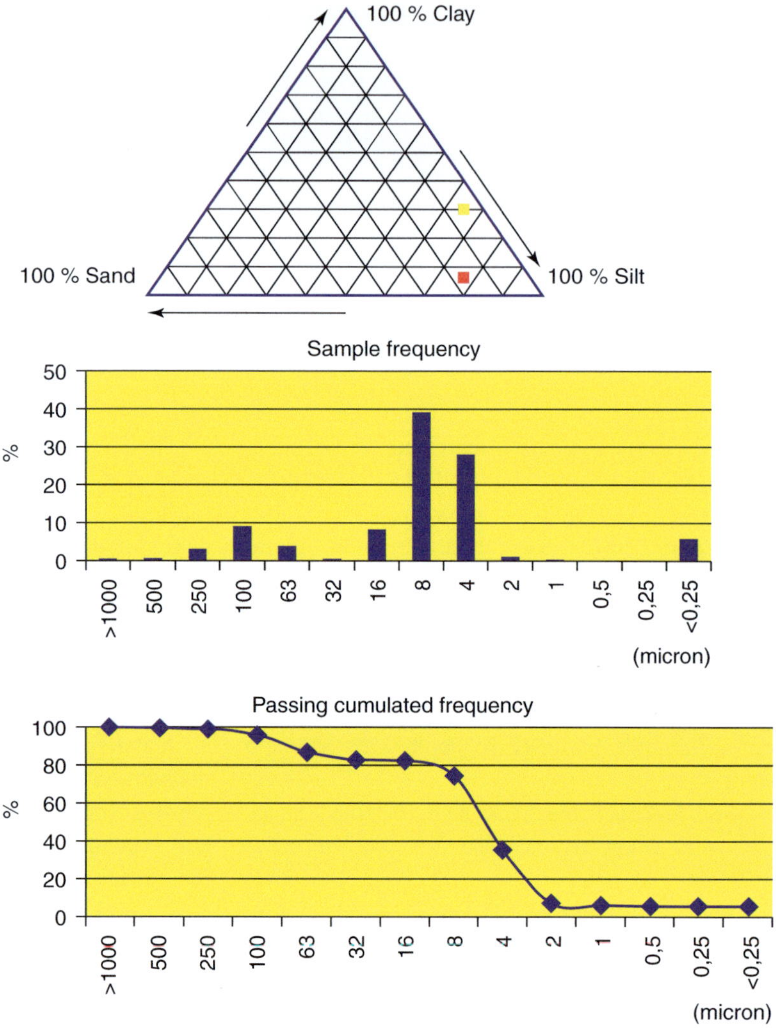

Fig. 7.12 Shepard (1964) diagram for siltstone grain size distribution (*red* is from sedigraph and *yellow* from hydrograph), classifying the material as sandy silt (*red*) and clay silt (*yellow*). Curves are from the sedigraph

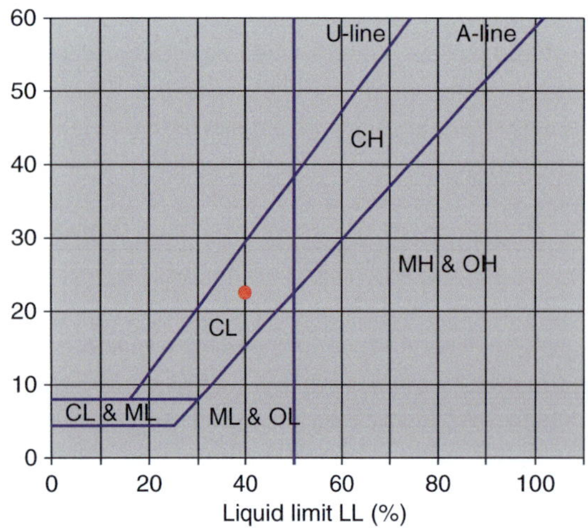

Fig. 7.13 Plasticity chart for Bamiyan siltstone

Keeping in mind the rough accuracy of the specific UCS value when coming from a field Schmidt hammer, it is interesting to note that siltstone is generally higher than conglomerate in dry conditions. When weathered, the situation is completely different, but it is likely that the presence of siltstone in the cliff is no more than 20–30 %.

In order to get a precise and reliable value of UCS, both siltstone and conglomerate have been investigated in a rock mechanic laboratory, with samples collected in the field and allowed to be taken from Afghanistan. The results are plotted in Fig. 7.16. Siltstone exhibits a UCS of 11.0 MPa, whereas conglomerate is, on average, 5.4 MPa. In general terms, it is possible to say that siltstone has a higher

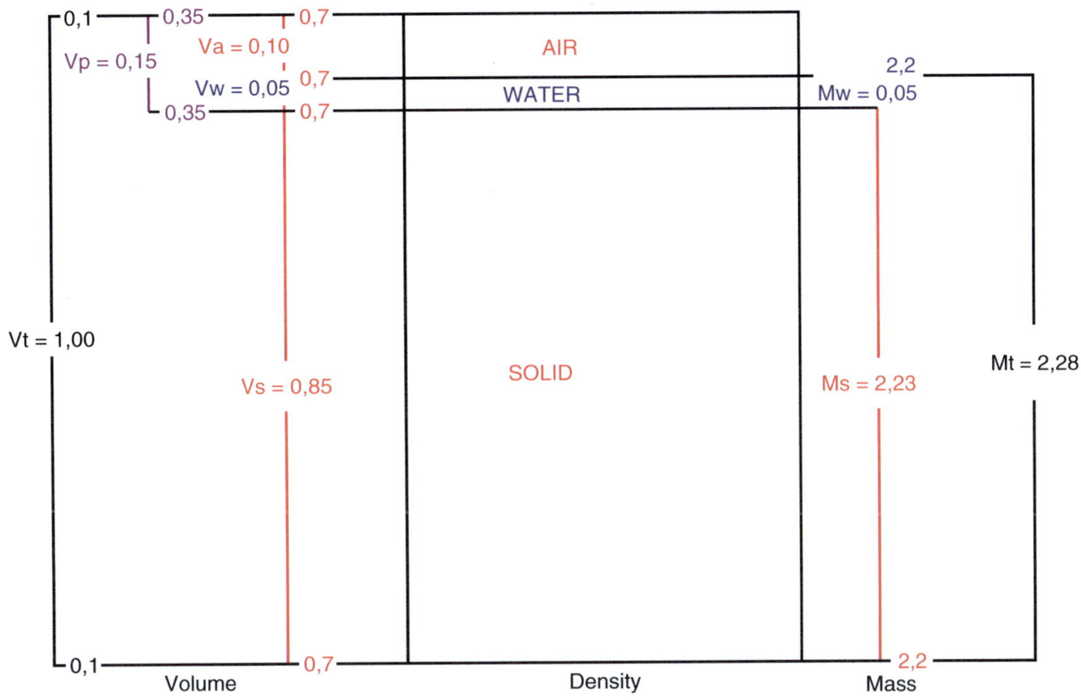

Fig. 7.14 Phase diagram for solid, water, and air constituting the siltstone lithotype of Bamiyan

Table 7.2 Major physical properties of siltstone

	Index parameters								Plasticity parameters			Grain size distribution		
Sample	Unit weight of soil mass Gr/cm³	Dry unit weight Gr/cm³	Unit weight of solids Gr/cm³	Water content %	Degree of saturation %	Porosity %	Void index	Consistency index	Liquid limit	Plastic limit	Plasticity index	Sand %	Silt %	Clay %
1	2.28	2.23	2.62	2.24	33	15	0.176	1.639	39.8	16.96	22.84	6	66	28
2												17	76	7

Sample 2 stands for sedigraph grain size analysis; 1 stands for hydrograph grain size analysis.

uniaxial compressive strength with respect to conglomerate.

From Fig. 7.16 it is also possible to note that conglomerate values are more concentrated in their mean value (standard deviation 1.70 MPa), whereas siltstone data are more widespread (standard deviation 2.60 MPa). As a first hypothesis on the investigated samples, these suggest the possibility of larger variation of mechanical properties of siltstone with respect to conglomerate, even when an unremarkable pattern is demonstrated by the field survey. Both materials can generally be defined as very weak rocks (ISRM, International Society for Rock Mechanics 1981).

In general terms, the uniaxial compressive strength, determined with a point load, is about 30 % of that exhibited on in situ tests, whereas the strength of conglomerate in a point load (5.6 MPa) is about 20 % of the in situ Schmidt hammer tests. This difference is not surprising because of the limitation of field tests in the definition of precise uniaxial compressive strength, the different

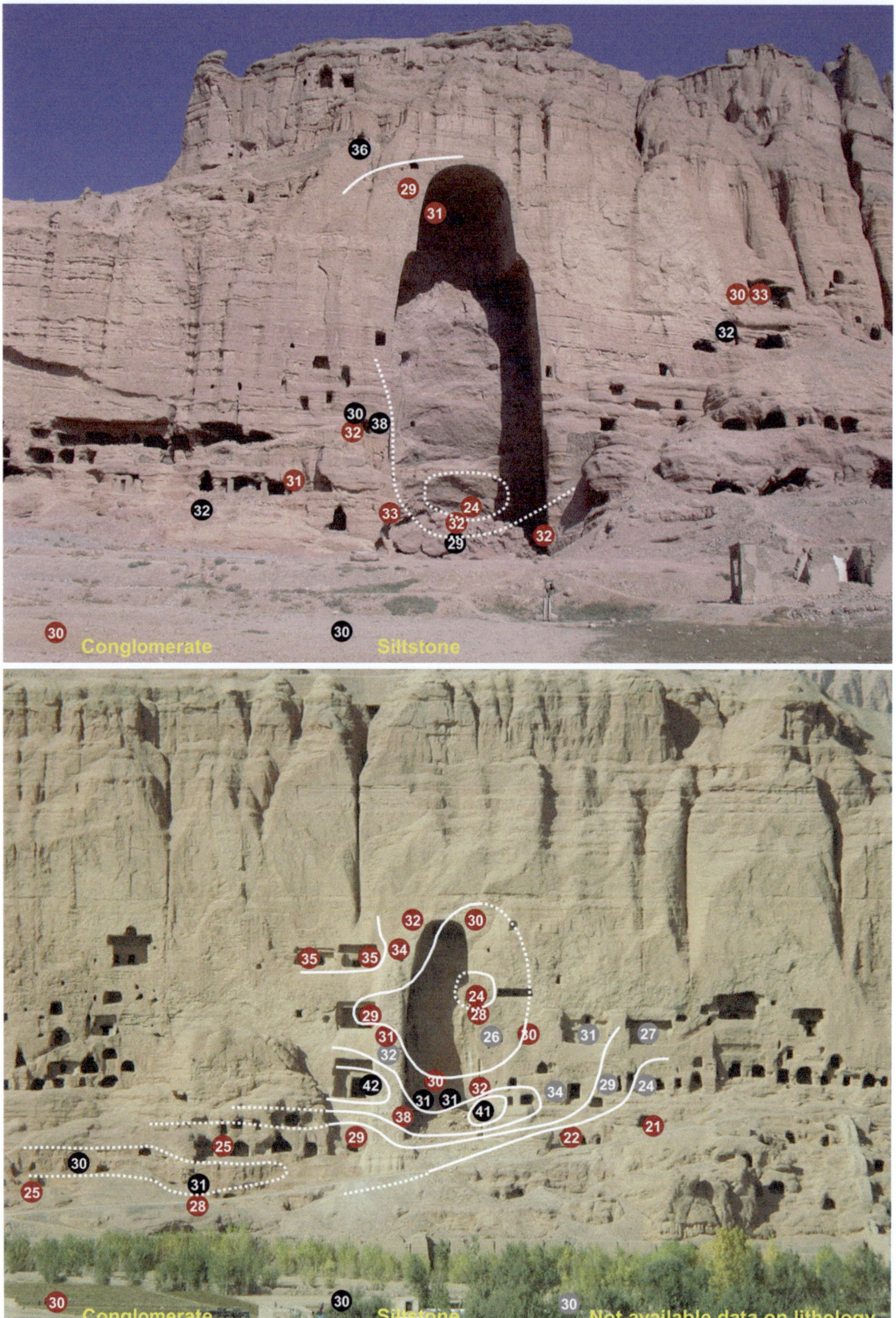

Fig. 7.15 Uniaxial compressive strength (MPa) from Schmidt hammer field survey, in the surrounding of both niches (*top* the Western giant Buddha niche and *bottom* the Eastern Buddha)

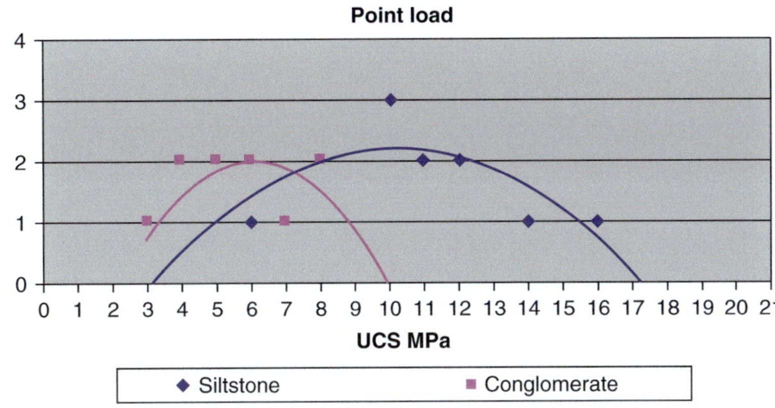

Fig. 7.16 Uniaxial compressive tests (MPa) from point load in both conglomerate and siltstone

confined condition of in situ samples with respect to the cubic samples tested in laboratory, the need of local calibration of the Schmidt hammer instead of using a general correlation chart between rebound of the hammer and UCS. Nevertheless, the field survey, even if less reliable with respect to point load, may yield useful information about the distribution of mechanical parameters along the entire investigated cliff.

Litotechnical classifications of materials (e.g., the Barton Q-system, Beniawsky RMR basic and modified for slope conditions, geological strength index [GSI]) (Barton et al. 1974, 1985; Beniawsky 1989) have been performed on the site (Fig. 7.17). The results mainly show different behavior between intact and blasted material. Furthermore, fractured siltstone exhibits generally low performance because of the presence of discontinuities spaced about 30 cm–1 m apart.

In situ density for conglomerate has been developed in Bamiyan, obtaining the value of 17.65 KN/m^3 and exhibiting a standard deviation of 0.03. This value is sensibly lower than the laboratory density obtained for siltstone (22.35 KN/m^3).

Shear strength parameters of both rock formations were considered to exhibit Hoek and Brown (1980) behavior, and for siltstone as well if it is not weathered. On the basis of this standard assumption, c′ and ϕ' have been determined by simulating the behavior of a sample in a triaxial test by means of an explicit difference finite code, known as FLAC (ITASCA Consulting Group 2000). The results obtained in the Eastern Giant Buddha site are friction angle of 32° and cohesion of 1.42 MPa for sandstone and friction angle of 52° and cohesion of 1.13 MPa for conglomerate (Fig. 7.18).

Hydrogeological Characters

According to Associazione Geotecnica Italiana (AGI) (1977) standards, a field permeability test was conducted on the site in order to acquire information on possible hydrogeological characters of rocks. The test, performed on conglomerate rock located not far from the Eastern Giant Buddha niche, exhibited a value of $K = 2\text{esp}10^{-4}$ m/s. Figure 7.19 reports the individual absorption values of conglomerate over time.

Geomechanic Characteristics of Discontinuities

Because discontinuities play a major role in rock stability, they have been investigated in terms of roughness and friction angle. The joint roughness coefficient (JRC) obtained from about 30 field tests exhibits a value of about 2 for sandstone and about 15 for conglomerate. From the roughness profile first- and second-order asperities have been also investigated (wavelength <50 cm and

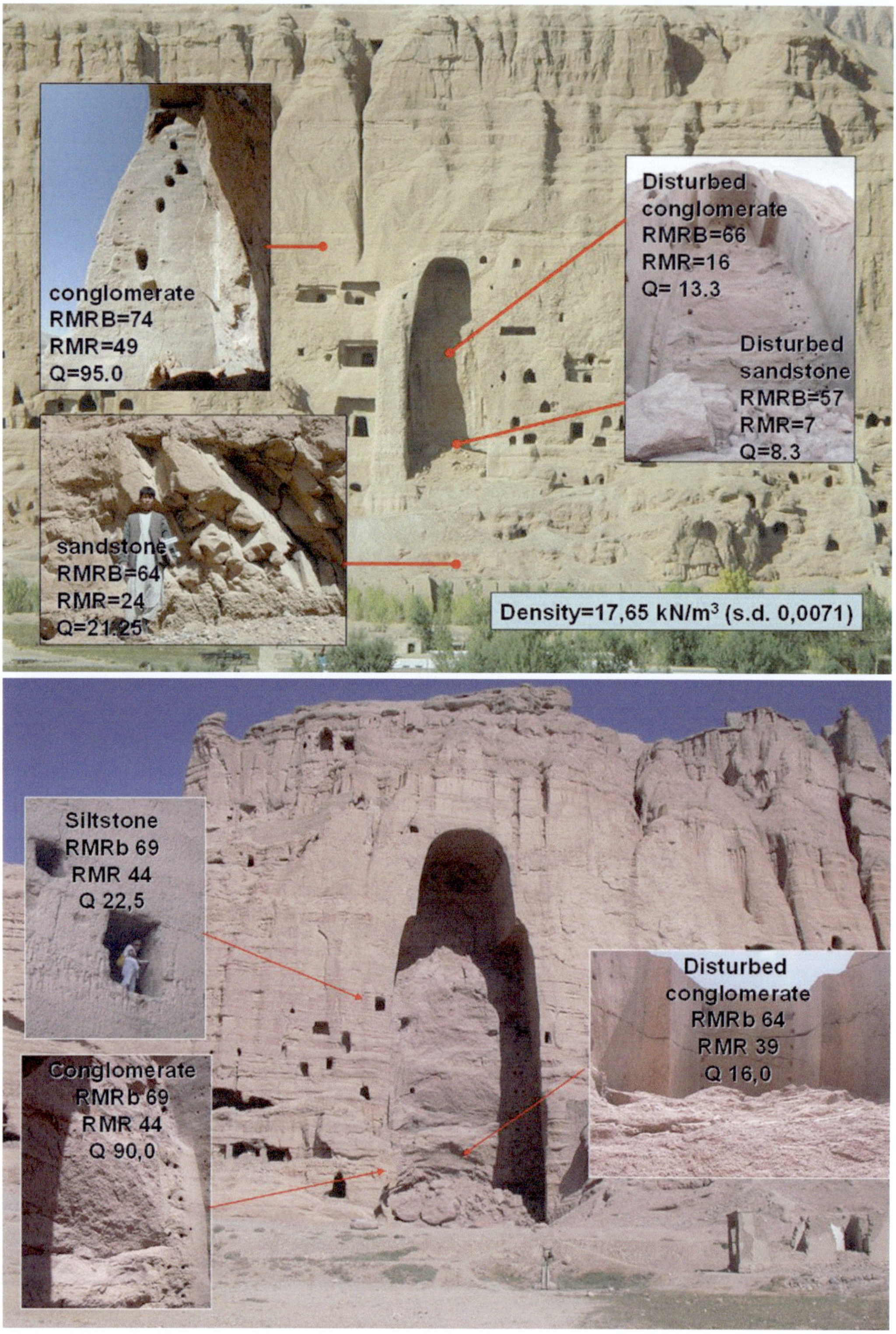

Fig. 7.17 Rock mechanic classification of local materials

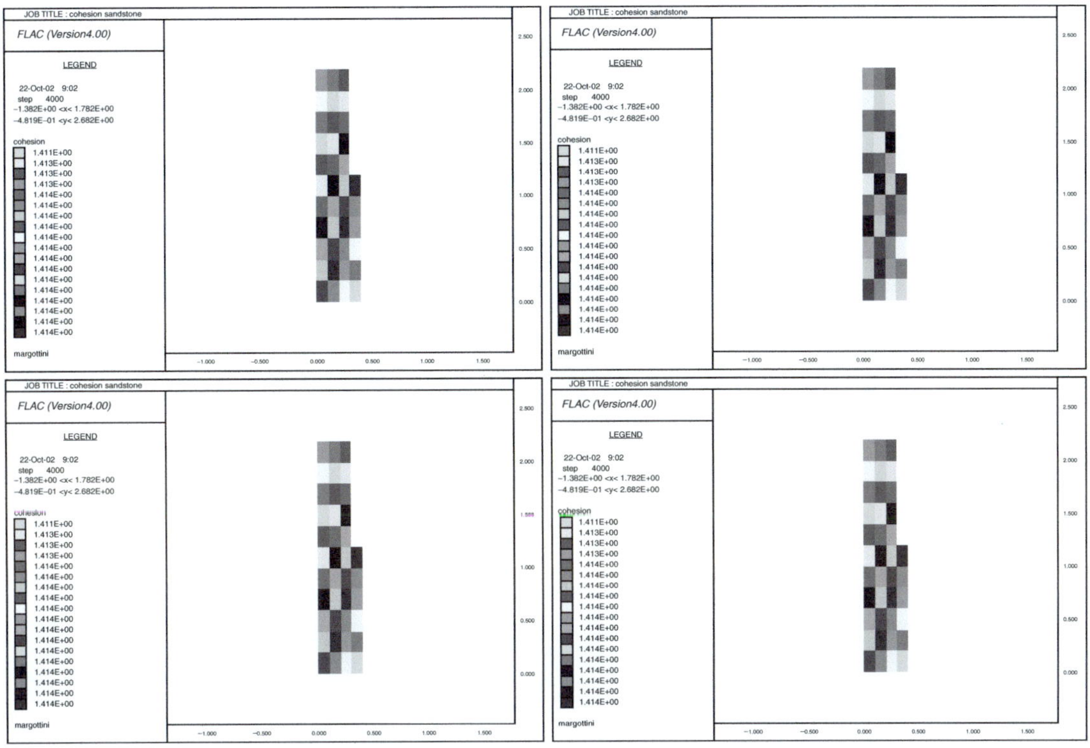

Fig. 7.18 Reconstruction the shear strength parameters (c', ϕ') according to the Hoek-Brown theory. Parameters are from a simulation of triaxial test within an explicit difference finite code (ITASCA Consulting Group 2000)

Fig. 7.19 Result of onsite permeability test on the Bamiyan conglomerate (Procedures according to AGI 1977)

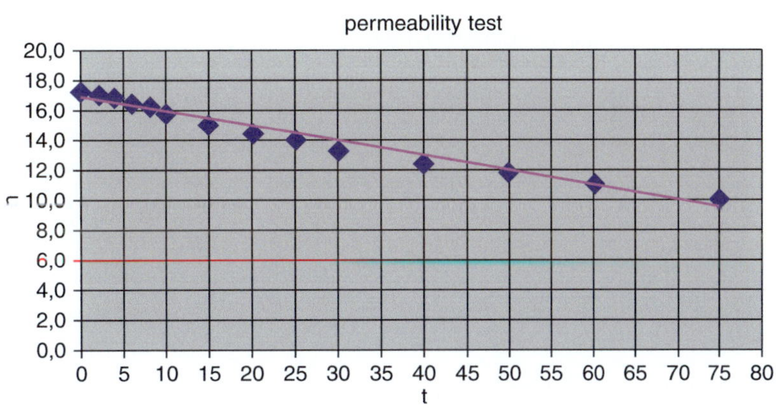

higher), in order to differentiate the contribution of asperities to the total friction angle ($\phi'_{basic} + i_{1st-2nd}$).

Reconstruction of the shear strength parameters of fractures (ϕ') was made by means of tiltometer tests. The results for the site of Eastern Giant Buddha are reported in Fig. 7.20. The obtained values are 42° (standard deviation 6°) and 38° (s.d. 4°) for conglomerate and 32° (s.d. 4°) for siltstone.

The tests have been performed in rough material; therefore, the result should partially include the contribution of asperities. This is more true for conglomerate, in which the JRC is about 15, and

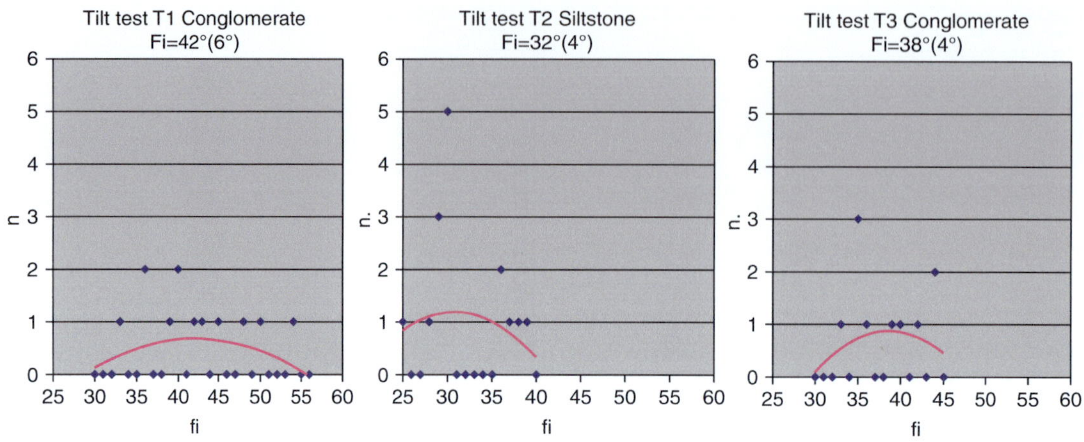

Fig. 7.20 Tilt test in the Eastern Giant Buddha site

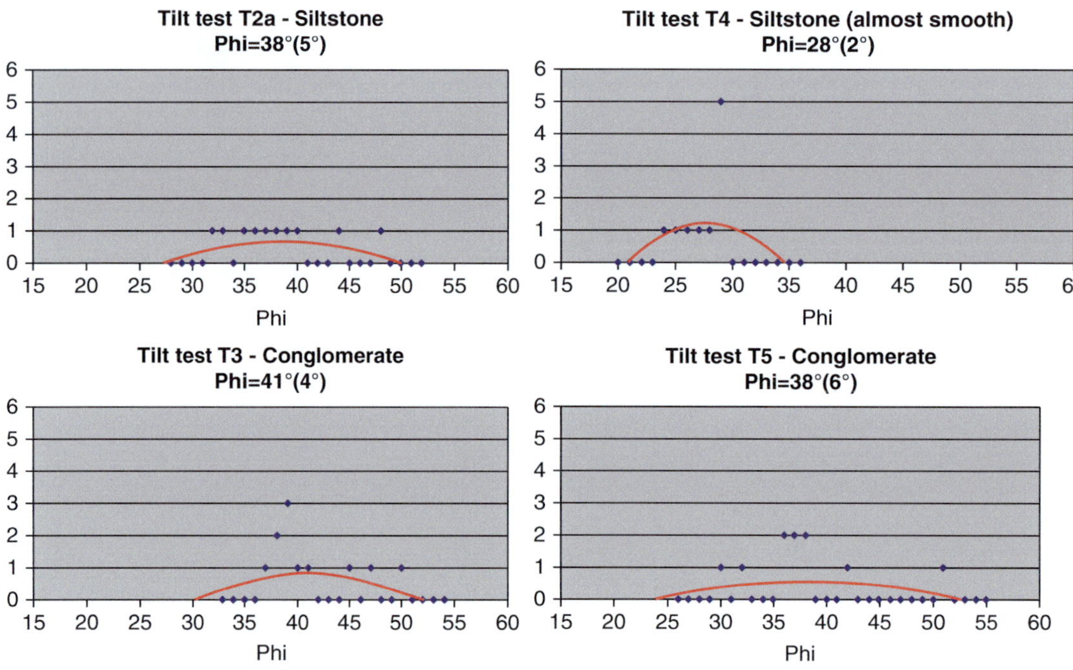

Fig. 7.21 Tilt test in the Western Giant Buddha site

less significant for sandstone, in which the JRC is about 2. In fact, the result of the tilt test for sandstone is almost coincident with that from simulation of a triaxial test using the Hoek and Brown (1980) approach. The result of ϕ' for conglomerate is relatively smaller than that obtained in the simulation of the triaxial test. This possibly demonstrates the importance of asperities in the total behavior of rock with respect to a tilt test that uses a small sample of material (about 20×20×20 cm).

Concerning the Western Giant Buddha niche, the discontinuity properties are not too different from that of the Eastern Buddha. The JRC obtained as an average of eight different sampling sites exhibits a value of 16 for conglomerate and 4 for siltstone.

Tiltometer tests for the purpose of identifying the shear strength parameters of discontinuities ($\phi' = \phi_{basic} + i_{1st-2nd}$), were developed on the site of the Western Giant Buddha, for both conglomerate and siltstone. For sandstone, the different behaviors of materials have also been investigated, with asperity ($\phi' = \phi_{basic} + i_{1st-2nd}$) and without (smoothed with carborundum stone; e.g., $\phi' = \phi_{basic}$). Conglomerate shows a value ranging from 38 to 41°. Siltstone shows a value of 28° for the smoothed sample and 38° for natural conditions (Fig. 7.21).

References

Associazione Geotecnica Italiana (1977) Raccomandazioni sulla programmazione ed esecuzione di indagini geognostiche

American Society for Testing of Materials (1977) Annual book of ASTM standards. Natural building stones: soil and rock; peats, mosses, and humus. Part 19. American Society for Testing Materials. Philadelphia, PA

Atkinson J (1997) Geotecnica. Meccanica delle terre e fondazioni. McGraw-Hill. Fizzonasco di Pieve Emanuele (MI)

Barton N, Lien R, Lunde J (1974) Engineering classification of rock masses for the design of tunnel support. Rock Mech 6:183–286

Barton N, Bandis S, Bakhtar K (1985) Strength deformation and conductivity coupling of rock joints. Int J Rock Mech Min Sci 22(3):121–140

Beniawsky ZT (1989) Engineering rock mass classification. Wiley, New York

Brown G (1972) The X-ray identification and crystal structures of clay minerals, 2nd edn. Mineralogical Society (Clay Minerals Group), London

Crosta G (1998) Slake durability vs. ultrasonic treatment for rock durability determination. Int J Rock Mech Min Sci 35(6):815–824

Franklin JA, Chandra R (1972) The slake durability test. Int J Rock Mech Min Sci 9:325–341

Goldstein J, Newbury DE, Echlin P, Joy DC, Roming AD Jr, Lyman CE, Fiori C, Lifshin E (1992) Scanning electron microscopy and X-ray microanalysis: a text for biologists, materials scientists, and geologists, 2nd edn. Plenum, New York, 820 pp

Hoek E, Brown ET (1980) Empirical strength criterion for rock masses. J Geotech Eng Div ASCE 106(GT9):1013–1035

ISRM, International Society for Rock Mechanics (1981) Rock characterisation, testing and monitoring ISRM suggested methods. Pergamon, London

ITASCA Consulting Group (2000) Fast lagrangian analysis of continua, vs. 4.0. Minneapolis, Minnesota

Le Roux A (1971) La lyophilisation, technique d'étude des textures des marnes et argiles. Bull Liaison Lab Ponts Chaussees 55:27–28

Pejon O, Zuquette L (2002) Analysis of cyclic swelling of mudrocks. Eng Geol 67:97–108

Shepard FP (1954) Nomenclature based on sand-silt-clay ratios. J Sediment Petrol 24:151–158

Shi B, Wu Z, Inyang H, Chen J, Wang B (1999) Preparation of soil specimens for SEM analysis using freeze-cut-drying. Bull Eng Geol Environ 58:1–7

Taylor RK, Spears DA (1970) The breakdown of British Coal measures rocks. Int J Rock Mech Min Sci 7: 481–501

Possible Evolution of Niches and Trends

8

Claudio Margottini

Structural Analysis of Discontinuities

Discontinuities play a major role in the slope stability of rocks. The geological material of Bamiyan presents a main pattern of discontinuities that were likely generated during excavation of the valley from the river, as a consequence of the absence of lateral support. More recently also, the explosion produced new fractures in rocks that were well localized in the niches, as well as enlarging pre-existing fractures. Concerning the stability of rock masses, the genesis of discontinuities is marginally relevant: If they exist they might play a role in possible collapse, otherwise they would not. The impact of the March 2001 explosion has been only marginally investigate because it caused rock falls in the niches and triggered a possible further collapse of the eastern side of the Eastern Giant Buddha niche. The collapse of the statues is not discussed in this chapter.

Distribution of discontinuities for the Eastern and Western Giant Buddha niches are reported in Fig. 8.1. From this it is possible to note only a major pattern in autochthonous materials and some orthogonal discontinuities. The lower sandstone formation, especially when belonging to the old sliding phenomenon laying left of the Eastern Giant Buddha niche, shows two major sets and some random patterns.

In the Western Giant Buddha, a major pattern exhibits two sets, which is evident in the cave top of the niche, suggesting potential long-term instability in this area.

Kinematic Analysis

The possible collapse mechanisms have been investigated through comparison of slope morphology and discontinuity patterns (e.g., wedge failure, planar sliding). General data on the entire cliff are in Chap. 5, whereas the present chapter offers information and figures related to the niches and surroundings. From the figures and related modeling it is possible to note the almost parallel incidence of slope face and major discontinuity patterns, not satisfying the condition for planar sliding or wedge failure (Turner and Schuster 1996), at least close to the niches. One must also consider that some discontinuities plotted in the figure represent the same fracture, detected in different vertical levels in the field survey (see Fig. 8.1 for localization of measurement sites) (Fig. 8.2).

The presence of many rock slides, also very ancient, as described in the geomorphological section along the cliff, are also manly related to the fault system with a general S quadrant dipping orientation with slope angles between 30 and 50°. This system of fault has not been detected surrounding of niches. Where this

C. Margottini
Department of Geological Survey of Italy,
ISPRA - Italian Institute for Protection and Environmental Research,
Via Brancati, 60, 00143 Rome, Italy
e-mail: claudio.margottini@isprambiente.it

C. Margottini (ed.), *After the Destruction of Giant Buddha Statues in Bamiyan (Afghanistan) in 2001*, Natural Science in Archaeology, DOI 10.1007/978-3-642-30051-6_8, © Springer-Verlag Berlin Heidelberg 2014

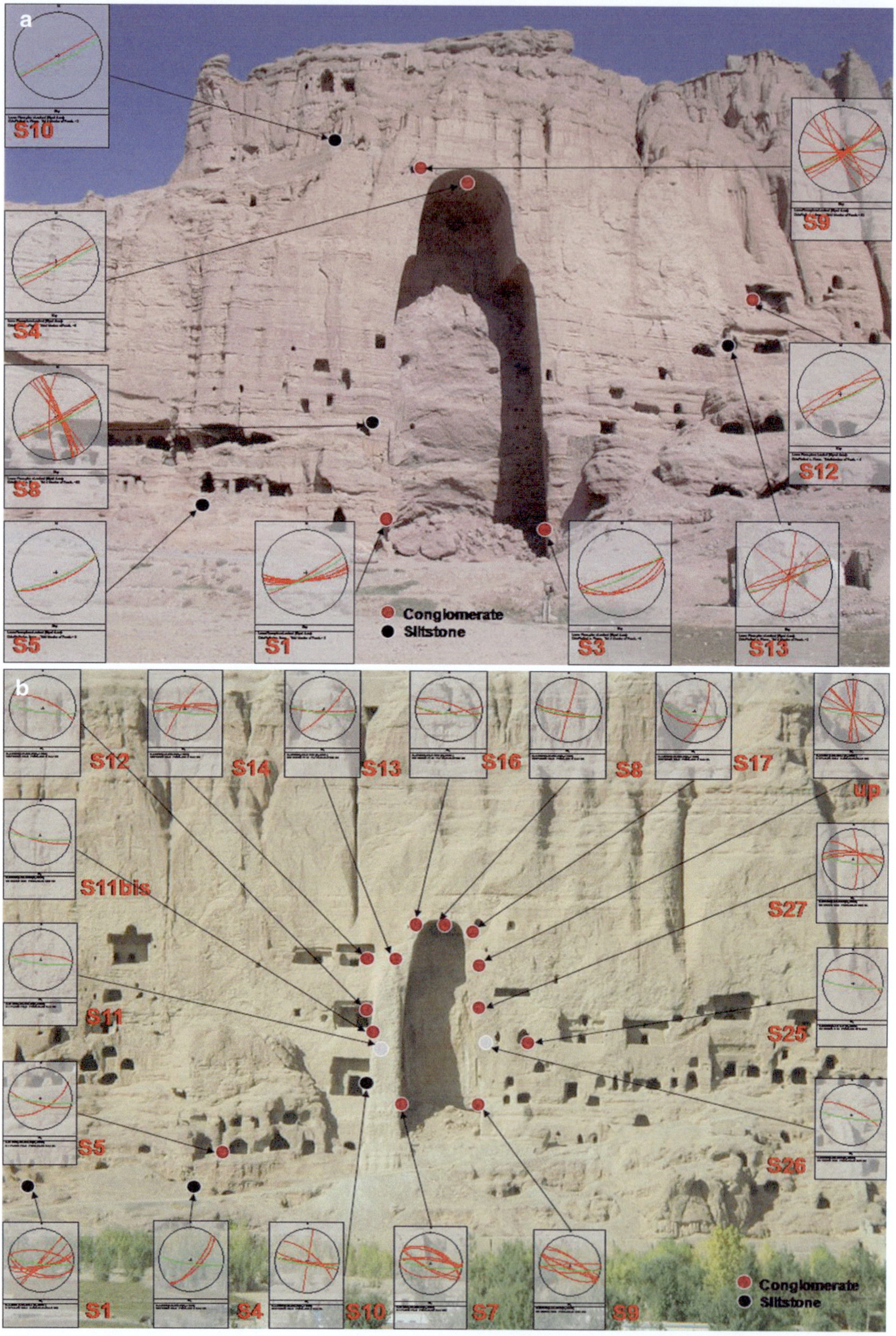

Fig. 8.1 Discontinuity trends for both the Western and Eastern Giant Buddha niches and surroundings

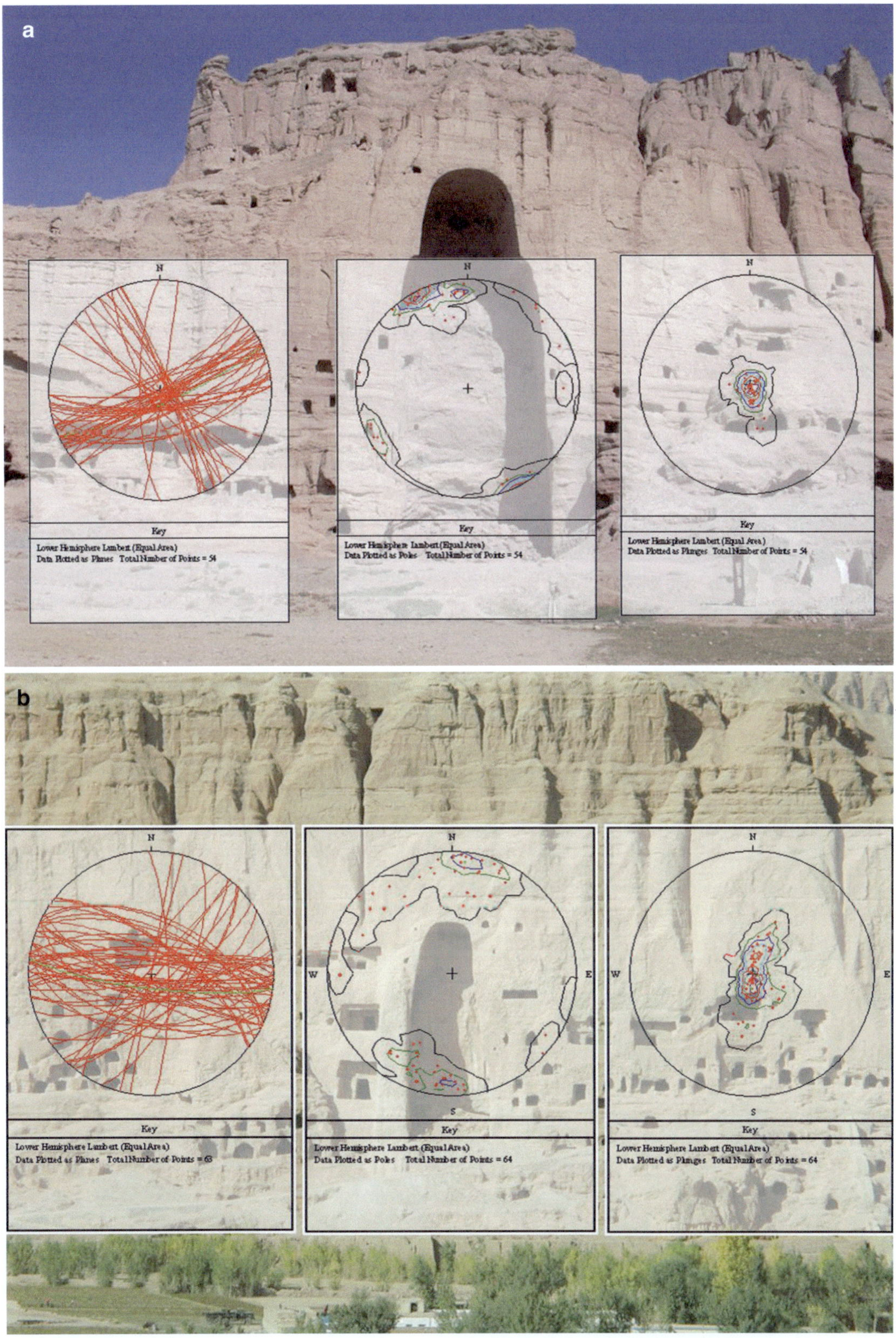

Fig. 8.2 Kinematic analysis for both the Western and Eastern Giant Buddha niches and surroundings

system still reaches the slope face, as occurred during the filling of the Bamiyan river valley with alluvial deposits, sliding is kinematically possible.

Stability Analysis

The investigation of possible stability conditions of the cliff in the surrounding niches was computed using the explicit-difference-finite code, FLAC (Fast Lagrangian Analysis of Continua in 3 Dimensions; ITASCA Consulting Group 2000). For a given element shape function, the set of algebraic equations solved by FLAC is identical to that solved with the finite element mode. However, in FLAC, this set of equations is solved using dynamic relaxation, an explicit time-marching procedure in which the full dynamic equations of motion are integrated step by step. Static solutions are obtained by including dumping terms that gradually remove kinetic energy from the system (Dawson et al. 1999).

In the area of Bamiyan niches, present-day stability conditions have been investigated according to both the Hoek and Brown theory of the shear strength of rocks and a set of modified parameters suitable to describe a finite element model: the possible consequences of a deepening of discontinuities today detected at the top of the niche and the lowering of cohesion for siltstone settled in the lower part of the cliff. The grid's mesh density has been computed to allow the best resolution in the area of possible movements (Fig. 8.3).

The results, for the small Buddha, can be summarized as follows.

- According to Hoek and Brown, shear strength parameters of conglomerate and sandstone, and a major discontinuity ranging from the middle of the cliff to the middle of the niche (only friction value for shear strength), the deformation of the cliff is relatively low and presently seems to be in stable condition (Fig. 8.4).
- Because the shear strength parameters can be modified with respect to the values obtained in the field, the cohesion of sandstone can be assumed to decrease at 25 % of the initial rate. This is a consequence of the fracturing degree that may generate possible shear planes in such rocks as integers of two or more family sets of joints. If a large crack is also detected

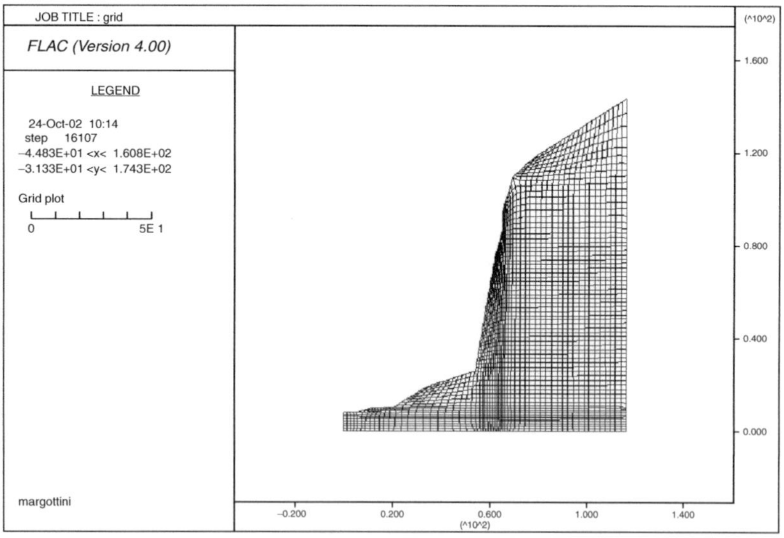

Fig. 8.3 The adopted mesh for the Eastern Giant Buddha niche stability analysis

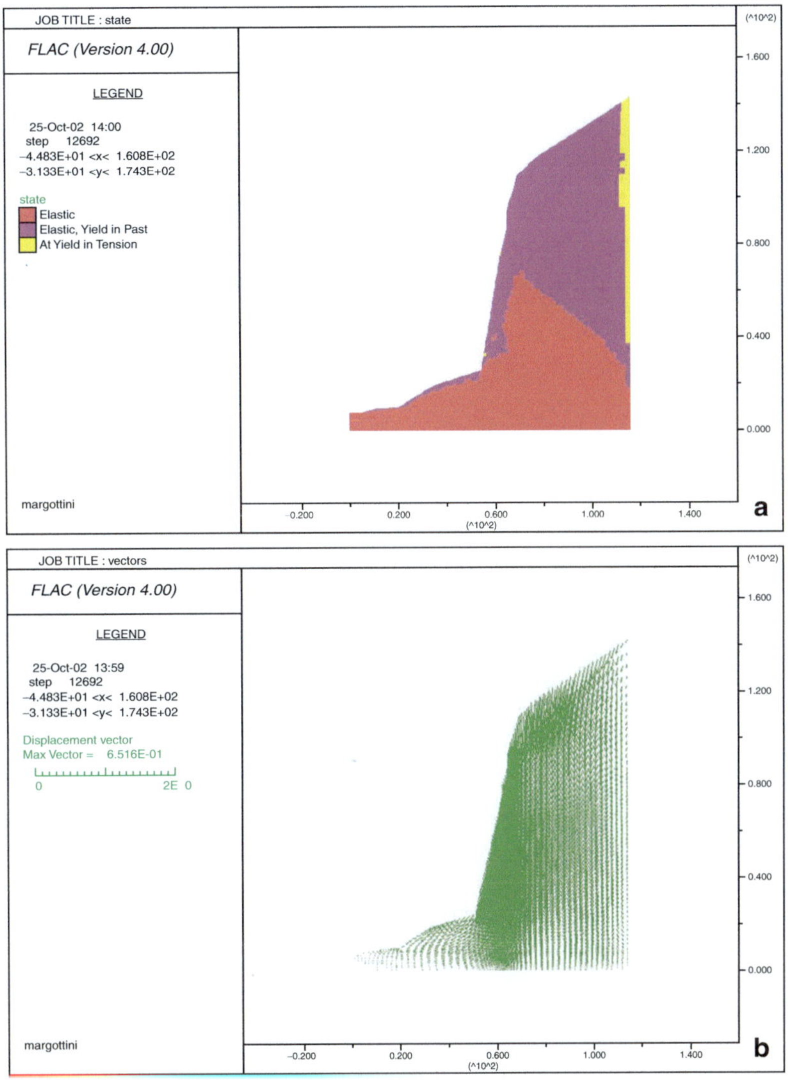

Fig. 8.4 (*Bottom*) State and displacement for the cliff of the Eastern Giant Buddha niche in present conditions

in the conglomerate at both back sides of the niches and reaches the top of the sandstone, the cliff will still be stable with factor of safety = 2.04, as shown in Fig. 8.5.

- Finally, if the cohesion of lower siltstone is considered near or equal to 0 because of the weathering process triggered by water infiltration, as well as from cracks in the conglomerate analyzed in the previous step, the cliff will become unstable, with maximum displacement and vector at the base of the niche (Fig. 8.6).
- This condition probably occurred in the past, as testified by the two rock slides on the left side of Eastern Giant Buddha niche and the right side of Western Buddha, respectively.

On the basis of slope stability analysis, it is possible to assume that the cliff will be generally stable until the discontinuities of

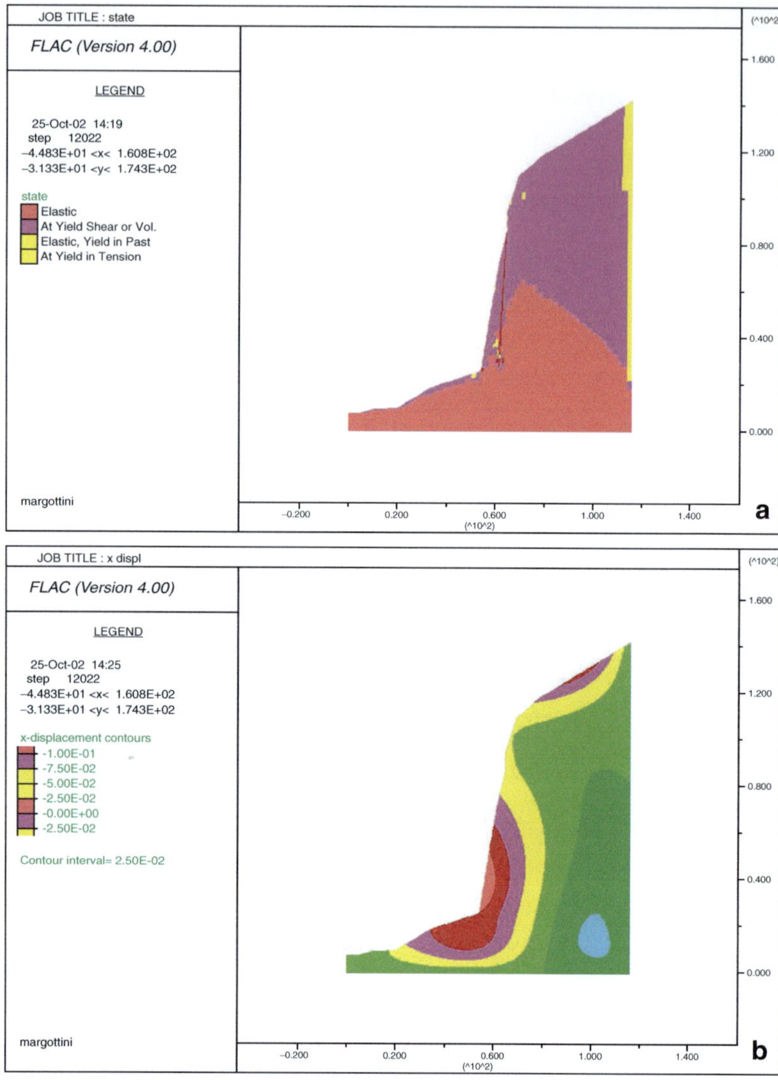

Fig. 8.5 State and displacement for the cliff of the Eastern Giant Buddha, with fracturing from the top reaching the base of the cliff and a 25 % reduction of cohesion in the lower siltstone

conglomerate from the top of the cliff reach the top of the lower sandstone. In such a situation, if the sandstone is losing cohesion, important sliding may occur until a value proxy reaches nil as consequence of internal high degree of fracturing and/or weathering for water infiltration during heavy rainfall and snowmelt. Clear historical examples exist left of the eastern niche and on the right of the western niche as well as in the lower part of the cliff covered by talus debris. These phenomena also occurred after the anthropization of the area and the construction of cavities; the sliding close to the western niche shows the floor of one of the cavities clearly tilted and in the eastern part of the cliff the sliding plane is dislocating the floor of the cavity (see Chap. 5). In general conclusion, because of the 2001 explosion and a possible global meteoclimatic change, negative evolution of the cliff has to at least be taken into consideration.

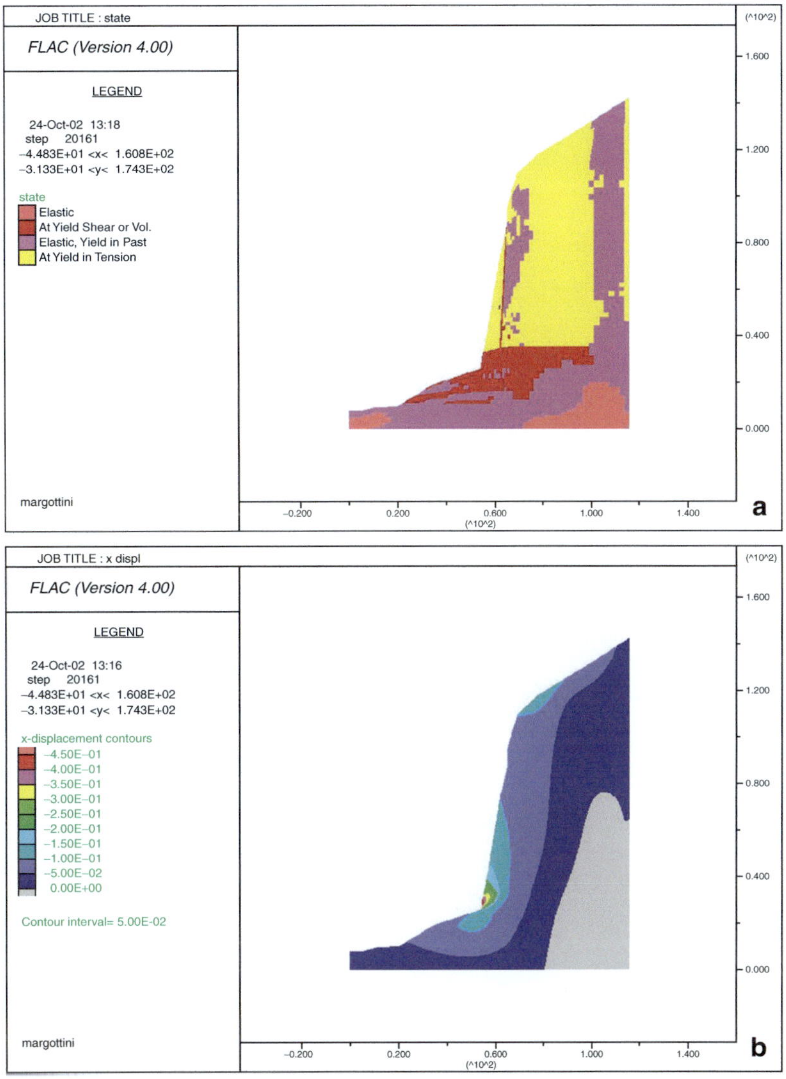

Fig. 8.6 State and displacement for the cliff of the Eastern Giant Buddha with fracturing reaching the lower part of the cliff; the lower siltstone exhibits no cohesion as a consequence of internal fracturing or weathering

References

Dawson EM, Roth WH, Drescher A (1999) Slope stability analysis by strength reduction. Géotechnique 49(6): 835–840

ITASCA Consulting Group (2000) Fast lagrangian analysis of continua, vs. 4.0. Minneapolis, Minnesota

Turner K, Schuster R (eds) (1996) Landslides, investigation and mitigation. Special report, transportation research board, vol 247. National Research Council, Washington

Past Experience in Conservation and Exploitation

9

Andrea Bruno and Claudio Margottini

Typologies of Works

The sequence of damage and restoration produced over the centuries since the construction of the statues is not clear. Even in the past, the contribution of human beings to maintain the integrity of the artistic sculpture was high. For example, in the period 1678–1707, Aurangzed, a Muslim ruler, shot a huge gun at the Buddha, likely damaging the foot of the Western Giant Buddha.

In recent times interventions have been developed. Their investigation may help us to understand the effects of future restoration works on local materials and their long-term duration.

Two major interventions have been recorded, but written documentation is rare: the intervention of the French Archaeological Mission of the late 1950s to the early 1960s and the well-known Indian Archaeological Survey of the late 1960s to early 1970s.

Following are reported some comments on previous restorations based on the comparison among the private photo collection of Andrea Bruno (September 1960) and Kurt Lambek (University of Canberra, Australia), taken in August 1967 and kindly provided for this work, and the photos from the Archaeological Survey of India (ca. 1968–1971, courtesy of Michel Jansen), and the present day. This discussion is certainly not complete and does not cover the huge number of interventions actually carried out during recent decades. It only describes the typology of works previously performed and their durability to the present.

1. The major remaining feature of the work developed by the French Archaeological Mission is the buttress, constructed to sustain the western side of the Eastern Giant Buddha niche. This demonstrates that, at least since the middle of the last century, the cliff was in a precarious condition, and probably at the limit of equilibrium.

The construction was performed in brick, and the result, shown in Fig. 9.1, demonstrates the typical intervention of that period: to solve the problem with limited attention paid to the environmental and aesthetical impact of construction. On the other hand, this side of the niche did not suffer too much as a consequence of the explosion of March 2001, confirming the efficacy of this intervention.

Unfortunately, the buttress was not well designed and the crack we would like to sustain started to re-open again. The reason is not clear even now, but there is a possibility that the soil foundation was not very well investigated, the buttress may be located over a cave, or the bearing capacity of the local soil may have been

A. Bruno (✉)
Studio Bruno, Via Asti, 17, Torino, Italy
e-mail: studiobruno@tin.it

C. Margottini
Department of Geological Survey of Italy, ISPRA - Italian Institute for Protection and Environmental Research,
Via Brancati, 60, 00143 Rome, Italy
e-mail: claudio.margottini@isprambiente.it

C. Margottini (ed.), *After the Destruction of Giant Buddha Statues in Bamiyan (Afghanistan) in 2001*,
Natural Science in Archaeology, DOI 10.1007/978-3-642-30051-6_9, © Springer-Verlag Berlin Heidelberg 2014

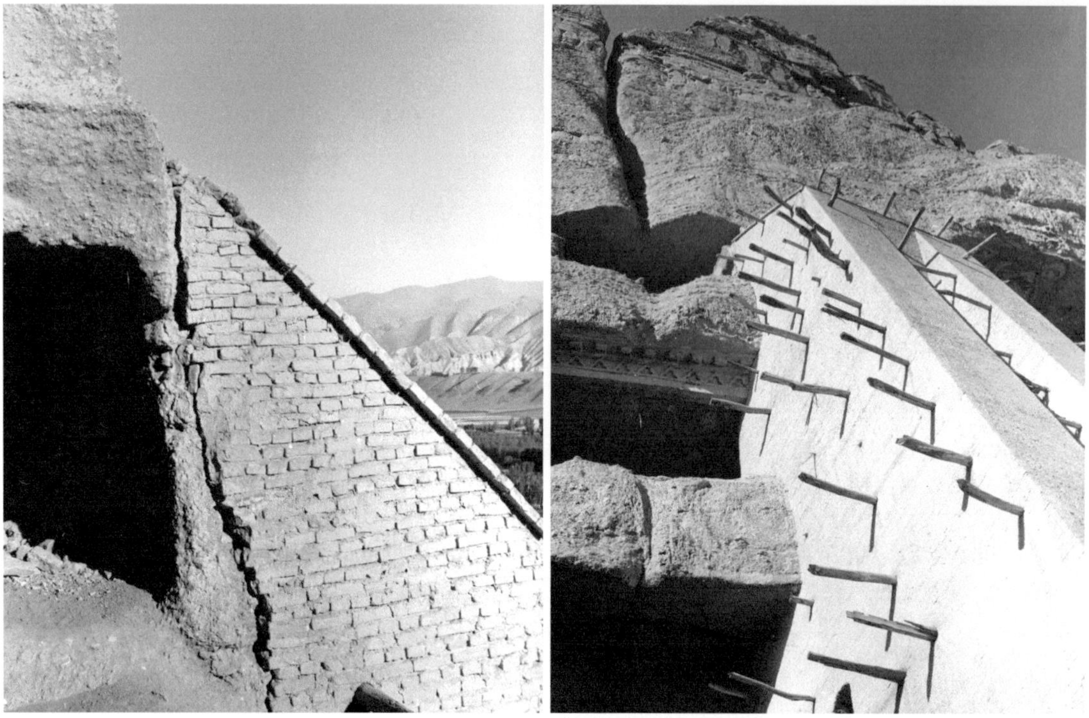

Fig. 9.1 The top of the buttress constructed by the French Archaeological mission (Photos A. Bruno, Sept. 1960)

underestimated. Even now, some evidences of deformation can be detected, suggesting the need for future intervention, at least to understand and recover the reason for such sinking. It is likely that the monitoring system of the main crack of the site did not reveal important widening in the period 2003–2007 in a small crack at the connection between the buttress and the cliff on the external side.

Finally, to overcome at least the heavy impact of the buttress, the Indian Archaeological Survey reshaped the structure and covered it with cement and natural earth, producing a camouflage not easy to recognize at first sight (Figs. 9.2 and 9.3).

2. Filling the fracture with concrete: A similar intervention was conducted, probably in 1960, in the Western Giant Buddha niche, as testified by K. Lambek's photos of August 1967, and likely aimed at avoiding water infiltration from the top of the cliff. Currently there is little evidence of this filling activity, suggesting a possible effect of the explosion or a natural enlargement of the fractures. This second hypothesis seems not to be confirmed by the evolution pattern of discontinuities.

Cliff deformation of the Eastern Giant Buddha site produced many fractures inside the niche (Fig. 9.4).

During the restoration of the buttress, designed from the Indian Archaeological Mission in the period ca. 1968–1971, many of the cracks were grouted with cement and mortar to protect them from water infiltration and stabilize them. That was necessary because, as mentioned, the foundation of the buttress was not properly designed, and hung onto the cliff itself. In consequence of this, horizontal deformation probably increased.

The result of this grouting from the Indian Archaeological Survey is still satisfactory, even if some fractures, widening by a few centimeters, are now visible. One of these cracks, located exactly in contact between the buttress and the cliff, is now under permanent manual monitoring.

The effect of the blasting on present-day widening of the fractures is not clear. Certainly, in the upper part, the collapse of an element of the wall separating the niche from the stairway, limited by the upward prolongation of such disconti-

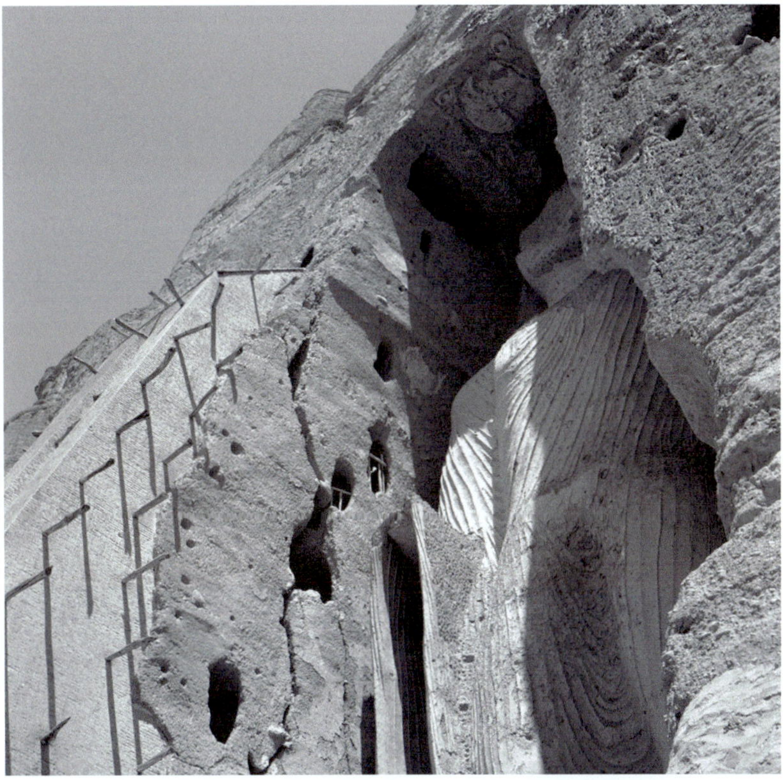

Fig. 9.2 The buttress and the most external part of the cliff showing the large crack and the detached block, which needs support. It is not clear whether the crack was moving at the time of the picture or not. Nevertheless, the crack was detected as slightly open in the 2002 survey by the authors (Photo A. Bruno, Sept. 1960)

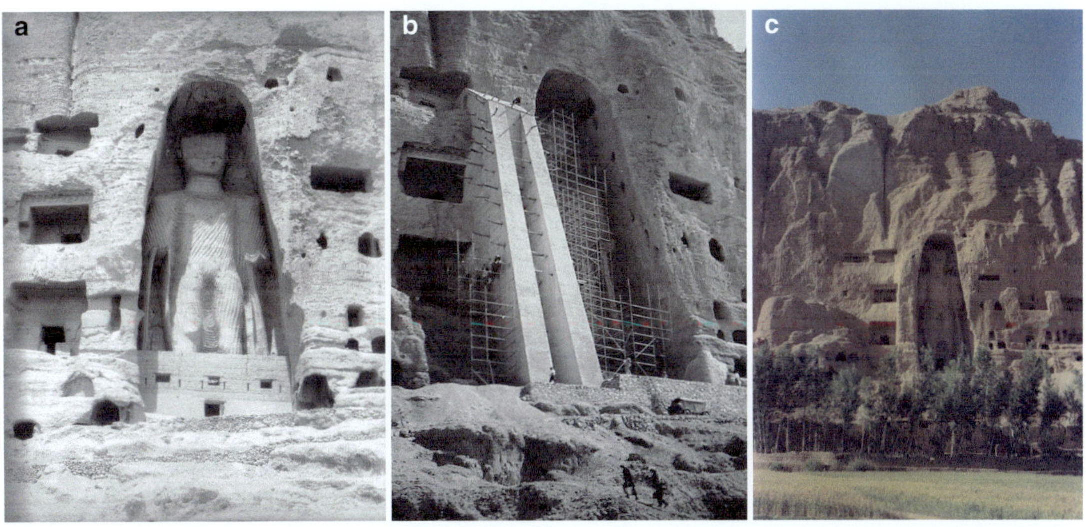

Fig. 9.3 The Eastern Giant Buddha before the French mission intervention, with the French buttress at the beginning of the Indian intervention (ca. 1968–1969) (Courtesy M. Jansen) after restoration by the Archaeological Survey of India in 1975 (Photo G. Arduino), respectively (**a**), (**b**) and (**c**)

nuities, poses some questions about the possible negative impact of the explosion (Fig. 9.5).

3. Reconstruction of missing parts: Looking at both the 1967 and present-day photos, it is clear that a small portion of the border niche of the Western Giant Buddha has been perfectly reconstructed. Also, the contact between the large slide in front of the Western

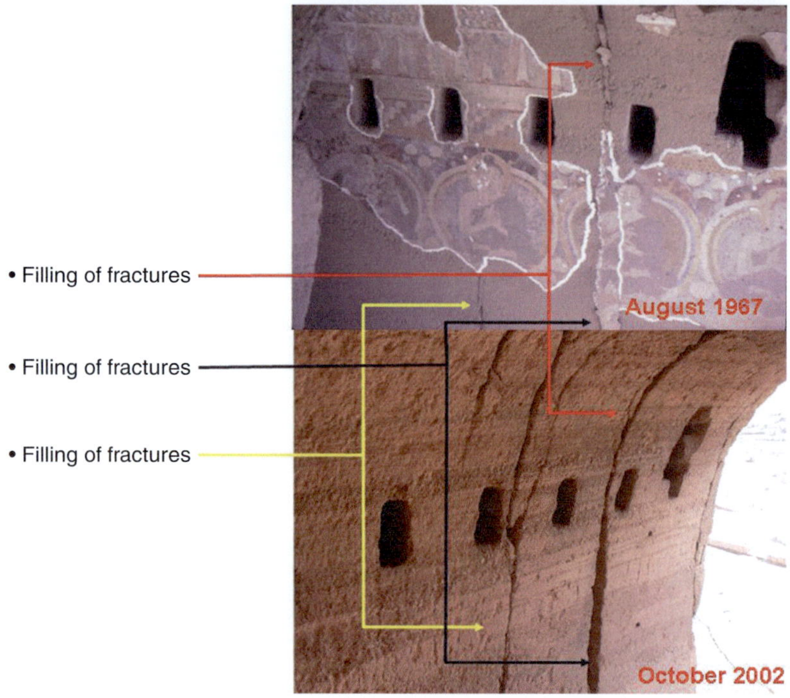

Fig. 9.4 Filling a fracture in the Western Giant Buddha with cement

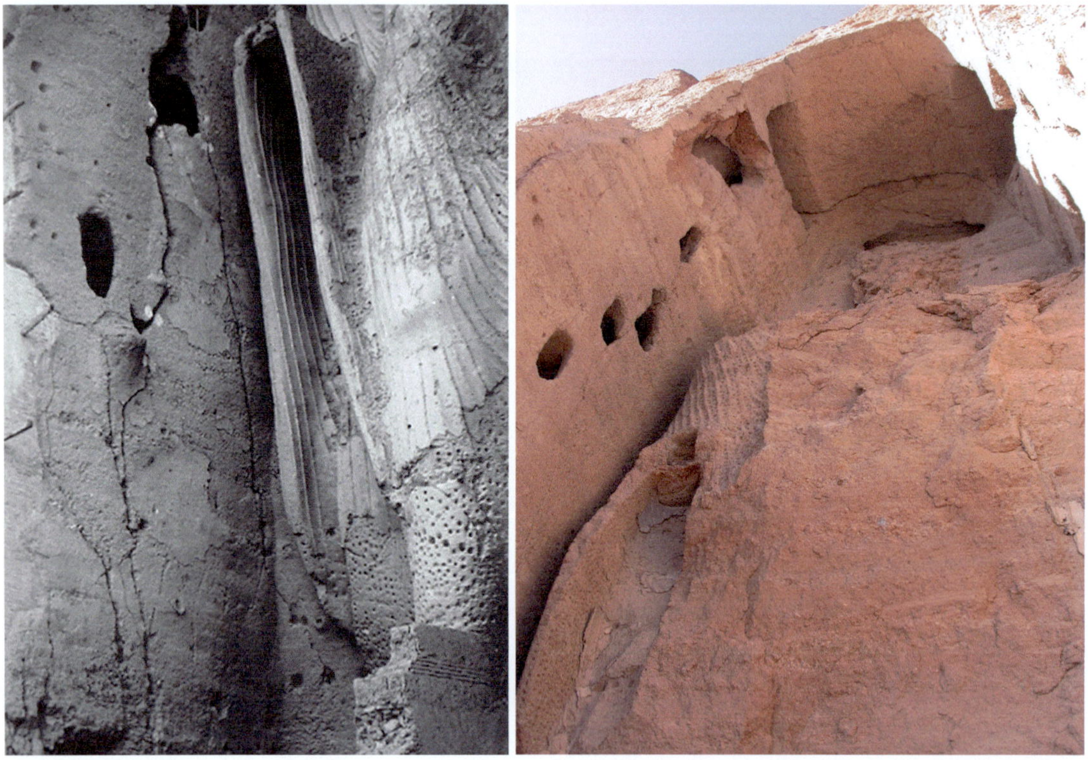

Fig. 9.5 Pattern of discontinuities in the Eastern Giant Buddha niche both before intervention of the Indian Archaeological Survey and currently (Courtesy Michel Jansen)

Giant Buddha and the cliff has been filled with concrete. A similar intervention of reconstruction of weak or missing parts has also been done for the Eastern Giant Buddha niche with excellent results (Fig. 9.6).

4. Protection from water runoff. This measure was considered extremely important in the past, and it still is. The Indian Archaeological Survey constructed two major protection channels above the Buddha niches. Their perfect maintenance and conservation is essential (Fig. 9.7).
5. Bolts have been used in the past to anchor small pieces of the statues to the cliff. The approach was correct and the result fully satisfactory until the blast occurred (Fig. 9.8).

Ideas for Exploitation

In the past only minor attempts were made to increase tourism and use the statues as a major attraction point in Afghanistan. This was mainly because of the safety issue of the country, which did not allow for a major exploitation plan and implementation for the site. Minor local private sector businesses were initiated, but unfortunately they did not reach any level of maturity. Similarly, the level of management and exploitation was mainly confined to the notes and scratch books of the experts who periodically visited Bamiyan. They immediately recognized the importance of the site and the huge potential for its development. But unfortunately history did not allow such improvement. Among the various experts, a major role was played by architect Andrea Bruno, who dedicated a large part of his life to the conservation of Afghan monuments. His first visit to Bamiyan was in September 1960. Since then, a large number of documents and reports have been published on the restoration and exploitation of Afghan monuments. The surveys performed by Andrea Bruno in the 1960s (Figs. 9.9, 9.10, and 9.11) are masterpieces for knowledge of the site, especially after the destruction of the statues in 2001. The work of Bruno was not only important for recording the site, but also for the first ideas on the exploitation of the Bamiyan Buddhas.

The project of a museum located just under the ground level in front to the Western Giant Buddha niche, where one might look at the head of the statue is a stimulating and intriguing concept. The idea can be put on the table when the scientific community and UNESCO start to think about what to do with the fragments of the statues that were destroyed in 2001.

Finally, the preceding notes report some of the first attempts to develop the status and improve the attractiveness of the Bamiyan valley. The case study demonstrates that there is still a possibility to enhance the scenery of the area and related landscape without creating false and modern manufacturing that disrespects the true spirituality of the site (Figs. 9.12, 9.13, 9.14, 9.15, and 9.16).

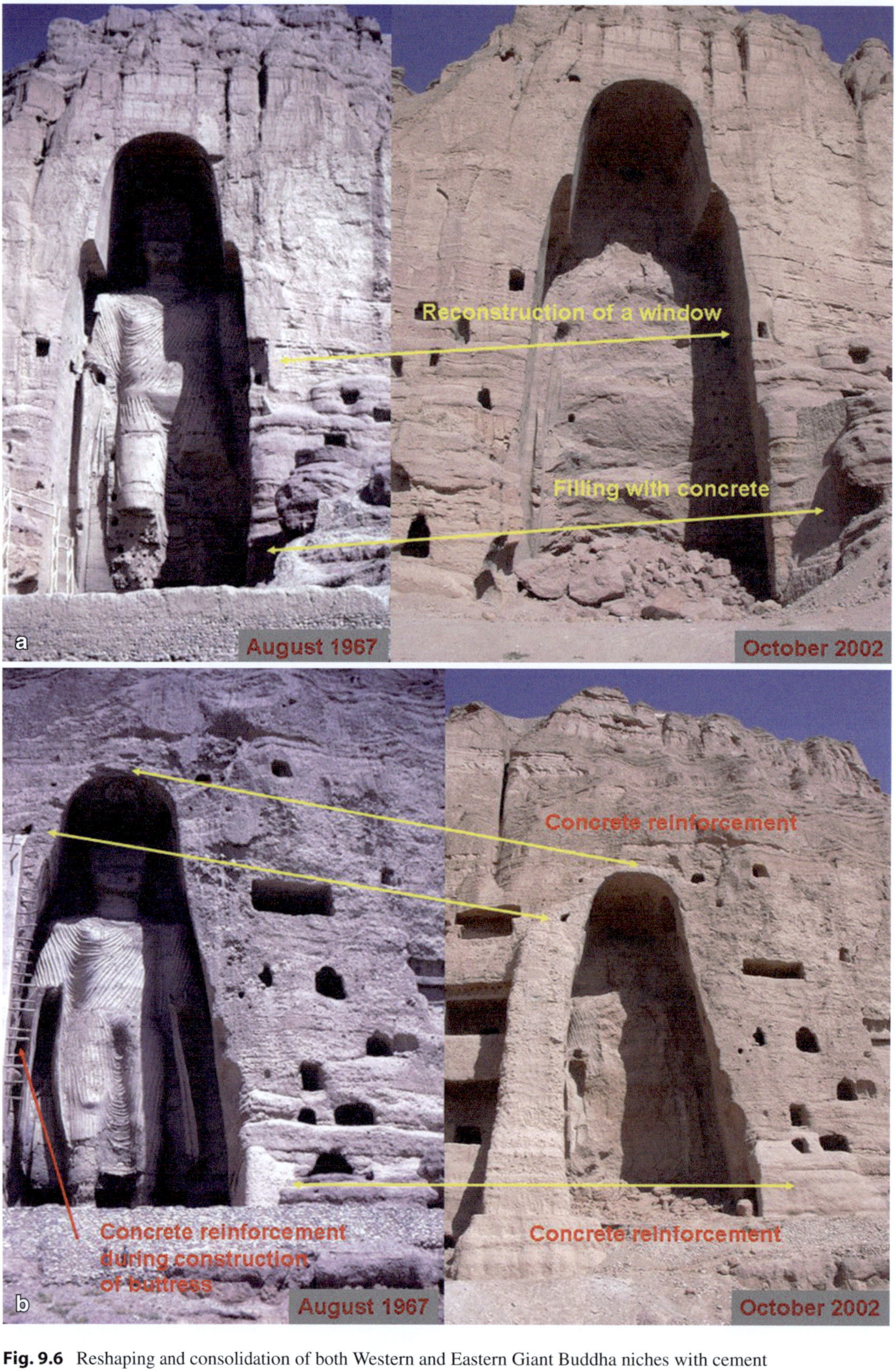

Fig. 9.6 Reshaping and consolidation of both Western and Eastern Giant Buddha niches with cement

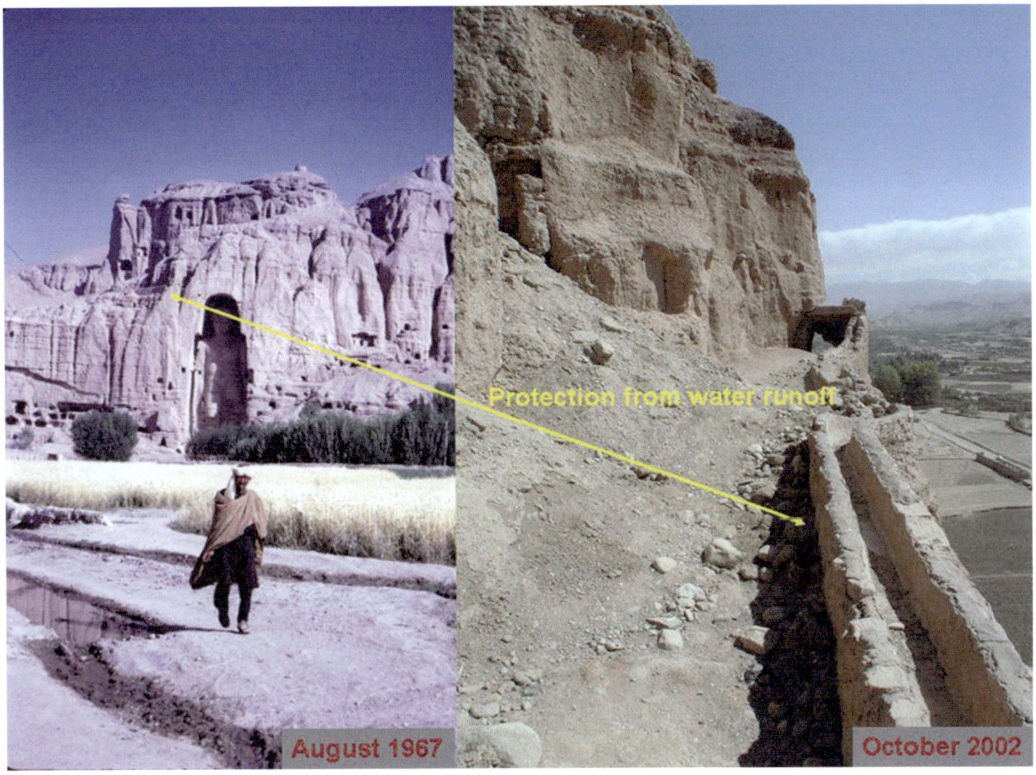

Fig. 9.7 Protection from water runoff and infiltration

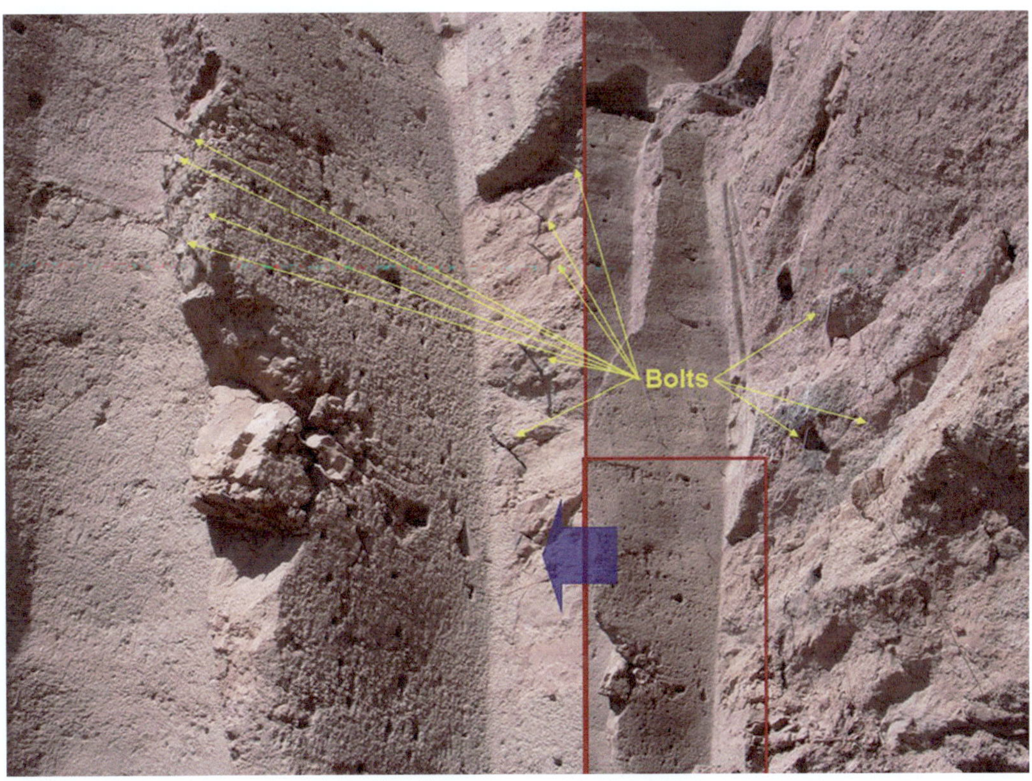

Fig. 9.8 Use of bolt for anchoring small pieces of the Western Giant Buddha

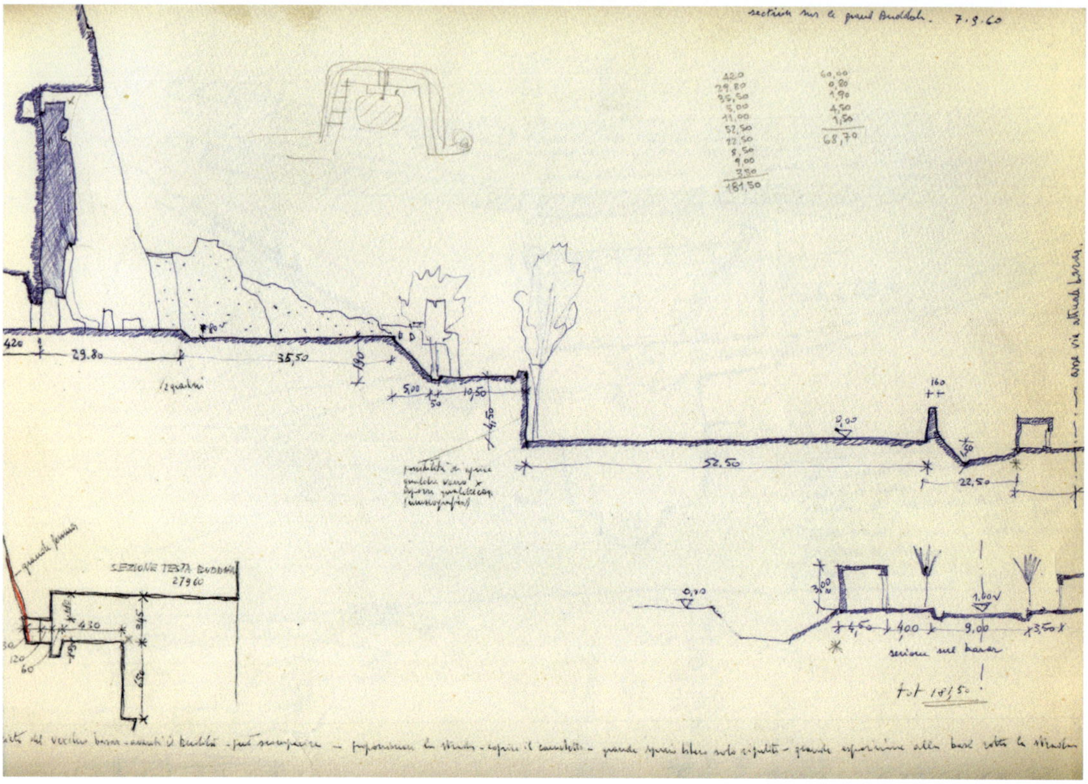

Fig. 9.9 Profile of the Western Giant Buddha niche and statue from the drawings of Andrea Bruno in 1960

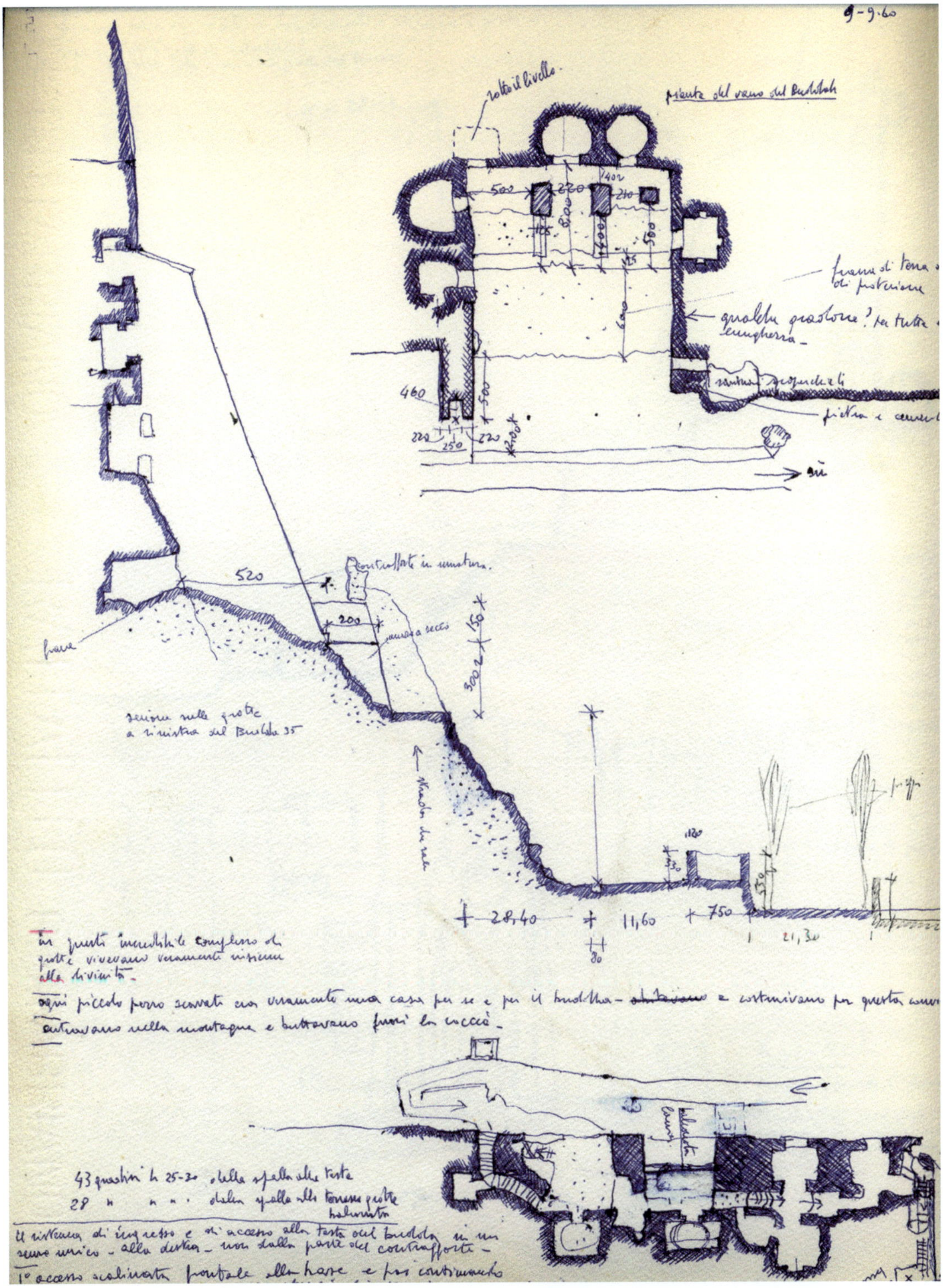

Fig. 9.10 Map of the Eastern Giant Buddha niche and profile and map of the cliff, western to the niche, in the drawings of Andrea Bruno in 1960 before the intervention by the Indian Archaeological Survey

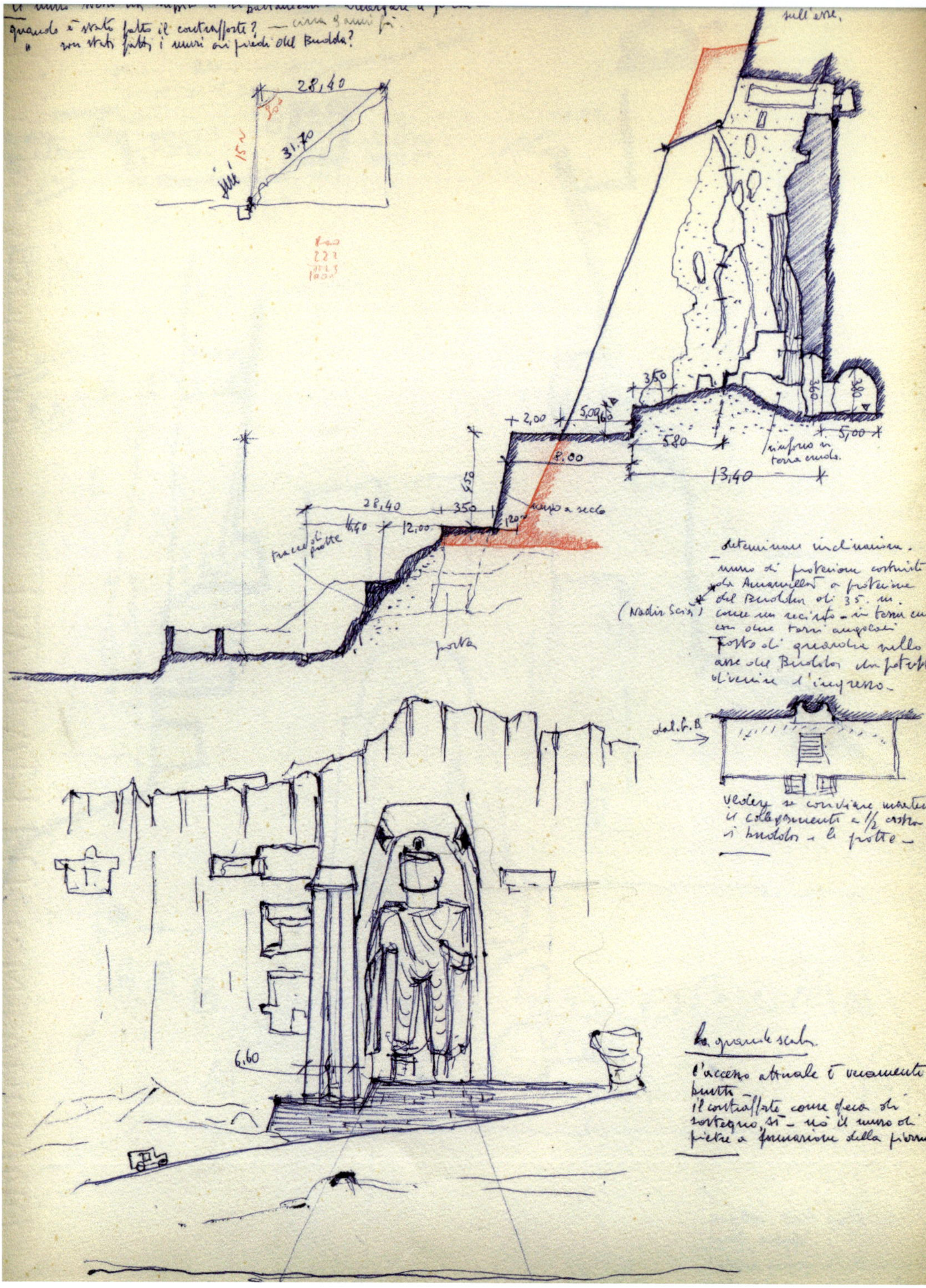

Fig. 9.11 Profile and front view of the Eastern Giant Buddha niche, in the drawings of Andrea Bruno in 1960 before the intervention of the Indian Archaeological Survey

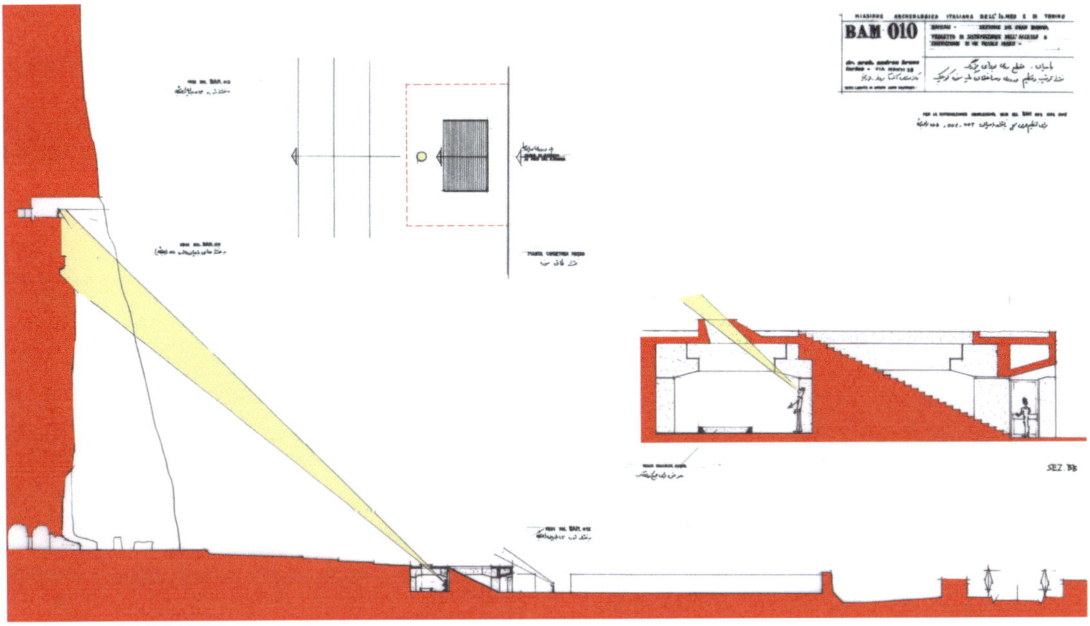

Fig. 9.12 The proposal of Andrea Bruno for the construction of a museum in front of the Western Giant Buddha niche in Bamiyan, 1960

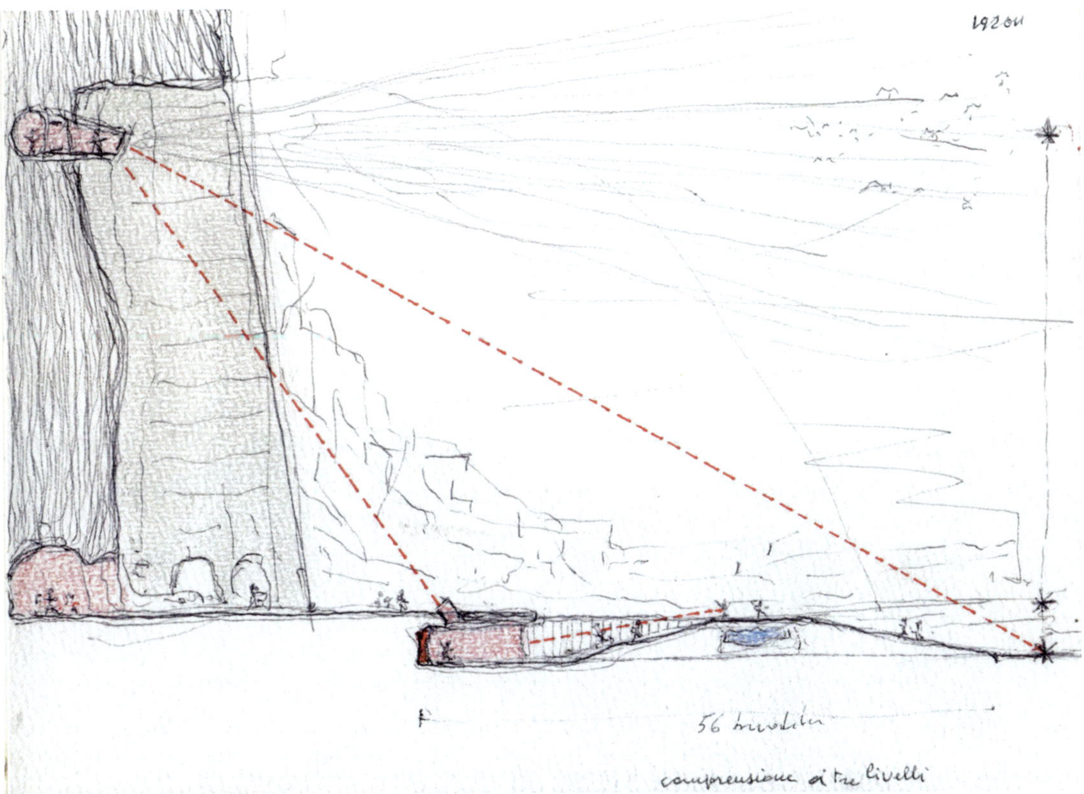

Fig. 9.13 The new project for the exploitation of the niche of the Great Buddha and the surrounding area, the observatory and the underground sanctuary-museum (Drawing by A. Bruno, 2010)

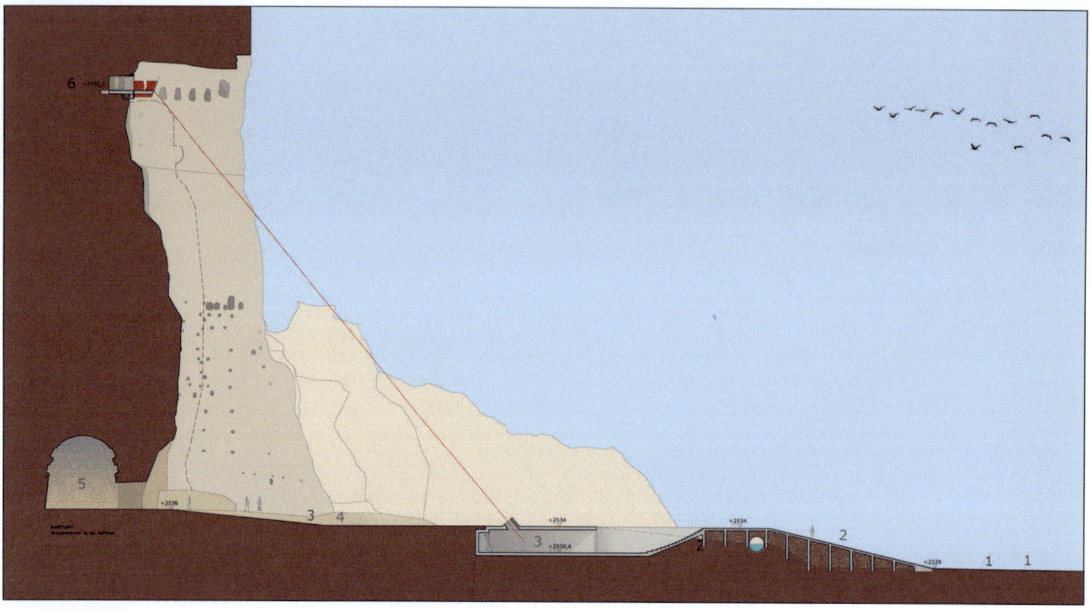

Fig. 9.14 General section of the project of valorisation of the niche in the Western Giant Buddha and requalification of the surrounding area, 2010. Project by Andrea Bruno: Legend for plans and sections: (*1*) "Great esplanade" - area in front of the niche. (*2*) Ramp. (*3*) Entrance to the new sanctuary. (*4*) Access to the tunnels. (*5*) Sanctuaries carved into the rock. (*6*) Observatory

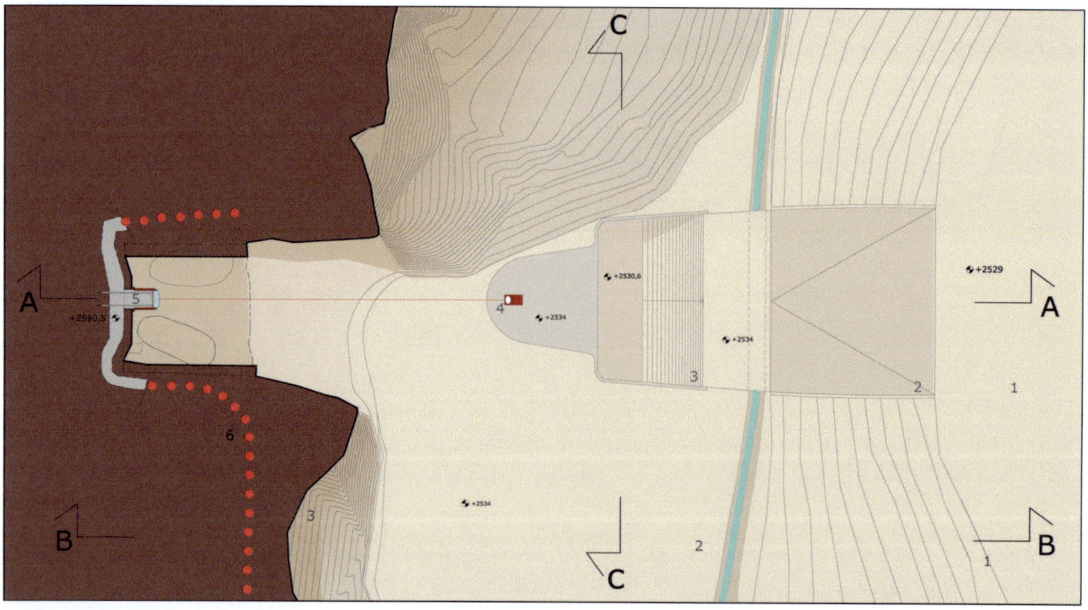

Fig. 9.15 General plan of the project at the level of the new Observatory (at approax. 65 m in height from the base of the Western Giant Buddha), 2010. Project by Andrea Bruno: Legend for plans and sections: (*1*) "Great esplanade" - area in front of the niche. (*2*) Ramp. (*3*) Entrance to the new sanctuary. (*4*) Oculo. (*5*) Observatory. (*6*) Access to the tunnels of the wall

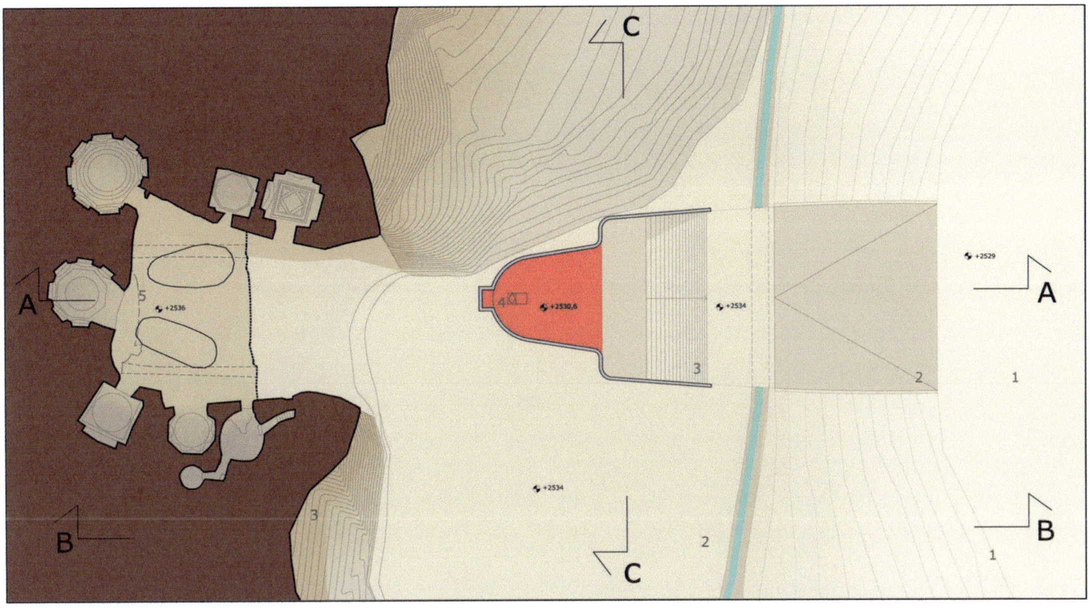

Fig. 9.16 General plan of the project at the level of the base of the Western Giant Buddha, 2010. Project by Andrea Bruno: Legend for plans and sections: (*1*) "Great esplanade" - area in front of the niche. (*2*) Ramp. (*3*) Entrance to the new sanctuary. (*4*) Reproduction of the Western Giant Buddha on a small scale. (*5*) Sanctuaries carved into the rock surrounding the niche of the Western Giant Buddha

10 Long-Term Conservation Strategy for Repair, Enhancement, Research, and Risk Preparedness for the Preservation of the Site

Claudio Margottini

Principles

Considering the previous described environmental processes, any stabilization strategy has to solve the causes of active processes and not focus only on mitigating the effects. Different stabilization measures have been investigated, considering also the need of implementing the work in an area of high cultural value but low technological support (at least at the moment). Also, consolidation works do not have to be restricted to the statues themselves, but must consider the entire area, especially for those processes that affect the whole cliff (e.g., rainwater drainage). In fact, as recognized by geomorphological investigation, many caves and part of the cliff are presently at risk of collapse and/or great erosion. Following are the essential points to be raised in a wider perspective and long-term view for the protection of cliff, caves, and niches. These elements are the foundations of a general stabilization plan for the consolidation of niches and cliffs, which should include the necessary geological, geomorphological, and hydrological studies as well as the structural condition of the caves and great cliff, and the monitoring and design of remedial works.

C. Margottini
Department of Geological Survey of Italy,
ISPRA – Italian Institute for Protection and Environmental Research,
Via Brancati, 60, 00143 Rome, Italy
e-mail: claudio.margottini@isprambiente.it

Eastern Giant Buddha

The Eastern Giant Buddha niche is now in a less risky condition. Complete stabilization of niches, as in the original plan, is probably necessary but is not an emergency (Fig. 10.1).

Western Giant Buddha

The risk of some collapse is now reduced in the Western Giant Buddha niche and grouting the top is also protecting it from water infiltration.

Back Wall of Both Niches

The fragile areas at the back of both niches are now the most important item to consider. There is the need to understand distribution, depth, and intensity of failures in the back side of both niches, which also affects the stability of the caves underneath. A preliminary proposal is described in the present volume. Any stabilization plan should be properly studied before execution. Possible investigation techniques include seismic noise, fragility tests on stones,

C. Margottini (ed.), *After the Destruction of Giant Buddha Statues in Bamiyan (Afghanistan) in 2001*, Natural Science in Archaeology, DOI 10.1007/978-3-642-30051-6_10, © Springer-Verlag Berlin Heidelberg 2014

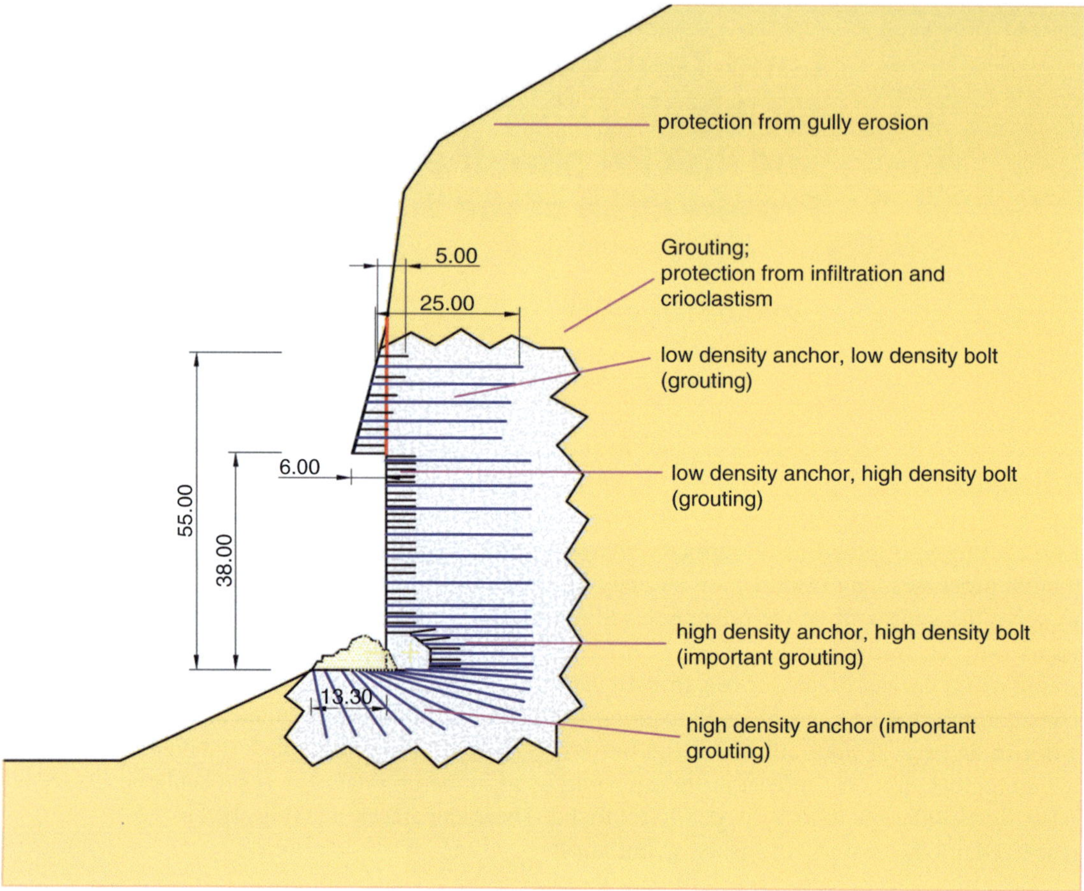

Fig. 10.1 Typology of stabilization techniques in different areas of the cliff and niche. Anchors and bolt can be seen in the cliff, outside of the niche, whereas mainly bolts and a low number of anchors may be proposed inside the niches

engineering geology tests, and ground penetration radar. At the moment both back sides are protected with a metal net that safeguards them from small rock fall, but it interferes with the possibility of investigating the consistency and deep propagation of shear failure behind the statue's original position (ground penetration radar prospecting). On the other hand, its on-site removal and replacement are quite difficult.

Buttress

1. The buttress of the Small Buddha niche was recognized as a fragile element of the site. A geotechnical study including boring the buttress foundation should be implemented. Sinking the buttress as well as opening and closing the cracks should be monitored. Accordingly, a mitigation plan should be stabilized after appropriate investigation.

Seismic Stability

1. Further study of the local seismic response to a potential earthquake at the Bamiyan site is recommended, particularly the cliffs and niches.
2. Vibratory ground motion monitoring should be carried out through installation of seismic ground motion recorders.
3. The effects of the Bamiyan earthquake of June 9, 1956 should be better investigated through local newspapers and documents as well as eyewitnesses and field survey.

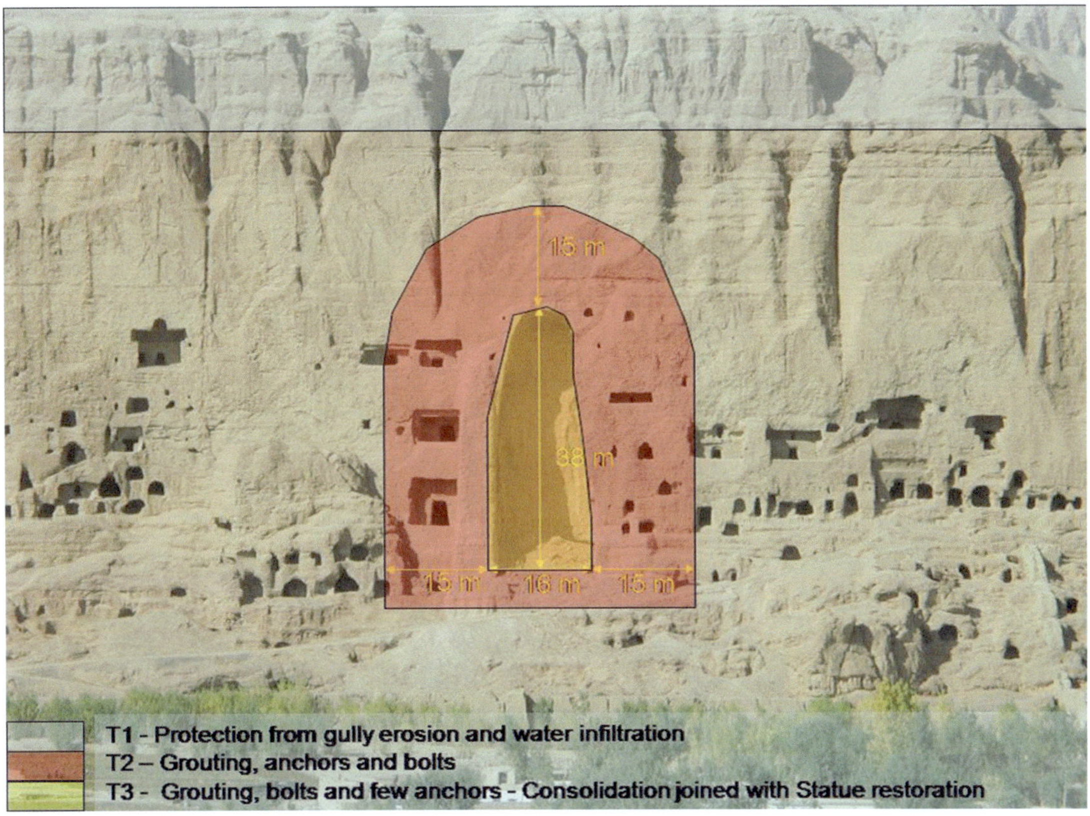

Fig. 10.2 Identification of major areas of intervention and related typology for the Eastern Giant Buddha niche

Caves and Cliff

Many caves are now at risk for incipient rock fall because of the morphological evolution of the cliff. The geomorphological map exhibits such sites under the term *active element*.

Experience suggests the use of passive anchors (stainless steel) and bolts for stabilization of the cliff and potential rock falls (Fig. 10.2).

This intervention, which was initially restricted to the Buddhas' site and surroundings, should be enlarged to the most endangered sites and caves. The appropriate design has to be planned directly on site in order to define exactly where to place individual anchors and bolts. Extreme care has to be used for the grouting mortar (i.e., cement with low water release) because of the slaking phenomenon affecting local material (Margottini 2003).

Drainage

Protection from water circulation and infiltration from the upper part of the cliff is one of the most important future actions. A geomorphologic as well as a drainage network map were established in order to ensure the description of the most relevant gully erosion phenomena and identify the most critical areas of the cliff. A preliminary work can consider the restoration of the existing channel constructed by the Indian Archaeological Survey in 1970, but a complete revision of the system has to be performed. In fact, the Indian channel correctly collects the water in the upper part of the cliff, but in some cases, it releases the water freely on the cliff, inducing considerable gully erosion (e.g., close to the Western Giant Buddha niche). Possibly, a more extensive channel has to be considered in order to also contribute to the preservation of most important cave

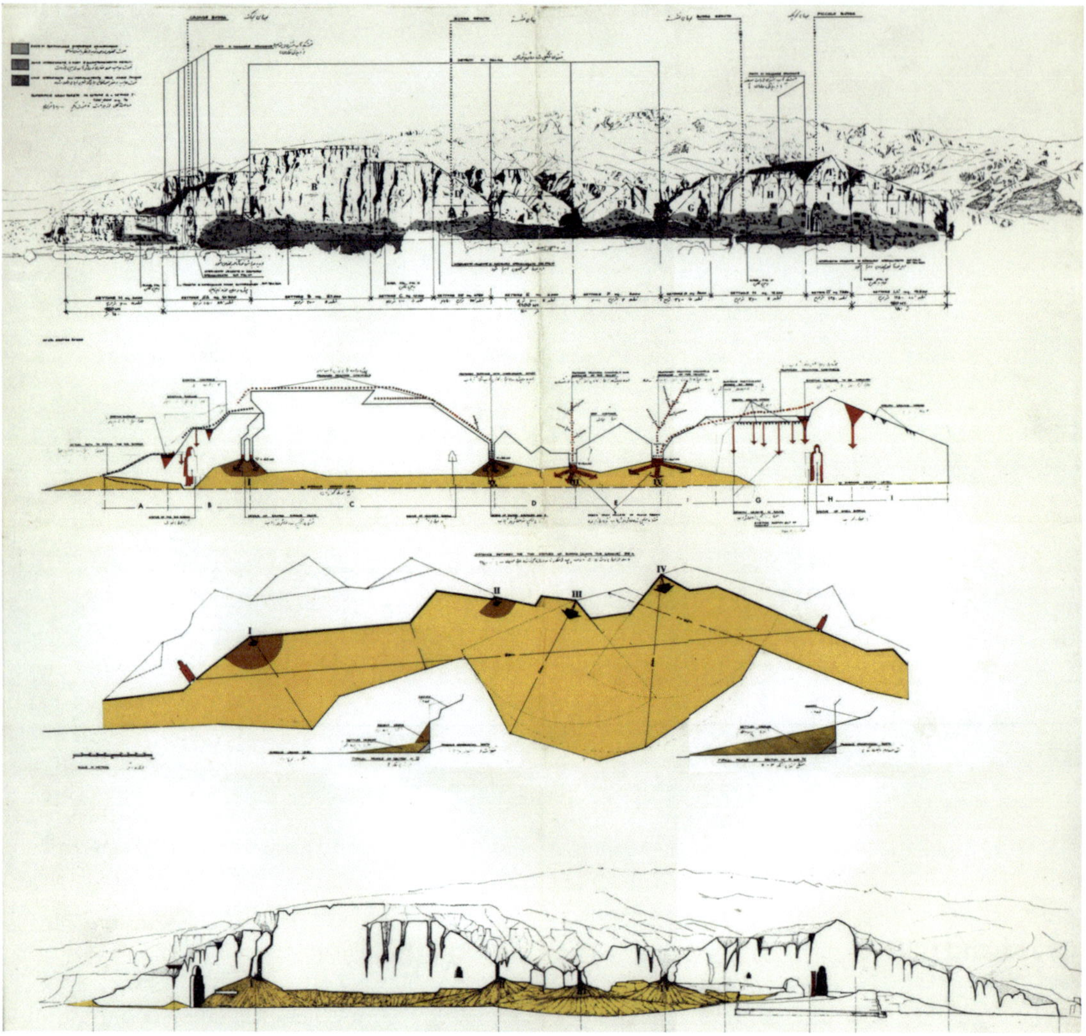

Fig. 10.3 Proposed canalization work form (Bruno 1966)

areas. This work consists mainly in canalization for the drainage of rainwater to avoid water infiltration, gully erosion, and debris deposition hiding part of the monumental complex (Fig. 10.3 from Bruno 1966). Finally, cracks have to be grouted and filled to avoid water infiltration from the surface through open discontinuities.

Documentation

1. All consolidation work carried out should be systematically documented (i.e., old and developing cracks). All technical reports and visual documentation should be translated and distributed to local Afghan officials.

Site Security

1. Regular de-mining units should be established on the site, especially at the top of the cliff.

Capacity-Building

1. Increased coordination and training should be established between the international experts and local experts and trainees.

2. Coordination with Afghan experts as well as the potential for capacity building is envisaged.

Publication

Because many international groups are working on the Bamiyan site, there is a need to collect all the produced material in order to ensure their wider distribution. The central point might be either UNESCO or the Minister for Information and Culture in Kabul.

Monitoring System: Manual Crack Gauge Monitoring System

Introduction

Field instrumentation is used for two primary purposes: to determine ground in situ conditions for design, and to monitor performance of critical elements of a project. During the investigation and design phase, instrumentation can be used to determine ground in situ conditions. Instrumentation monitoring can be extended over several months or years as needed to measure seasonal effects. During construction, instruments can be used to monitor in situ conditions to verify design assumptions and warn of possible changed conditions or impending hazards.

Geotechnical instruments are also used to characterize site conditions, verify design assumptions, monitor the effects of construction, enforce the quality of workmanship, and provide early warning of impending failures. In these regards, they are used to augment standard investigation practices and visual observations, conditions that would otherwise be difficult to evaluate or quantify because of their location, magnitude, or rate of change.

The need of a monitoring system in the Bamiyan cliff was recognized from the beginning of the investigation. This is because the understanding of crack and cliff evolution is a basic element for understanding present-day trends and potential future failures. Simple equipment was requested that would have no impact on the cultural heritage site of Bamiyan and was suitable to be managed by local technicians.

Cracks on rock slopes, buildings, walls, and slides can be monitored using a strain measuring tool. A large range of crackmeters (jointmeters) exist to suit all needs. Tools include a transparent graduated grid/scale, calipers, survey tape, micrometer, dial gauge, mechanical strain gauge (deformometer), and electrical crack gauge. One simple monitoring method is to firmly place a stake/hub pin on each side of a crack and then measure the distance between them. The magnitude and time rate of movement is monitored. Commonly, as the rate of movement accelerates, the likelihood of an imminent failure increases and concern should be raised.

The Crackmeter Monitoring System in the Bamiyan Cliff

Considering the outlined needs (simple equipment, easy to manage by local technicians, no impact on cultural heritage) the choice was made to use a removable deformometer (Fig. 10.4).

This equipment allows the measurement of relative movements between two reference pins positioned across the joint (fissure). Supplied with a carrying case, it has a precise sliding rail with adjustable base length. The movements between the two anchors are obtained with a mechanical dial gauge and measures exhibit a theoretical accuracy of 0.001 mm.

A crack gauge monitoring network was initially established in September 2003 (Colombini and Margottini 2003) in both niches and further implemented in June 2007 along many other places of the cliff. Figure 10.5 reports all the monitored cracks. In Table 10.1 the "initial reading" (zero value) of each sensor is reported.

As mentioned, a monitoring system was first established by the author in 2003. The position of this sensor is reported in Figs. 10.6 and 10.7 according to the original label. Presently, they have been re-named as in following the table, which also report all the collected data. For the date June 25, 2007, these readings are slightly different from the previous table because they are measured with different equipment.

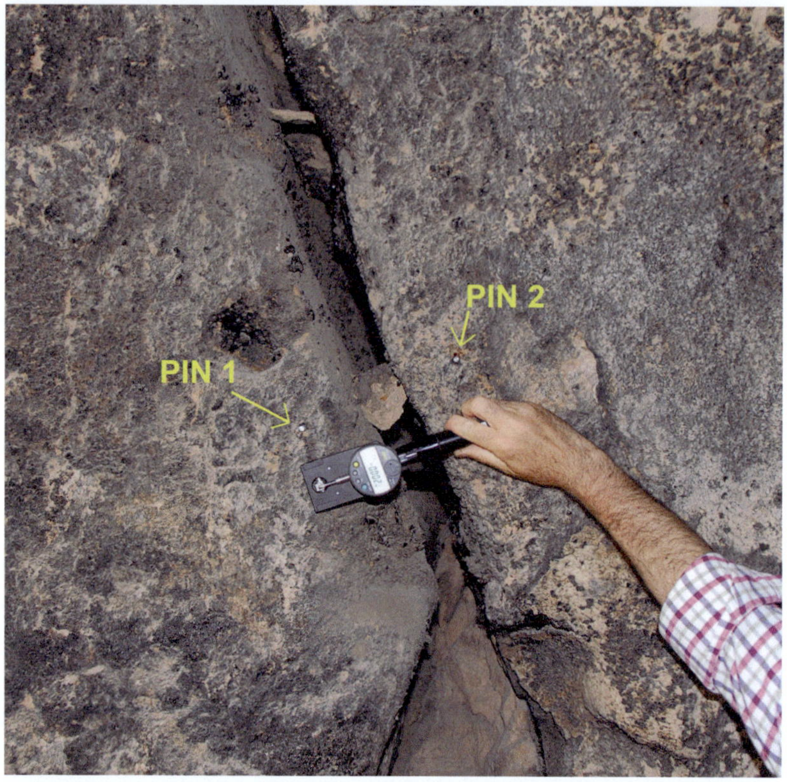

Fig. 10.4 The selected tool for the manual crack gauge monitoring network

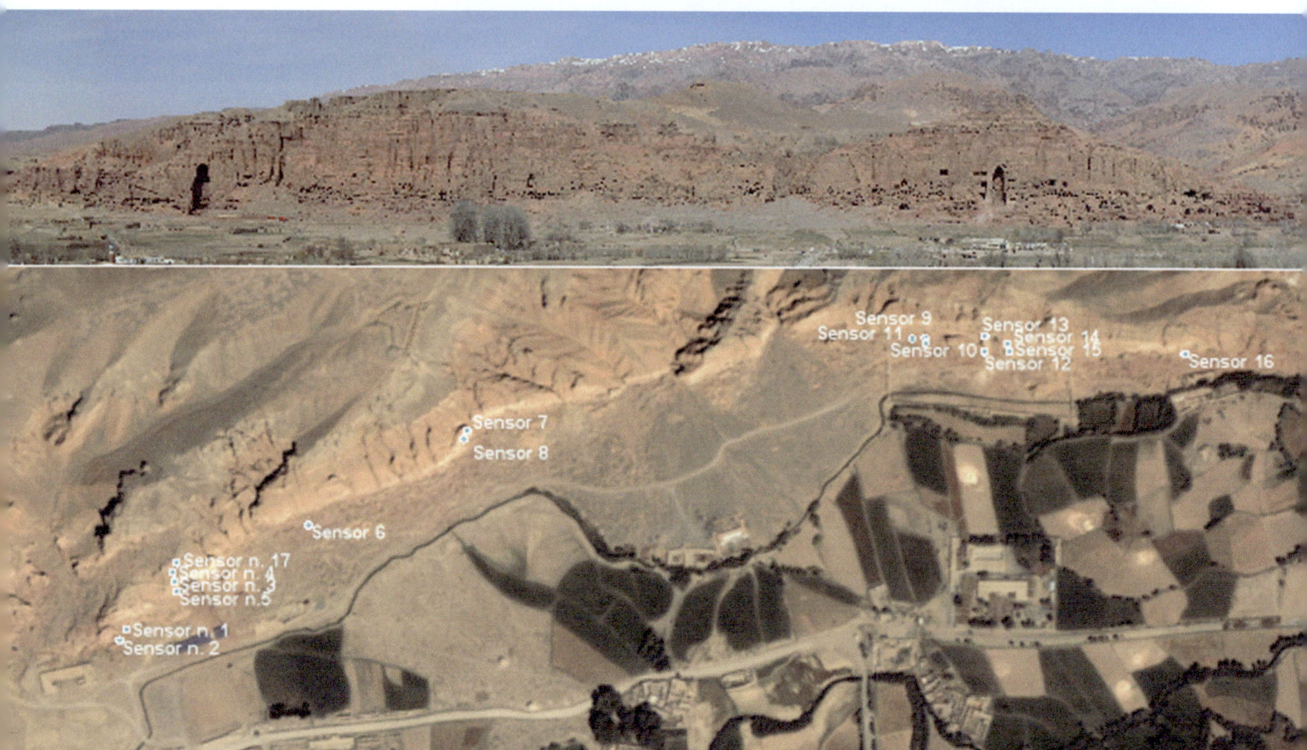

Fig. 10.5 The manual monitoring network for crack gauge measurement

Table 10.1 Initial values of monitoring sensors

Date	Sensor 1	Sensor 2	Sensor 3 (former 6)	Sensor 4 (reinstalled in 2012)	Sensor 5 (reinstalled in 2012)	Sensor 6	Sensor 7 (reinstalled in 2012)	Sensor 8	Sensor 9	Sensor 10 (reinstalled in 2012)	Sensor 11 (reinstalled in 2012)	Sensor 12 (former 1)	Sensor 13 (former 2)	Sensor 14 (former 3)	Sensor 15 (former 4)	Sensor 16	Sensor 17 (former 5)
June 25, 2007	1.865	0.885	3.208	2.316	0.525	3.639	1.389	1.451	1.177	2.56	1.177	1.182	0.032	4.694	0.91	1.288	2.205
November 23, 2010	2.03	1.394	3.494	[redacted]	4.483	4.015	[redacted]	1.53	3.064	[redacted]	[redacted]	1.317	0.048	4.805	1.055	1.467	2.307
June 20, 2012	1.968	1.137	3.333	2.085	2.091	3.945	2.116	1.368	3.017	2.043	2.676	1.297	0.005	4.775	1.063	1.453	2.182

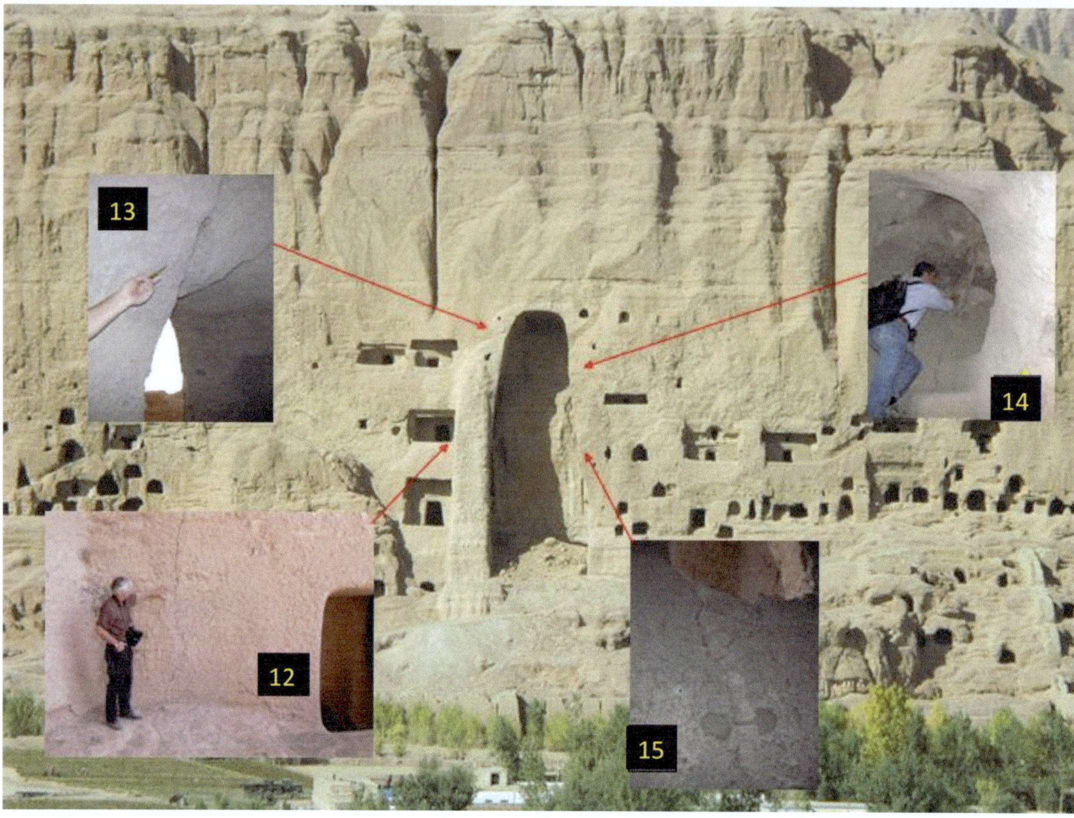

Fig. 10.6 Identification of monitored discontinuities in the Eastern Giant Buddha niche. Numbers (*12–15*) indicate the sensors

From the existing sensors, and referring to the two niches only, it is possible to say that (Fig. 10.8):

1. No deformation occurred until now in the site of Western Giant Buddha
2. Minor enlargement was observed in stations 3 and 4, occurring during the execution of emergency consolidation work on the eastern side of Eastern Giant Buddha niche in spring 2004. This minor deformation was also confirmed by an automatic sensor applied during implementation of the work. After completion of the work in the spring of 2004, no deformation was recorded. On the western side of the Eastern Giant Buddha niche, monitoring of the buttress (station 1) and monitoring of a large crack on the top (station 2, probably the same recovered in the vault of the lower cave behind the feet of the statue), do not reveal any remarkable widening.

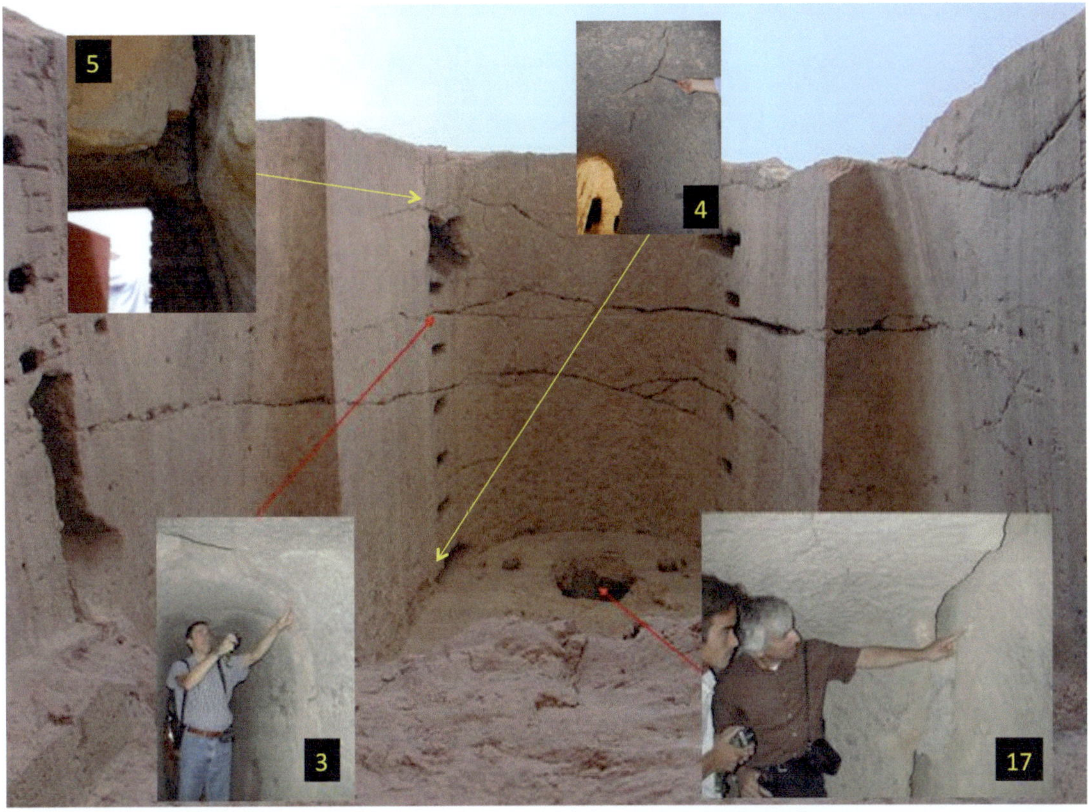

Fig. 10.7 Identification of monitored discontinuities in the Western Giant Buddha niche. Numbers (*3–5, 17*) indicate the sensors

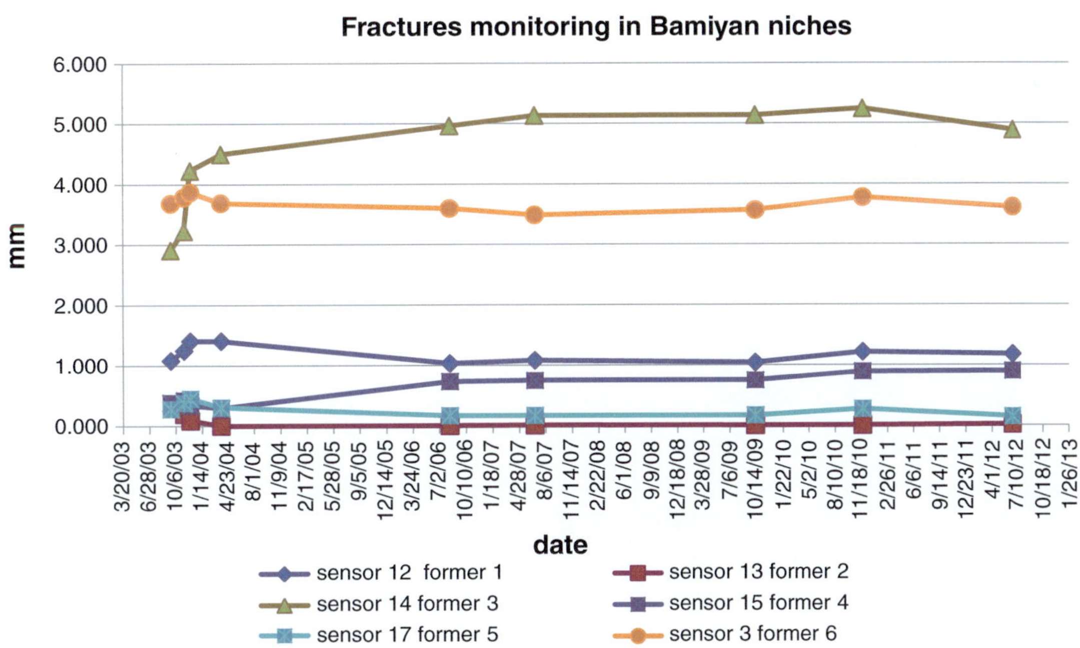

Fig. 10.8 Observed deformation on selected discontinuities on cliff and niches

Table 10.2 Series of existing monitoring sensors

Data	Sensor 12 (former 1)	Sensor 13 (former 2)	Sensor 14 (former 3)	Sensor 15 (former 4)	Sensor 17 (former 5)	Sensor 3 (former 6)
September 16, 2003	1.085	0.200	2.907	0.386	0.281	3.685
November 06, 2003	1.250	0.090	3.225	0.425	0.380	3.785
November 30, 2003	1.408[a]	0.002[a]	4.225[b]	0.340[b]	0.450[a]	3.870[a]
March 26, 2004	1.409	0.000	4.498	0.291	0.305	3.691
August 10, 2006	1.035	0.005	4.962	0.730	0.166	3.600
June 25, 2007	1.082	0.000	5.131	0.746	[c]	3.489
October 10, 2009	1.043	0.000	5.135	0.748	[c]	3.568
November 23, 2010	1.217	0.016	5.420	0.891	0.268	7.447
June 20, 2012	1.178	−0.027	4.886	0.899	0.143	2.964

[a]Interpolate values
[b]Manually increased after comparison with the automatic monitoring system operating in the same cracks in October–November 2003
[c]Covered by grouting of crack. To be reactivated

Appendix: Location of Individual Manual Monitoring Sensors, for Crack Gauge Measurement

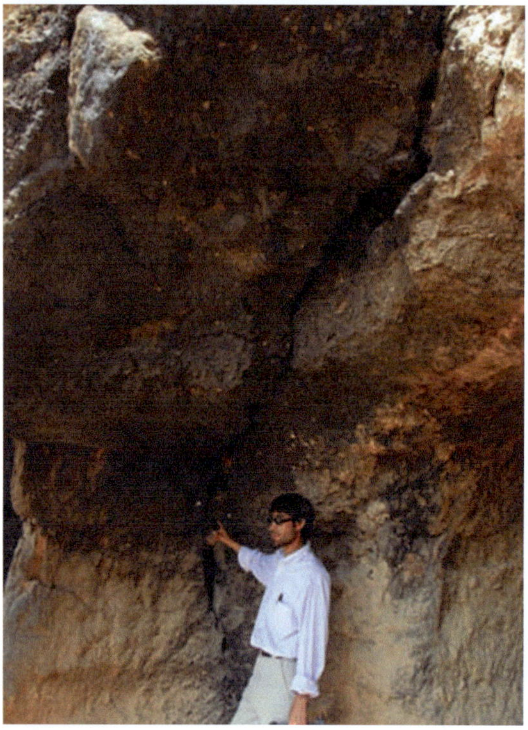

Sensor n. 1

Sensor n. 2

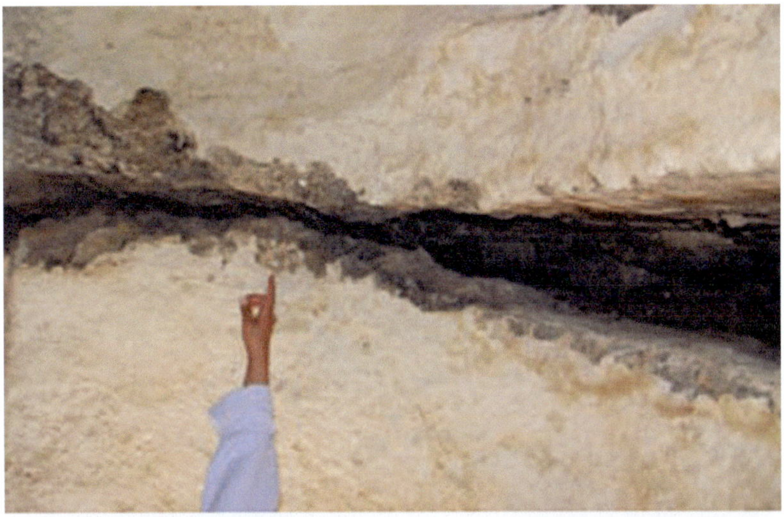

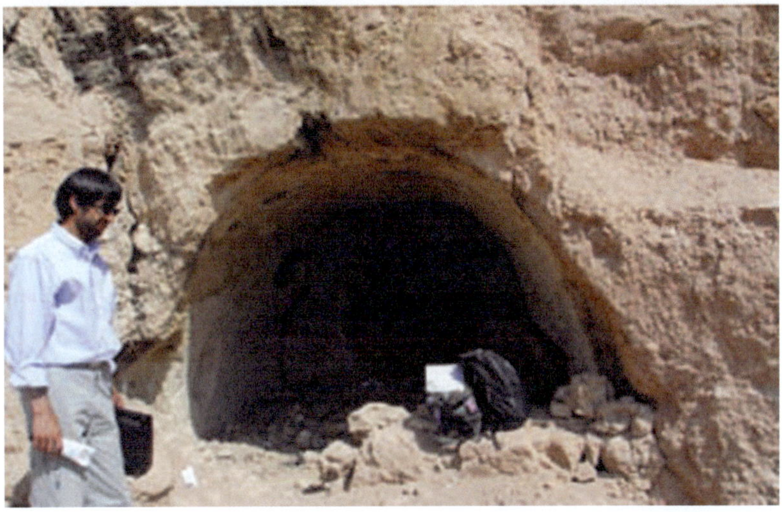

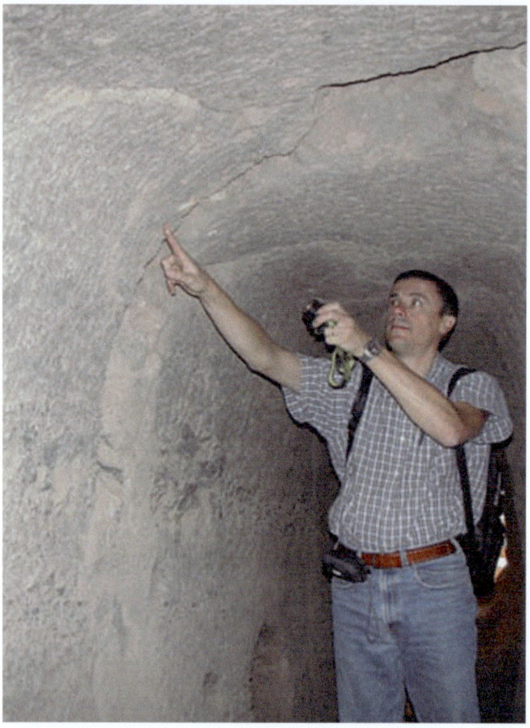

Sensor n. 3 (former n. 6)

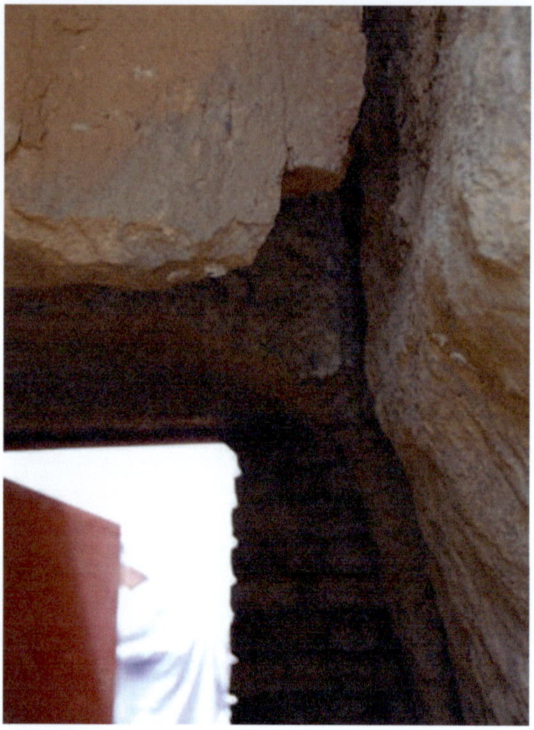

Sensor n. 5 (entrance top level corridor)

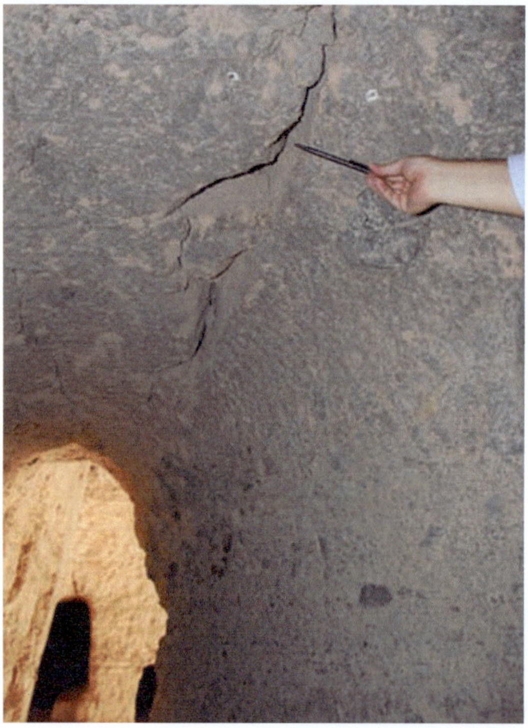

Sensor n. 4 (almost back side)

Sensor n. 6

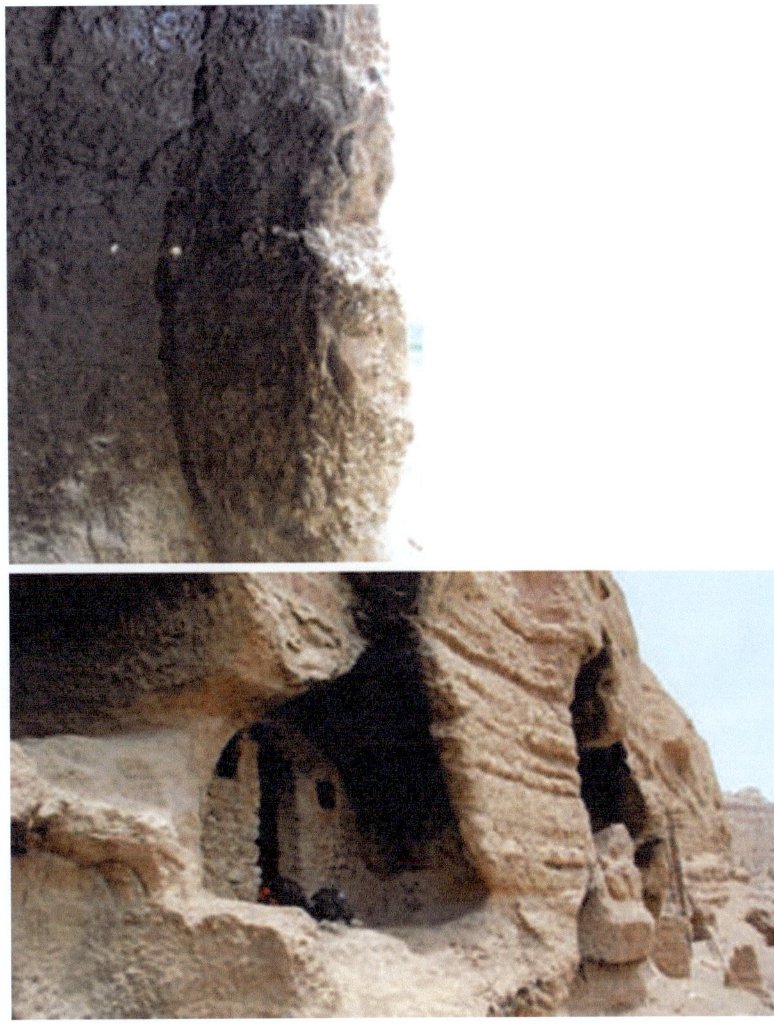

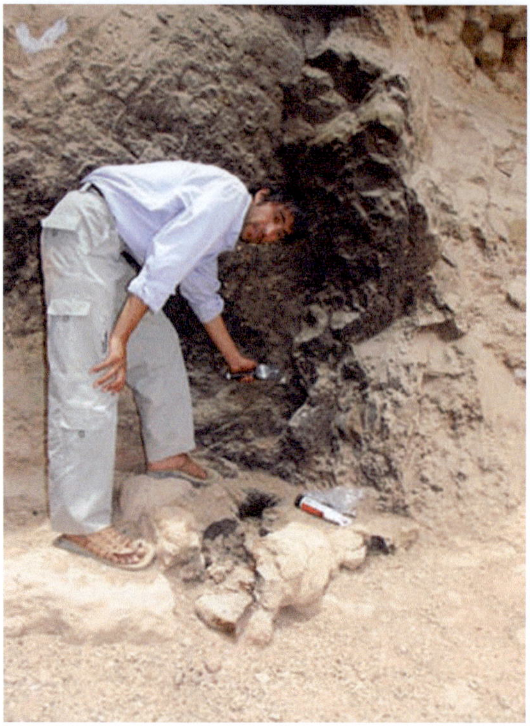

Sensbor n. 7

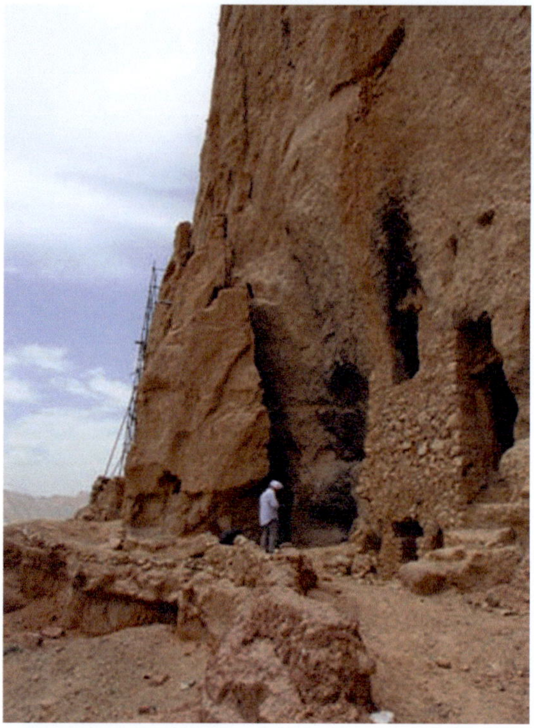

Area of sensors n. 7 and 8

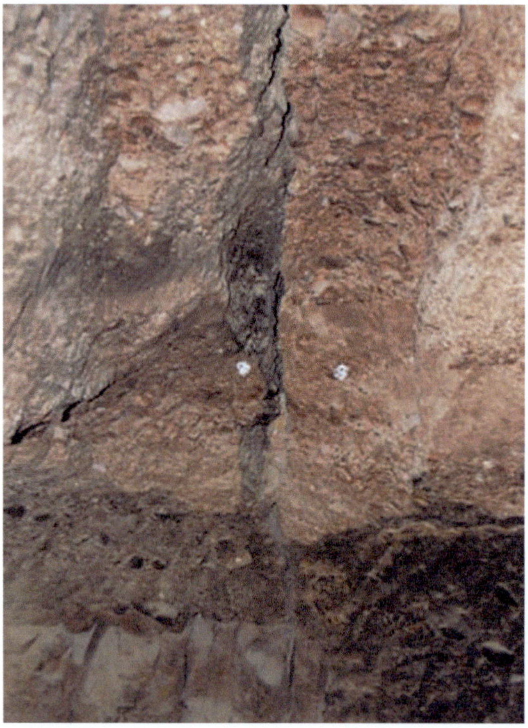

Sensor n. 8

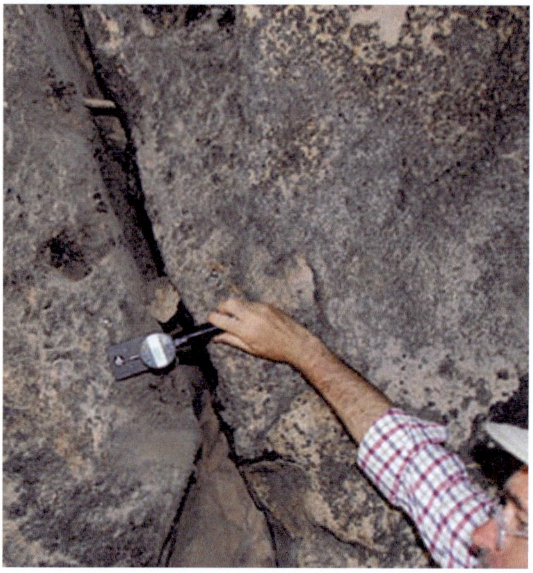

Sensor n. 9

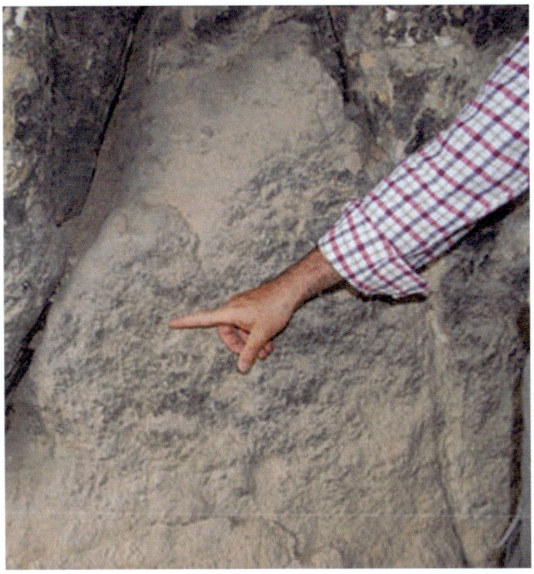

Sensor n. 10

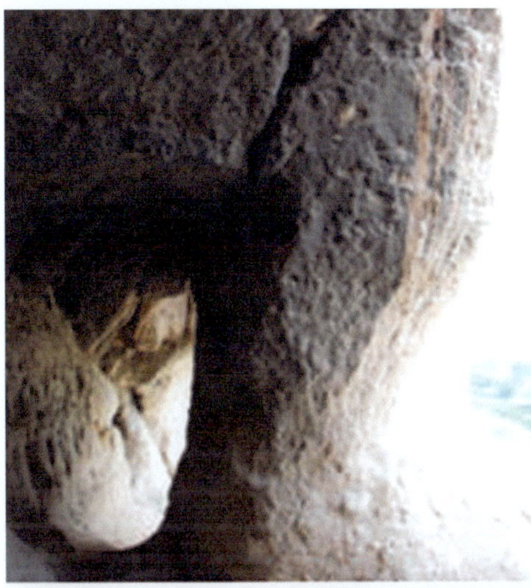

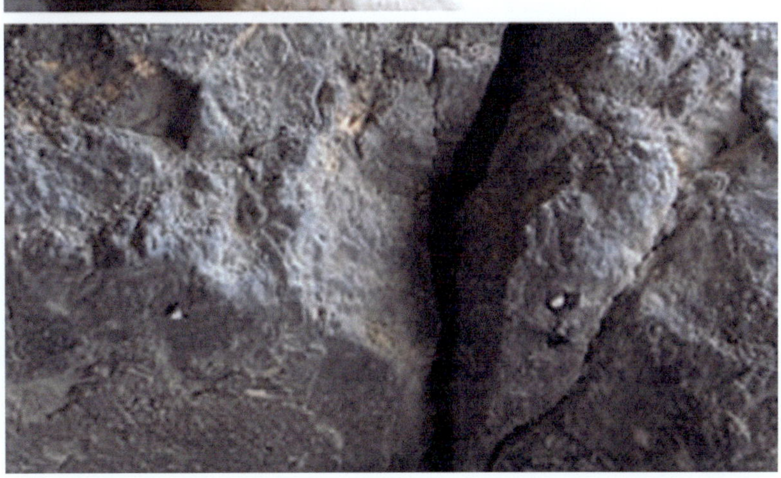

Sensor n. 11

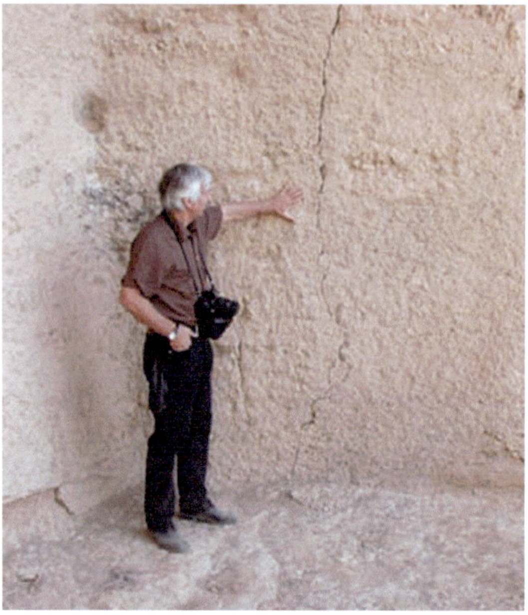

Sensor n. 12 (former n. 1)

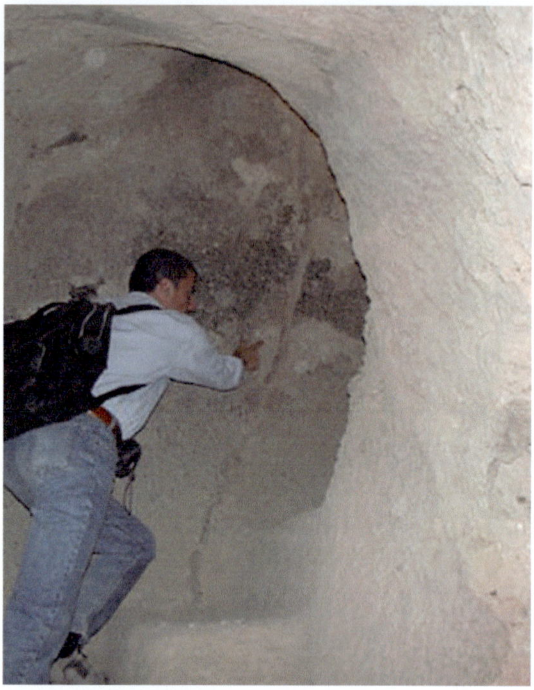

Sensor n. 14 (former n. 3) middle of the stairway

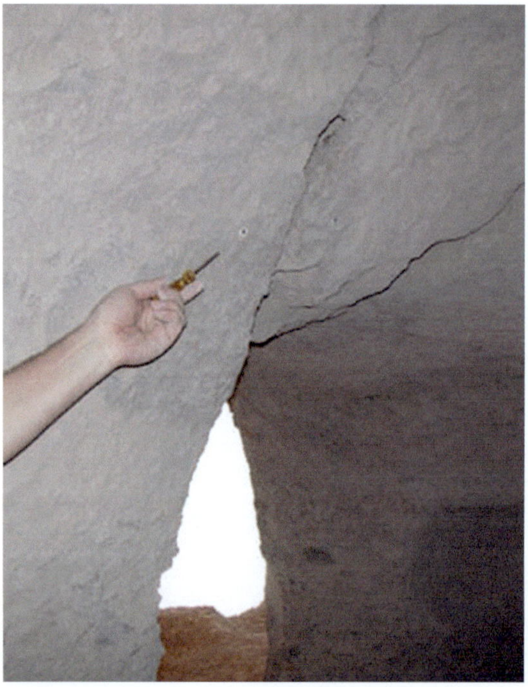

Sensor n. 13 (former n. 2) top of niche

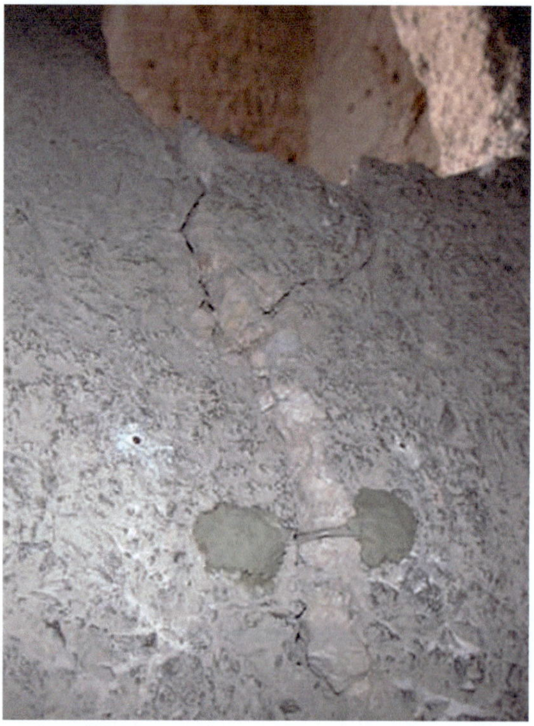

Sensor n. 15 (former n. 4)

Sensor n. 16

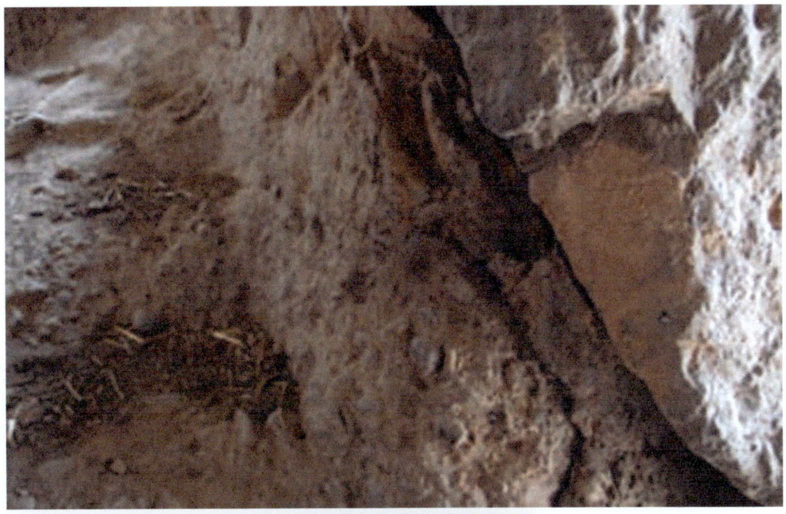

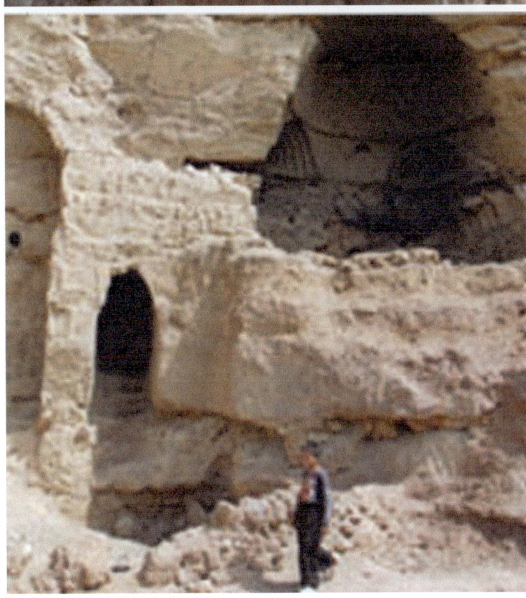

Sensor n. 17 (former n. 5) presently covered by mortar

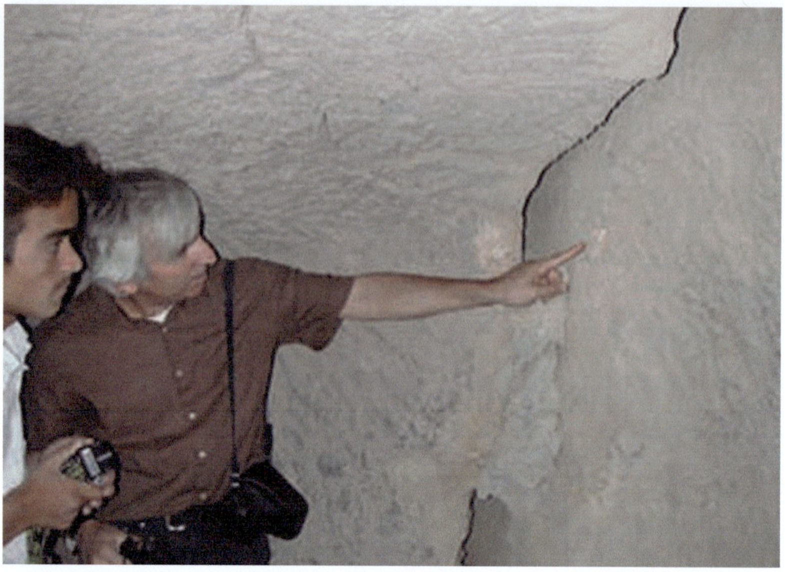

References

Bruno A (1966) Projects and works by Andrea Bruno (1960–1995). In: Mastropietro M (ed) Restoration and behind. Architecture from conservation to conversion. Edizioni Lybra imagine, Torino

Colombini V, Margottini C (2003) Buddhas niches in Bamiyan valley: emergency measures in autumn 2003, to permit future consolidation. UNESCO Report, Paris, Sept 2003

Margottini C (2003) The geomorphological instability of the Buddha niches and surrounding cliff in Bamiyan valley (Central Afghanistan). UNESCO Report, Paris, June 2003

11 Emergency Intervention

Vittorio Colombini, Carlo Crippa,
Claudio Margottini, and Gedeone Tonoli

Identification of Most Critical Areas

The explosions of March 2001, besides the demolition of the statues, reduced the stability of the shallower parts of the niches (Margottini 2003, 2004a). All the investigations described in previous chapters, together with many field surveys, allowed the identification of the most unstable sites. In the Eastern Giant Buddha niche, as well as the collapse of the statue, three minor rock falls occurred from the top of the niche. Blasting also degraded the strength of the rear of the highest right part of the niche, where a stairway is located inside the cliff and the wall between the stairs and the niche is quite thin (about 30–50 cm). This part presently has the most critical instability (A3 in Fig. 11.1). As a consequence of an existing buttress, the left side did not suffer as much damage, although in the upper part a rock fall occurred and some instabilities are now evident.

In the Western Giant Buddha niche, the major blast effects were the collapse of the statue and the consequent instability of the rear of the niche. A small rock fall occurred from the top of the niche (left side). Probably, the greater thickness of the wall between the stairway going up into the cliff and the niche (about 1 m) inhibited the propagation of the blasting effects, and resulted in less severe damage. A large crack, about 20–30 cm wide, is present in the corridor at the back of the head of the statue.

Figure 11.2 shows the most critical areas found in the field inspection and/or identified by analysing the different geological aspects investigated in this paper.

Emergency Measures Taken in 2003–2006

Overall Strategy on the Eastern Giant Buddha Niche

After the general strategy for stabilization, a follow-up of activities was performed in September 2003, aimed at the identification of potential negative evolution of the cliff and niches during winter 2003–2004 (Margottini 2003, 2004a; Margottini et al. 2005). The results of a field mission suggested an immediate response to the upper east side of the Eastern Giant Buddha niche where the existing large fissures were widening and the risk of an immediate rock fall was estimated to be very high. This collapse could involve a large

V. Colombini
Studio Colombini,
Via Merulana, 272, 00185 Rome, Italy
e-mail: v.colombini@gmail.com

C. Crippa • G. Tonoli
Trevi Division Rodio, Via Pandina, 5,
26831 Casalmaiocco (Lodi), Italy
e-mail: ccrippa@trevispa.com; gedeone.tonoli@tiscali.it

C. Margottini (✉)
Department of Geological Survey of Italy,
ISPRA – Italian Institute for Protection and Environmental Research,
Via Brancati, 60, 00143 Rome, Italy
e-mail: claudio.margottini@isprambiente.it

C. Margottini (ed.), *After the Destruction of Giant Buddha Statues in Bamiyan (Afghanistan) in 2001*, Natural Science in Archaeology, DOI 10.1007/978-3-642-30051-6_11,

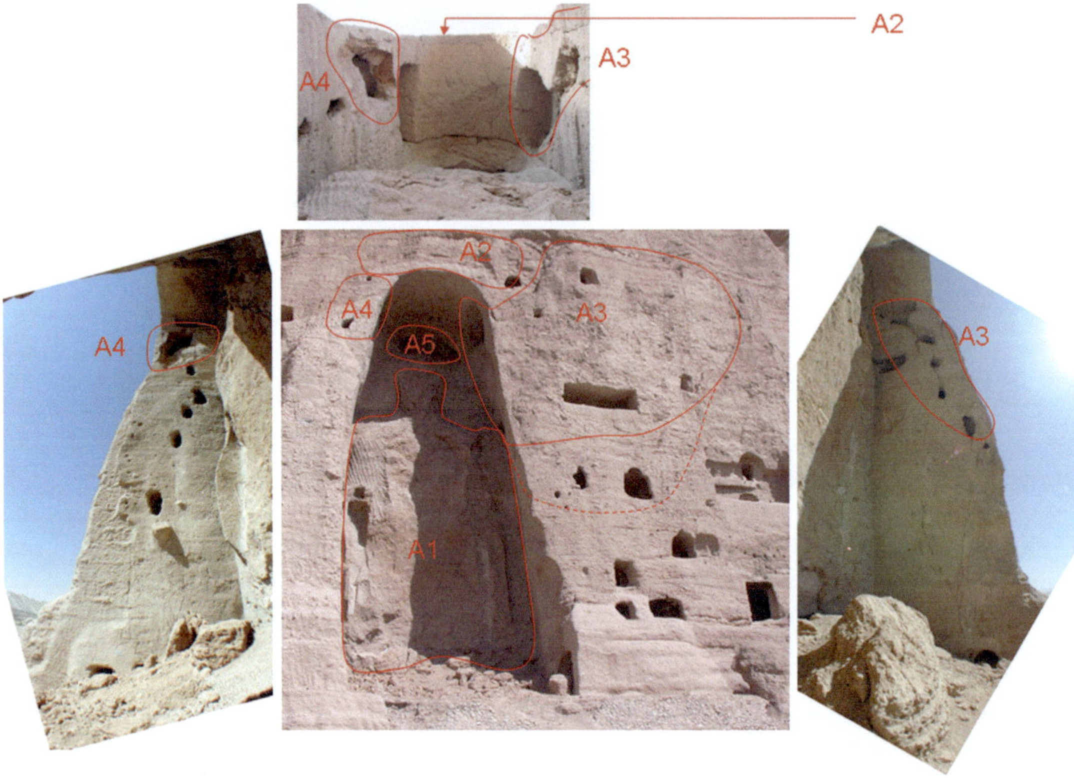

Fig. 11.1 Identification of the most critical instability areas in the Eastern Giant Buddha niche. The *A3* block in the Eastern Giant Buddha niche exhibits the most acute instability

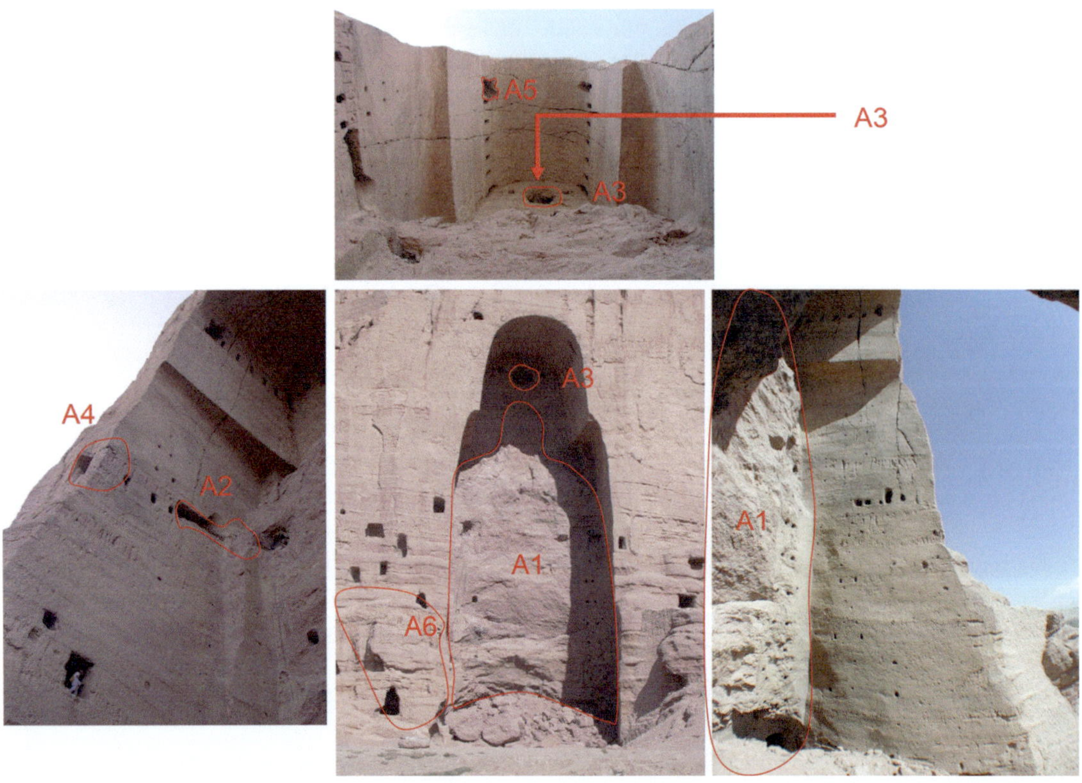

part of the upper eastern part of the cliff an then totally destroy the niche (Fig. 11.3).

Emergency consolidation work was immediately planned and carried out in this most critical of rock fall-prone areas to avoid any further collapse in the coming winter season, but also to enable archaeologists the safe cataloguing and recovering of the Buddha statues' remains, still lying on the floor of the niches. The stabilization activities started in October 2003 and continued until the beginning of December 2003 (eastern side). A second operational phase was implemented in the period April–June 2004 (eastern side) and the final one in the period September–November 2006 (western side and top). Figure 11.4 shows all the study areas and the sites for intervention. Without considering the study phases, the practical activities included four different steps:

1. The *installation of a monitoring system*, to evaluate in real time any possible deformation of the cliff. Sensors were designed to monitor the entire working area, connected with an alarm system, to make working conditions safer.
2. The *realization of temporary protection* includes steel cables and two iron beams suitable to avoid lateral deformation inside the niche from an unstable cliff and blocks. Among the temporary work, a wire net was installed on the back side of both niches to allow archaeologists to work on the ground floor in safe conditions, just after the consolidation of the niche's wall.
3. The *final stabilization* of the east side of the niche, west upper side and top. In these areas anchors, nails, and grouting were executed, in order to reduce the risk of rock fall and collapse. Particular care was addressed to the problem of grouting material because of the very high slaking capability of siltstone. The anchors placed in 2003 were pregrouted to avoid any oxidation and then percolation inside the niche. From 2004 it was decide to use only stainless steel materials, even if not pregrouted.
4. *Minimization of intervention* (anchor/nail head finishing) completes the execution of work. Anchor and nail heads were designed to be placed slightly inside the rock and then covered by a mortar allowing a total camouflage of the work. A number of tests on the better mixture, between cement, local clay/silt and water, to be used for covering the anchor/bolt heads, were also designed and developed in 2003, in cooperation with ICOMOS experts. The results highlight the better chromatic stability and robustness of the mixture.

Implementation on the Eastern Wall

In the eastern side of the niche a large external block was prone to collapse (Fig. 11.3), as were many others in the inner part.

A real-time monitoring network was planned and realized to monitor the most remarkable cracks and discontinuities. Eleven potentiometric crack gauges (0–50 mm, 4–20 mA), fitted with couplings and connecting cable (total length 350 m) were supplied in the first phase in 2003, with acquisition system (data logger) and data management software. An alarm system to detect any deformation (movement) possibly induced by the work on the main cracks present in this part of the cliff was also installed. The accuracy of the gauges was requested to 0.01 mm, to allow an accurate measurement of even small deformation. The position of the sensors is reported in Fig. 11.5.

The temporary protection includes a network of 0.6″ diameter steel cables with a light pretensioning, to sustain the most unstable block from possible collapse; steel cables were fixed to short nails, irregularly placed to avoid any stress concentration in a given line or area. Two temporary beams, located laterally to support the cliff deformation were designed and executed. Each beam

Fig. 11.2 Identification of the most critical instability areas in the Western Giant Buddha niche. The *arrow* points to a serious problem inside the niche. Other important areas to secure are *A4* and the top of *A6*

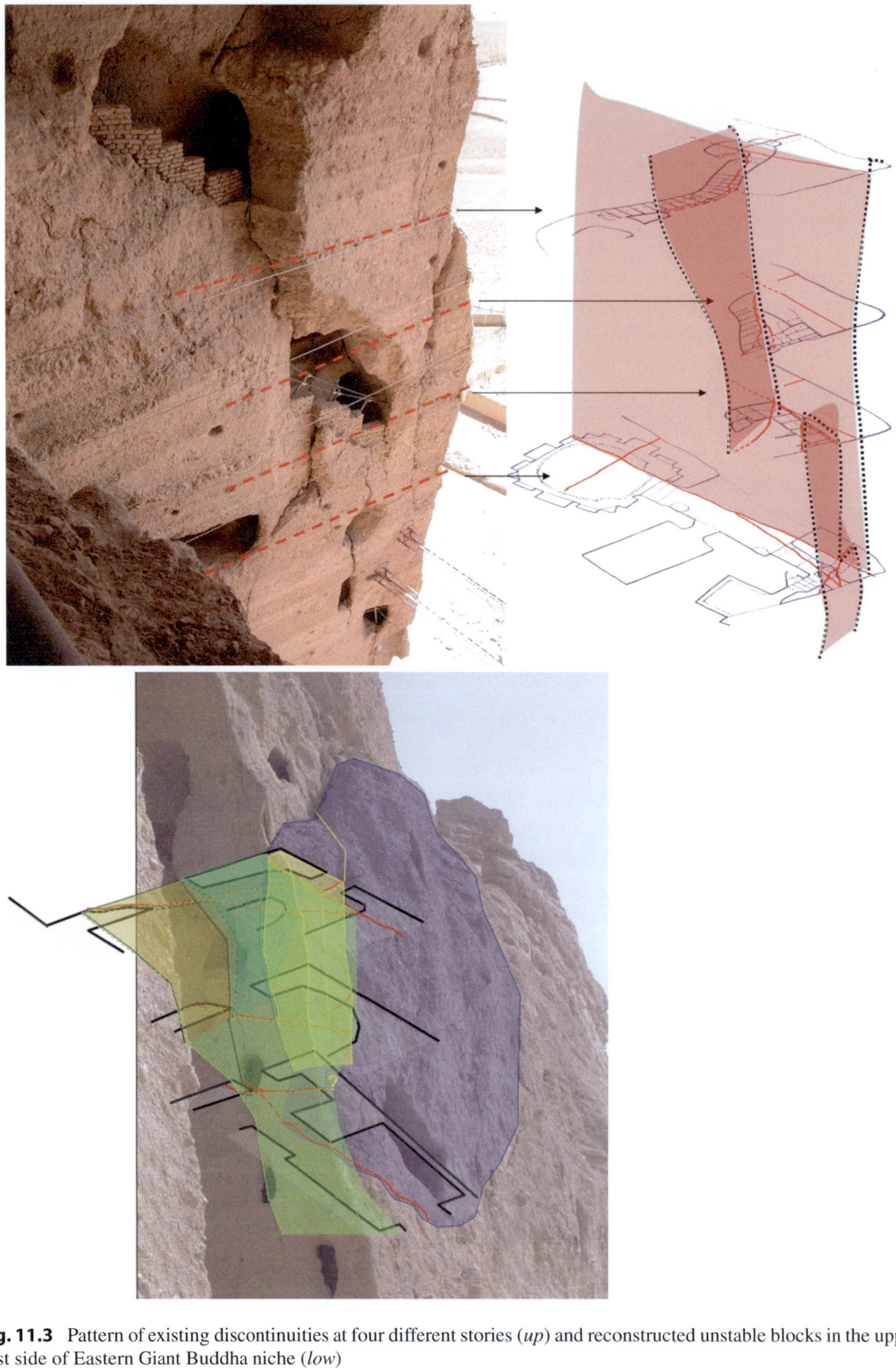

Fig. 11.3 Pattern of existing discontinuities at four different stories (*up*) and reconstructed unstable blocks in the upper east side of Eastern Giant Buddha niche (*low*)

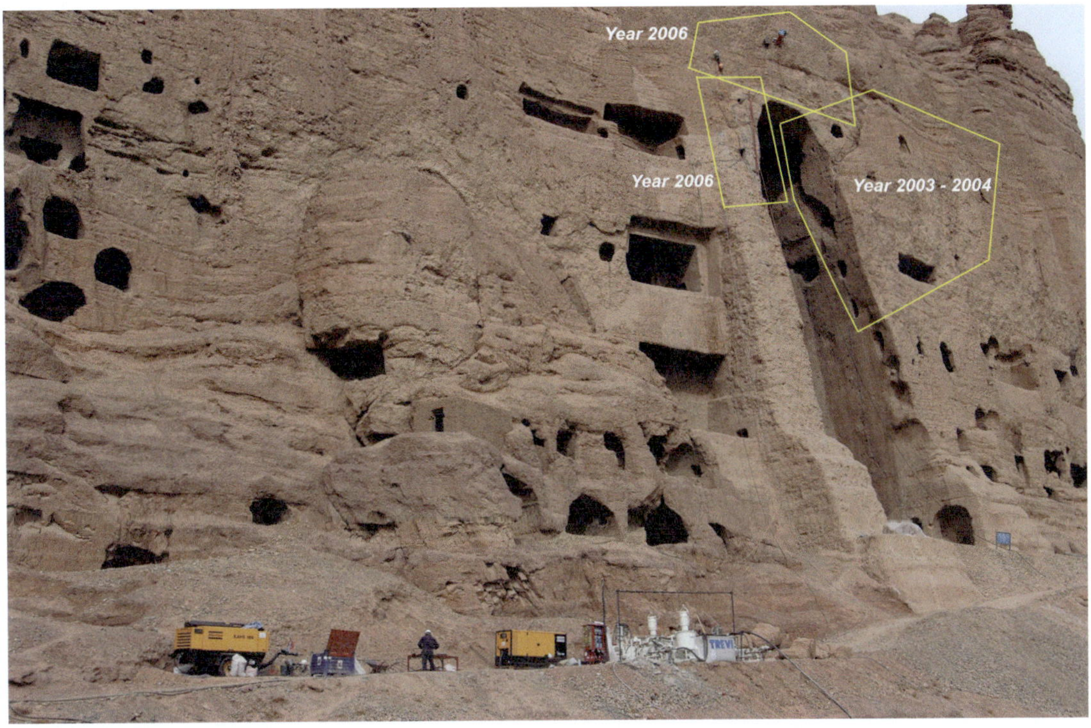

Fig. 11.4 Localization of the three areas of intervention in the Eastern Giant Buddha niche

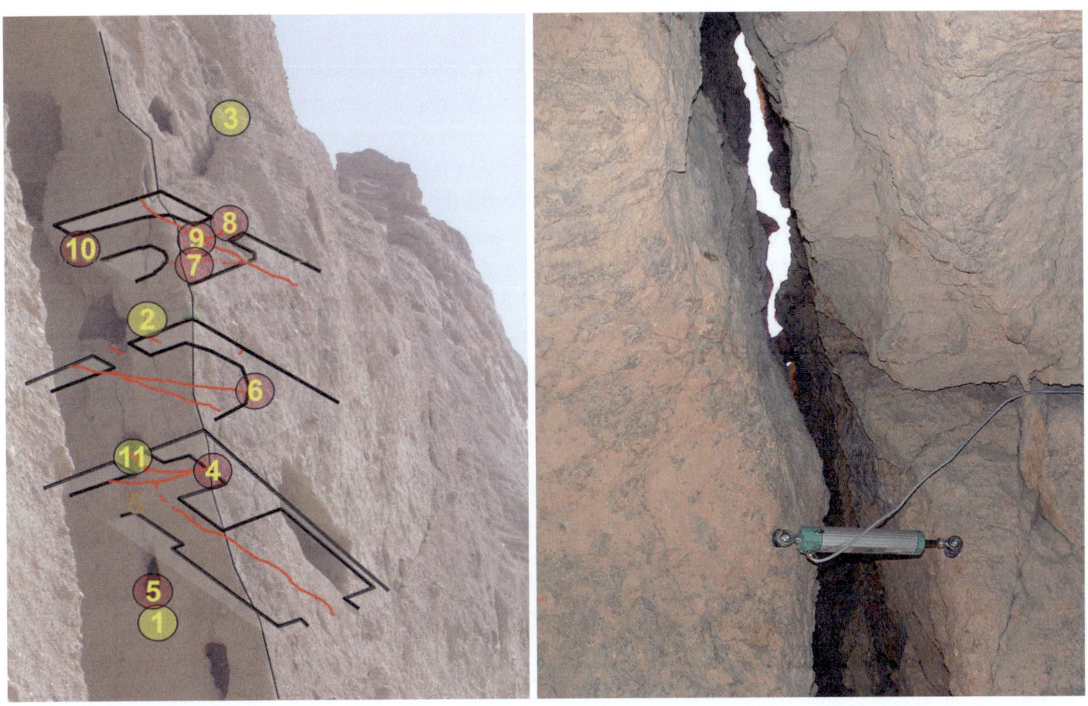

Fig. 11.5 Distribution of the 11 sensors monitoring the cracks underlying the most unstable block. On the right, a detail of such sensors

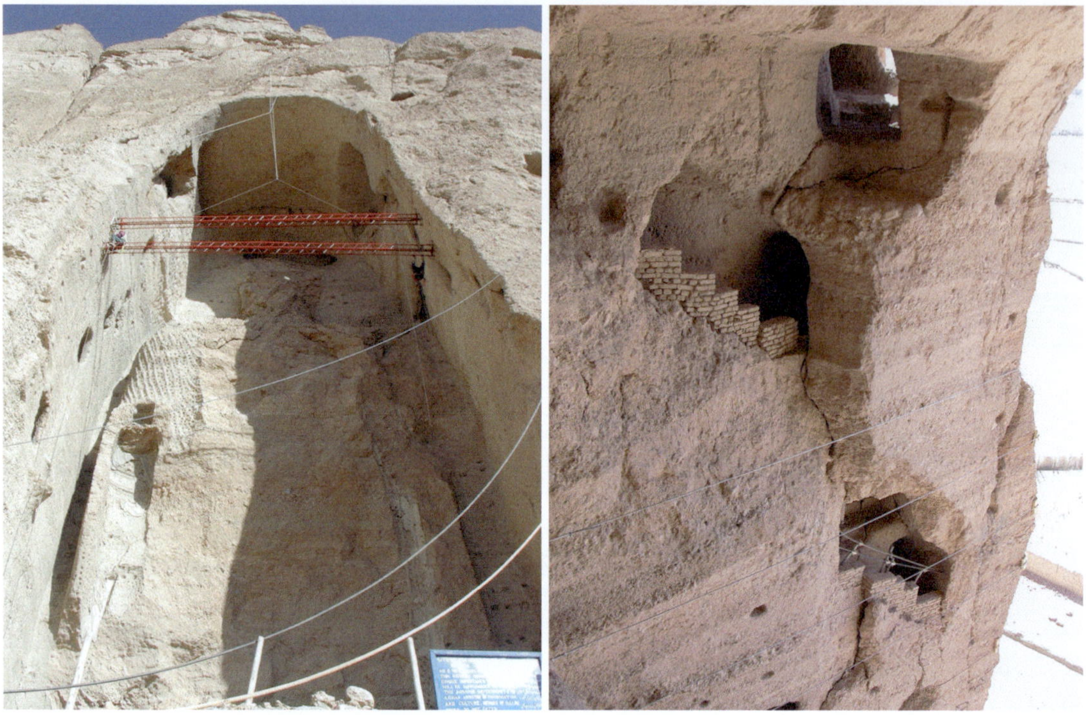

Fig. 11.6 The temporary beams (*left*) and steel cables (*right*) for the temporary support of upper eastern part of Eastern Giant Buddha niche

was calculated to offer a resistance of about 40 t, similar to two designed long anchors. Figure 11.6 shows the steel cables and the two iron beams. The temporary protection elements were removed in 2006.

The consolidation was designed by means of passive anchors and nails, correctly grouted. Long anchors are spaced about 4 m apart because they exhibit, in this configuration, a safety factor equal to two, without considering the contribution of nails. Nails will not follow a precise configuration because they have to be designed on site to strengthen the shallower part of the block. Even anchors may have some nonhomogeneous distribution, a function of internal cavities. Details on calculation are reported in Fig. 11.7. In these we have:

- The geometrical distribution of load and the assumptions for calculation, based on mechanics of a rigid body, and the related moment
- The assumptions for moment calculation of anchors and the related safety factor
- A comparison test about the possibility to generate toppling according to the static loads and the uniaxial compressive strength of the material

A major concern at the very beginning was certainly the problem of adherence between grouting material and silt-stone, a very slaking material. For this reason the choice was made to low water-release grouting. This can be achieved mixing water and cement with a superplasticizer, a chemical additive suitable to maintain the water inside the mortar. The adopted composition was: $W/C = 1/2.0$ + superplasticizer. A comparison between the standard strength for anchors and the possible mobilized one was investigated. Because anchors are designed to provide 20 t each, the borehole has a diameter of $f = 9$ cm, and the active length was limited to only 5 m, we have: $2 \times \pi \times 4.5 \times 500 = 14{,}131$ cm^2.

Without direct tests, adherence between the mortar and the rock is generally calculated between 6 and 8 kg/cm^2; assuming 5 kg/cm^2 we have about 70 t. Then, assumed strength of an anchor is 20 t that, divided by 70 gives a result of about 28 % of normal standard; alternatively, the real obtained adherence is about 1.41 kg/cm^2, which is much less than the design value of 5 kg/cm^2. The resulting safety factor is about five, suggesting a reasonable security with the designed loads. Anyway, because of missing information on detailed geomechanical distribution of discontinuities deep in the rock, these feasibility assumptions were considered in favor of security.

The correctness of the adopted solution and also of the bounding capacity of grouting mortar is given from the anchor suitability tests (Fig. 11.8), performed in 2004 to understand the bonding capacity of anchors in both siltstone and conglomerate. The design strength of passive anchors was assumed as 20 t, for a bounded length after the major discontinuity at least of 5 m (about 4 t per linear meter). The anchor suitability test was performed for 1 m length to 40 t, close to the yield capacity of steel. Up to this value no remarkable permanent elongation was detected, to demonstrate the correct bounding effect between siltstone and conglomerate and the anchors (Fig. 11.9). These data confirm, once more, the appropriate choice of superplasticizer as an additive suitable to avoid any slaking phenomena in the siltstone.

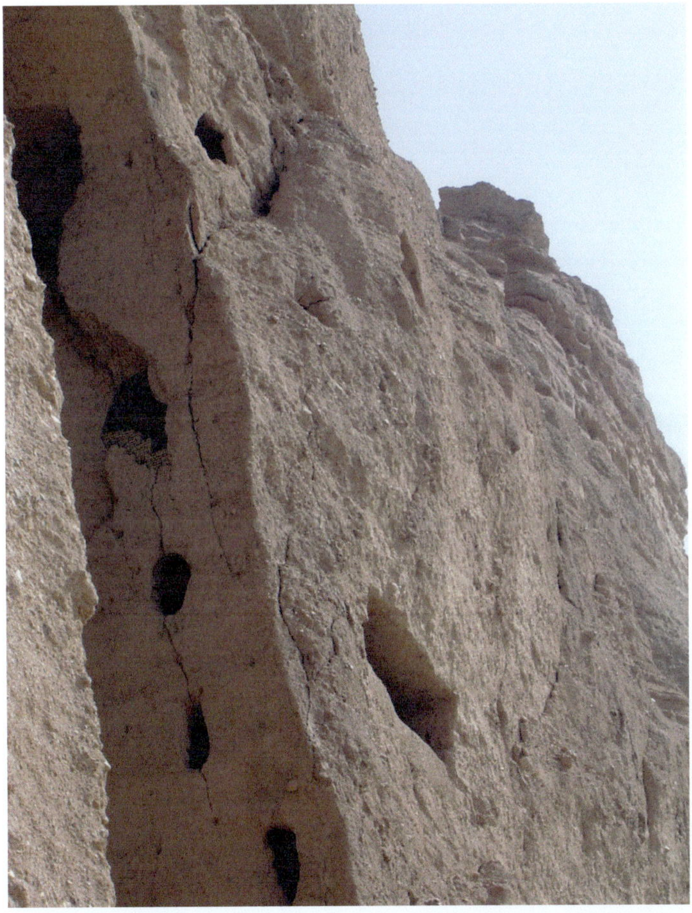

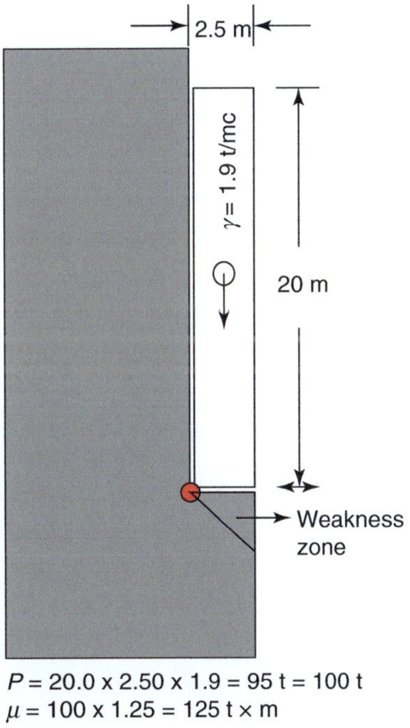

Fig. 11.7 The external block prone to collapse, (*left*) the geometrical distribution of load and the assumption for calculation (*right*), moment calculation of anchors and the related factor of safety (*next page left*) and an evaluation test about the possibility to generate toppling according to the static loads and the uniaxial compressive strength of material (*next page right*)

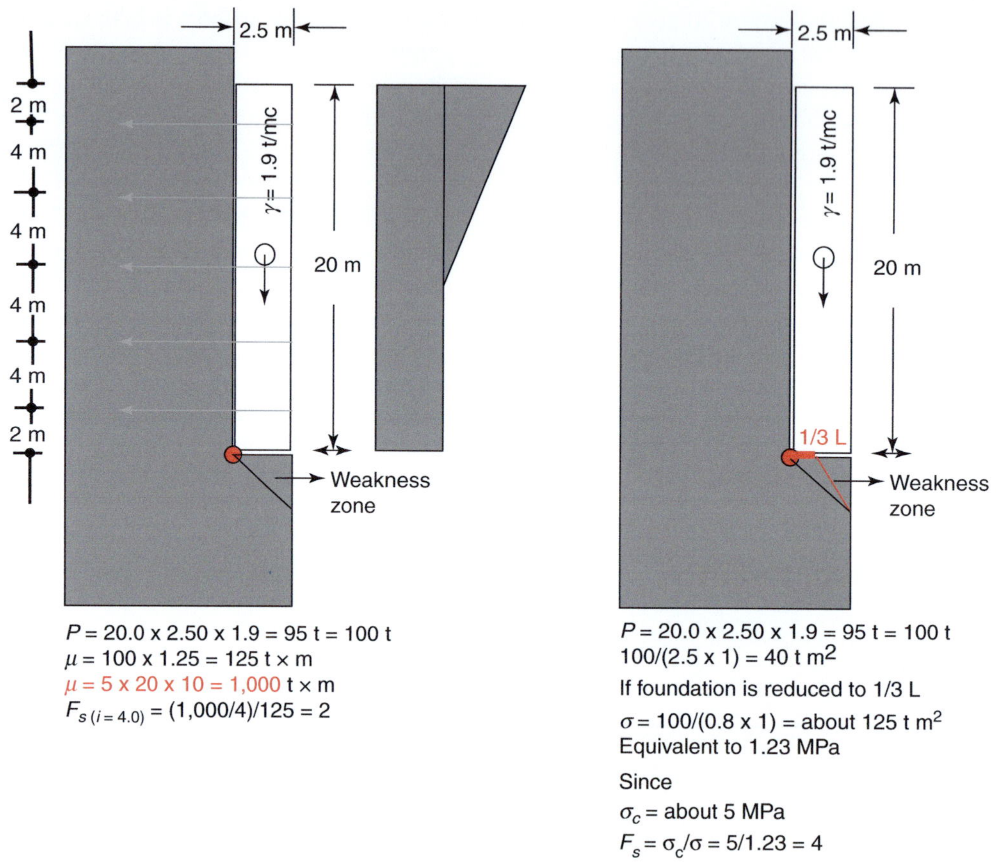

Fig. 11.7 (continued)

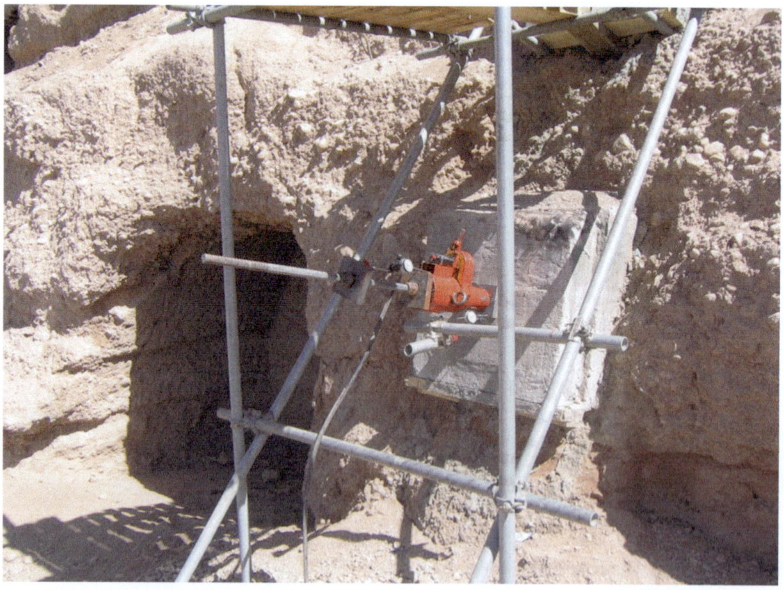

Fig. 11.8 The anchor suitability test site

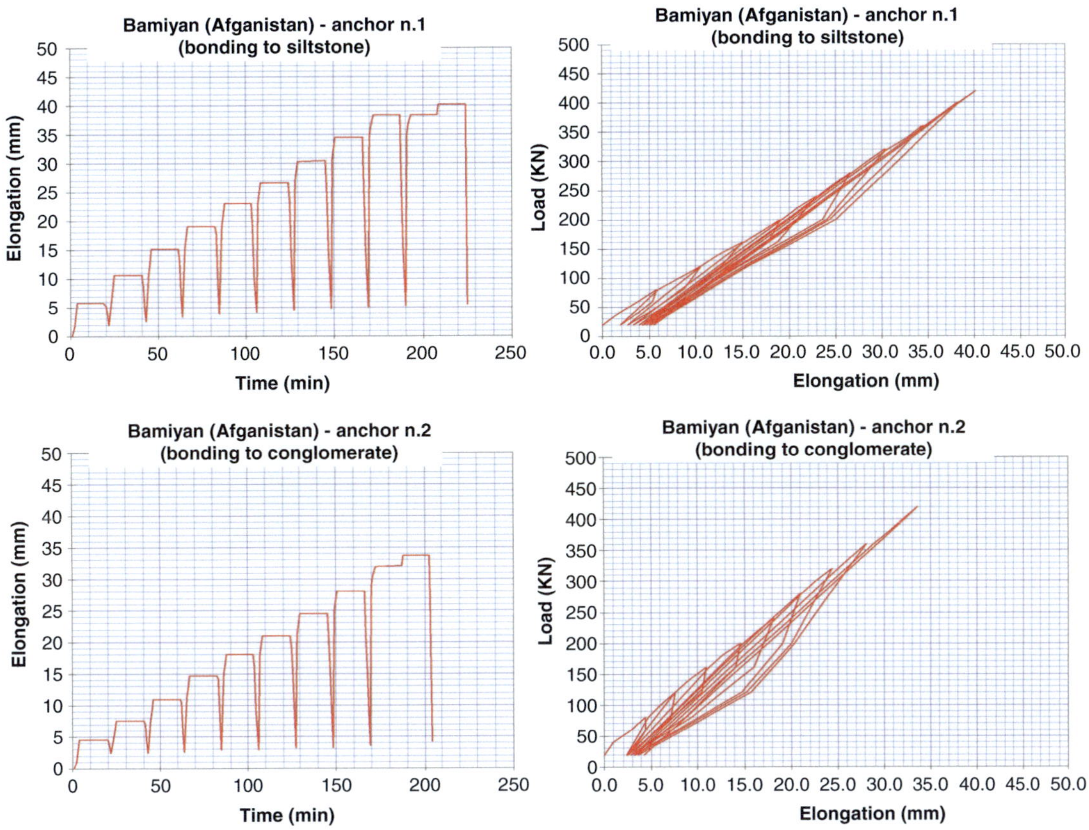

Fig. 11.9 Anchor suitability tests for siltstone and conglomerate, in 1-m length anchors. The load (kN) and respective time (min) and elongation (mm) are reported, showing, up to 40 t, upholding of the elastic domain and the lack of any permanent deformation for the tested anchor

In detail, in the eastern part of the niche have been placed:

1. Six short passive anchors (steel nails, dia 16 mm, FeB 44 K, threaded, with couplers, anchor plates, and nuts—single bar length $L=2.5$ m) with diameter of 36 mm and length about 5 m, placed on the internal side of the niche (diamond head rotary machine)
2. Twenty-nine stainless steel passive anchors with a diameter of 26 mm and a length of 5–10 m (in any case double the last encountered fissure from surface), 20 on the internal side and 9 on the external
3. Seventeen passive anchors, pregrouted to avoid oxidation, with a diameter of 90 mm and length of 15 m, for a total length of 200 m, placed on the external part of the cliff (Anchor bars VSL, dia 26.5 mm, st 835/1,030, preinjected, with external corrugated sheathing, including plates and nuts)

Apart from the above technical aspects, the main difficulties in this project were not only the typology of intervention and the materials used but also how to execute the work in a country like Afghanistan with low availability of equipment. Certainly, the first idea was the construction of scaffolding. However, because of the very high probability of rock fall destroying it, with additional risk for the workers staying below hanging rocks, as well as the economic cost of the scaffolding itself, and the approach of winter season in 2003–2004, the need to find an alternative solution came up. After careful investigation and evaluation of possible alternatives for implementing the job, the choice fell on the use of professional climbers.

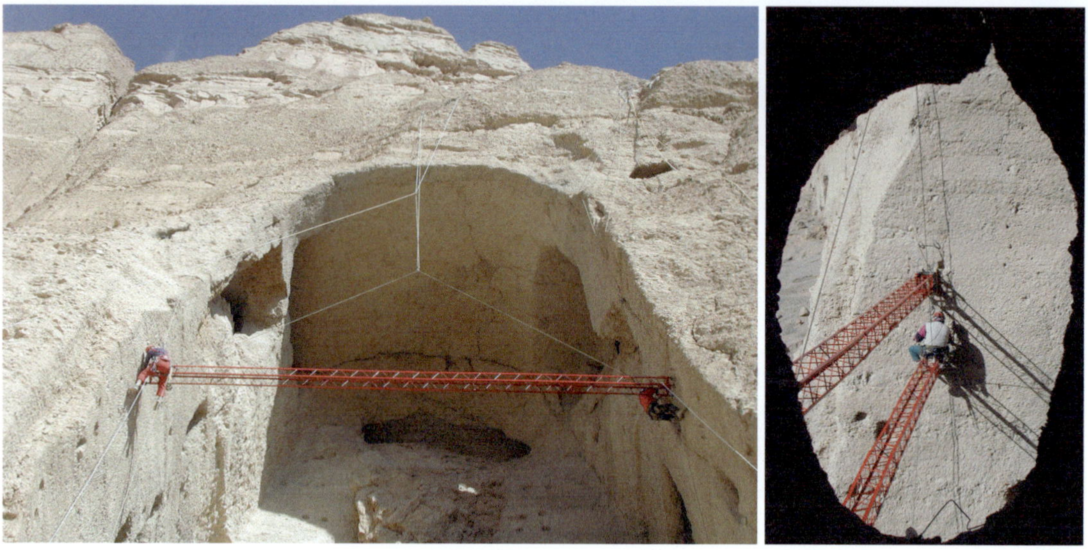

Fig. 11.10 Installation of temporary struts. Note that during the installation, the strut is fixed on top of the cliff in the safe zone

Climbers, also supported by ground staff, operated directly on the surface, hanging from the top of the cliff, in a safe area, moving from top down and then in safe conditions with respect to any potential rock fall (Fig. 11.10). Also a major difficulty was the calibration of drillings with respect to the existing cavities. In fact, a large number of caves (around 800) and tunnels are located on the cliff, constituting a unique example of rupestrian settlement. The selection of drilling then required a detailed investigation into their orientation and inclination to avoid drilling and grouting into the archaeological caves. As mentioned previously, great attention was paid to the methodology for consolidation. Short- (16 mm) and medium- (20 mm) length passive anchors (stainless steel) have been used, with rotary drilling machines with diamond heads to avoid any possible vibration. Use of cooling fluid faced the occurrence of slaking-prone siltstone in the presence of water: because of this, the usage of water was limited when drilling the conglomerate and a mix of compressed air and water was adopted when discontinuities were detected and when a possible level of siltstone was encountered. Pregrouted long passive anchors, used only in the first phase in 2003 (26 mm) have been used with a roto-percussion machine and use of air as a flushing medium. From a temporal point of view they have been drilled only after the use of shorter ones and from the further part of the unstable blocks toward the most critical one. The purpose of small anchors is to join together all the unstable masses and fix them to the proxy stable geological background. The long anchors have to unite this part to the most internal and stable geological material. Direction and inclination of anchors have been defined on site but, in any case, direction of deformation and perpendicularity to discontinuities have been taken into consideration. Temporal execution considered the principle of starting from the most stable place to the most unstable. This is to start consolidation from the part where disturbance can better be restrained. In particular, with respect to the internal side of the niche, the lowermost unstable block has been approached from the bottom to the top. In fact, in the top of the niche there is a hanging block that cannot be touched without having stabilized the lower part. Figures 11.11, 11.12, 11.13, 11.14, 11.15, and 11.16 describe different adopted technologies and equipment in relation to the target

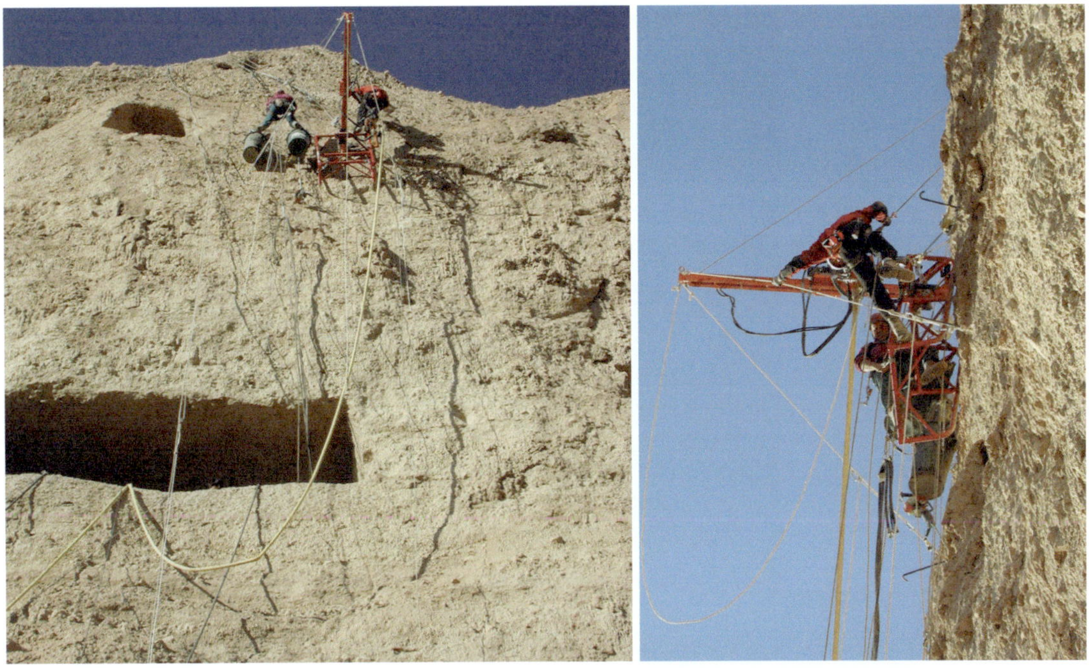

Fig. 11.11 Final consolidation with the use of professional climbers and large rotary machine

Fig. 11.12 Execution of nails for the stabilization of unstable blocks

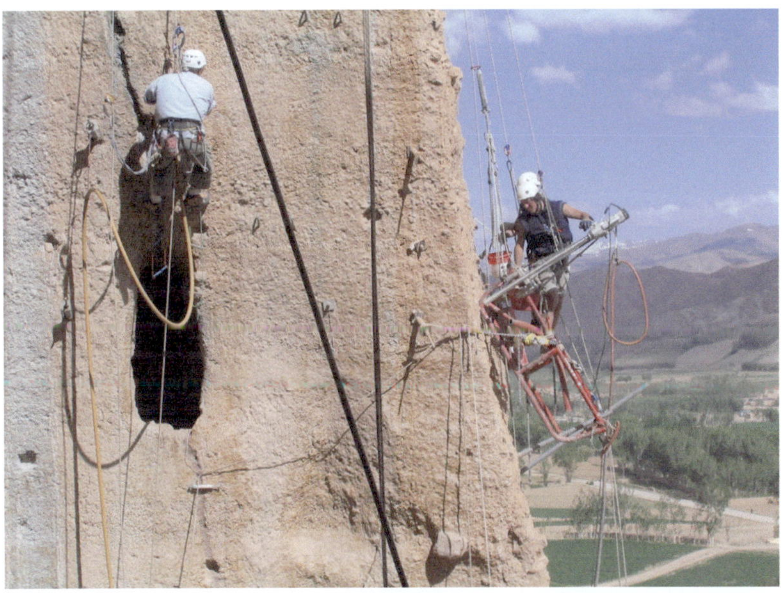

problem. Grouting was made with cement with superplasticizer added to avoid any water release, capable of interference with the slaking siltstone as well as to get the best possible adherence between bar and rock, composed as in Table 11.1.

In total, for grouting and filling in the eastern wall of the niche, approx. 17 m^3 were injected, with 19,000 kg of cement divided into:

1. Short anchors grouting as 1,200 kg
2. Anchors grouting as 8,200 kg
3. Crack filling (from top) as 9,600 kg

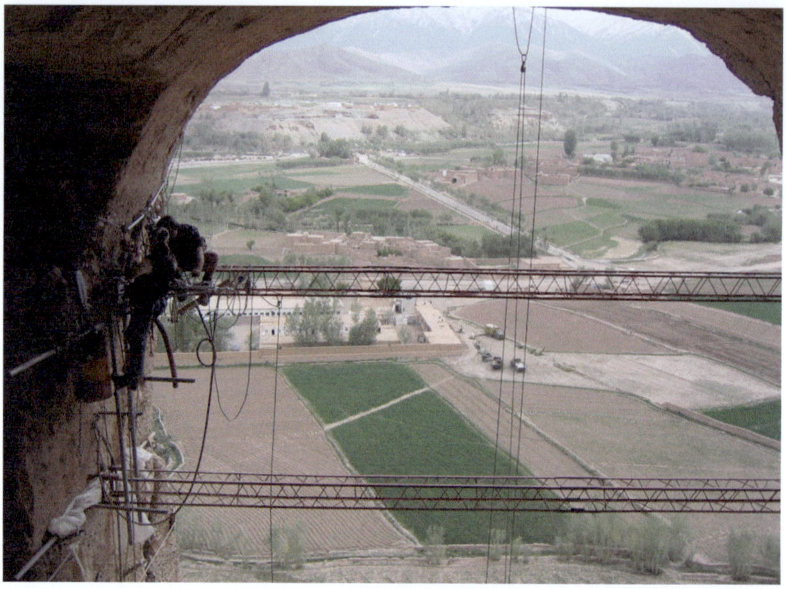

Fig. 11.13 Execution of nails on the roof of the niche

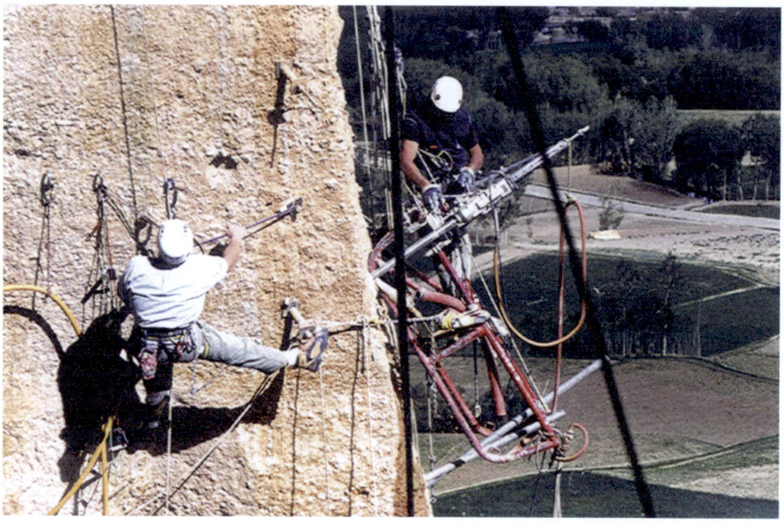

Fig. 11.14 Drilling for nails at the corner of niche

Minimization of impact was implemented by covering with mortar of suitable color all the anchor steel plates. In this way it is now very difficult to identify the places where anchors and nails were settled. The composition and color of mortar was established with the support of technicians from the International Council of Monuments and Sites. A final arrangement should be provided by a conservator. Figure 11.17 presents the results of the executed activity.

The solution and the techniques adopted as well as the four steps for improvement of activities proved quite satisfactory because the monitoring system did not record any remarkable deformation in the unstable blocks through the working period (Fig. 11.18).

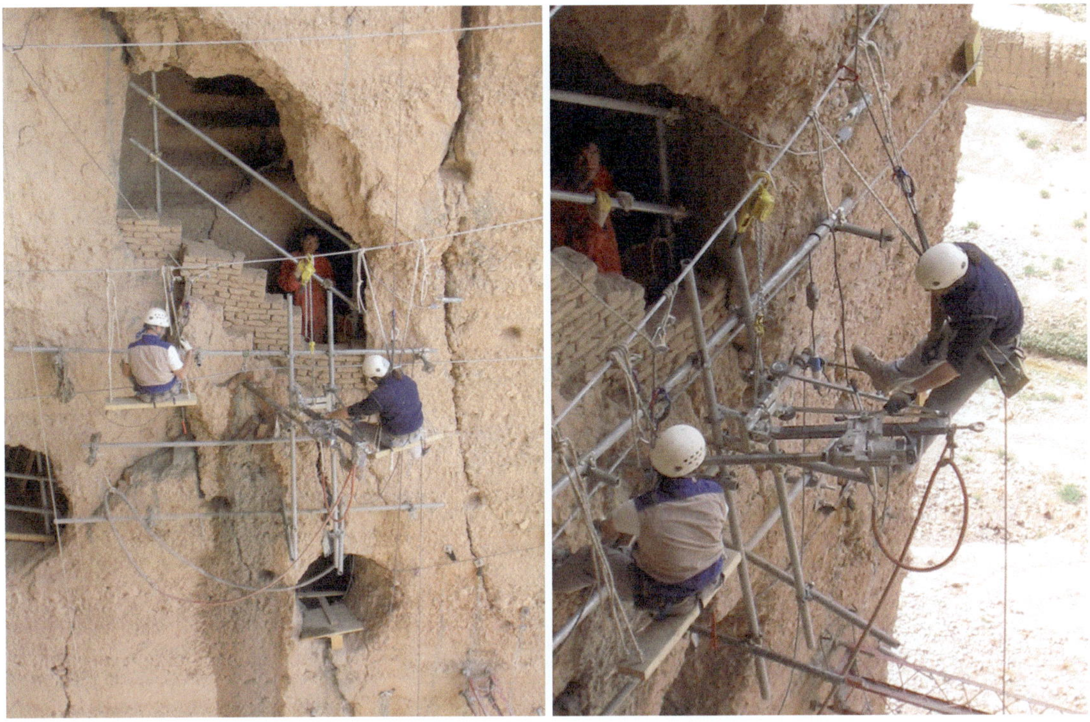

Fig. 11.15 Execution of nails with rotary machine and diamond head, at the inner side of the niche

Implementation on the Western Wall

The western side of the niche is also suffering the effects of explosion as well as the sinking of the existing buttress. The buttress was probably constructed to reduce the risk of collapse of this flank, which was considered extremely unstable, to justify a very massive intervention by a French archaeological expedition in the late1950s and early 1960s; the structure was finally strengthened and mitigated in the impact by the Archaeological Survey of India in 1970. Because the buttress seems to be connected with bolts to the cliff it is possible that the sinking of this structure may produce a horizontal stress, toward the external, inducing additional instability as testified by the intervention of the Archaeological Survey of India (Fig. 11.19, Courtesy Prof. Maeda). There is now some evidence (e.g., widening of small cracks) from which it is possible to hypothesize that the buttress is hanging from the cliff, more than sustaining it. This situation might increase the existing damage.

Effects of the March 2001 explosion are mainly evident at the top of the niche, probably where there is a maximum concentration of stress in consequence of the morphology of the niche (arch and pillar, as described in Colombini and Margottini 2003a). In particular (Margottini 2004b, 2006) there is a small pillar (Fig. 11.20) that needs immediate emergency intervention before collapse, possibly inducing large deformation processes to the whole western part of the niche. This part was also completely restored by the Archaeological Survey of India in the late 1960s to early 1970s. Apart from the planned minor emergency intervention, any large intervention in this area should include geotechnical investigation on the present buttress foundation and, later on, the complete stabilization of the niche. Likely, the manual monitoring system installed in 2003 does not exhibit presently any further deformation of the most severe cracks. The present emergency intervention, planned in the upper part of niche, was designed in order not to fix any part of the buttress to the cliff, because its possible evolution has not been investigated.

Fig. 11.16 Detail of execution of nails from inside the caves by means of small and very small rotary machines and diamond heads

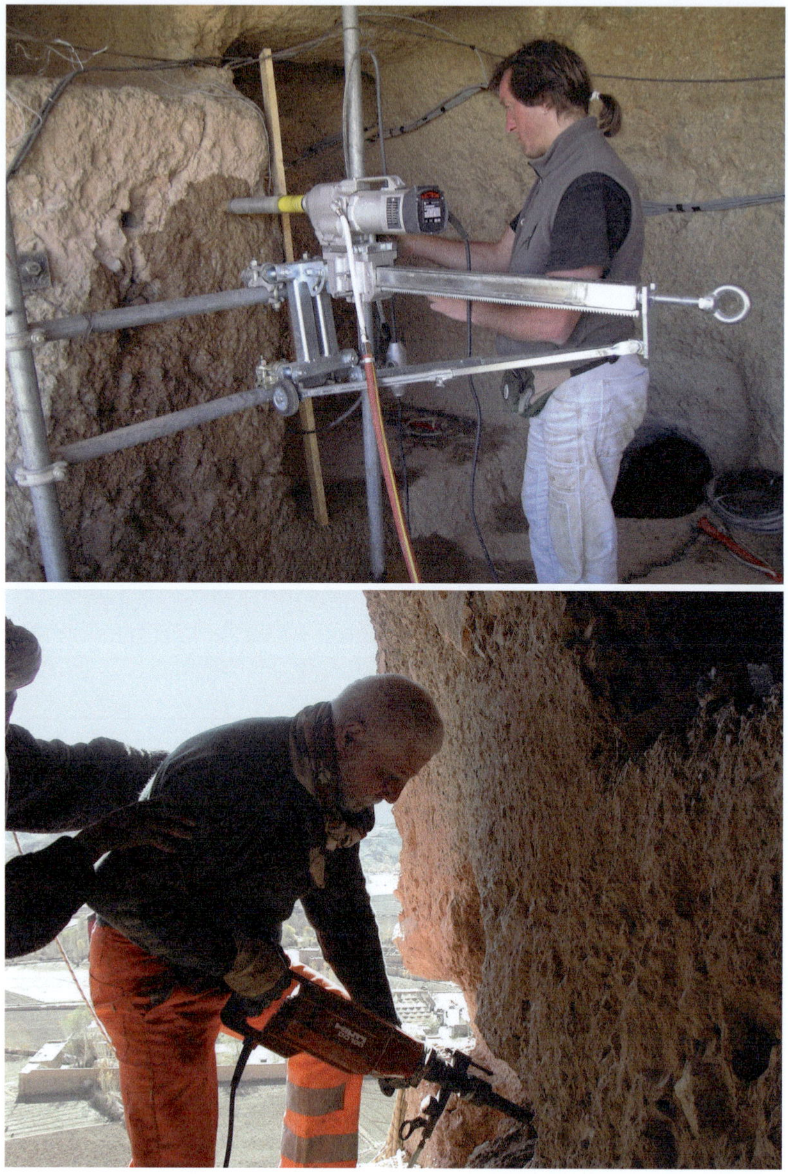

Table 11.1 Composition of grouting materials for both anchoring and cracks filling

	Anchoring grout (kg/m³)	Cracks filling mortar (kg/m³)
Water	540	300
Cement	1,360	610
Sand		1,270
Additive superplasticizer	7	7

Also in this situation the general strategy of an emergency intervention was developed in four steps:

1. A monitoring system on the most relevant discontinuities. No. 6 potentiometric crack gauges were newly installed and tested in the west wall of the niche of the Eastern Giant Buddha to monitor the cracks identified as most dangerous in the area of the drilling and

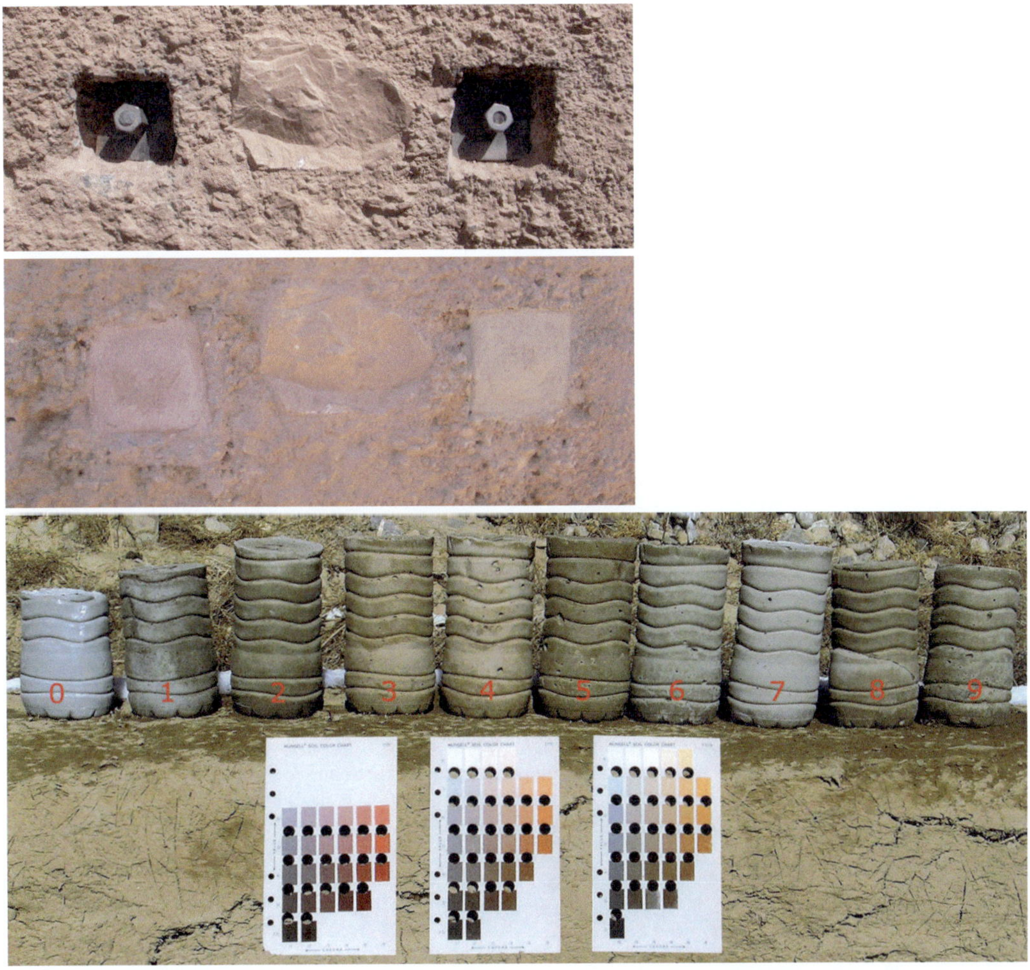

Fig. 11.17 Covering of anchor heads with proper mortar (test site)

grouting work. The scheme of installation is reported in Fig. 11.21.

2. A temporary support was created (Fig. 11.22), by means of:
   (a) The two existing long iron beams were moved into the upper part of the niche, to provide some lateral support to the niche.
   (b) Two iron/wood beams capable of supporting any lateral deformation of the small pillar; The construction details of the beam have been finally adapted to the materials actually available in Afghanistan.
   (c) Steel cables binding completely the pillar and cliff; 13 temporary steel cables have also been installed on the western side of the niche. Four of them were fixed through steel bolts on the inner/outer wall of the niche, whereas nine were circular cables embracing horizontally (six) or vertically (three) the rock pillar and the septum at an elevation of 2,570 m.
3. Emergency intervention includes nails and grouting as follows:
   (a) Prior to starting the drilling works, the large cracks in the area of the pillar at the left wall of the Eastern Giant Buddha were thoroughly filled in and grouted with cement grout. After the preliminary

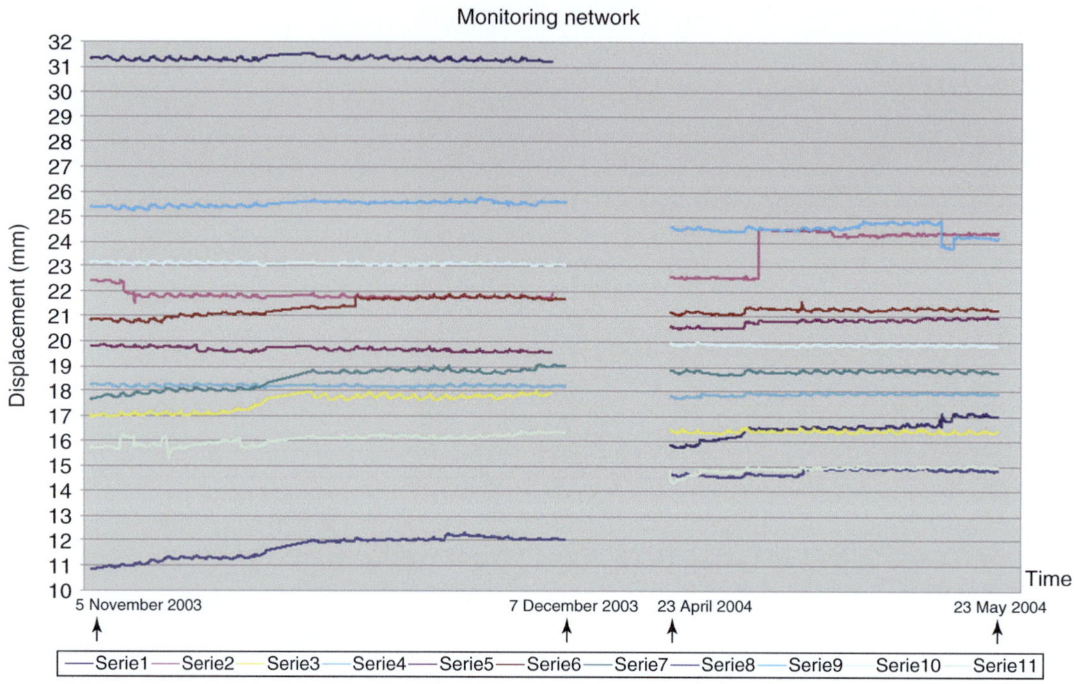

Fig. 11.18 Time evolution of the 11 extensometers operating in the period November 5th–December 7th, 2003 and April 23rd–May 23rd, 2004 in the eastern wall of the niche, and showing no remarkable movements on the cliff. Some minor steps have been caused by climbers who hit the gauge placed on the cliff

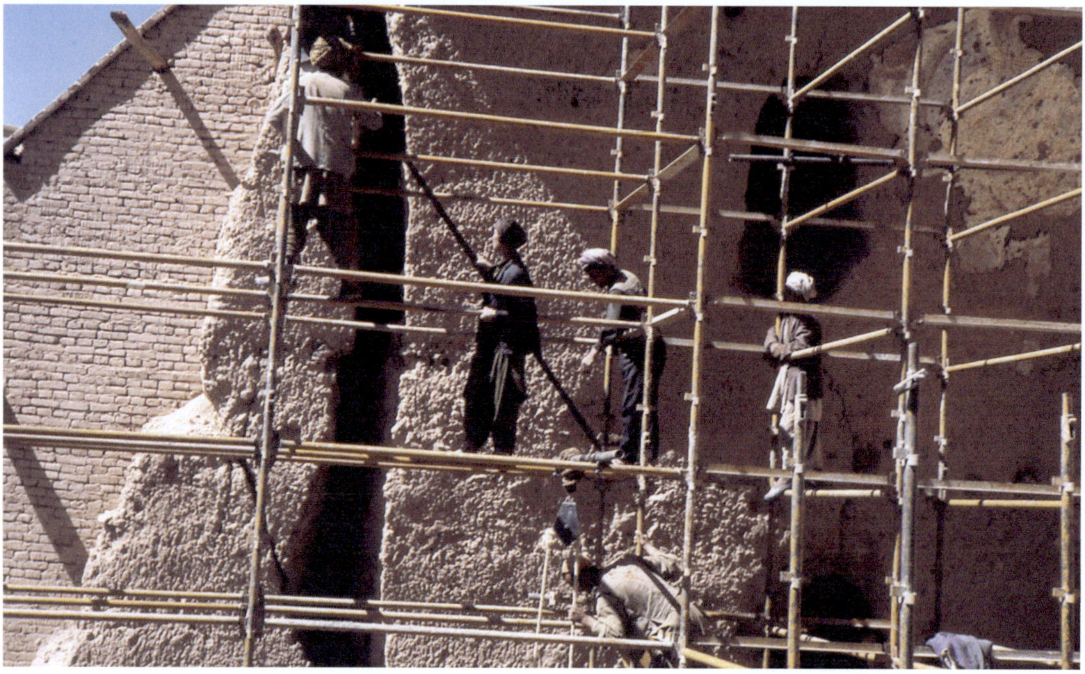

Fig. 11.19 Consolidation work from the Archaeological Survey of India in 1969 (Courtesy of Prof. Maeda)

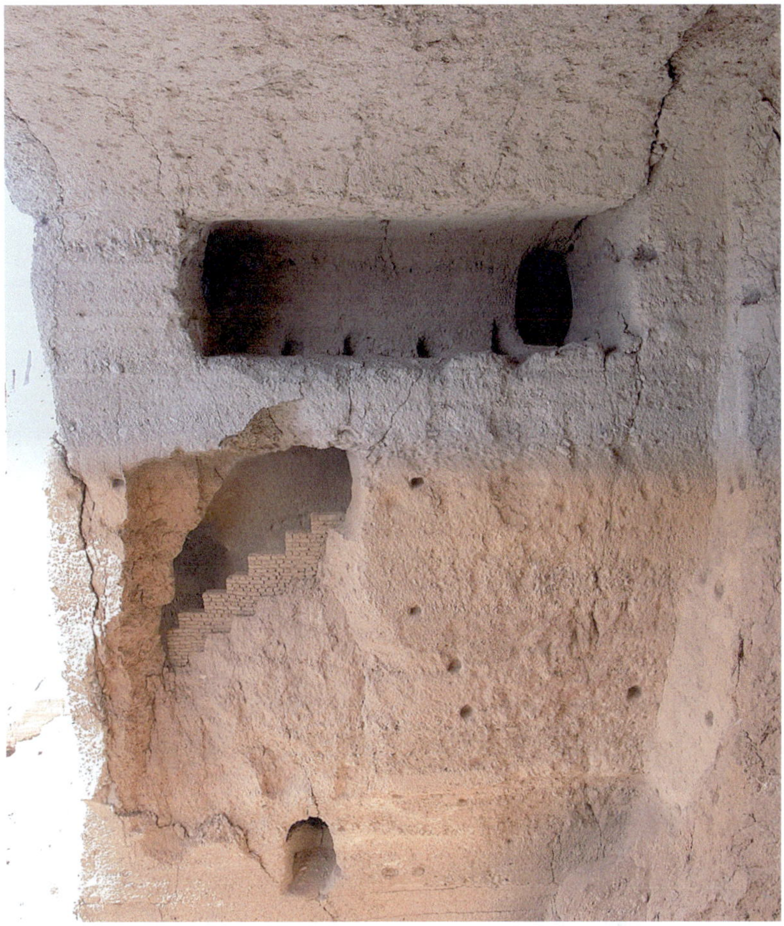

Fig. 11.20 The most unstable element (pillar) in the western side of Eastern Giant Buddha niche

caulking, the main fissures were filled in using 0.96 m3 of low water-release cement grout, with some 1,200 kg of cement. Grout composition utilized was, as usual, C:W=0.5 with superplasticizer.

(b) For the drilling operations, a diamond rotary system, 50 mm dia., has been adopted with the aim of limiting as much as possible interferences, produced by vibratory effects, to the limited stability of the structure in this area. A total of 12 stainless steel passive anchors, dia 20.0 mm, have been installed, with a total drilled length for these 12 anchors of 52.4 m. In detail, nine of these nails connect the pillar internally, in both directions parallel to the face and perpendicular to it to create a robust net; two short passive nails (located at below and above the critical pillar, with a depth less than the back side plane where a large crack has been detected after the removal of fragments in the lower caves behind the feet of the statue. One passive anchor parallel to the surface aimed at stabilizing the upper gallery where a large fissure is present;

(c) Low water-release grouting in the boreholes, maintaining the composition of water and cement successfully adopted in the eastern side, which includes: W/C=1/2.0+superplasticizer. Approx. 2.0 m3 were grouted for the nails installation with 2,400 kg of cement.

4. Minimization of impact follows the same criteria established for the eastern side.

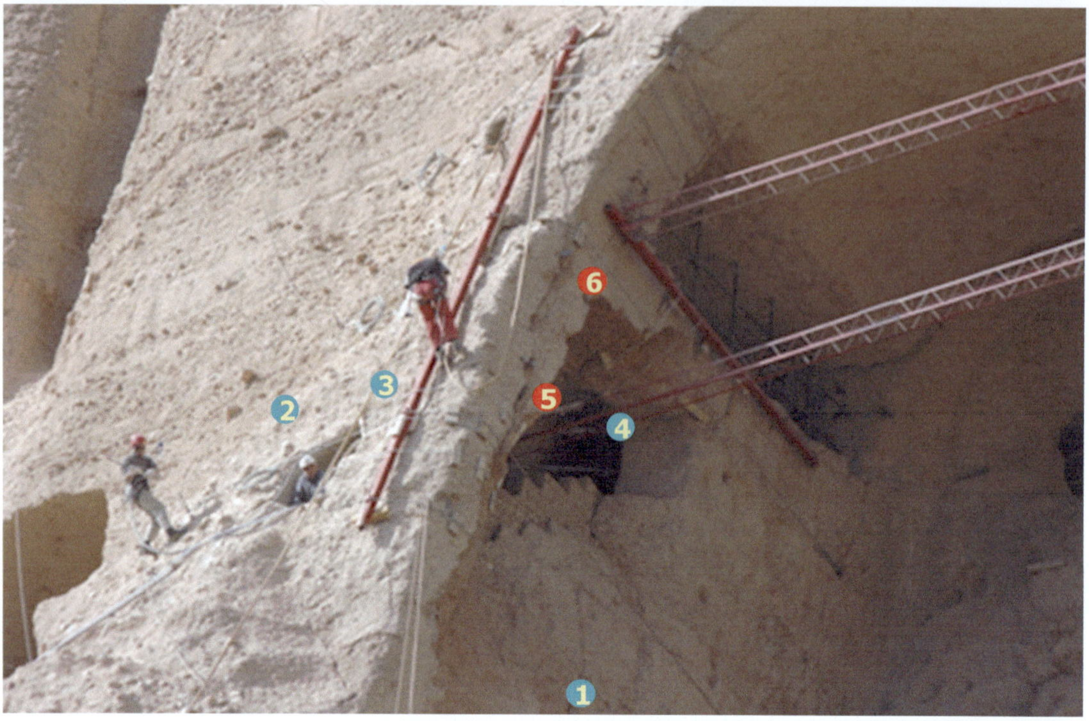

Fig. 11.21 The location or potentiometric crack gauges for discontinuity monitoring and alarm; external sensors are in green and red are internal to the cave

In the western wall of the niche, the total grout for grouting and filling was estimated at approx. 3 m^3, with 3,600 kg cement (cracks and anchors). As mentioned before, the correctness of grouting was demonstrated by the suitability test for the anchors that does not differentiate the anchor bounded to conglomerate from the one bounded to siltstone, in which slaking is highly possible.

Figure 11.23 provides the distribution of the anchors as well as their chronological sequence. This last is quite important to avoid disturbance to the most critical part of the cliff without having reinforced in advance the less unstable portion of it. In addition, consolidation work and grouting provide a further protection against water infiltration; in fact, water circulation may potentially cause slaking of the siltstone as well as generating additional pore pressure. Execution of the work is indicated in Figs. 11.24, 11.25, 11.26, 11.27, and 11.28.

Implementation on the Upper Part

In the upper part of the niche it was decided to install three permanent stainless steel passive anchors, 12.0 m long, subhorizontally, with the following purposes:

To monitor the tensional state of the rock masses, by means of ten strain gauges placed in two of the anchors

To grout the medium part of the cliff to avoid water infiltration within the niche during snowmelt or prolonged rainfall

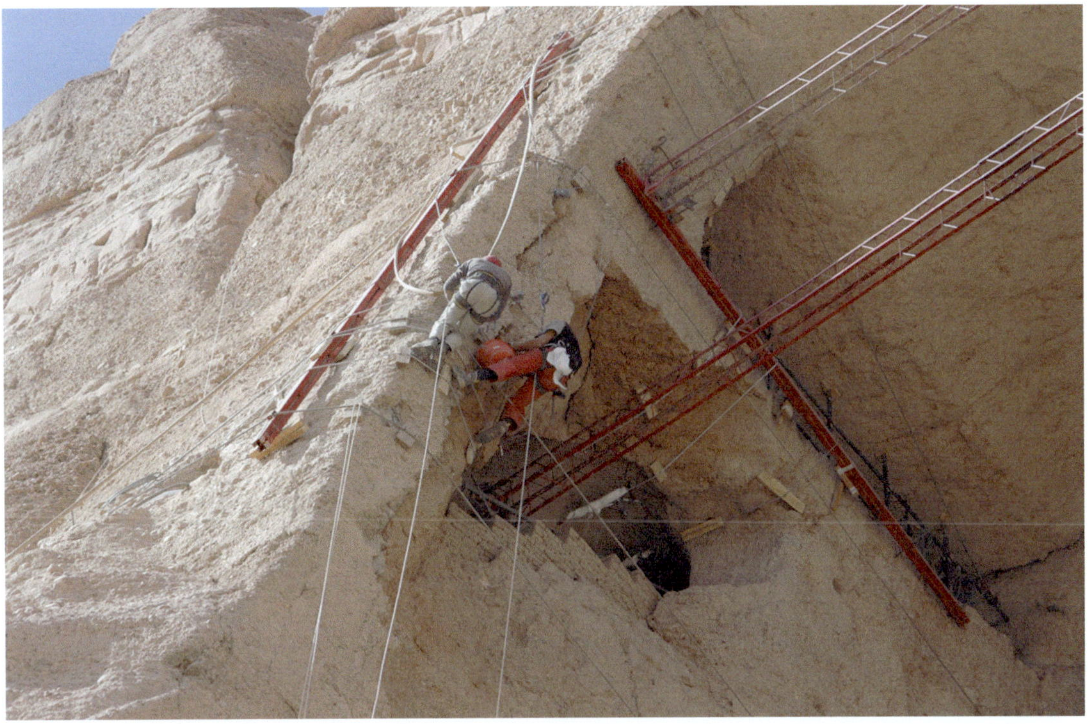

Fig. 11.22 The temporary support necessary for secure execution of the work in the western wall: two long iron beams replaced in the upper part of the niche, the short iron/wood struts for local support and the steel cables

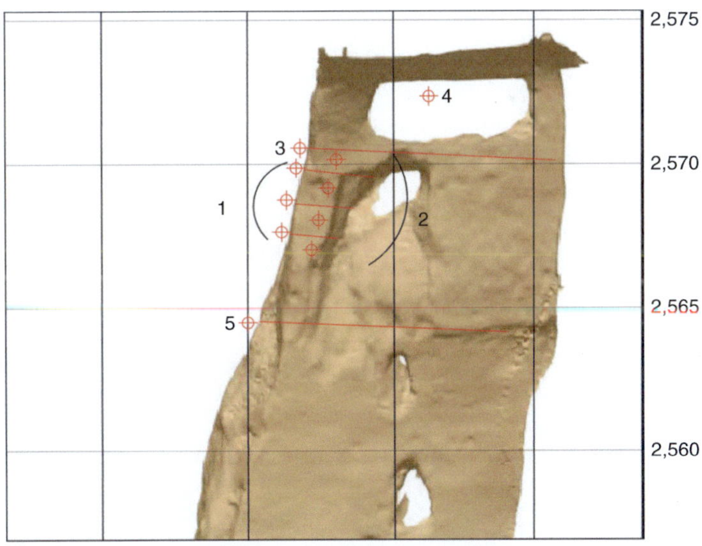

Fig. 11.23 Distribution of the anchors and chronological sequence of their installation

Fig. 11.23 (continued)

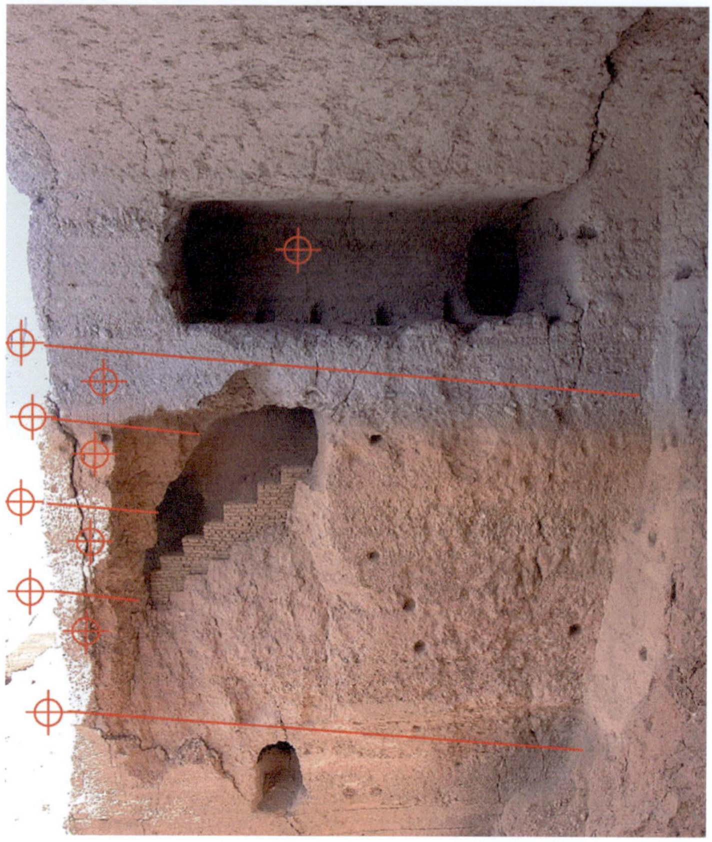

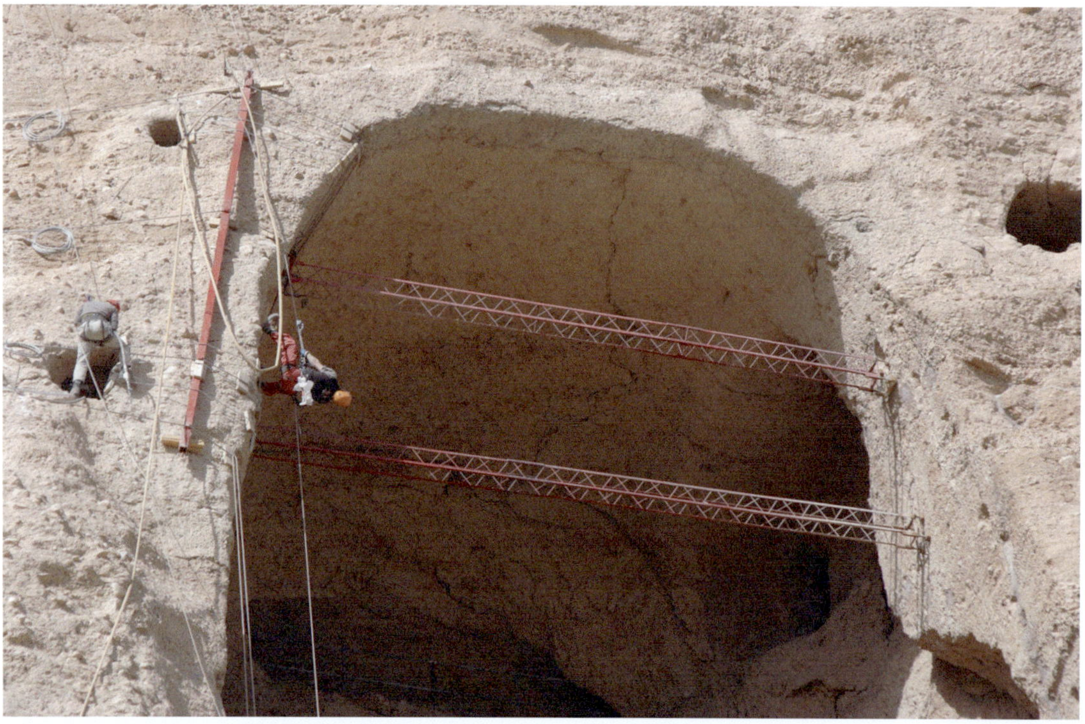

Fig. 11.24 Temporary protection on the west side of the niche

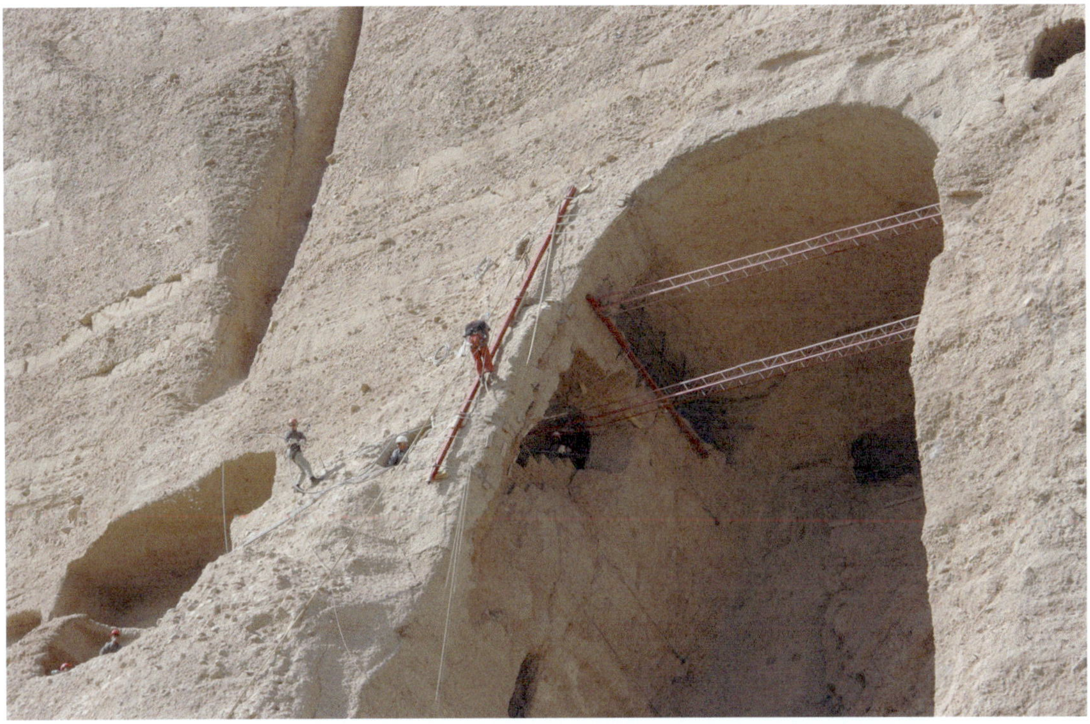

Fig. 11.25 Climbers at work

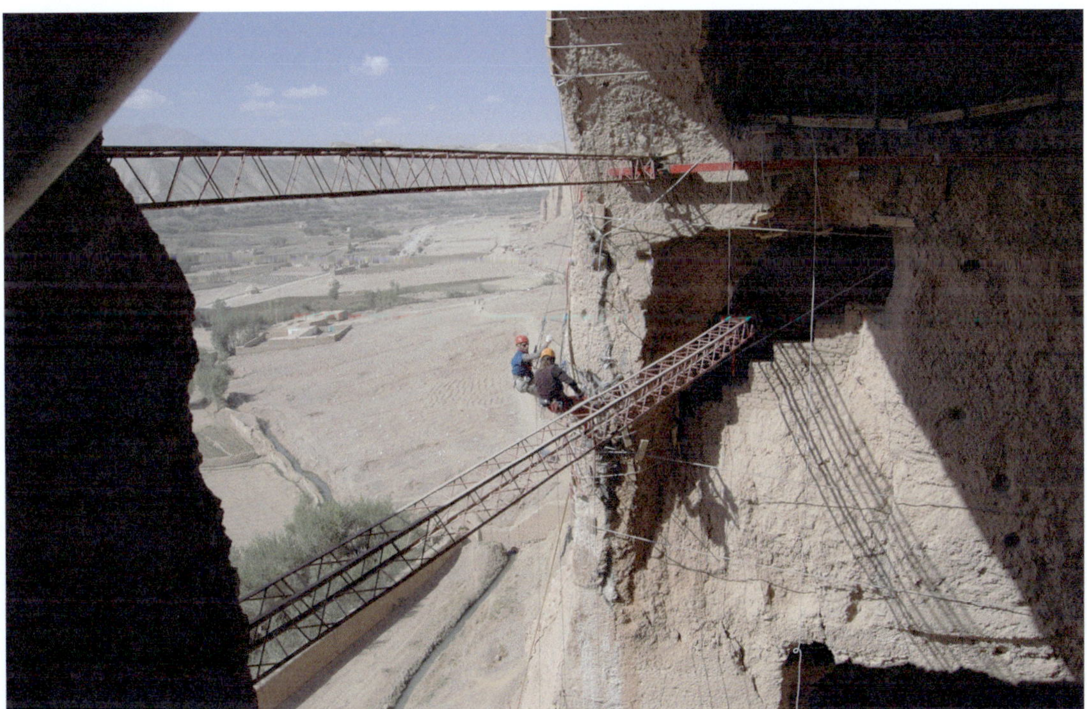

Fig. 11.26 Installation of steel cables as temporary protection

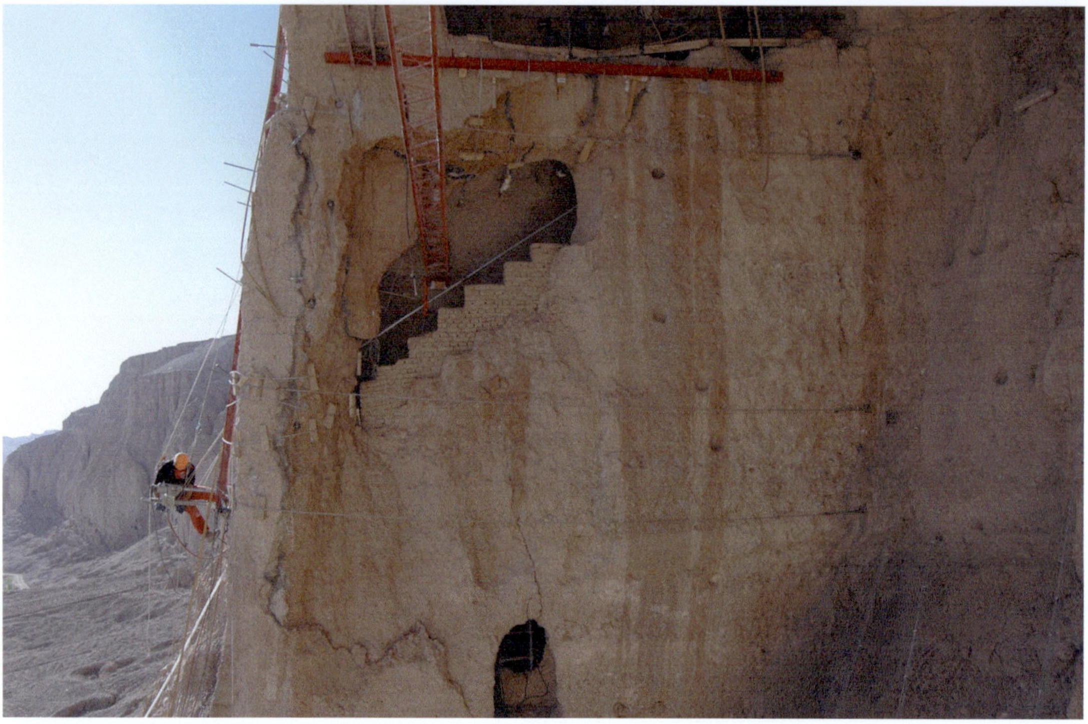

Fig. 11.27 Drilling for nails on the west side, with intermediate power machine and diamond head

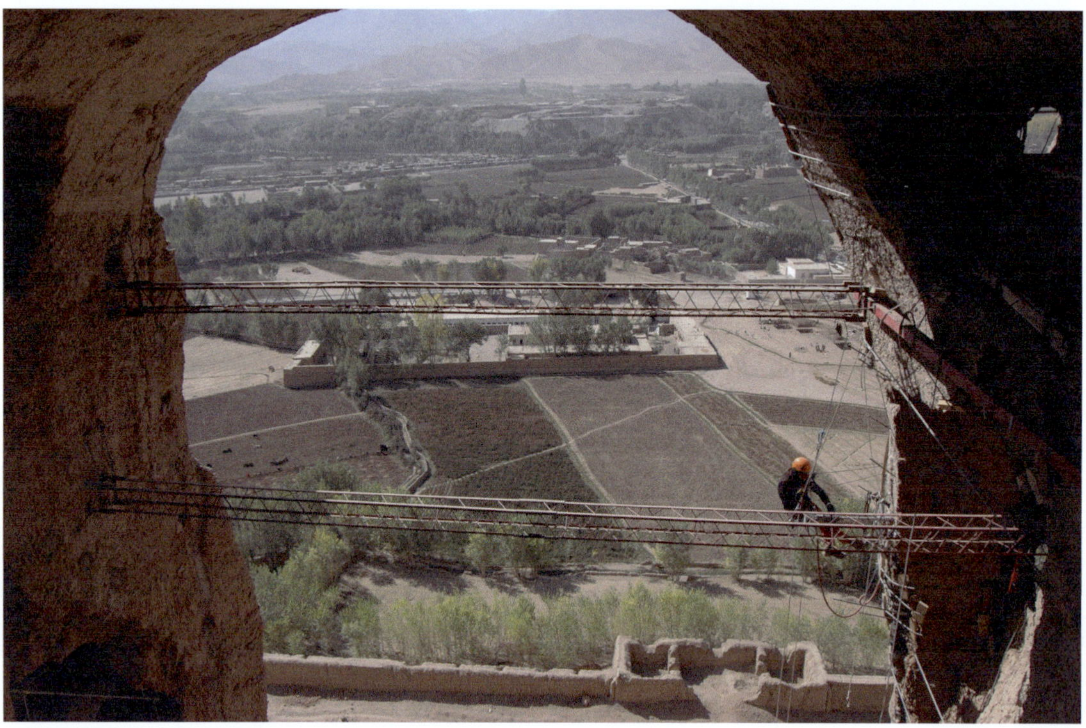

Fig. 11.28 Drilling for nails

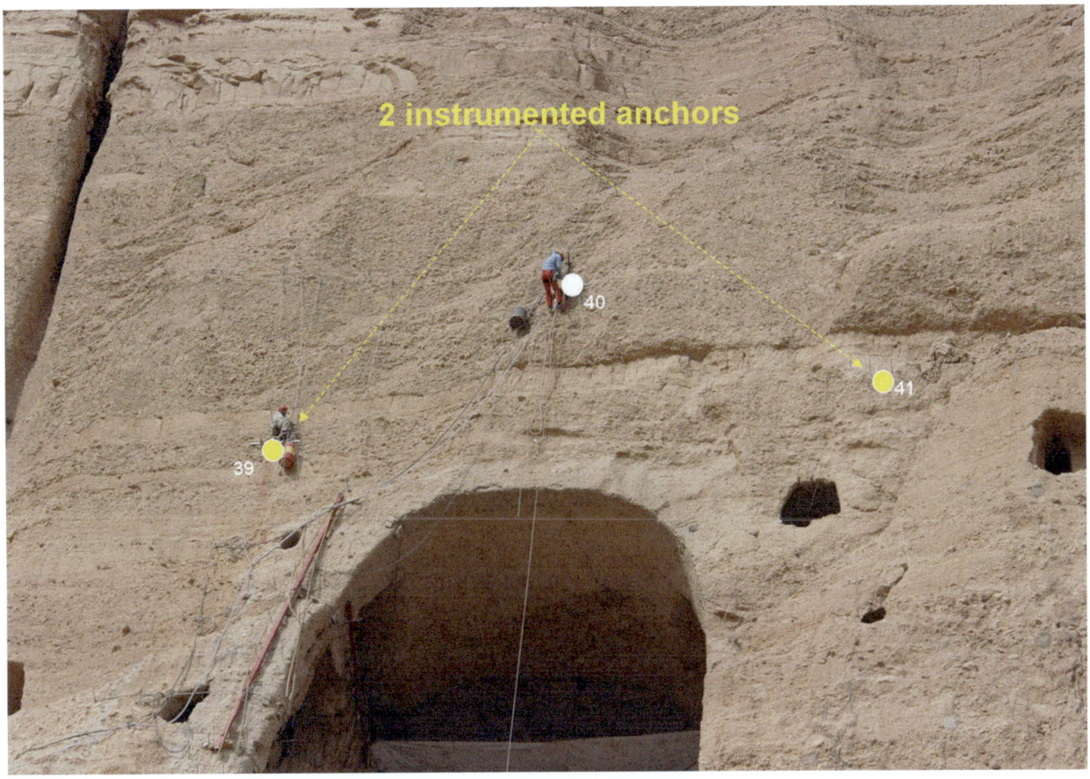

Fig. 11.29 Position of anchors on the top of niche. The *yellow* ones are monitored with 10 strain gauges

The position of these long anchors is reported in Fig. 11.29.

Five hundred and forty liters of cement mix were utilized for the grouting of the anchors (600 kg of cement). Open vertical fissures, reaching downward in the niche, were intercepted in the three boreholes at depths up to approximately 8 m (anchor no 39) and 9 m (anchor 41).

Anchors no. 39 and 41 were instrumented with the installation of a series of five spot-weldable strain gauges (Fig. 11.30) in each anchor. The characteristics of these instruments are described below. The cables connecting the sensors have been conveyed into small grooves to suitable steel boxes located in the niche, where readings can be taken utilizing the portable data logger with LCD display (Table 11.2; Figs. 11.31 and 11.32).

Location and Type of Anchors and Nails on the Eastern Giant Buddha Niche

In conclusion, in the Eastern Giant Buddha niche 64 passive anchors and nails were installed, for a total length of 443.5 m. The total amount of grouting was established at 19.7 m^3 of cement grout with 24,000 kg of cement. The next figures 10.33 to summarize all the long passive anchors and nails and the related location. Table 11.3 reports type and length of each anchor and nail previously described.

In order to simplify the reading of previous figures, the following pictures have been provided, also displaying the position of anchors and nails over a real photo of the site (Figs. 11.34, 11.35, 11.36, and 11.37).

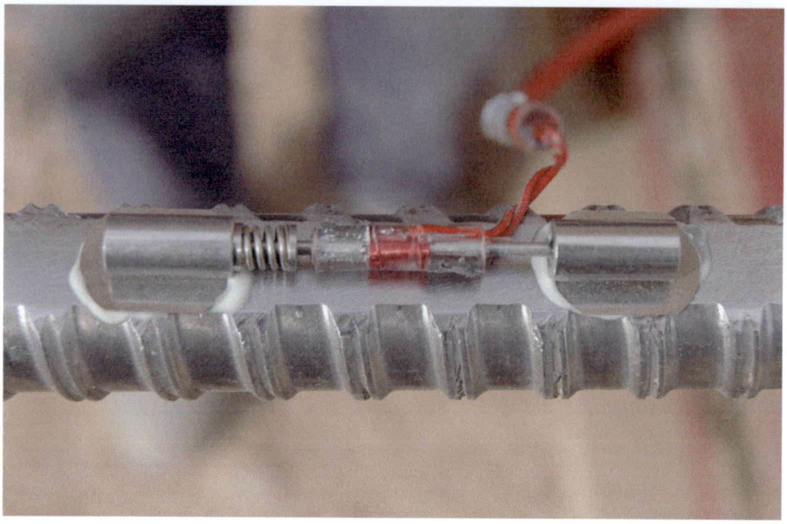

Fig. 11.30 Detail of a strain gauge

Table 11.2 Characteristics of the installed strain gauges

Transducer type	Vibrating wire
Standard range	3,500 microstrain
Sensitivity	1.0 microstrain
Accuracy	0.1 % F.S.
Non linearity	Less than 0.5 % F.S.
Temperature range	−30 °C to +80 °C
Gauge length	50.8 mm

First Interventions in the Western Giant Buddha Niche

Despite the destruction of the statue, the Western Giant Buddha niche did not suffer extensively as a consequence of the explosion (Margottini 2006). Emergency intervention includes (Fig. 11.38):

1. Grouting of the large fissure placed in the corridor, back side of the niche
2. Other minor sites to consolidate

The grouting of the large crack in the back side of the niche has been done from inside as well as from outside (top of the cliff).

Initially, the fissure was grouted and closed in the internal part of the niche, in order to protect the niche from cement infiltration and leaching from the top. Small pipes were required inside the cement to avoid internal overpressure.

From the top of the cliff, inclined drills were used and, when the fissure was encountered in the perforation, it was grouted with the same mixture of cement and superplasticizer described in Margottini (2003).

Major attention was required for the execution of drilling on top of the cliff, because of the possible existence of land mines even after a complete demining of the site as a result of rainfall runoff (Figs. 11.39 and 11.40).

Other minor interventions were required in two small sites, as reported in Figs. 11.41, 11.42, 11.43, and 11.44. The possible risk of collapse, even for small pieces of rock, was completely avoided.

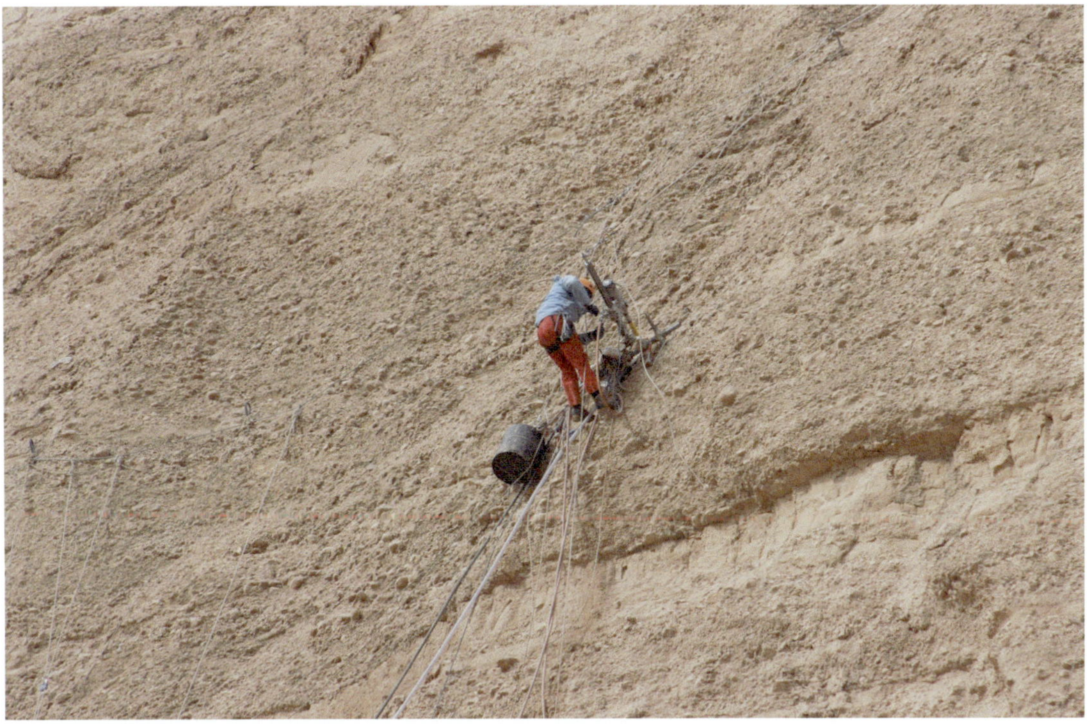

Fig. 11.31 Drilling for the installation of long anchors on top of niche

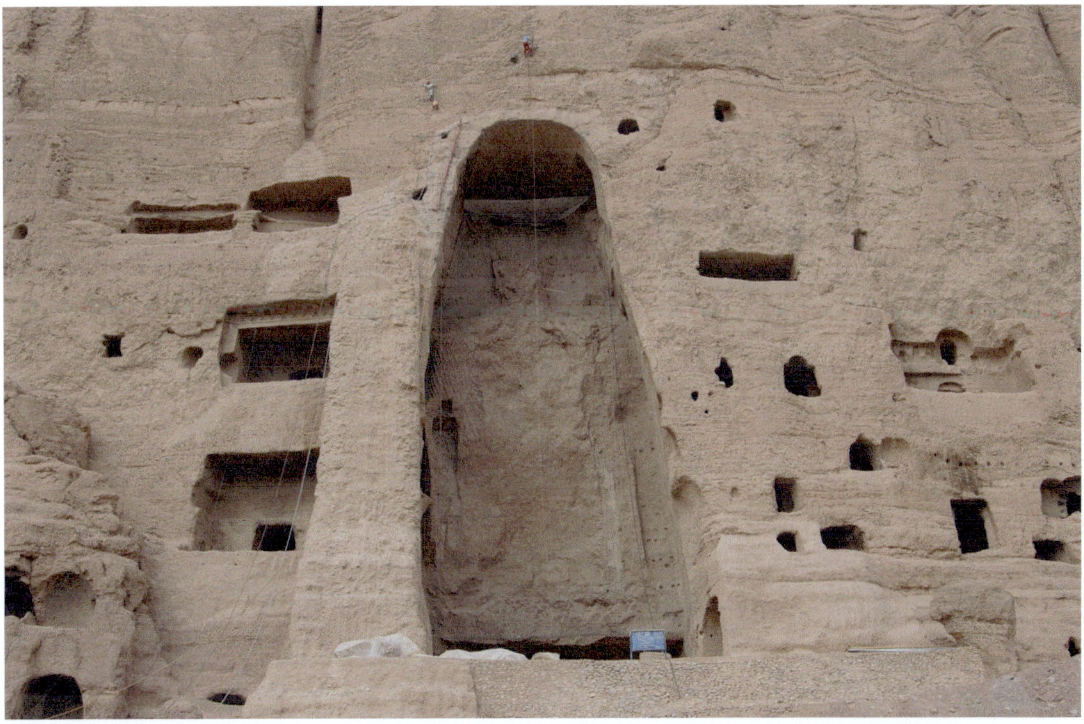

Fig. 11.32 View during execution of the work

Table 11.3 Type and length of each installed anchor and nail (for numbers refer to Fig. 11.33)

Anchor no.	Drill Ø 90 mm Dywidag Ø 26/50 mm (m)
A	7.50
B	7.50
C	7.50
D	15.00
E	15.00
F	15.00
G	15.00
H	7.50
I	7.50
L	7.50
M	7.50
N	15.00
O	15.00
P	15.00
Q	7.50
R	7.50
S	7.50
Anchor no.	**Drill Ø 50 mm Gewi Ø 16 mm (m)**
no.	*m*
1	4.20
2	5.15
3	5.20
4	5.30
5	4.85
6	4.50
Anchor no.	**Drill Ø 50 mm stainless steel Ø 20 mm (m)**
no.	*m*
7	4.50
8	4.50
9	4.50
10	4.60
11	3.30

Table 11.3 (continued)

12	1.30 (Gewi Ø 16mm)
12 bis	5.00
13	5.00
14	7.00
15	7.50
16	7.50
17	7.50
18	7.50
19	7.50
20	7.50
21	3.00
21 bis	3.50
22	5.50
23	6.50
24	3.70
24 bis	5.00
25	4.50
26	2.50
26 bis	7.00
Anchor no.	**Drill Ø 50 mm stainless steel Ø 20 mm (m)**
27	1.50
28	1.60
29	2.10
30	4.00
31	4.30
32	4.60
33	5.00
34	6.00
35	8.30
36	5.00
37	2.00
38	8.00
39	12.00
40	12.00
41	12.00

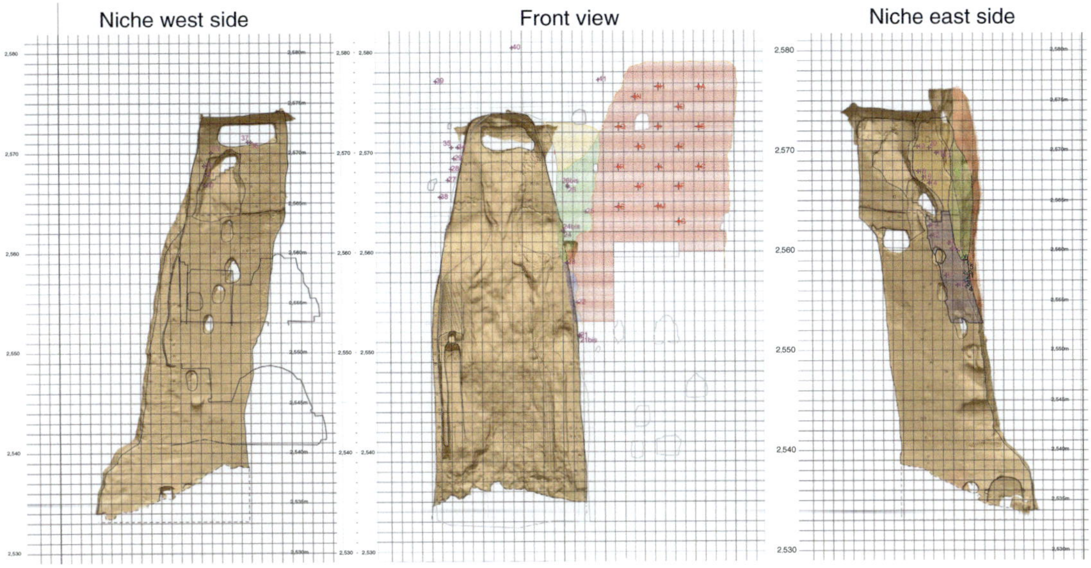

Fig. 11.33 Typology, position, and length of executed anchors and nails (*red* is for passive anchors, pregrouted, with diameter = 26 mm and length = 15 m; *violet* is for stainless steel passive anchors with diameter = 26 mm and length = 5–10 m; *green* is for short passive anchors with diameter = 16 mm and length about 5 m, placed on the internal side of the niche)

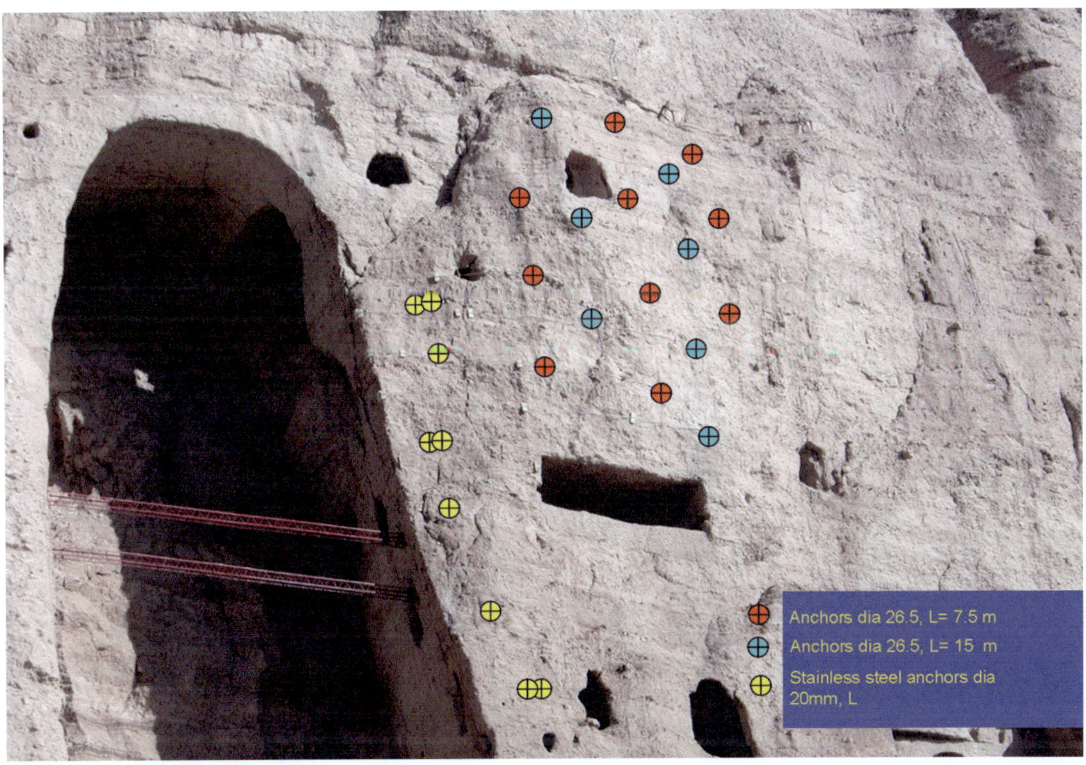

Fig. 11.34 Positioning of anchors on the east wall

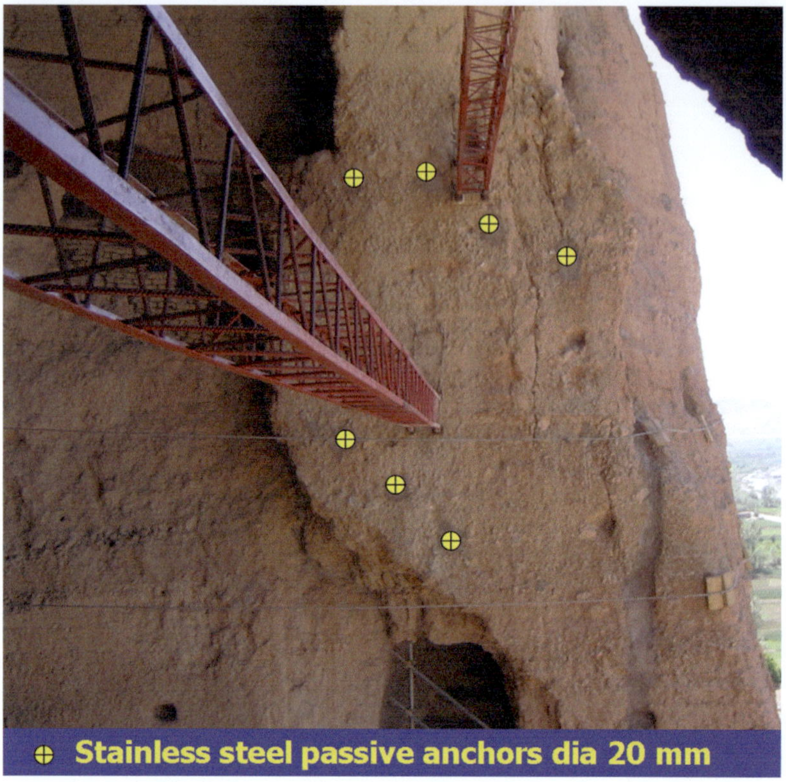

Fig. 11.35 Positioning of anchors on the high inner part of east wall

Fig. 11.36 Positioning of anchors on the low inner part of east wall

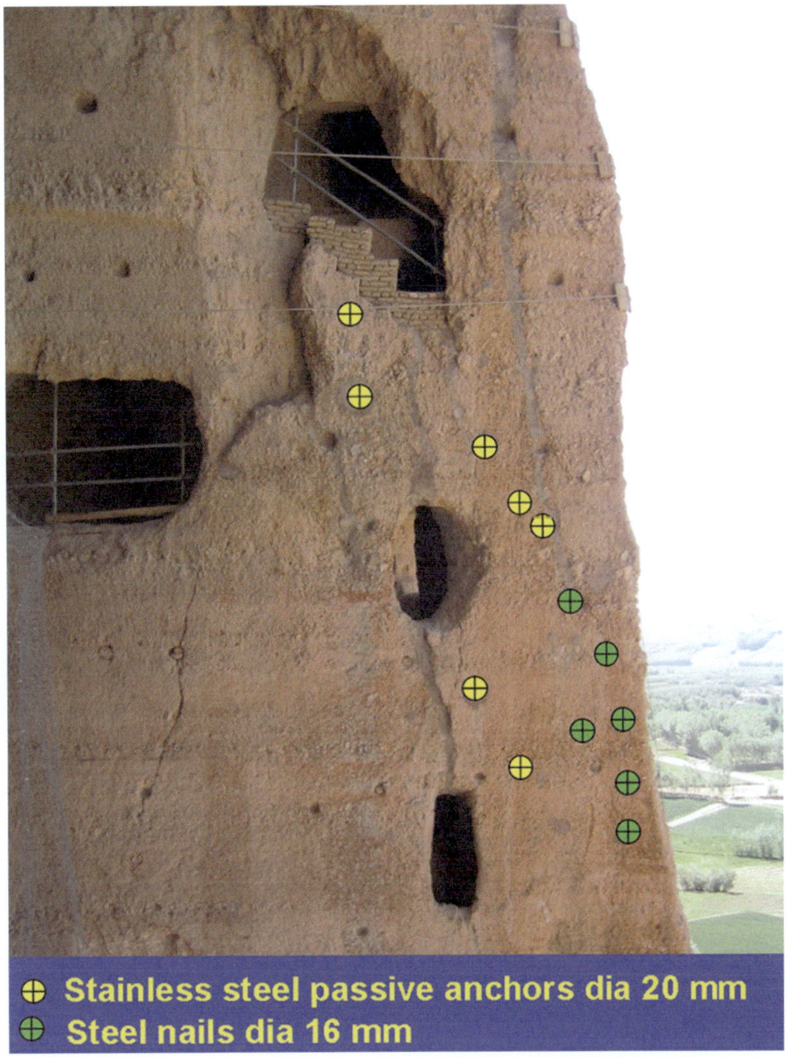

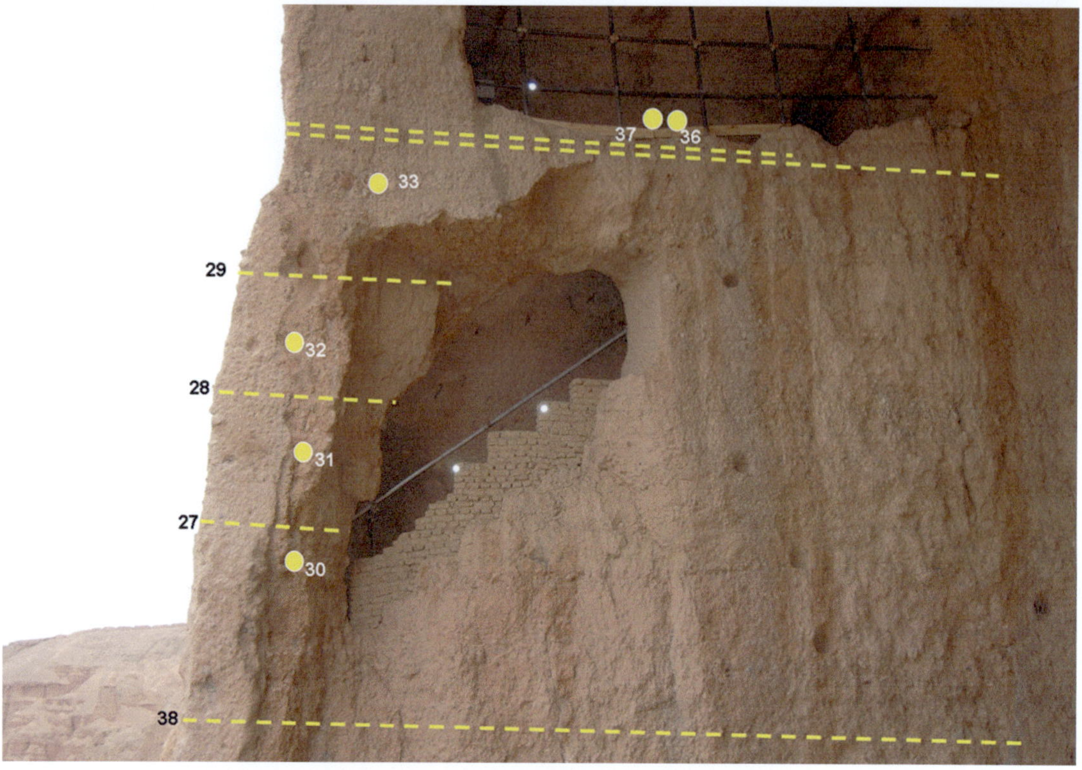

Fig. 11.37 Positioning of anchors on the west wall (Eastern Giant Buddha niche)

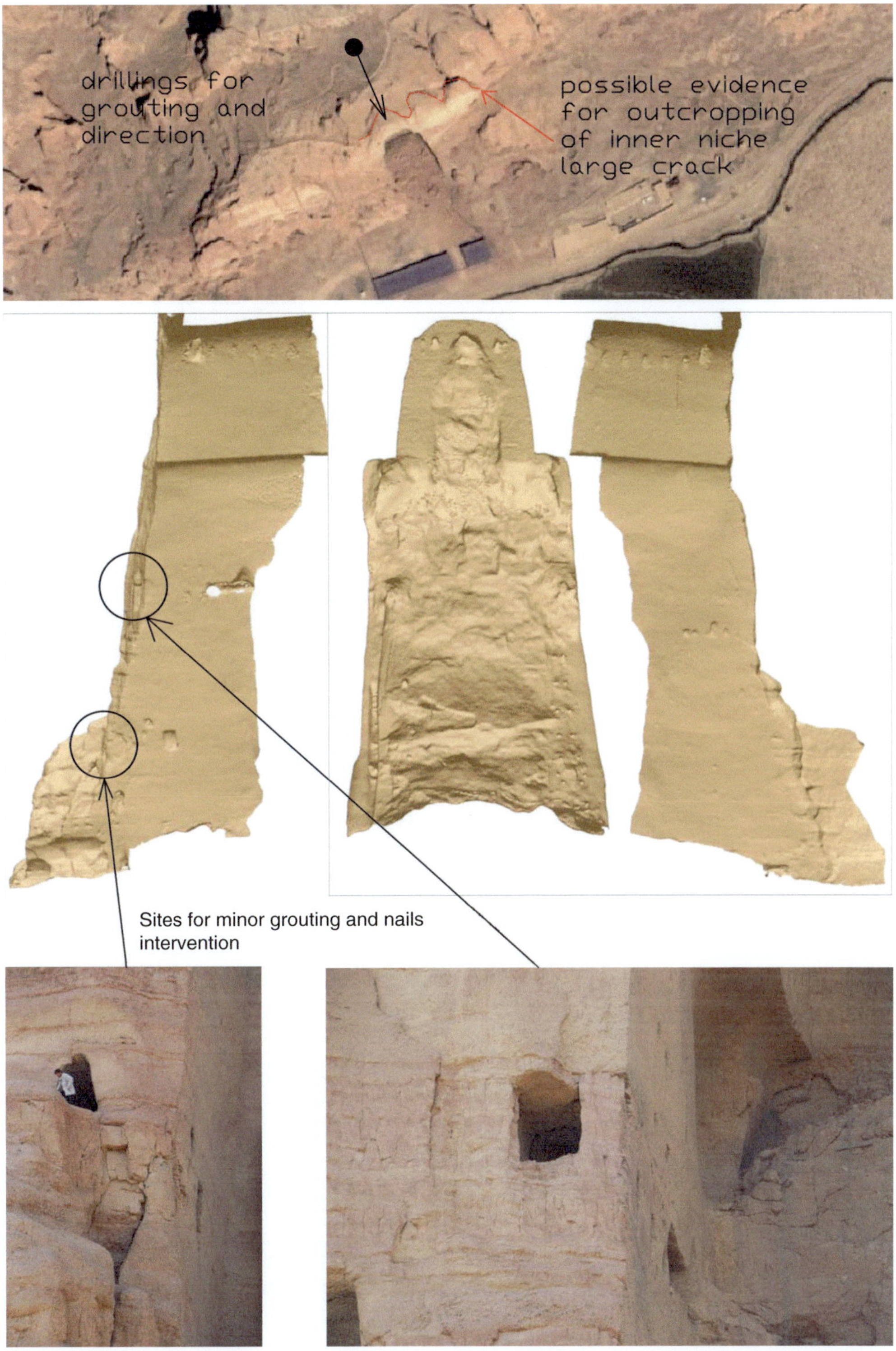

Fig. 11.38 Emergency intervention in the Western Giant Buddha niche (Topographic data from PASCO 2003)

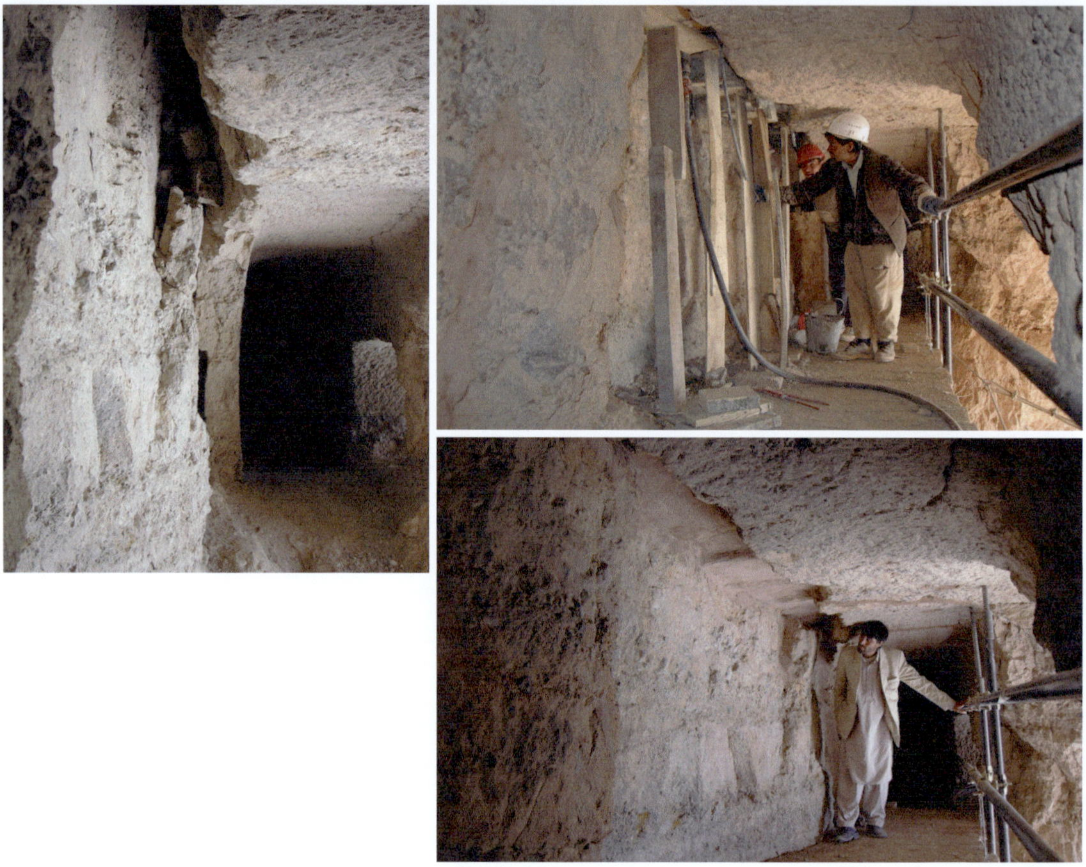

Fig. 11.39 (*The three figures*) The large crack in the corridor on the back side of the niche before, during, and after the grouting

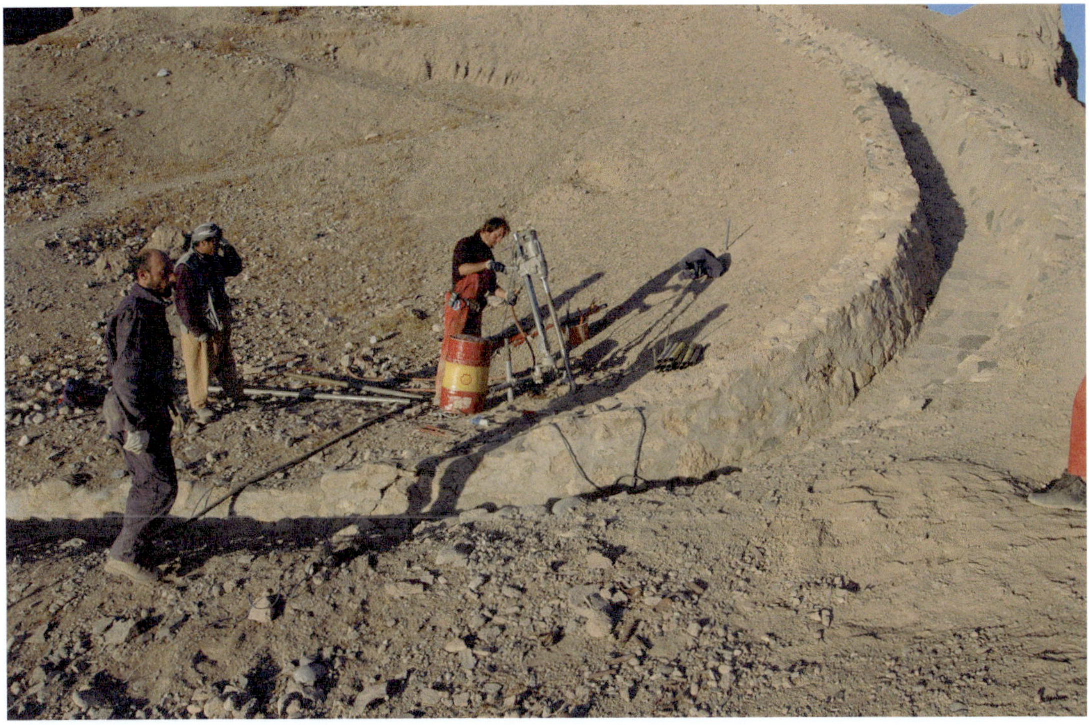

Fig. 11.40 Detail of the grouting from the top of cliff

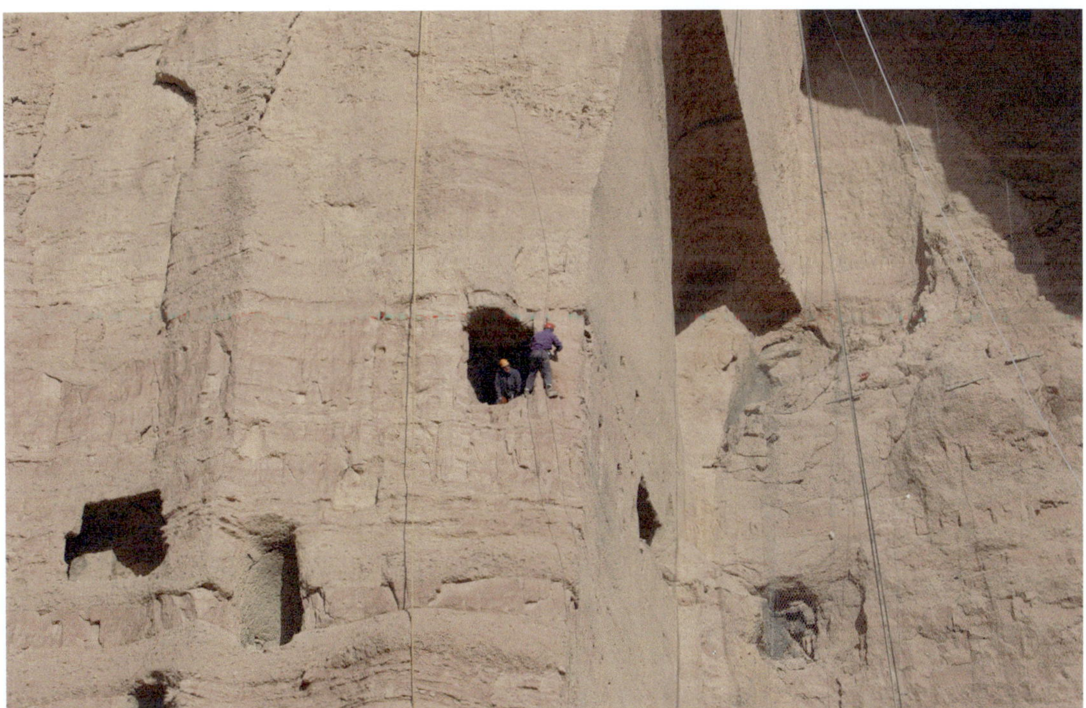

Fig. 11.41 Preparation for consolidating the potential small rock fall at middle level of the niche

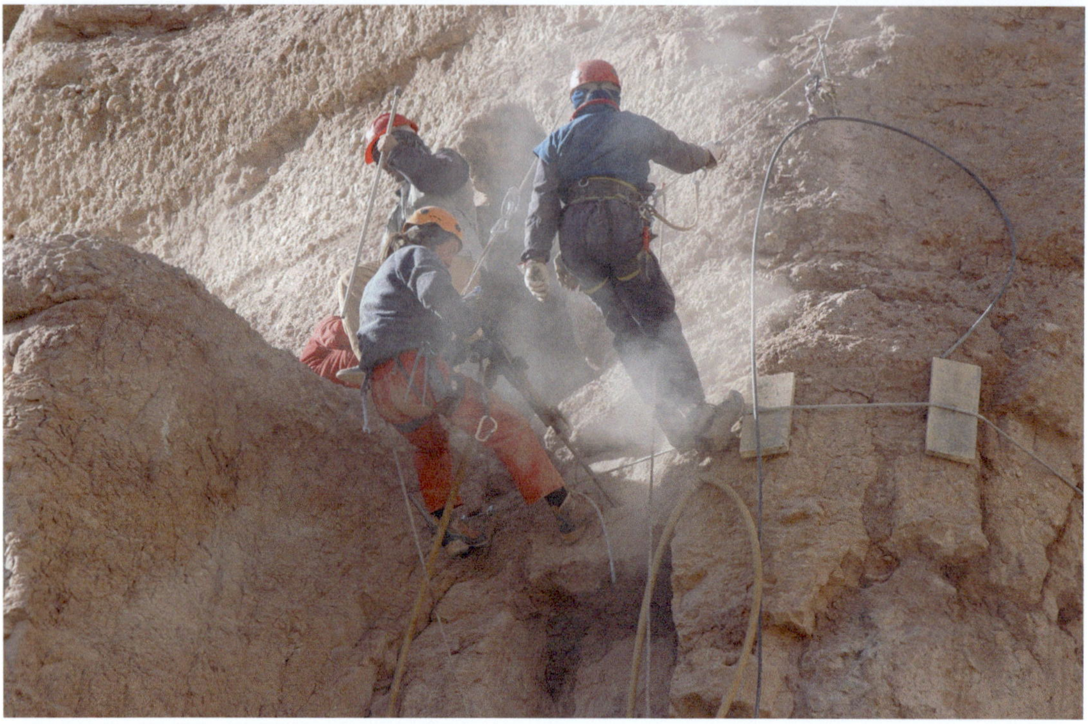

Fig. 11.42 Drilling in the lower part of the niche

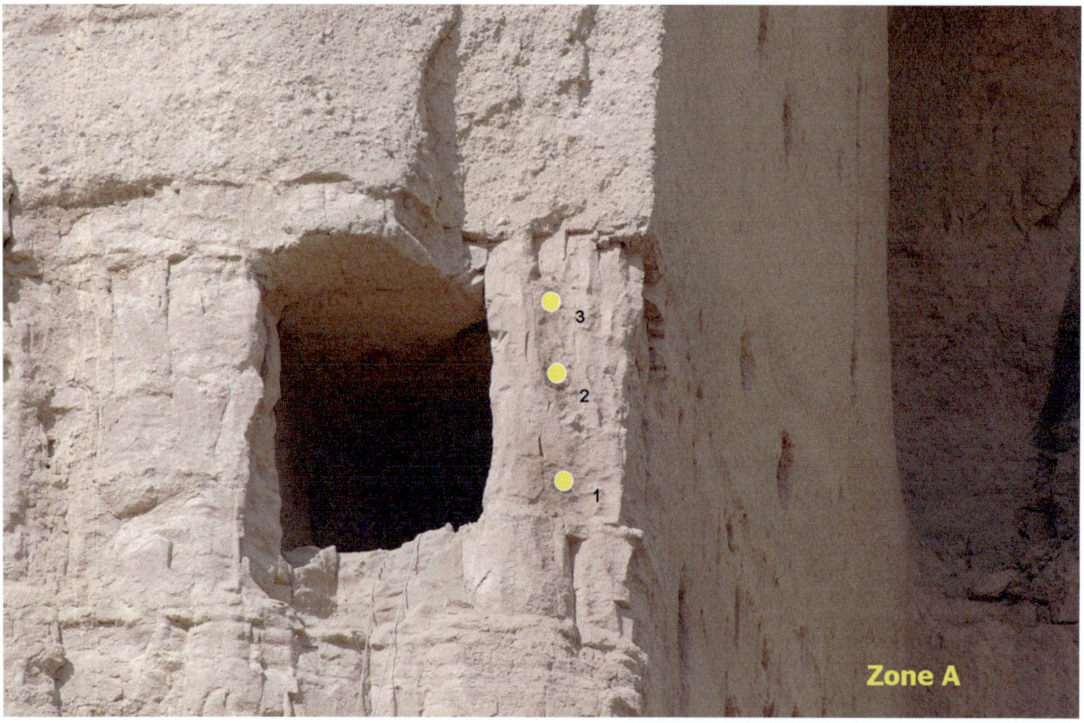

Fig. 11.43 Position of nails in the unstable small rock masses at middle level of the niche

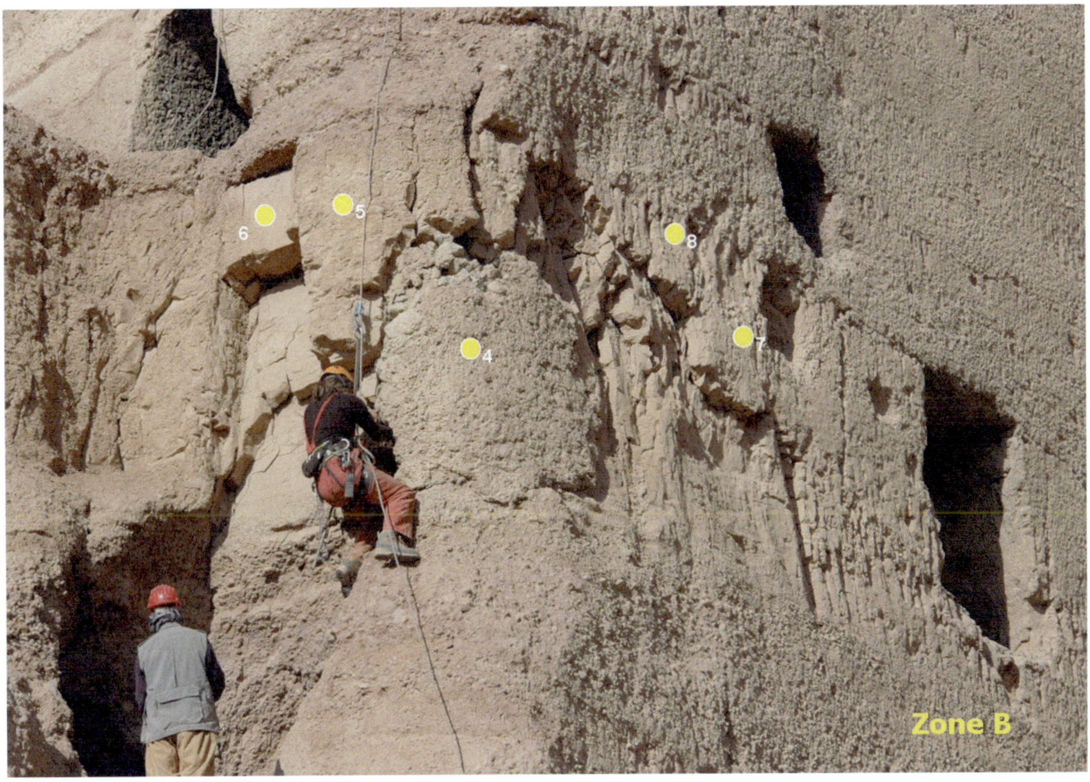

Fig. 11.44 Position of nails in the lower part of the niche

References

Colombini V, Margottini C (2003) Reducing the most critical rock fall-prone areas in the Buddhas niches in Bamiyan (Central Afghanistan). UNESCO report, July 2003

Margottini C (2003) The Buddha niches in the Bamiyan valley (Central Afghanistan): instability problems and restoration plans in the UNESCO intervention. J Jpn Landslide Soc 450(3):246–249 (in Japanese)

Margottini C (2004a) Instability and geotechnical problems of the Buddha niches and surrounding cliff in Bamiyan valley, Central Afghanistan. Landslides 1(1):41–52

Margottini C (2004b) Definition of 2004 activities for implementing the emergency consolidation of Eastern Giant Buddha and for beginning the preservation of Western Giant Buddha from geological hazards. UNESCO report, Mar 2004

Margottini C (2006) Definition of 2006 activities for completing the emergency consolidation of Large Eastern Buddha and for the preservation of Large Western Buddha. UNESCO report, Aug 2006

Margottini C, Colombini V, Crippa C, Tonoli G (2005) Emergency intervention for the geo-mechanical conservation of the niches of Bamiyan Buddhas (Northern Afghanistan). In: Sassa K, Fukoka H, Wang F, Wang G (eds) Landslides – risk analysis and sustainable disaster management. Springer, Berlin/Heidelberg, pp 75–79

PASCO (2003) Buddhas laser scanner project. UNESCO report

12 The Back Wall of the Niches: Preliminary Considerations

Claudio Margottini

Introduction

The explosion of March 2001 produced the collapse of the two statues, the instability all over the cliff, and a shear zone of hanging blocks in the back side of both niches. This part, despite of the destruction, is still very important because some plasters and excavated elements from the original outline are still on site. Because of this, the back side of the niches are sites to investigate in detail, to explore the connection between civil engineering, earth science, and restoration.

Considering the positive experience developed in consolidating the Eastern Giant Buddha niche, it is reasonable to proceed, also in this area, with a similar plan of activity.

Following are some preliminary notes on the different parts of the back side of the niches, to be used as first input to the final design.

Endangered Vaults at the Base of the Niches

The removal of fragments from the ground of the niches, mainly in the Eastern Giant Buddha, has confirmed the large instability of blocks settled in the lower part of the back side, and the risk of collapse, in the vault. Additionally, in the Eastern Giant Buddha niche, a master crack was discovered just behind the fragments and parallel to the cliff, increasing the risk of fall also for a large portion of the back side. A numerical evaluation of the stability conditions of the cliff is not presently available, but the overall shear strength of rock mass is relatively low, in consideration of the wide opening (about 10 cm) of the fissure. Nevertheless, the persistence of this large crack was investigated in the nearby caves, likely showing a probable limited penetration into the lateral sides (Fig. 12.1).

It is interesting to note that the crack probably represents the internal boundary of the statue's rock excavation, because in the upper part of the niche a fracture is clearly delimiting the statue from the background massive material. Figure 12.2 shows the section of the Eastern Giant Buddha niche and the detected discontinuities either into the niche or in the close proximity cavities and temple, only in the western side. From this it is possible to verify the possible continuity of this crack along the entire section of the niche. The discontinuity belongs to the master system of fissures parallel to the front cliff, likely caused by the lateral detensioning of the most external part of the slope during the deepening of the Bamiyan valley. An additional important support could derive from the reconstruction of the rock sections destroyed by the explosion, to be decided together with restorers and archaeologists.

In the vault the following problems still persist: collapse of small pieces of rock (minimum

C. Margottini
Department of Geological Survey of Italy
ISPRA – Italian Institute for Protection and Environmental Research,
Via Brancati, 60, 00143 Rome, Italy
e-mail: claudio.margottini@isprambiente.it

C. Margottini (ed.), *After the Destruction of Giant Buddha Statues in Bamiyan (Afghanistan) in 2001*,
Natural Science in Archaeology, DOI 10.1007/978-3-642-30051-6_12, © Springer-Verlag Berlin Heidelberg 2014

Fig. 12.1 The silhouette of statues on the back side of both niches (Western on *left* and Eastern on *right*). At the ground of Western Buddha are the remains of the March 2001 explosion. Note the height of the western niche (*left*) is about 55 m and the eastern niche (*right*) is about 41 m (From PASCO 2003b)

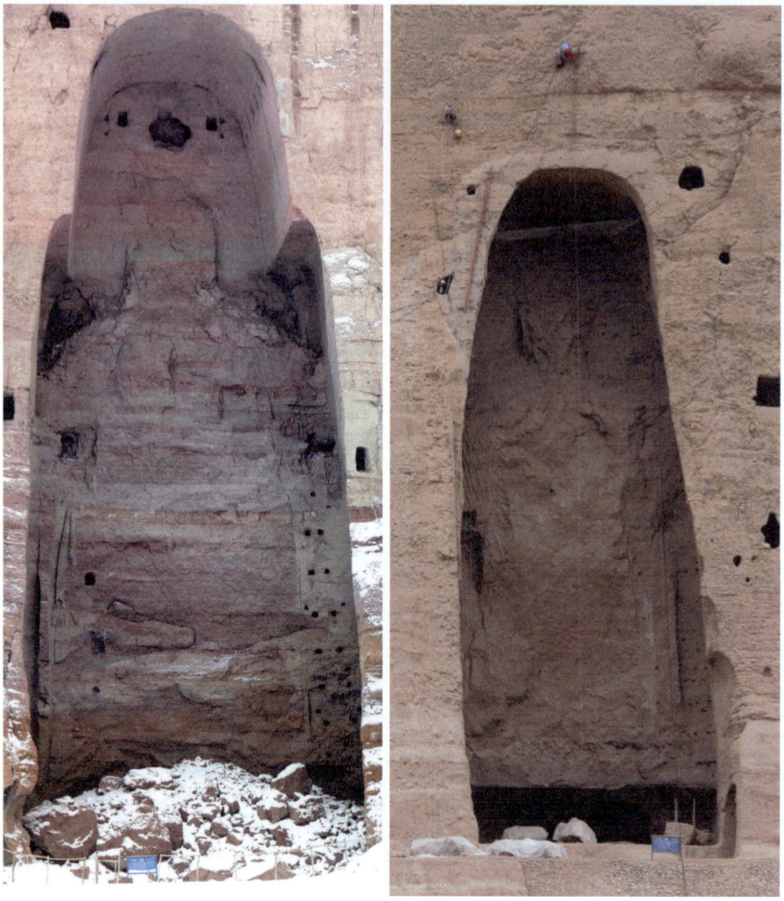

size ca. 30×30×30 cm), fall of large blocks of rock (around 1–2 m^3), and possible instability of the entire back side because of the large discovered fissure. In such a condition it is strongly recommended to install temporary support by means of wooden pillars. They should be settled perpendicular to the vault, and stably fixed to the ground. When the pillars are not perpendicular to the ground surface, an appropriate small excavation to host it must be created.

The Back Wall of Both Niches

During the field work, a possible strategy for the consolidation of the back side of the niches was investigated. In such a way some preliminary ideas are presented here, to be further confirmed with some mathematical modelling and compared with the experience of restorers. This area is, in fact, a connection point between engineering geology specialists and experts of restoration. Nevertheless, according to some preliminary estimation based on a laser scanning map produced by PASCO (2003b) and photo interpretation, the surface still maintains the original feature in only about 7 % in the western niche (apart from an area of 10 % destroyed in 1707 that remained untouched from the recent explosion) and about 20 % in the eastern niche. The remaining part is now fresh rock, weakened by the explosion with the typical morphology and consistency of a "shear band." In this area, any solution for securing the cliff has to be developed according to rock mechanics principles; at the contact between the original statue's surface and fresh rock, the proposed solution has

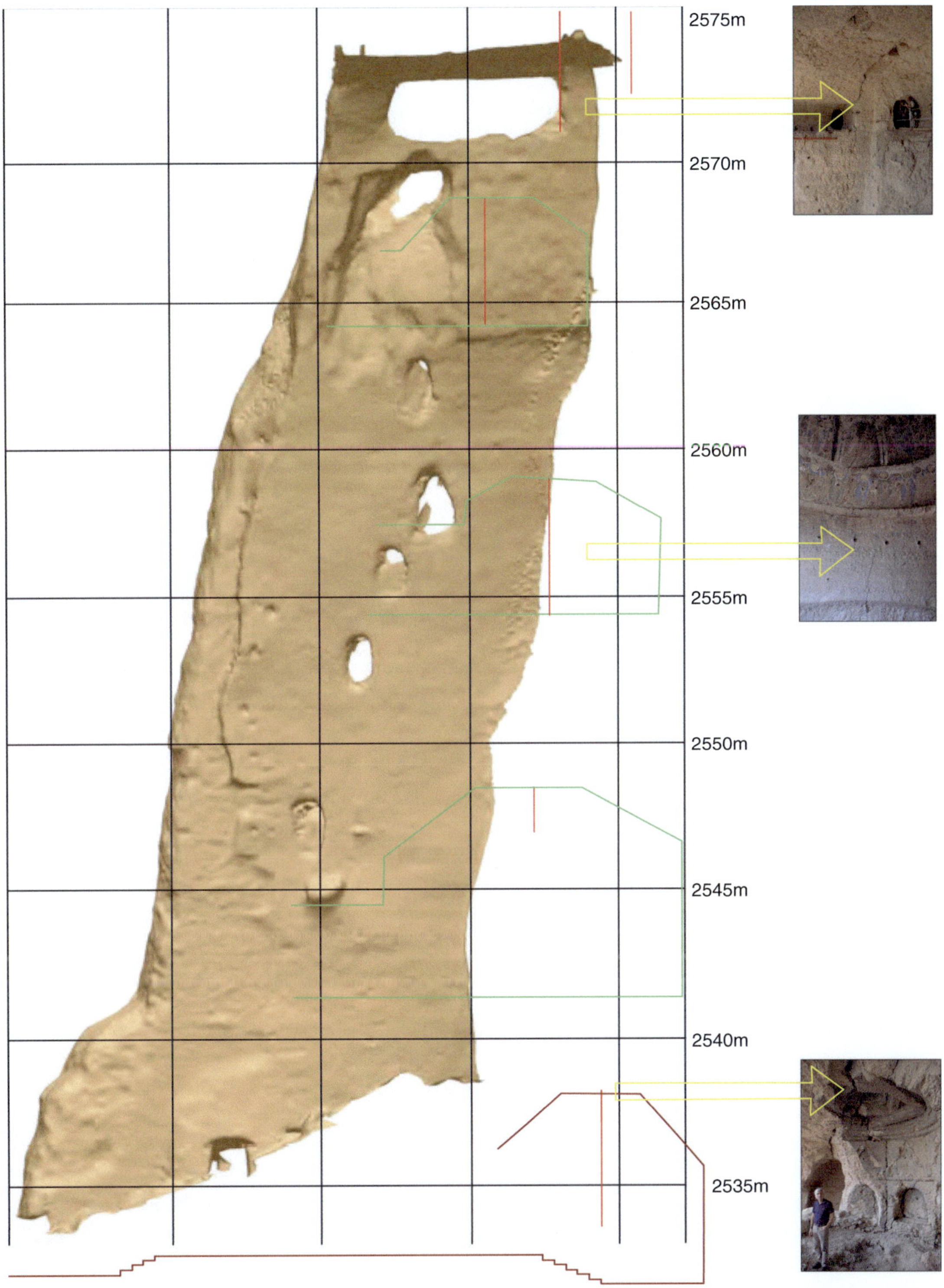

Fig. 12.2 Reconstruction of upward and lateral propagation of the large crack, in *red*, discovered at the *bottom* of the vault, in the ground level, after the removal of the statue's fragments (Topographic data from PASCO 2003a). The different thickness of the line stands for widening >10 cm (*bold*) and <10 cm (*light*). In *green* is the projection of lateral caves that were investigated to evaluate the propagation of the crack

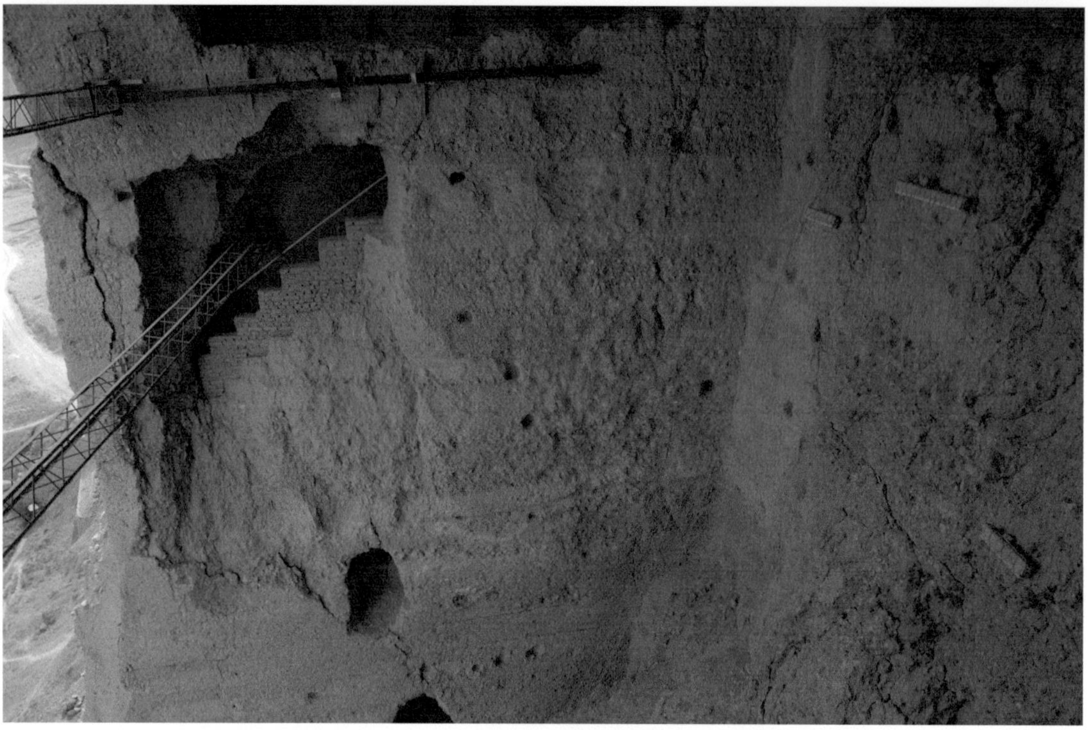

Fig. 12.3 The metallic net protecting from the detachment of small rocks from the shear band of destroyed statue. The photo also shows the wooden spacers, covered by nonwoven fabrics, placed on the niche to avoid any direct contact between the net and the remainder of the statue

to fit with restoration principles, adapted to rock mechanics stability.

As a general principle, any possible strategy has to begin with the identification of the final requirement. In other words, whether the niche will be preserved as it is, there is the need for an accurate stabilization, avoiding block collapse and maintaining as much as possible its present state. If the final solution will consider, for example, the reconstruction of a new Buddha statue, a supporting structure to sustain some of the original fragments not totally destroyed yet is required. Such a construction must be adequately anchored to the intact rock mass and then the consolidation of the back side has to be designed having this in mind. The following notes have been elaborated considering, as a target, the need to maintain the present state of the niches, avoiding artificial reconstruction. In this light, the destroyed niches are the emblem of human incapacity to preserve properly its heritage, to take care of the traces of our passage on the earth.

According to the positive experience already conducted in the Eastern Giant Buddha niche, the general strategy for the back side conservation should include:

1. *Temporary support for the unstable blocks including the vault in the ground level.* The existing net is certainly operating properly (Fig. 12.3), even if a better fixation at the underneath of the back side of the niche is required. This is to allow the safety of workers at the ground level. Presently the net is not completely fixed to the cliff and some falling rock may reach the ground. In the execution of a possible consolidation of the back side by means of scaffolding, the installation of more temporary steel cables, horizontally placed, is required.
2. *Damage assessment, in terms of surficial degradation and crack pattern distribution.* The back side is heavily damaged, because the six contemporary dynamite explosions of March 2001 produced the total detachment of the rock, originating a typical "shear band" (e.g., a thickness of rock deformed by the laying

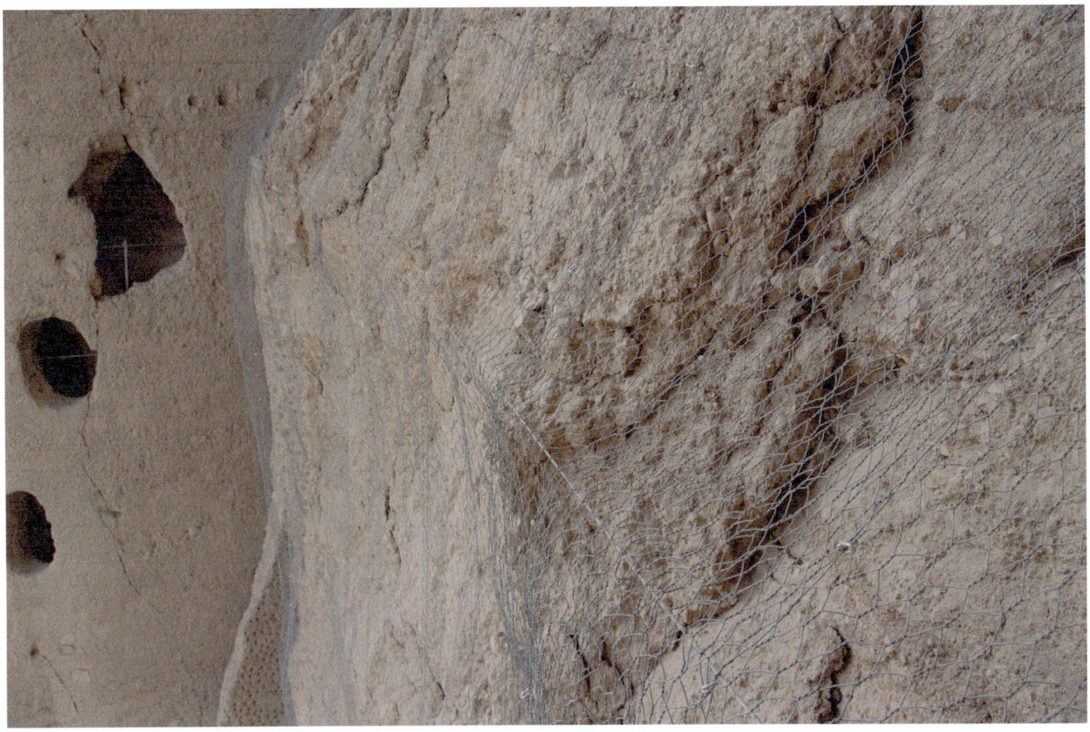

Fig. 12.4 Detail of temporary net and of the shear zone on the back side of Eastern Giant Buddha niche

Fig. 12.5 The shoulders of the Western Giant Buddha

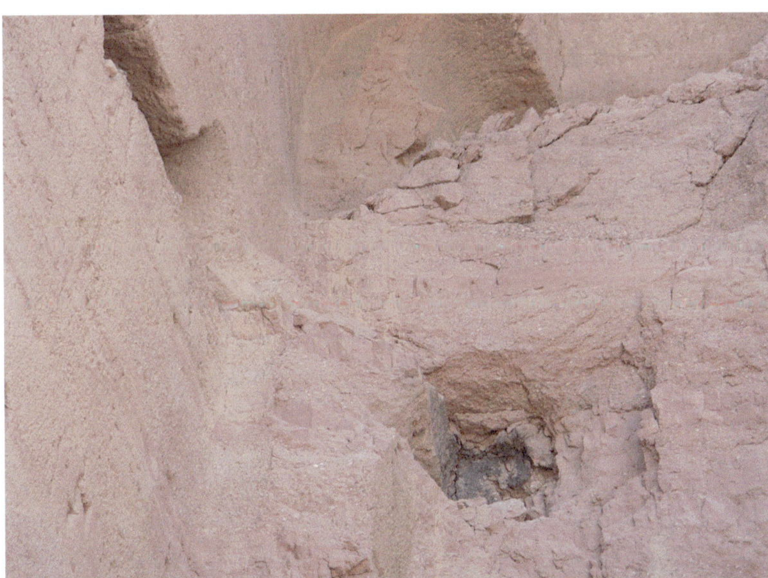

down movement of a large portion of material). The real depth of the shear band into the rock mass is not very clear. Some estimation can be made from superficial inspection, but this can result in a large underestimation of the real situation. Figures 12.4, 12.5, 12.6, 12.7, 12.8, 12.9, and 12.10 provide an overview of the current status.

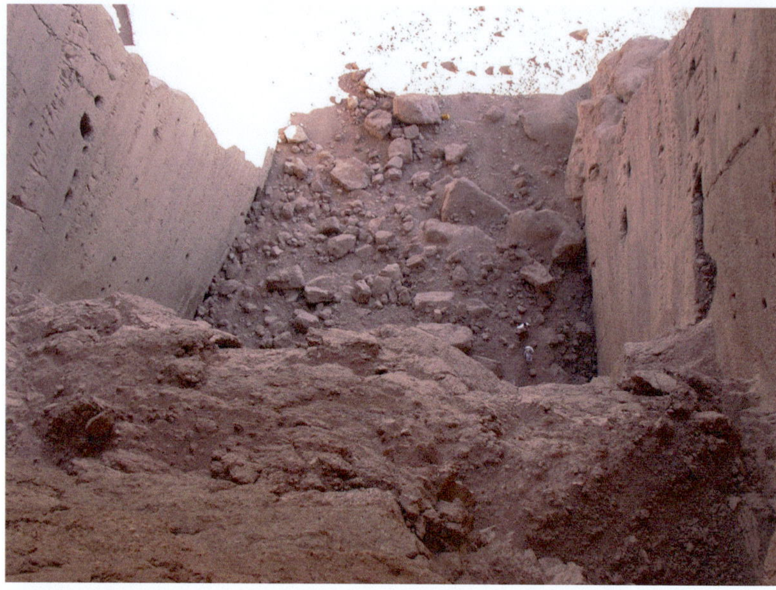

Fig. 12.6 (*Top* on the *right*) The shoulders of the Western Giant Buddha from the top

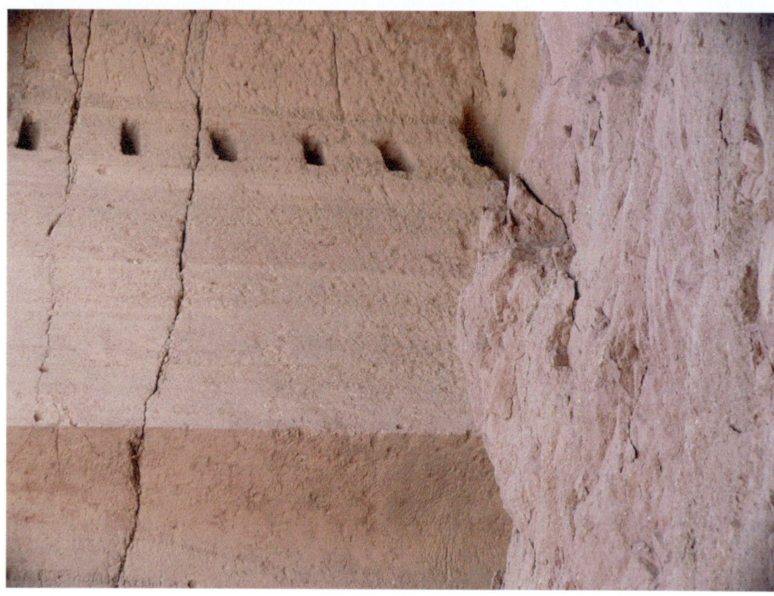

Fig. 12.7 (*Center* on the *left*) Detail of detachment in the Western Giant Buddha

Damage assessment is an essential part of the work because any solution has to be designed in accordance with the volume of surface and detached blocks, persistence of cracks, and spacing of them into the rock mass. There are many methods to assess such damage assessment.

The best solution would require a direct inspection supported by geophysical prospecting, such as ground penetrating radar (GPR), to understand the exact distribution in the depth of detached material (Roch et al. 2006). Unfortunately, such a technique is not suitable in this case with the already installed metallic net. In fact, the electromagnetic waves emitted by GPR antennas are usually reflected by such metallic barriers, misleading the signal and then the resulting outcome. A surface inspection, from the scaffolding, may

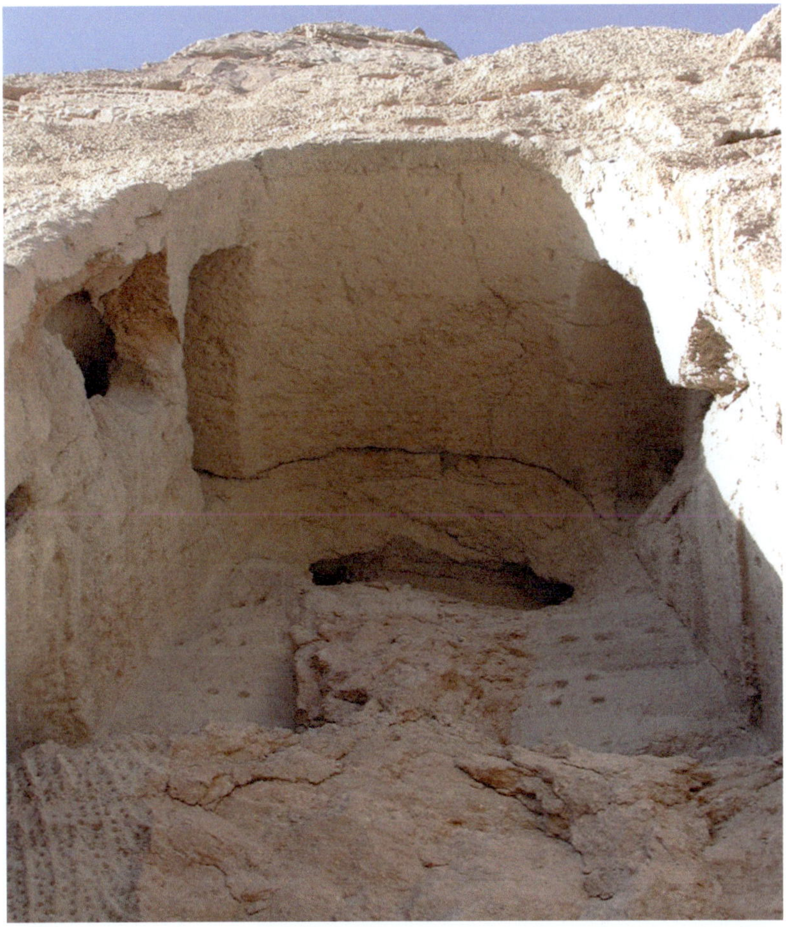

Fig. 12.8 (*Center* on the *right*) The shoulders of the Eastern Giant Buddha

provide an accurate estimation at least in the external part of the surface. An estimation of density of cracks and persistence can be provided by means of traditional rock mechanics techniques such as the evaluation of block size and then the assessment of volumetric joint count (Jv; Palmstrom 1974, 1982, 2005). This approach has to be conducted directly on site, by means of proper scaffolding.

Nevertheless, in order to provide a preliminary estimation of the damage suffered from the back side of the niches from the explosions, the evaluation of the block size on the back side of the Western Giant Buddha has been conducted from laser scanning maps produced by PASCO (2003b) and available rectified photos. The aforementioned niche was chosen because an undeformed picture taken by climbers is available. Also an anaglyph image is available for this niche. An anaglyph mixes into one image a stereoscopic view using the complementarity of colors in the RGB channels. With coloured glasses, one can then filter the image and see the depth information of the model (Courtesy of Pierre Smars, ICOMOS).

Being an important parameter, the block size is represented, either explicitly or implicitly, in the main quantitative rock mass engineering classification systems used in the design of rock support, such as:

(a) The ratio between RQD (Rock Quality Designation) and a factor for the number of joint sets (Jn) in the Q system
(b) RQD and joint spacing (*S*) in the RMR system

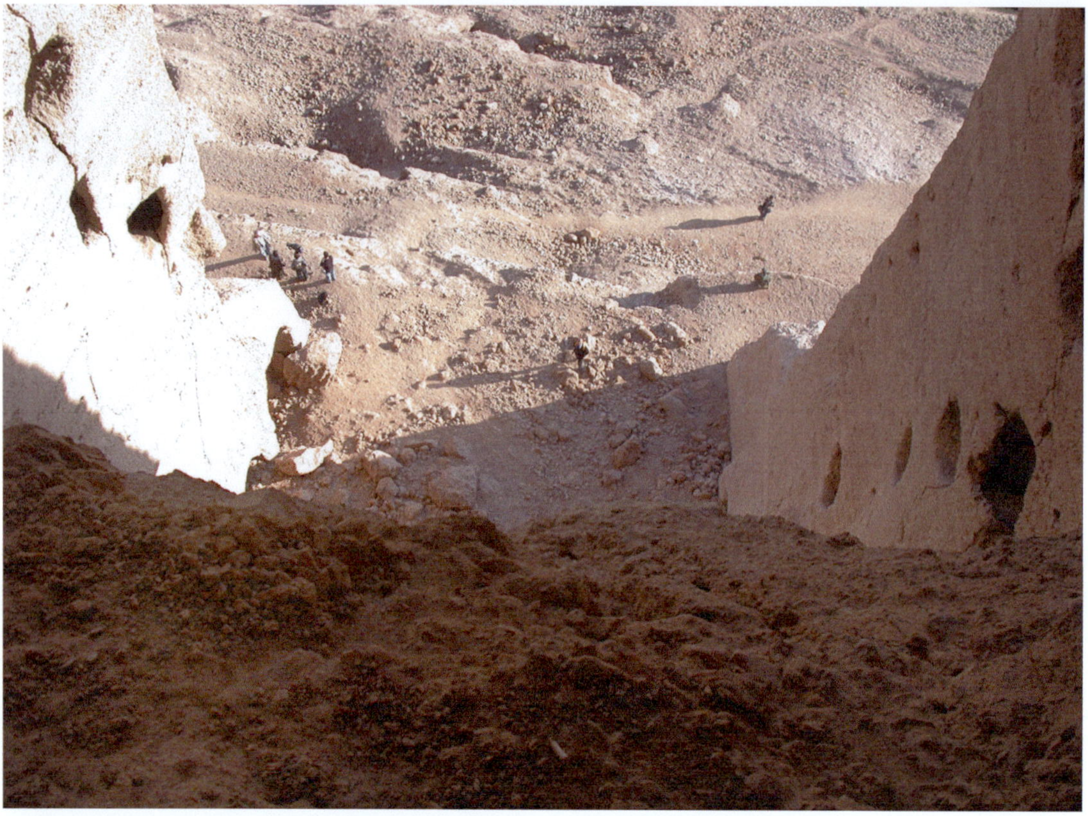

Fig. 12.9 (*Low* on the *left*) The shoulders of Eastern Giant Buddha from the top

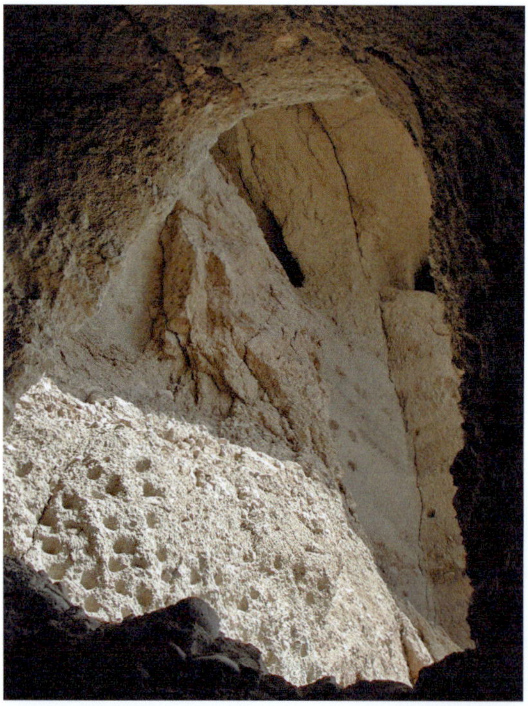

(c) Block volume (Vb) in the RMi (rock mass index), and the number of joints sets (nj) when RMi is applied in rock support evaluation

Also, the qualitative geological strength index system applies block size expressed as various degrees of blocky and broken rock masses in the determination of its values for rock mass strength (Palmstrom 2005).

The volumetric joint count (Jv) was introduced by Palmstrom in 1974. Earlier, a similar expression for joint density measurements was applied by Bergh-Christensen (1968) as the number of joints in a blast round. Being a three-dimensional measurement for the density of joints, Jv applies best where well-defined joint sets occur.

Fig. 12.10 (*Low* to *right*) Detail of the detachment in the Eastern Giant Buddha

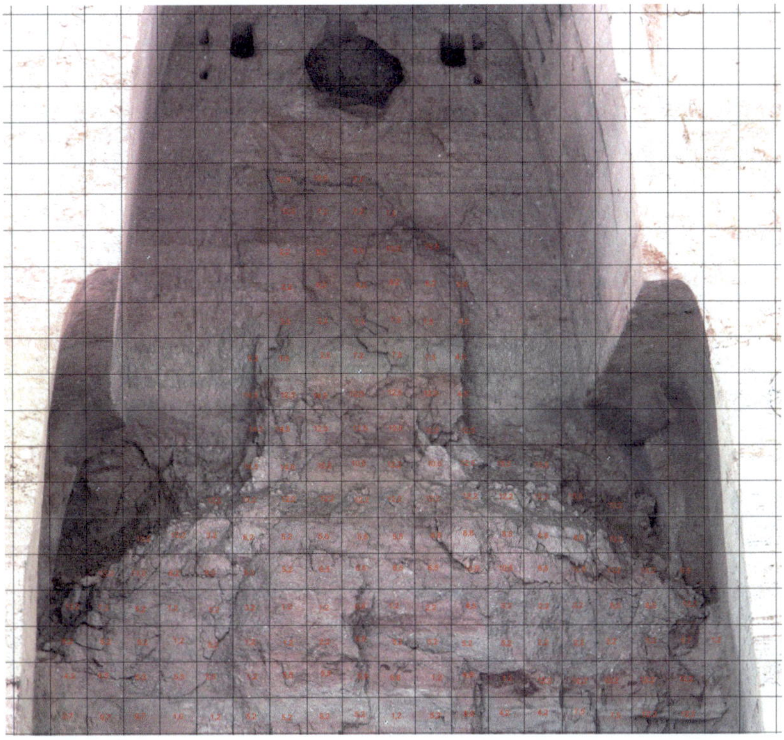

Fig. 12.11 Distribution of the volumetric joint count (Jv) in a portion of the niche

Jv is defined as the number of joints intersecting a volume of 1 m³. Where the jointing occurs mainly as joint sets

$$Jv = 1/S1 + 1/S2 + 1/S3 + \ldots 1/Sn$$

where *S*1, *S*2 and *S*3 are the average spacings for the joint sets.

As has been shown by Palmstrom (1995, 1996, 2005), it is possible to correlate block volume (Vb) and the volumetric joint count (Jv). This is:

$$Vb = \beta \times Jv^{-3}$$

where β is the block shape factor, having the following characterization:

- For equidimensional (cubical or compact) blocks β=27
- For slightly long (prismatic) and for slightly flat (tabular) blocks β=28–32
- For moderately long and for moderately flat blocks β=33–59
- For long and for flat blocks β=60–200.
- For very long and for very flat blocks β>200

A common value for β=36.

Palmstrom (1995) has shown that the block shape factor (β) may crudely be estimated from

$$\beta \approx 20 + 7\frac{a3}{a1}$$

where *a*1 and *a*3 are the shortest and longest dimensions of the block.

A chart for block estimation from Jv is reported in Palmstrom (1982).

The above concept has been applied to the Western Giant Buddha niche, for selected 2-D areas of 1 m², were a rectified image is available. Given that the 3-D application is not possible to develop because the scaffolding is not yet installed, and the information on the propagation inside the rock of the effect of the explosion is not available, the result is only a first approximation. Figure 12.11 is a preliminary report of the distribution of the Jv for a detail of the Western Giant Buddha niche. In Fig. 12.12 all the niche is reported, showing the preliminary

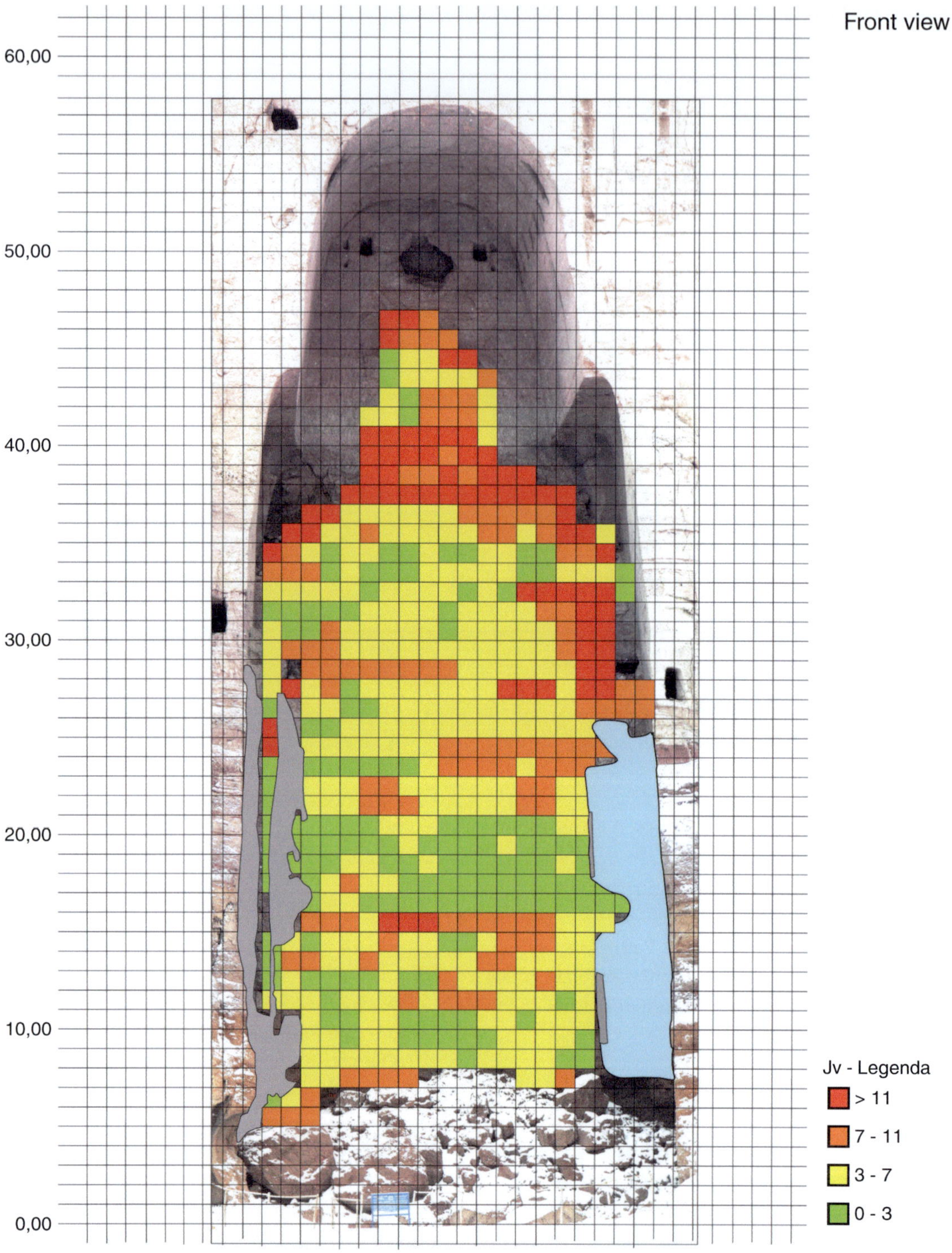

Fig. 12.12 Joint volumetric count for the Western Giant Buddha niche. Legend has been readapted from traditional classification (Palmstrom 2005) to fit the available data. In *grey* are original surfaces still preserved and in *sky* is the area resulting from the 1707 destruction

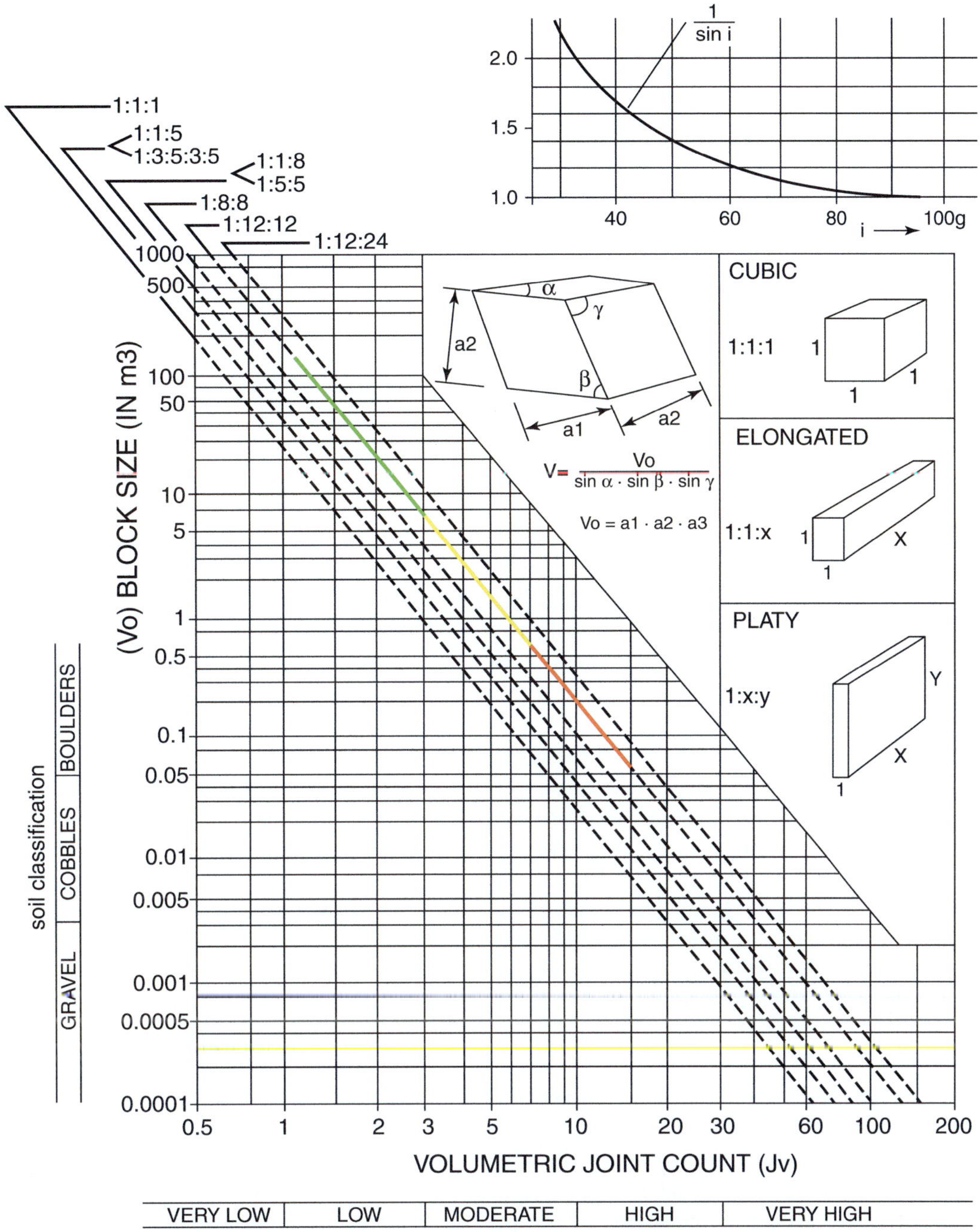

Fig. 12.13 Comparison of average values of block size for the classes of volumetric joint count (Jv) (Palmstrom 1982). In color are the classes selected in the present study, with the same legend as in Fig. 12.12

assessment of many areas of instability. Comparing these data with the chart of Fig. 12.13 in which the relationship between Jv and block size is reported, it is possible to estimate that 12 % of the surface exhibits block size smaller than 0.17 m^3, 20 % exhibit block size between 0.17 and 0.6 m^3, 43 % between 0.6 and 6.0 m^3, and 25 % approximately less than 6.0 m^3.

3. *Establishing a proper consolidation strategy*. Because it is not correct to define a unique solution all over the niche but there is the need to adapt different techniques at the different parts of the niche, as in previous works, it is absolutely necessary to perform a zoning of different problems (e.g., shallow detachments, cracks in the depth) in order to establish the most appropriate consolidation procedure. The damage assessment based on rock mechanics methodologies, possibly joined with some geophysical investigation, may allow the understanding of the thickness of the shear band to consolidate and the dimension of block size. In general, according to the preliminary damage assessment of Fig. 12.12, it is possible to say that such problems require the avoidance of collapse of portions of the rock face by recovering the original rock mass strength. The positive experience already conducted in the Eastern Giant Buddha niche suggests the use of similar techniques (e.g., passive anchors, nails, and grouting) but with different methodologies. Because this part of the niche still maintains the silhouette of the statue, it should be as little disturbed as possible. Likely, the removal of very few totally displaced small blocks would be required. Nevertheless, there is a need to tie together the single pieces of rock, originated by the laying down movement along the shear band. For small fragments, many types of nails can possibly be evaluated: pregrouted, stainless steel, carbon fiber, fiberglass. All these do not alter in time and provide a reliable strength. Mobilized shear strength has to be properly evaluated to assess density and distribution. This last is dependent on the grouting material that is another important element to consider; generally, it depends on the selected type of nail and the target (e.g., nailing an archaeological remain or very deep rock). In principle, the grouting material should provide the required bounding capacity to nail in order to: fit with the slaking problem of the rock, exhibit a similar (or slightly lower) mechanical strength and elasticity of original material, and not damage the archaeological remains. Considering also that the proposed solution requires a long time of execution and should possibly be implemented by a local firm, it is advisable to use a simple but efficacious solution. In this first hypothesis, three phases can be foreseen, likely with the use of stainless steel nails and passive anchors, making for a very simple procedure for installation:
   (a) The first phase has to be conducted in very blocky areas, with extremely low vibration and low water pressure cooling equipment (diamond head), for a depth of about 50 cm and a diameter less than 20 mm (nail 10 mm). This provides the opportunity to grout and nail the most superficial crust of the back side. In such a way a first resistant surface can be ensured. The grouting material to inject should be identified in coordination between restorers and consolidation experts, likely through field tests. A mortar already approved by Cultural Heritage Ministries could provide the most appropriate solution, avoiding chemical material in favor of a more natural product. In any case the low water release in the mixture, as a consequence of slaking for some siltstone strata, is a fundamental requisite. The proposed technique is similar to the one in use for the consolidation of masonry with microinjection. The spacing of such small grouting/injection can depend on the surface condition of materials and, as a first approximation, falls within the order of 50 cm.
   (b) The second phase is aimed at stabilizing the remaining shape of the figure till the master crack identified in 2006, in the underneath caves behind the feet of the Eastern Giant Buddha. In this phase it is expected to consolidate the material with grouting and nails, in order to obtain a homogeneous and stable rock mass, to a depth of no more than 3 m. In this case the drilling will be conducted with low vibration equipment, with diameters of about 32 mm (including the stainless steel nail of 16 mm and grouting). The spacing can here be considered relatively constant, on the order of about 1.5–2 m.
   (c) The third phase is aimed at fixing the consolidated mass to the stable rock of the cliff,

and the grouting of the master crack identified in 2006. The stabilization will be conducted with passive anchors and grouting for a depth of about 10–15 m. The depth of anchors should be, in this case, twice the distance from the last detected large fissure encountered during the drilling. The diameters can be, as first sought, about 90 mm, in order to allow a correct placing of anchors. The spacing of such anchors will be evaluated through a stability analysis of the back side of the niche under present conditions. It is possible that anchors should be concentrated in the lower part of the niche, whereas in the upper part where the thickness of material is lower, the extension to 4–5 m. of passive anchors/nails foreseen in phase 2 can have satisfactory results.

Figure 12.14 shows the theoretical distribution of such nail/anchors in the Eastern Giant Buddha niche, and the relationship with the large crack detected at the inner back side (*dashed red line*). Real damage assessment is not considered in this sketch.

It is now clear that the proposed solution, with the three steps of drilling as previously described, is likely suitable to solve the target problem. The only uncertainty still to clarify includes the real nature of grouting material to inject for each of the three phases. Nevertheless, no final solution can be defined without a joint and fruitful cooperation within restorers and rock mechanics specialists. Even if the original part of the Buddha Statues is now a minor percentage with respect to fresh rock, there is no doubt that in this kind of work that is at the border among many disciplines, only a real collaboration can overcome misinterpretation and incorrect understanding of the real problems and needs (Plate 12.1).

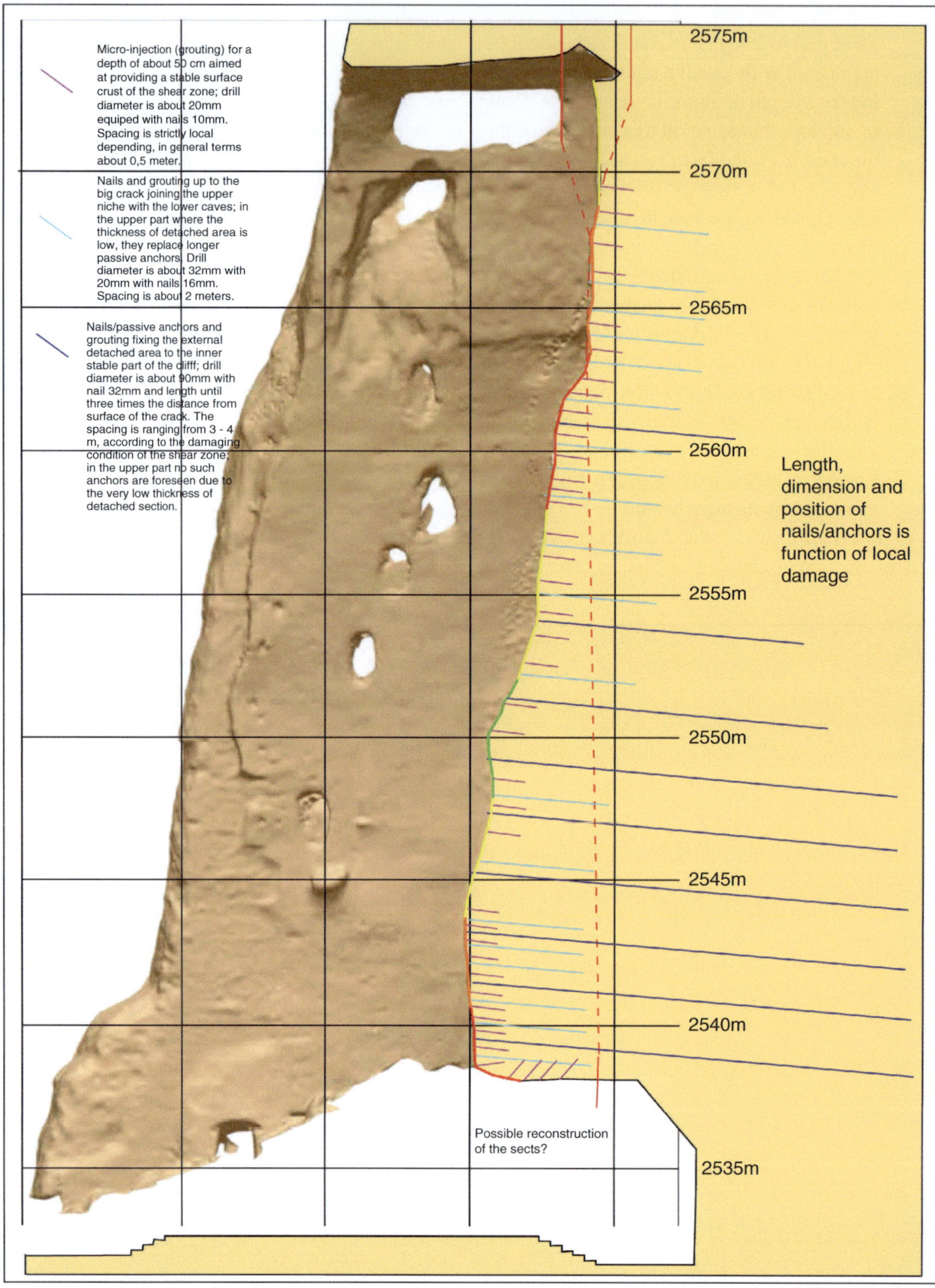

Fig. 12.14 Tentative sketch for the stabilization of the back side of both niches (case study developed for the Eastern Giant Buddha). *Pink*, *green*, and *blue* stand for shallow nails, intermediate nails, and deep anchors, respectively; the *dashed red line* is the probable reconstruction of a large crack detected at the inner back side of the niche.

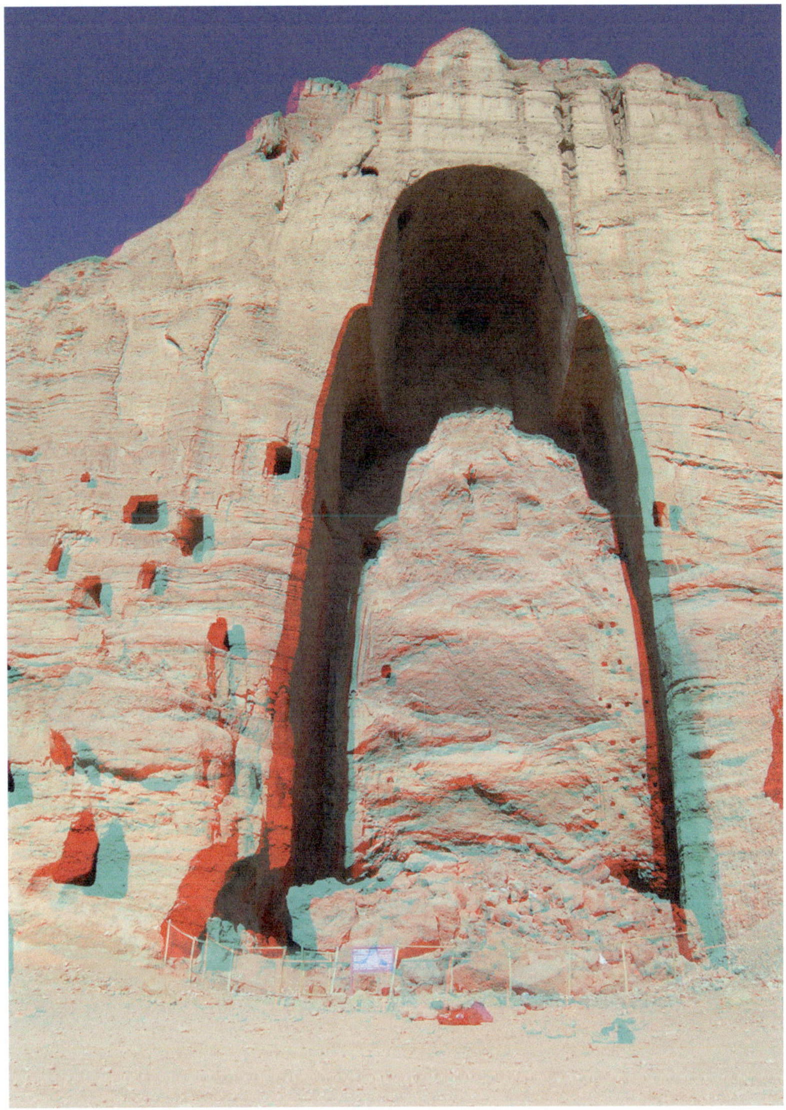

Plate 12.1 The anaglyph image of the Western Giant Buddha niche. This technique is quite useful to display a static view of 3-D models. An anaglyph mixes into one image a stereoscopic view using the complementarity of colors in the RGB channels. With colored glasses, one can then filter the image and see the depth information of the model (Courtesy of Pierre Smars, ICOMOS)

References

Bergh-Christensen J (1968) On the blastability of rocks (in Norwegian). Lic Techn thesis, Geological Institute, Technical University of Norway, Trondheim. 320 p

Palmstrom A (1974) Characterization of jointing density and the quality of rock masses (in Norwegian). Internal report, A.B. Berdal, Norway, 26 p

Palmstrom A (1982) The volumetric joint count – a useful and simple measure of the degree rock mass jointing. In: Proceedings of the IV congress international association of engineering geology. vol 2. New Delhi

Palmstrom A (1995) RMi – a rock mass characterization system for rock engineering purposes. PhD thesis, University of Oslo, Department of Geology, 400 p

Palmstrom A (1996) The weighted joint density method leads to improved characterization of jointing. In: International conference on recent advances in tunnelling technology, New Delhi, 6 p

Palmstrom A (2005) Measurements of and correlations between block size and rock quality designation (RQD). Tunnels Underground Space Technol 20(2005): 362–377

PASCO (2003a) Bamiyan mapping project. UNESCO report

PASCO (2003b) Buddhas laser scanner project. UNESCO report

Roch KH, Werner C, Bruckl E (2006) Potentials of monitoring rock fall hazards by GPR: considering as example the results of Salzburg. Landslide 3: 87–94

13 Safeguarding the Clay Plaster Remains of the Eastern Buddha Statues and the Rear Side of the Niche

Bert Praxenthaler

Introduction

The back wall of the Eastern Giant Buddha niche is characterized by important remains of the original surface and plaster. Despite the explosion, there is still 20 % of the original plaster and this make the consolidation of the back wall of the Eastern niche an important task for archaeological conservation. On the other side, the Western Buddha niche contains only 7 % of the original surface and plaster.

Safeguarding the plaster remains means not only fixing the mortar to the bedrock but also reinforcing the latter, in order reconstitute the original integrity of the rock altered by the explosion.

The potential solutions, methodologies and techniques for achieving such an important task have been discussed among the many involved experts. As a matter of fact, it was recognized that no unique solution can be defined for all over the niche. There is a need to adapt different approaches to the different problems in different parts of the niche. In detail: original plaster detachment, small blocks originated by explosion, cracks in the depth, have been recognized as major threats. Essential support to implement the job was the availability of scaffolding, kindly donated by the Messerschmitt Foundation (Germany).

B. Praxenthaler
Epfenhausen, Bahnhofstr. 12,
D-86929 Penzing, Germany
e-mail: bert@praxenthaler.de

The adopted approach is considering:

- Fixing the remaining part of the original clay plaster and the very blocky areas with a proper gluing mortar.
- In very blocky areas, shallow nails will be installed, with extremely low-vibration and low-pressure water cooling equipment (diamond head) and possibly air. The expected depth of drilling is no more than 50 cm and a diameter less than 20 mm (nail 6 mm). This provides the opportunity to grout and nail the most superficial crust of the back side. In such a way a first resistant surface can be ensured.
- A following phase is aimed at stabilizing the remaining shape of the figure until the master crack identified in 2006 can be repaired, in the underneath caves behind the feet of the Eastern Giant Buddha. In this phase it is expected to consolidate the material with grouting and nails, in order to obtain a homogeneous and stable rock mass to a depth of no more than 3 m. In this case the drilling will be conducted with low-vibration equipment, with a diameter of about 32 mm (including the stainless steel nail of 16 mm and grouting).
- The last phase is aimed at fixing the consolidated mass to the stable rock of the cliff, and grouting of the master crack identified in 2006. The stabilization will be conducted with passive anchors and grouting for a depth of about 2.5 m behind such a master joint. The diameters can be about 90 mm, in order to allow a correct placing of anchors.

C. Margottini (ed.), *After the Destruction of Giant Buddha Statues in Bamiyan (Afghanistan) in 2001*,
Natural Science in Archaeology, DOI 10.1007/978-3-642-30051-6_13,

The preliminary position of anchors/nails was developed on the basis of damage assessment of the cliff, but their actual position was defined directly on site. As already developed in the emergency intervention for cliff and niche (see previous chapters), it is evident that final position, density, and length can only be decided on site according to the real condition of the cliff. Apart from many details, it is possible to say that longer anchors will be mainly concentrated on the lower part of the niche, to ensure global stability to the vault niche. This is also in consequences of some tectonic faults detected during the field inspection that are producing a kinematic context prone to fall. These kinds of tectonic faults were detected also in some other parts of the cliff (Margottini 2009) as well as, and likely produced the collapse of some part of the leg and foot of the Western Giant Buddha. Also, long anchors were designed to fix the remaining part of the shoulder and the head to the bedrock identified behind the master crack previously described.

The grouting was another important issue. Because of the positive experience of the consolidation performed in the past and the verified stability over time (Margottini 2009), it was decided to use the same cement and mortar composition as previously defined, including the amount of superplasticizer. Following are the details on the executed intervention.

Safeguarding the Clay Plaster Remains

The clay surfaces of the Eastern Buddha that survived the blowing-up of the statue in 2001 had been spot-bonded to the back wall in 2004 as part of an emergency stabilization measure.[1] This work could only be carried out by abseiling with a rope, because a scaffold had not yet been put up. In 2008, it was possible to fix these original clay plaster surfaces. For this purpose the scaffold was extended with bracket-shaped cantilevers to get close enough to the clay plaster fragments. These two fragment areas can be found on the left side under the former arm of the Buddha, at a height of 9.6 m and measuring 15 m^2, and on the right side along a fold of the gown, slightly inside the other arm, at a height of 9.0 m and measuring 6 m^2 (Fig. 13.1).

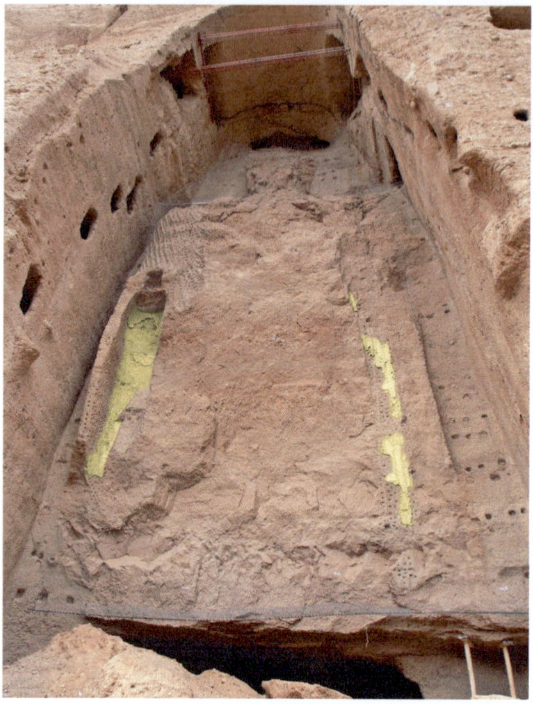

Fig. 13.1 Eastern Buddha, preserved original clay plaster layers (marked in *yellow*), in total about 21 m^2, pre served in 2008 by gluing and needling

Composition of the Original Clay Plaster

The original clay plasters on the Eastern Buddha are all made up of three layers. The lowest layer immediately on the stone, the slush coat, contains chaff und was applied to the surface very wet and then spread with the fingers. These finger traces are still clearly visible today. To improve adhesion, clay lumps were pressed into previously

[1]In 2004, a mixture of ochre-coloured clay and gypsum was used, mix ratio 3:1.

mortised holes (diameters of 6–8 cm) with the help of a suitable stone. The actual plaster material was applied onto this layer, a clay mixture containing chaff and animal hair, c. 2–3-cm thick. Onto this second layer a fine layer of 2–3 cm was applied, which served as support for the color coat. In several areas only the undercoating remains (Figs. 13.2 and 13.3).

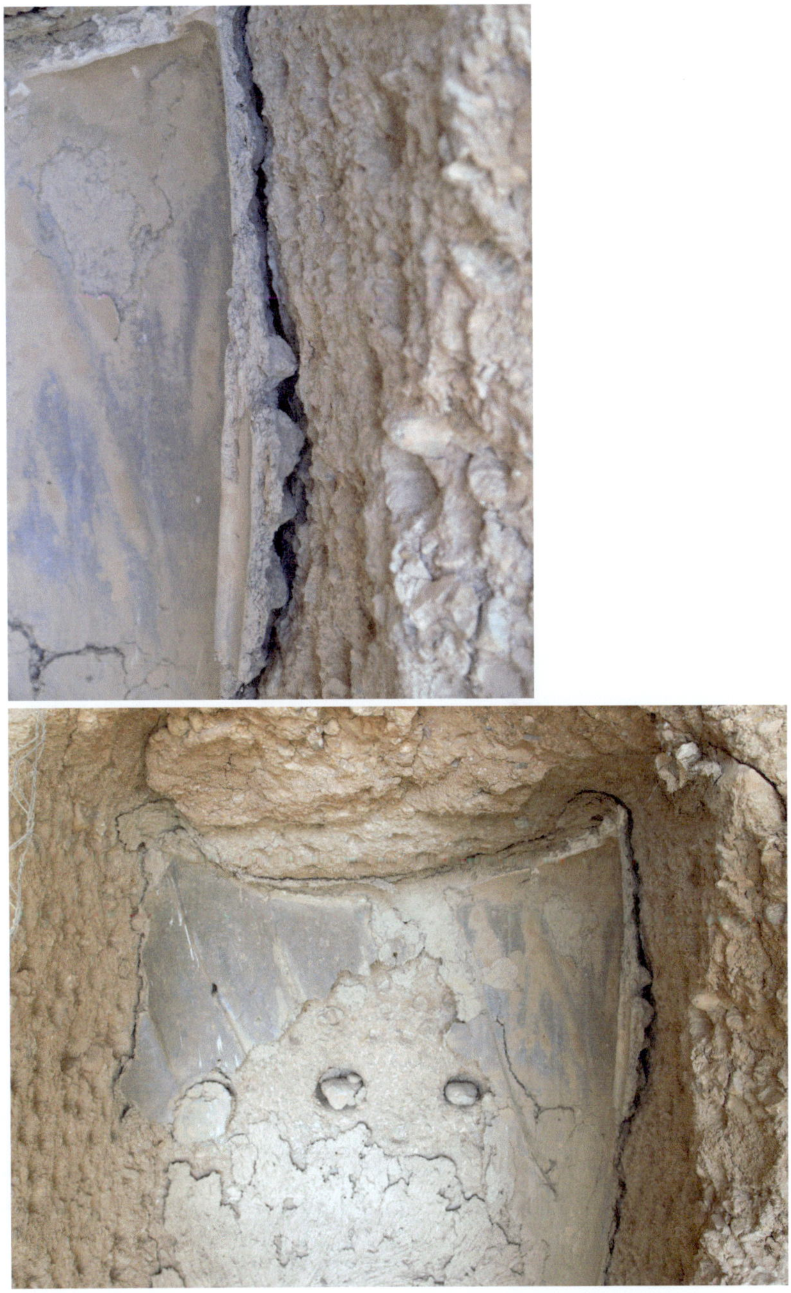

Figs. 13.2 and 13.3 Eastern Buddha: remains of clay plaster on the left, with lapis lazuli. Photo shows condition before treatment. All fragments are slightly detached from the rock face and are only kept together by being wedged on the left and right with the "clay-stone burling"

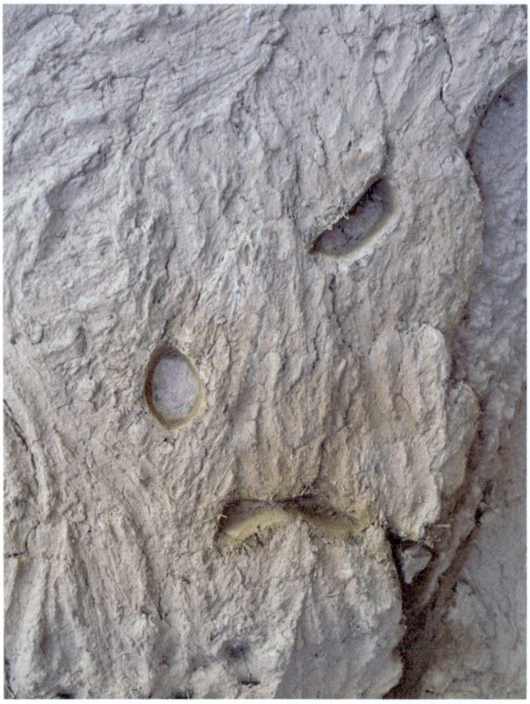

Fig. 13.4 Eastern Buddha, finger traces in the lowest clay plaster layer

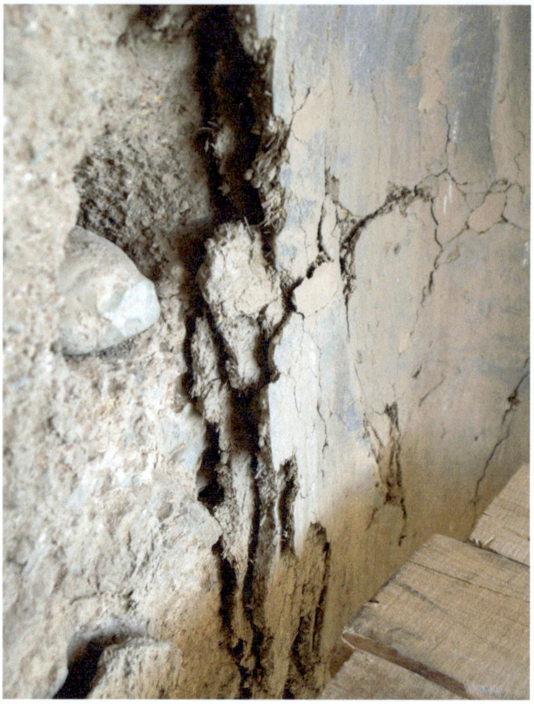

Fig. 13.6 Three layers of clay plaster, partly detached from one another

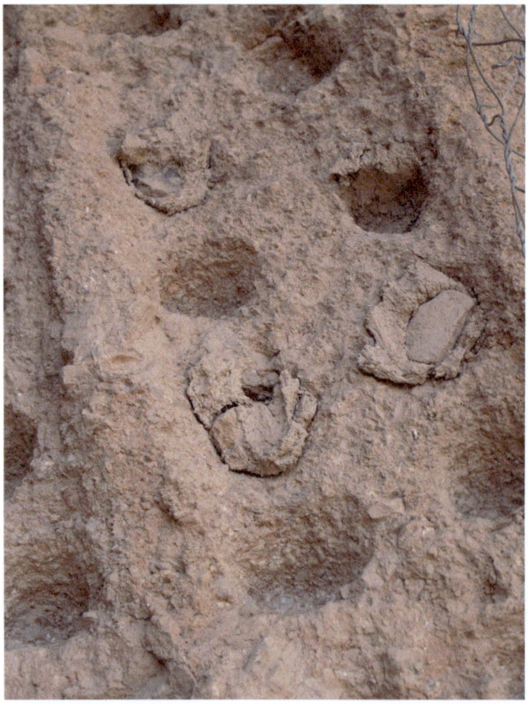

Fig. 13.5 Clay lumps pressed into the holes with stones

Bonding, Edging, and Needling the Clay Plaster Fragments

For the 2008 stabilization work on the clay fragments, tests with various mortars were carried out in advance. Upon the advice of Prof. Emmerling (Chair of Conservation Sciences at the Technical University Munich) Ledan® was used as component.[2] After adhesion tests a mixture of Ledan® TA 1, red clay, and Dralon fiber seemed suitable for back wall bonding. For fixing the edges yellow clay, Ledan® TB 1 and when required Scotchlite K1 were employed as mortar.[3] At appropriate

[2]Ledan® TA 1 Leit 03, Ledan® TB 1, both are binding agents based on lime with aggregates of Terra Pozzuoli. Purchased from Fa. Dr. Kremer.

[3]The mixture of the bonding mortar: 1 VT Ledan® TA 1 and 3 VT red clay, c. 0.3 VT Dralon fiber. Due to its better bonding adhesive power the red clay was selected. For fixing the edges 1 VT Ledan® TB 1 and 3 VT yellow clay were mixed, depending on layer thickness aggregates of Scotchlite K1. Yellow clay for colour adaptation. All materials purchased from Fa. Dr. Kremer.

spots where the mortar was particularly thick, for instance at elevations of folds, darts of V4A ribbed bars or sanded glass fiber beams were used.[4] These darts were bonded with a mortar of Ledan® and yellow clay. Whenever possible, armature bores were carried out in areas of the clay plaster that already had surface damages or holes in order not to reduce the original substance unnecessarily. The spots where the darts were placed were marked with a small nail.

In order to be able to move those clay plasters, which were very detached from the rock face, back a little to their old position and thus to avoid the mortar for bonding and the stabilization of the edges becoming too thick, these loose plaster layers were carefully moistened on their reverse side to make the material a little more malleable. By means of a splaying apparatus, which we had made by metal craftsmen at the Bamiyan bazaar, the clay plaster fragments could be repositioned millimeter by millimeter and were battered afterward.

At first, this work was carried out partly by Bert Praxenthaler and by stone restorer Mujtabah Mirzai, together with up to five Afghan workers. In the course of the work, the reverse-side bonding and the fixing of the edges were mostly executed by the Afghan workers Qurban, Eshaq, Karim, Haidar, and Abdul Ali. However, drilling and needling were only done by Praxenthaler and Mirzai (Fig. 13.7).[5]

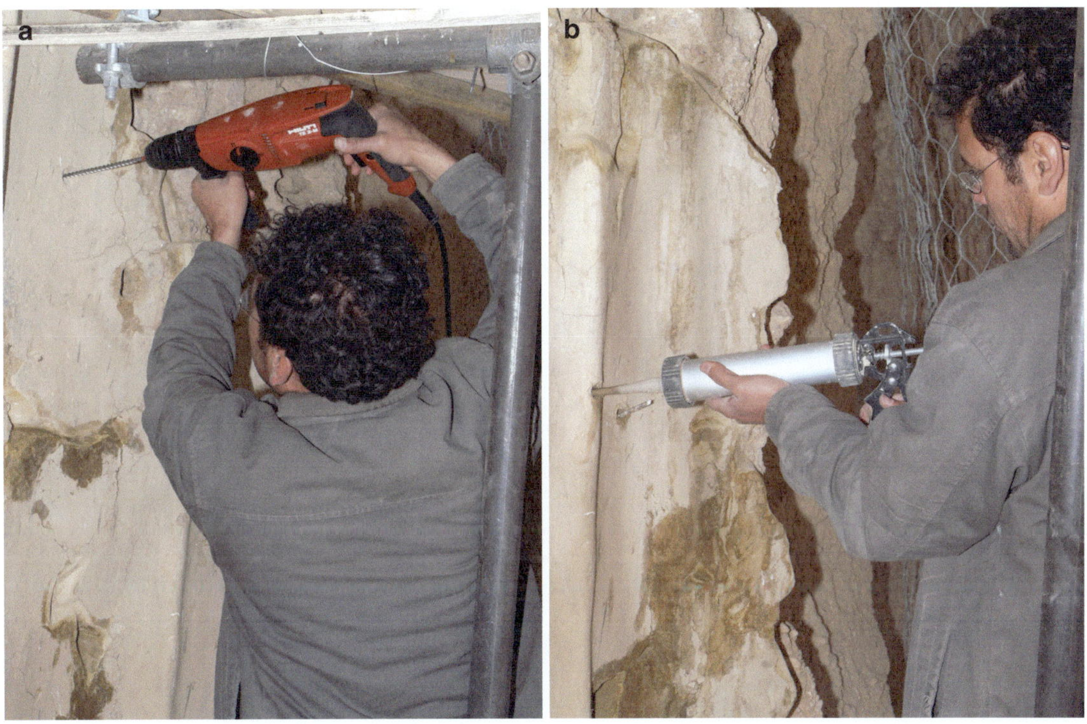

Fig. 13.7 (**a**, **b**) Original clay plaster fragments (*right side* of the Eastern Buddha). Preparation for the setting of needles: Mirzai drilling a hole into an existing bullet hole

[4]Ribbed bars V4A 6 mm from Kummetat Stahl, Frankfurt, glass fibre beams 5 mm from Fa. Fibrolux, Hofheim.

[5]Particularly qualified workers were integrated into the works step by step. They were not merely made familiar with the work techniques, but also with the basics of restoration.

Fig. 13.8 (**a–c**) Bonding of the clay plaster fragments. By means of a splaying apparatus, pressure is carefully put on the clay plaster fragments. In most cases this apparatus was hinged to the scaffold. The surface of this apparatus was adapted to the texture of the clay plaster and was made of small wooden boards. On the left underneath the arm hole

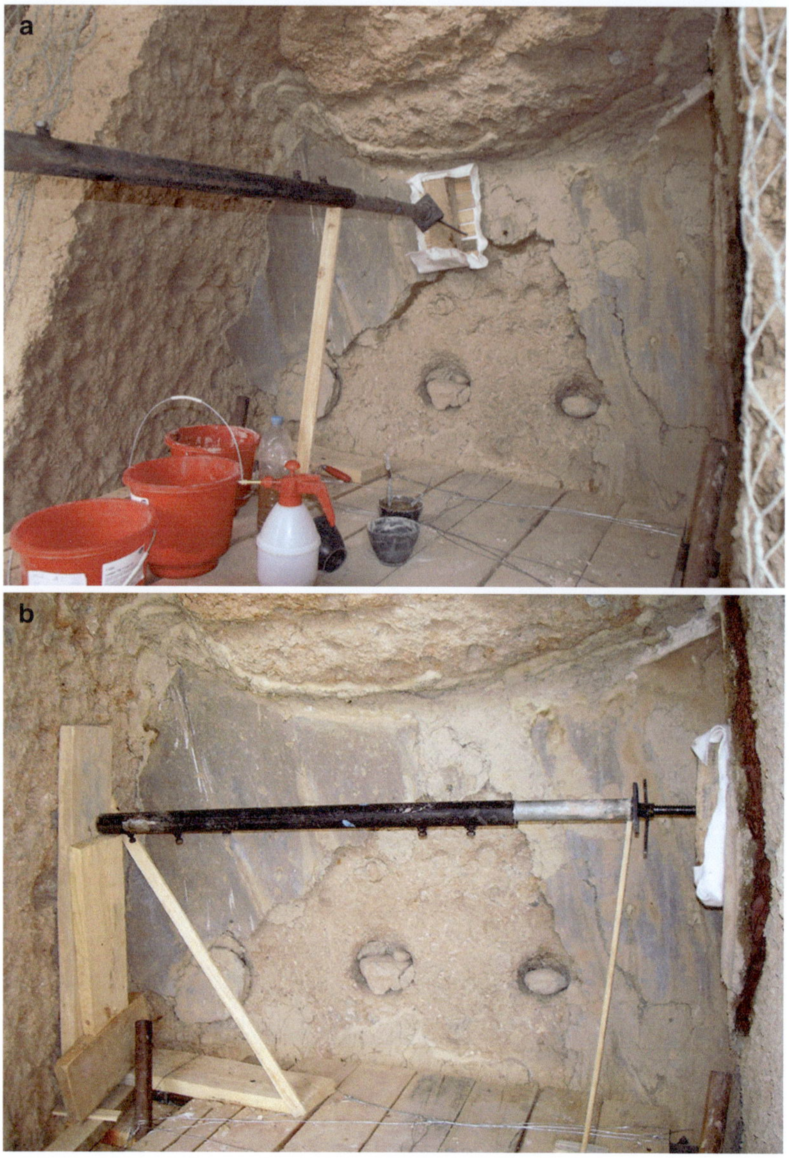

Fig. 13.8 (continued)

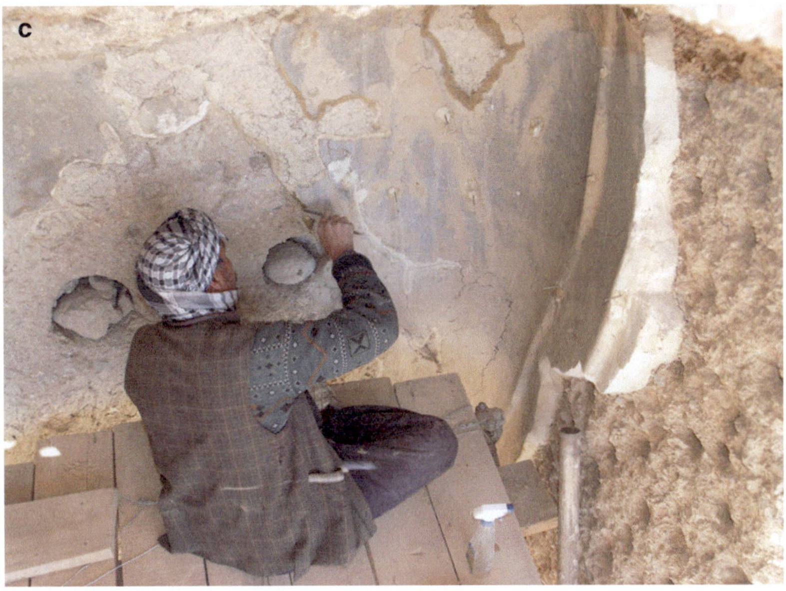

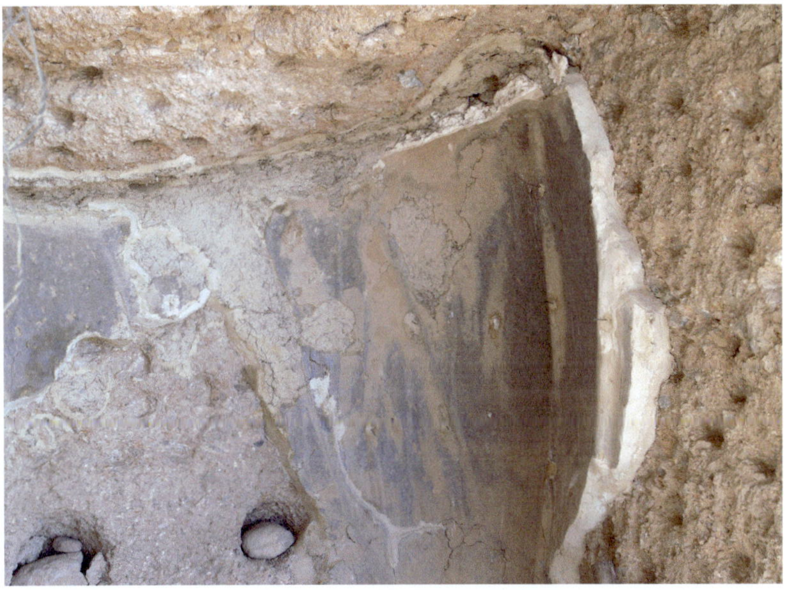

Fig. 13.9 Bonded and secured clay plaster fragment underneath the arm hole

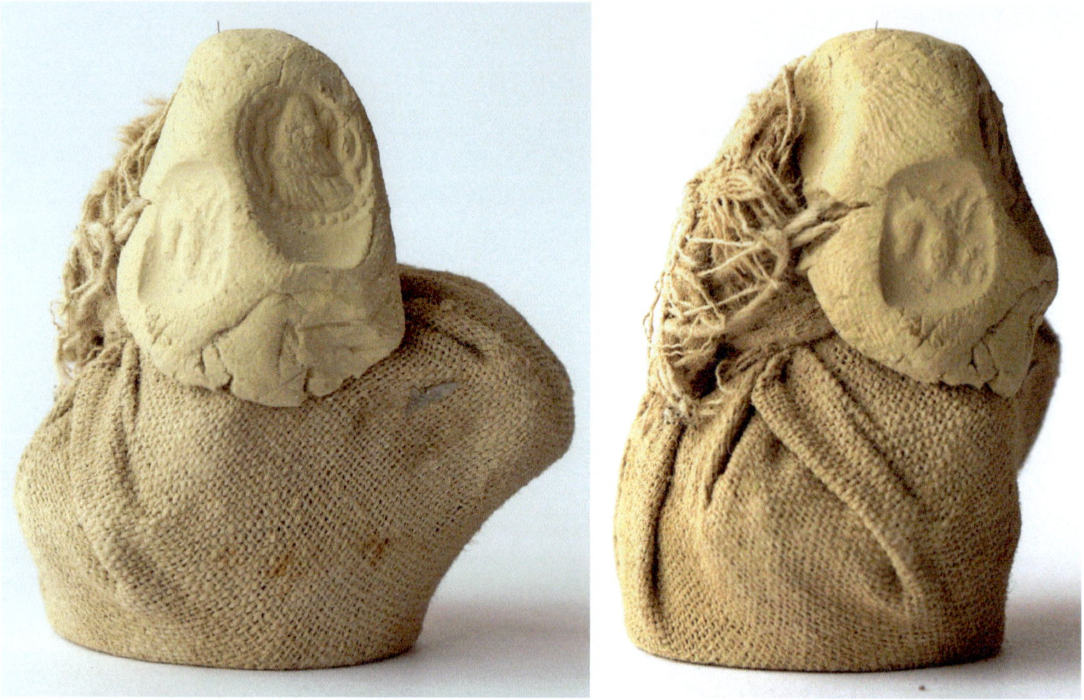

Figs. 13.10 and 13.11 Eastern Buddha. Buddhist relic found on October 3, 2008

Relic Find

During the work on the Eastern Buddha an object was found on 3 October 2008 in a cavity about 17 m above the ground level of the niche and above the clay plaster fragments on the left. This is most likely a Buddhist relic, which apparently was deposited while the statue was erected in a 10-cm hole together with parts of plants and then closed with a stone. This cavity is the backmost base of the hole that used to hold the beam of the Buddha's right arm. It is located below the undestroyed rock parts of the Buddha's right shoulder. Effectively, the relic was hidden at the place of the Buddha's elbow joint.

The object consists of a cloth sack measuring 4 × 6 × 3.5 cm, tied with a thread and sealed with a clay seal. The base is circular with a diameter of 3 cm. The clay seal has two different oval stamps. Some of the dried plant parts could be identified as belonging to a local plant by the name of "esfand." The find was handed over to the Governor of Bamiyan, Dr Habiba Sorabi, on Friday, 3 October and the relic is now in the care of the provincial government of Bamiyan (Figs. 13.10 and 13.11).

Stabilization Measures for the Rear Side of the Niche

After the remains of the clay surfaces of the Eastern Buddha still in situ had already been secured in 2008, in June/July 2009 backfilling the cracks and needling the brittle back wall of the niche was implemented. In this context especially the areas of the statue that had survived the explosions had to be taken into consideration, such as the left shoulder and left fold of the cloak, parts of the head, and smaller areas of the right fold of the cloak.

Concept for Rock Stabilization

Effectively, for the stabilization of the back wall of the niche a plan of four stages was intended. Before the work, the scaffold was adapted to the individual situation and the protective net against rockfall was taken down.

1. Stabilization with heavy-duty tension belts (temporary support)

   At first, on both sides of particularly brittle spots anchors with ring eyelets were inserted. At these anchors tension belts were attached, which together with timber wedges as interlayers could prevent rock pieces from falling down. These tension belts are only meant as an interim measure and can be removed once the stabilization is complete.
2. Backfilling the cracks and fissures

   For backfilling of the cracks, first of all the vertical crack grooves were sealed from outside with a mixture of yellow clay and Ledan®. After this grouting the mortar mixture was filled in through an opening. Mortar mixture for fissures of up to 20 mm: Ledan® TA 1 with washed and sifted river sand; for wider fissures: cement and river sand with Rheobuild 1,000 as aggregate (after Tonoli/Crippa). This backfilling was carried out in several phases to avoid the unstable, moisture-sensitive stone being partially dissolved and to reduce the danger of shearing. To enable the backfilled adhesive mortar to harden, work was carried out at several places at the same time. Once a partial improvement of the consolidation was achieved, the grouting could continue at a higher level and then be backfilled.
3. Placing of steel anchors

   After the hardening of the mortar, the drilling could begin. The backfilling of the cracks led to some degree of stabilization of the brittle parts; for an additional consolidation, steel anchors of 1 m length and 12 mm diameter (stainless V4A ribbed bars) were used. With conventional drilling apparatuses (Hilti T76), the drilling was carried out in several steps (8, 12, and 16 mm). The holes were freed from dust with compressed air and afterward the steel bars were fixed with cement mortar.

The question of application and measures had been considered at the site in 2009 in collaboration with Prof. Fecker. Further support with contributions and suggestions was delivered by Prof. Margottini in October 2009 during his visit to Bamiyan (see report Margottini 2009), so very exact dosage of treatments could be achieved. Every problem zone of the Buddha got its special application and therapy (Fig. 13.12).

Anchor Drilling

As a drilling unit a core drilling machine with diamond heads and combined air and water cooling was used (Schützeichel, d41 diameter). For smaller needles and anchors, rotopercussion machines were used.

The cooling in general was done with water. For special areas like the left fold of the clothing at the left side of the Buddha air cooling was the choice, as water would have damaged the original parts of in situ mud plaster. The ingress of water to the fissures inside the cliff would have degraded the shear strength and the adhesion of the cliff.

After 5 weeks of drilling and 18 drillings with a total of drilled 80 m, just two drill pipes had been left. With new pipes with modified teeth design, better adapted to the rock material, more drillings could be performed. In total 45 core drillings with a length of 200 m had been done, not counting the additional ones that had been necessary after grouting the holes, to be able to set the steel bars (Figs. 13.16, 13.17, and 13.18).

Fissures and Cracks, Grouting the Bore Holes

When drilling, we came again and again to cracks and fissures. Before the setting of the steel bars, the bore holes could be used to fill up fissures with mortar. Into some of those holes we had to fill in tons of mortar, for hours. In total 19 tons of mortar had been filled in (machine: Comdrill infusion jet pump).

Setting of Anchors

The material for anchors and nails has been 22-, 12-, and 10-mm girded and threaded steel in stainless quality (V4A, V2A). For setting the

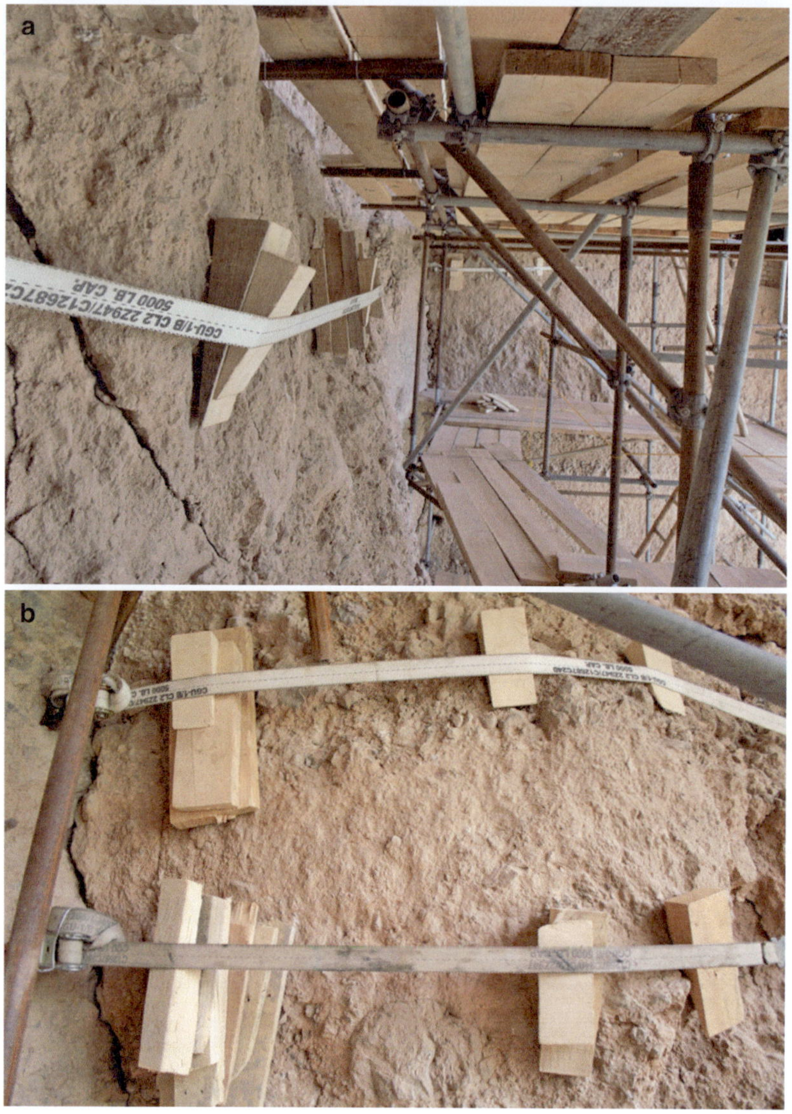

Fig. 13.12 (**a**–**d**) Temporary covering of the brittle parts at the Eastern Buddha by means of heavy duty tension belts as a first step of securing the back wall

Fig. 13.12 (continued)

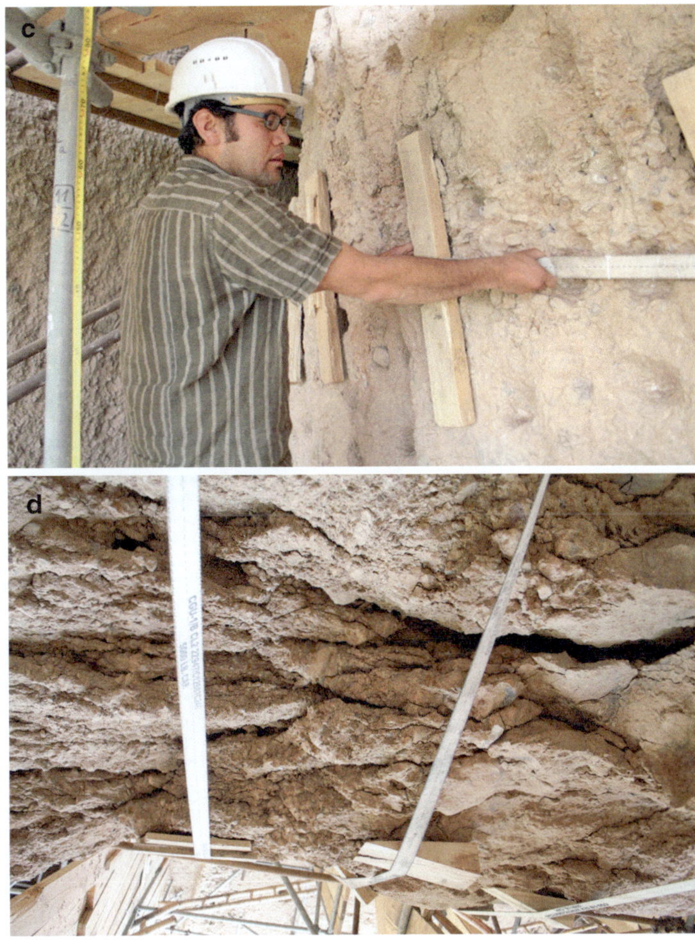

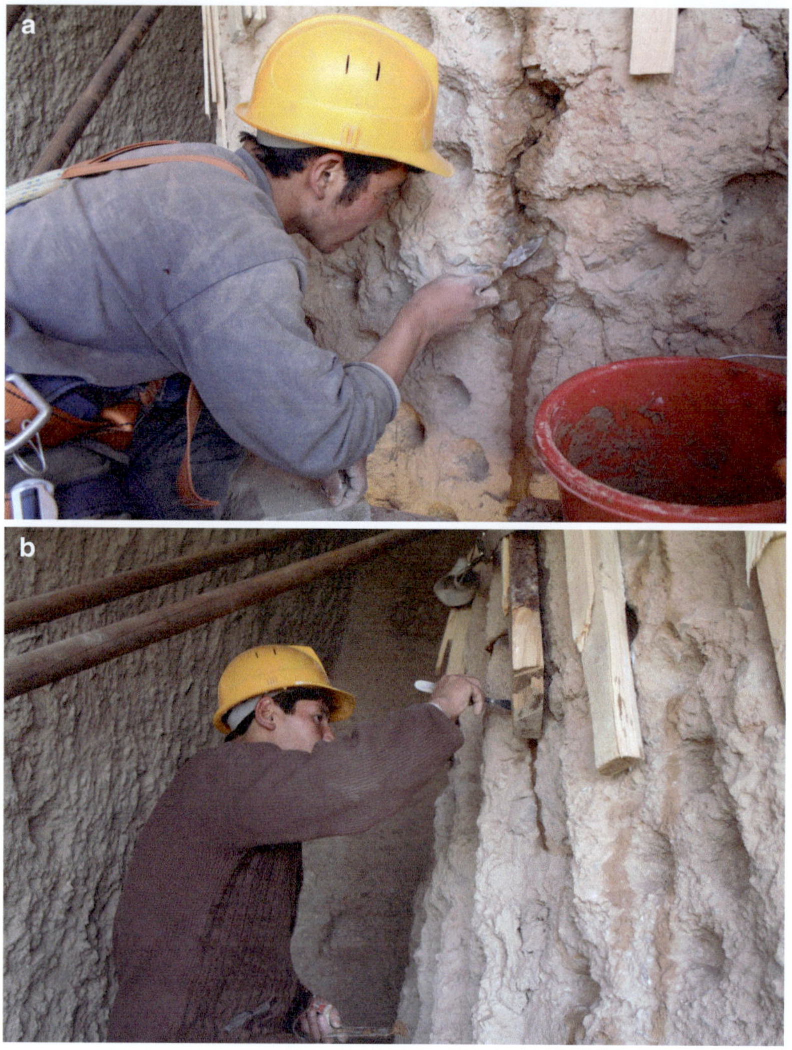

Fig. 13.13 (**a**, **b**) Sealing the cracks and backfilling step by step with mortar

anchors first the material was prepared in suitable size. The bore hole was filled up with mortar (cement mortar with BASF super plasticizer). Then the steel bar was pushed in; an excess length of 15 cm was left for further use.

Filling Up Micro Fissures with Ledan®

Some areas showed a very porous conglomerate rock, cut by very fine fissures. There, with medical injectors Ledan® were put into the gaps so a certain basic stability for later drilling could be obtained.

Nails and Needles

With 10- and 6-mm V2A steel bolts, smaller stable or stabilized portions of rock could be fixed to the rear side. As mortar in those cases mostly Ledan® was the choice, but sometimes cement mortar with super plasticizer (BASF).

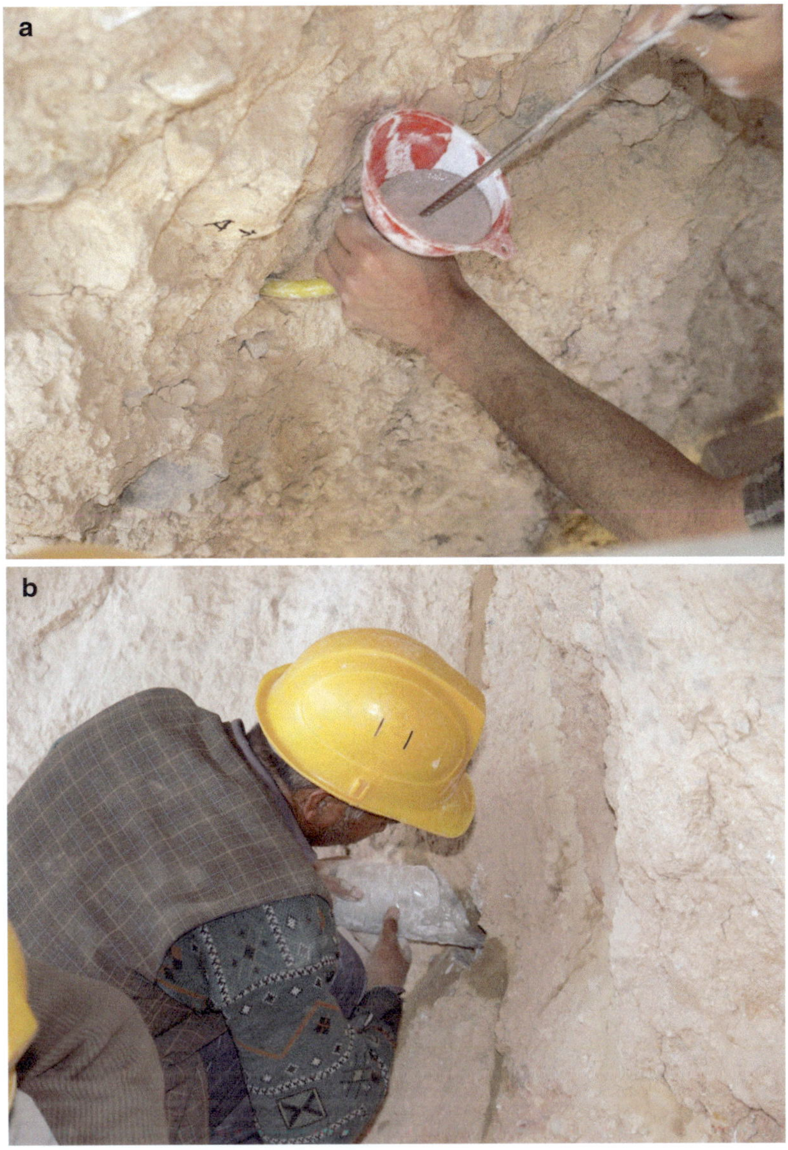

Fig. 13.14 (a, b) Backfilling the cracks with mortar

Case Studies Described in Detail

Large Fissure in the Caves

On level 3/4, three 7.5-m length drillings had been performed. Those bores had been used to fill up the major tectonic gap, which is situated about 4 m behind the rear side of the niche (on the proposition of Prof. Margottini). That gap (see Fig. 12.2) is open to the caves underneath the rear side and clearly visible. Before the grouting process those gaps had to be sealed provisionally with rags and foamed rubber.

Just that single grouting action used 1.2 tons of mortar in an afternoon. Our machine was a mortar jet pump with mixer, able to get the material up to a height of 40 m. After the grouting process some days later, those bore holes had to be drilled again to push in the anchor steel bars, as the mortar was hardened now.

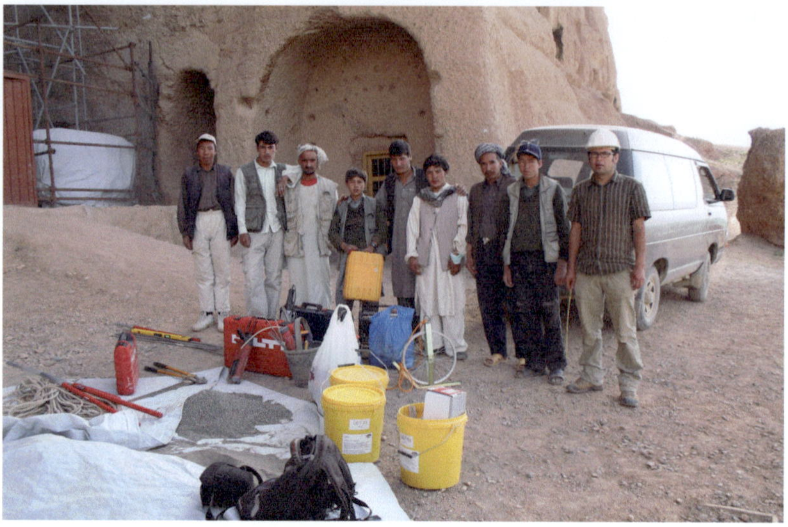

Fig. 13.15 The team at the Eastern Buddha

Mantle Plait of the Buddha

The remains of the mantle plait of the Buddha, whose basic part had been blown off and was now overhanging and showed a lot of fine cracks going parallel to the plait. That part was stabilized with five anchors d20/1,400, according to the proposal of Profs. Fecker and Margottini.

Fragments of the Head of the Buddha

The in situ remains of the head of the Buddha had been most unstable. To remove the loosened debris of the head further and further would have meant to give up the fragments. To keep the fragments, thousands of tiny micro injections had to prepare stable fields, which could be nailed to the stable material inside the rock cliff. Additionally a small buttress was constructed, in order to bear the weight of some 3-ton rocks situated above.

Conclusion

The complete rear side of the niche of the Eastern Buddha with all in situ remains of shoulder, mantle plait, and fragments of the head now appear to be stable and safe.

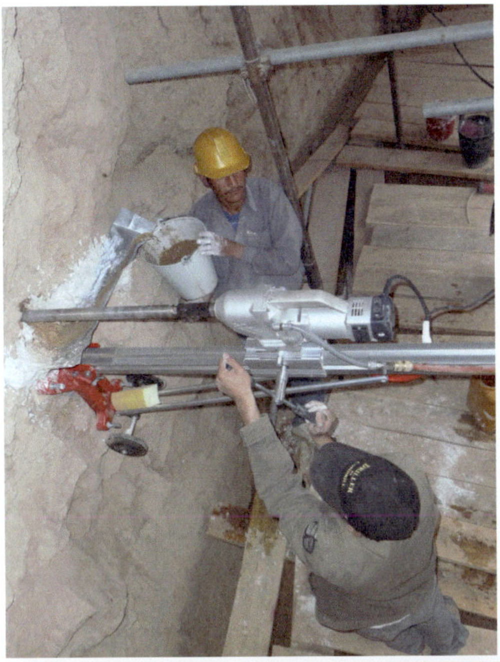

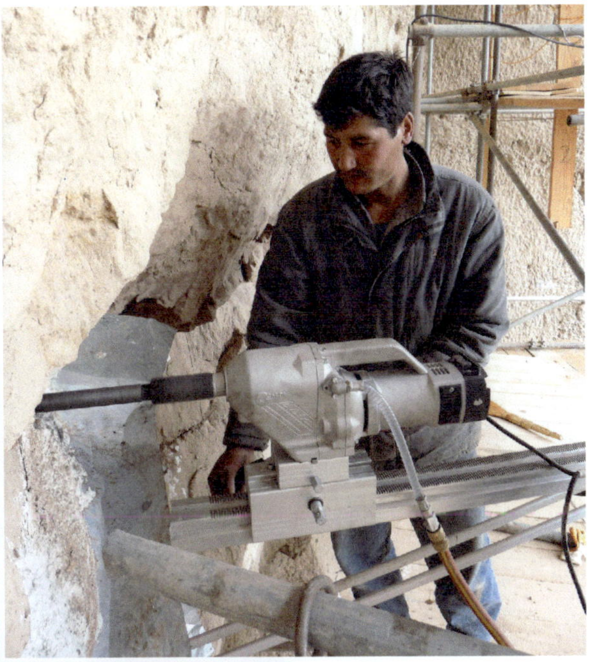

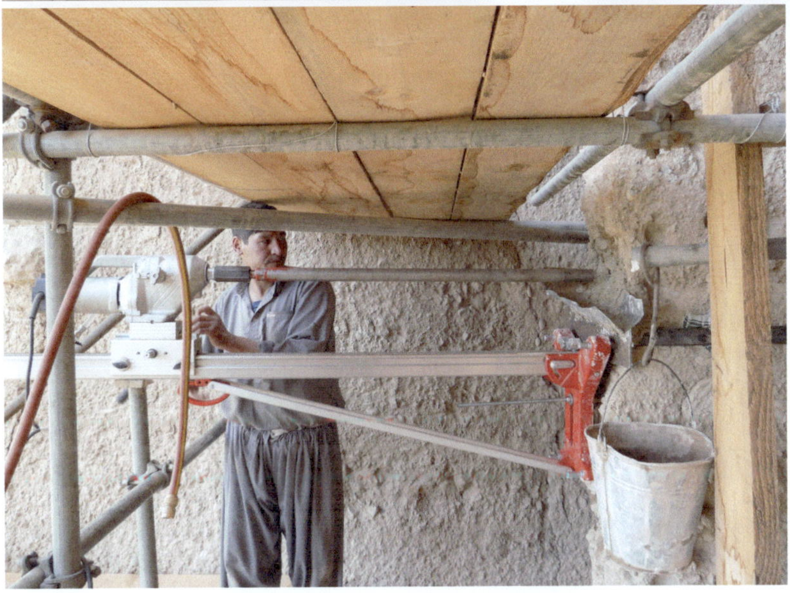

Figs. 13.16, 13.17, and 13.18 Diamond core head drilling in different situations. The water used for cooling is collected by a suitable folded panel in buckets, so the rock surface was protected of water

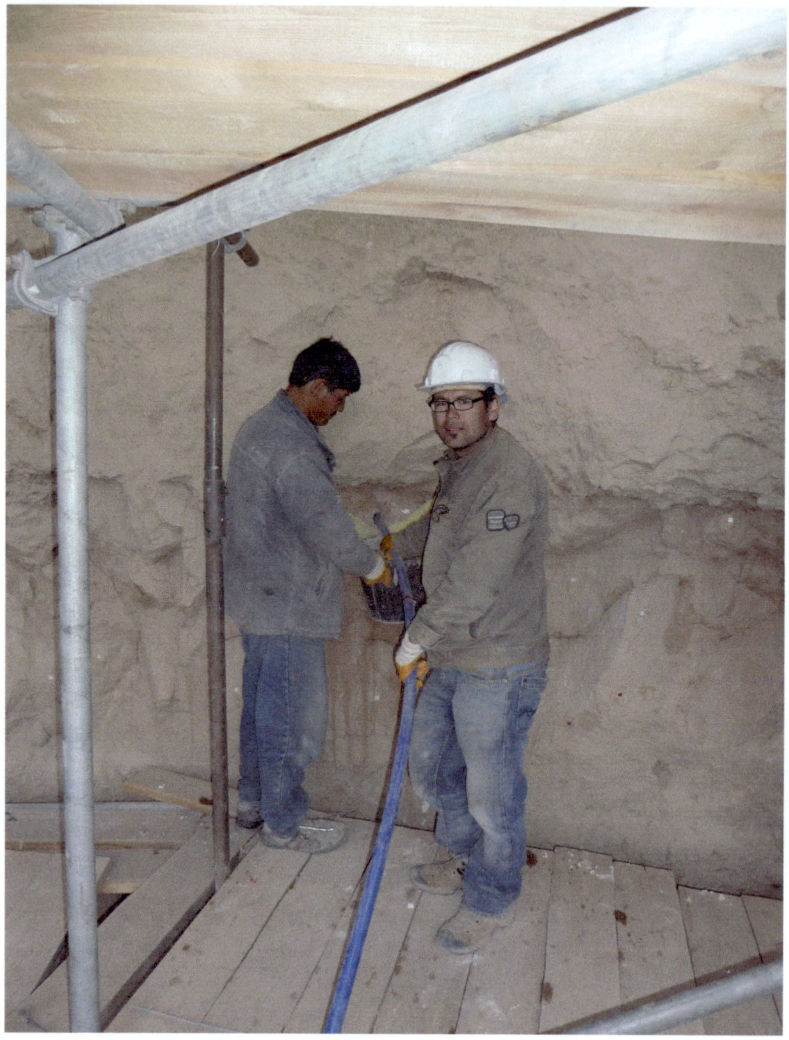

Fig. 13.19 Filling in the mortar in the bore holes. After filling up the fissures with mortar, a second drilling had to be done to position the anchors

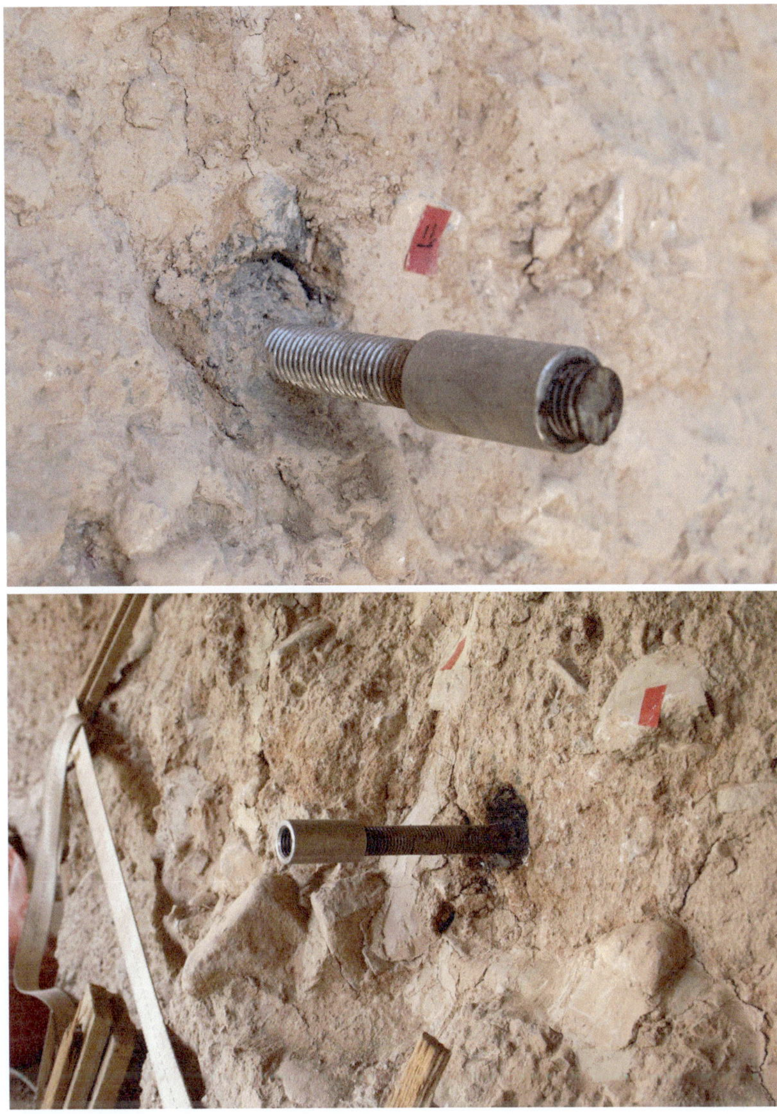

Figs. 13.20 and 13.21 Anchor of stainless steel (V2A) with a thickness of 22 mm

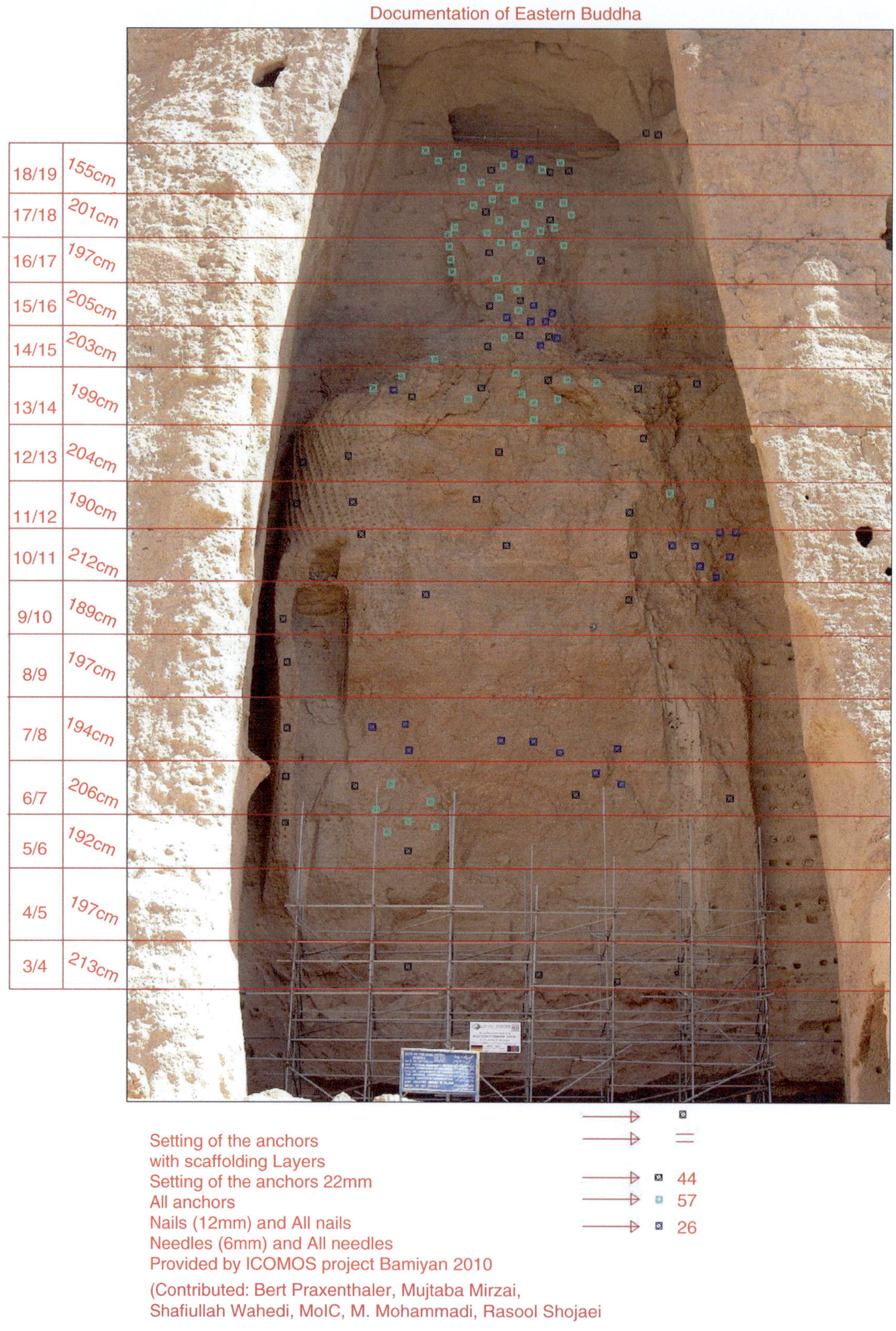

Fig. 13.22 Eastern Buddha: A map of anchors and needles

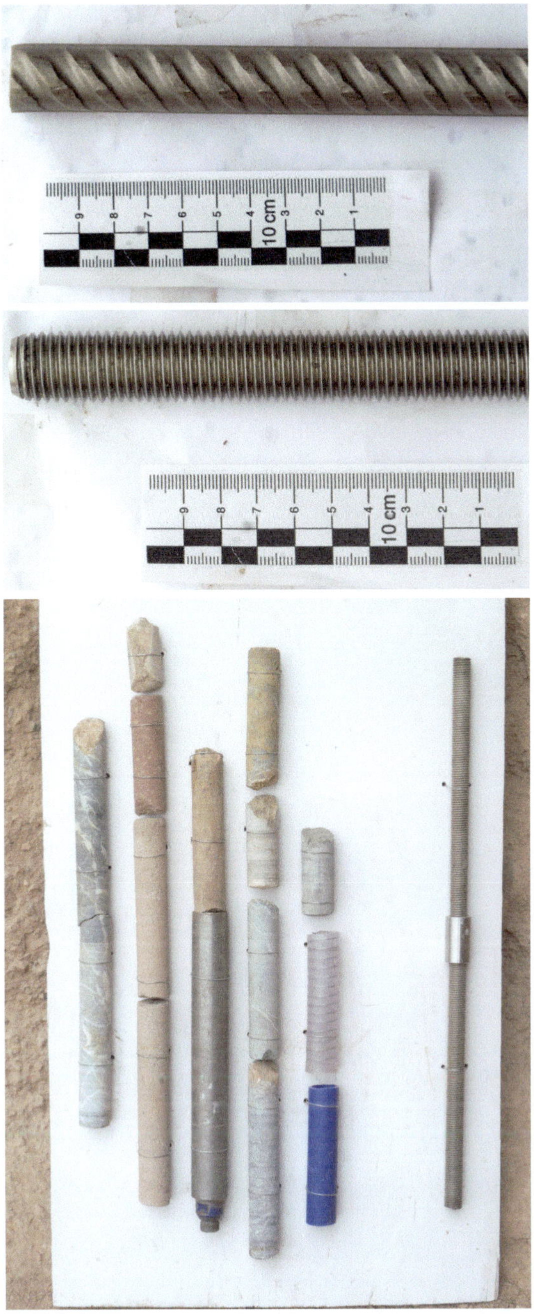

Figs. 13.23, 13.24, and 13.25 Anchor material of V2A-steel, typical bore cores, and different materials used for the stabilization

Reference

Margottini C. ISPRA, Geological survey of italy & university of modena and reggio emilia the consolidation of back wall of east giant buddha niche (Bamiyan-Afghanistan) report on mission 6–12 oct 2009.

Conclusions

Carlo Crippa and Claudio Margottini

This section describes all the emergency intervention performed in Bamiyan (Central Afghanistan) for the consolidation of niches and unstable blocks resulting from the explosions executed by the Taliban in March 2001, aimed at destroying the V–VI c. A.D. giant statues of Buddhas.

The effects of explosions were quite dramatic: the two statues totally collapsed; also some small parts of the niches fell down and, mainly, a large part of the Eastern Giant Buddha niche was close to collapse. UNESCO immediately undertook an emergency intervention for securing the remaining elements of such wonderful cultural heritages and, thanks to the generous Government of Japan's financial support, the work started in November 2003.

The activities were developed according to the following general scheme:

1. Engineering geological study of the site, including laboratory tests and field studies (laboratory tests were conducted in Europe on selected samples; site analysis started in 2002 and lasted for the entire execution of the consolidation work. The major constraint to field investigation was the presence of land mines).
2. Installation of a high precision monitoring system
3. Implementation of temporary support infrastructure, to stabilize the blocks that were at limit equilibrium and also during the execution of work
4. Execution of the consolidation work, with professional climbers to avoid any activities below the hanging and unstable blocks, by means of a system of small and long passive nails and anchors, properly grouted
5. Minimization of impact of anchor heads, with a mixture of special mortar, investigated in detail with the support of an ICOMOS expert

The result was quite satisfactory, in an area that is slowly recovering from decades of war. For this reason, due to lack of knowledge, it was necessary to adopt the best professional judgment in identifying the most hazardous situations. In the meantime, the most appropriate technologies and solutions were carried out to stabilize the cliff in a very short time and without any risk for workers.

After the engineering geological studies, started in September 2002, and the work of October–December 2003, March 2004, and October–December 2006, the cliff and niche of the Eastern Giant Buddha, the most critical part, are now more stable and the risk of collapse almost avoided. Also, the niche of the Western

C. Crippa
Trevi Division Rodio, Via Pandina, 5, 26831 Casalmaiocco (Lodi), Italy
e-mail: ccrippa@trevispa.com

C. Margottini
Department of Geological Survey of Italy, ISPRA – Italian Institute for Environmental Protection and Research, Via Vitaliano Brancati, 60, Rome 00144 , Italy
e-mail: claudio.margottini@isprambiente.it

C. Margottini (ed.), *After the Destruction of Giant Buddha Statues in Bamiyan (Afghanistan) in 2001*, Natural Science in Archaeology, DOI 10.1007/978-3-642-30051-6, © Springer-Verlag Berlin Heidelberg 2014

Giant Buddha has been protected from water infiltration and the risk of minor rock falls avoided. Further work will be needed in the future, especially in the back wall of both niches, but at least the major risk of a collapse involving also the few remains not destroyed by the Taliban is now turned away.

Nevertheless, other initiatives are now necessary, and the securing of the cliff has to be enlarged to the many sites where the old caves are excavated, especially those affected by severe geomorphological processes. Although they do not have the same dimension and importance as the Giant Buddha Statues, the caves represent a unique heritage that testifies to the old civilization of the valley, contributing to depict the incomparable "Bamiyan Cultural Landscape."

The cooperation and mutual support among the different types of expertise involved in the Bamiyan project, such as earth science, archaeology, restoration, architecture, civil engineering, and related technologies, can now be considered a successful experience. It is hopeful that such a working approach can be transferred to those places where the conservation of cultural heritage is strongly related with environmental and geological hazards. It is a "grey" area of competences where only the synergic contribution of a multidisciplinary approach, and not interdisciplinary, can properly support the long-term conservation of worldwide cultural heritage.

Index

MIX
Papier aus verantwortungsvollen Quellen
Paper from responsible sources
FSC® C105338

If you have any concerns about our products,
you can contact us on
ProductSafety@springernature.com

In case Publisher is established outside the EU,
the EU authorized representative is:
Springer Nature Customer Service Center GmbH
Europaplatz 3, 69115 Heidelberg, Germany

Printed by Libri Plureos GmbH
in Hamburg, Germany